KEITH ROWNEY

RESPIRATORY PROTECTION HANDBOOK

RESPIRATORY PROTECTION HANDBOOK

WILLIAM H. REVOIR
CHING-TSEN BIEN

LEWIS PUBLISHERS
Boca Raton New York

Acquiring Editor:	Ken McCombs
Project Editor:	Erica Orloff
Marketing Manager:	Greg Daurelle
Direct Marketing Manager:	Arline Massey
Cover:	Dawn Boyd
Prepress:	Kevin Luong
Manufacturing:	Sheri Schwartz

Library of Congress Cataloging-in-Publication Data

Revoir, William H., 1924–1991.
Respiratory protection handbook / William H. Revoir, Ching-tsen Bien.
p. cm.
Includes bibliographical references and index.
ISBN 0-87371-281-1 (alk. paper)
1. Respirators—Handbooks, manuals, etc. 2. Respiratory organs--Protection—Handbooks, manuals, etc. 3. Industrial safety--Equipment and supplies—Handbooks, manuals, etc. I. Bien, Ching-tsen. II. Title.
T55.3.G3R48 1997
681′.76—dc20 96-47080
CIP

Direct all inquiries to CRC Press LLC, 2000 Corporate Blvd., N.W., Boca Raton, Florida 33431.

Lewis Publishers is an imprint of CRC Press

International Standard Book Number 0-87371-281-1
Library of Congress Card Number 96-47080
Printed in the United States of America 1 2 3 4 5 6 7 8 9 0
Printed on acid-free paper

To Lois and Jean

Preface

I met Bill Revoir when I became a member of the ANSI Z88.2 Subcommittee which developed the 1980 ANSI standard for the Practices of Respiratory Protection and Bill served as the Chairman. I received my industrial hygiene training prior to the enactment of the Occupational Safety and Health Act. During that time, no lectures on respiratory protection were provided since engineering controls was considered the method in controlling workplace hazards. I received my assignment as the OSHA liaison to NIOSH for respiratory testing and certification shortly before I joined the ANSI Z88.2 Subcommittee. The participation in the ANSI Subcommittee gave me a rare opportunity to learn the trade from masters like Ed Hyatt, John O'Neill, and Bill Revoir. Bill assumed the major responsibility of developing the ANSI Z88.2-1980 standard, which provides very detailed instructions for establishing an effective respiratory protection program. Although Bill worked for the respirator manufacturer, he demonstrated that he really cared for the safety and health of workers. He pointed out to me what was wrong with the respirator certification process, and why certain approved respirators would not provide adequate protection. Since I worked for OSHA, I often received calls from Bill asking me why OSHA did not inspect certain workplaces which be believed were hazardous to workers.

After NIOSH announced the revision of respirator testing and certification regulations in 1979, Bill understood that NIOSH needed help. He organized an ANSI Ad Hoc Subcommittee on Respirator Testing and Approval and also served as the leader of the air-purifying respirator section. I had the privilege to work with him again. To demonstrate that the current filter certification regulations were grossly inadequate, Bill proposed that the Ad Hoc Subcommittee conduct tests. With the agreement of major respirator manufacturers such as MSA, Norton and 3M, extensive filter testing was conducted to demonstrate that the current filter testing methods were insensitive to indicate the performance of filters, and that a submicrometer sodium chloride aerosol should be the choice for filter testing. The 1987 NIOSH proposed revision of the respirator testing and certification regulations adopted most of the recommendations proposed in the Ad Hoc Subcommittee's report.

After Bill Revoir retired and moved to the Washington D.C. area, we had more time to see each other and exchange views on respiratory protection. He mentioned that he was asked to write a book on respiratory protection, and I encouraged him to accept the invitation since he was very qualified in writing the subject. Unfortunately, Bill was not able to complete the book before his untimely death. I was asked by the editor of the publisher to complete Bill's book, and also adding a section on how to comply with the OSHA respiratory protection standard. Since I was involved in developing standard interpretations and compliance policies on respiratory protection for OSHA, it would be inappropriate for me to write about this subject. There are many government sponsored projects on respiratory protection. However, reports of these projects do not have wide distribution. There are also no reviews of many articles published in the journals. A complement to Bill's manuscript should be the addition of a section addressing the developments of respiratory protection since OSHA's existence. It should include reviews and summaries of studies of both unpublished and published studies or reports. The articles which I selected are mainly limited to U.S. articles published after 1970, and I also take the liberty of selecting the articles which I feel are of general interest.

The first part of the book addresses subjects such as the establishment of an effective respiratory protection program, elements of respiration, classification of air contaminants and its effects on humans, oxygen deficiency, particulate and gas filtration, classifications of respirators and their advantages and limitations, definitions of fit testing terms and application of fit factors, and special problems. The second part of the book includes the development of respiratory protection in the last twenty years, respirator certification tests, filter testing and factors that would affect the performance of filters, the prediction of service life of the sorbent, factors affecting the performance of the sorbent, a compilation of cartridge and canister service life data, developments of the fit testing methods, protection factors resulted from quantitative fit testing, simulated workplace testing, workplace testing, and other related issues such as faceseal and exhalation valve leakages.

I would like to express my appreciation to the assistance provided by the Mine Safety Appliances Company and other respirator manufacturers. They include Bullard, Cabot, Draeger, ISI, North, Racal, Scott, Survivair, and 3M. I would like to thank Klaus Willeke for providing art works from his studies. I am also grateful to Jeff Birkner and John Steelnack for reviewing the manuscript and making many helpful suggestions. The staff of the Technical Data Center of OSHA has also provided assistance. Finally, I am indebted to my wife for her encouragement in writing this article and to my daughters for their tireless efforts in proofreading.

Ching-tsen Bien

About the Authors

The late Mr. Revoir, a consultant in respiratory protection, received a B.S. degree in Chemical Engineering from the Massachusetts Institute of Technology and a M.S. degree in Physics from the Northeastern University.

Mr. Revoir spent over thirty years in the development of respiratory protection devices. He published many articles on respiratory protection and received four patents on respiratory protection technology.

Mr. Revoir was a member of many professional organizations such as the American Industrial Hygiene Association (AIHA), American Society for Testing and Materials (ASTM) and the American National Standards Institute (ANSI).

Mr. Revoir's major contribution in respiratory protection was that he served as a major contributor and Chairman of two ANSI Z88 Subcommittees on respiratory protection. The standard on the Practices for Respiratory Protection was adopted by several regulatory agencies, and the standard on Respirator Testing and Certification was adopted by the National Institute for Occupational Safety and Health (NIOSH).

Ching-tsen Bien is a consultant in industrial hygiene. He received his B.S. and M.S. degrees in Chemical Engineering from the Cheng Kung University and the University of New Brunswick respectively. He also received a M.S. degree in Industrial Hygiene from the University of Pittsburgh.

Mr. Bien has worked for the Occupational Safety and Health Administration (OSHA) for more than twenty years. During his tenure at OSHA, his major responsibility was providing technical assistance on respiratory protection, protective clothing, and industrial hygiene. Mr. Bien is a certified industrial hygienist, registered professional engineer, and also a member of various American National Standards Institute (ANSI) Z88 Subcommittees on respiratory protection.

In Memory of the Senior Author

1924–1991

Bill Revoir, the senior author of this publication, was a friend, professional colleague, bench mark for defining professional responsibility, and an example of how to integrate professional concern for work health/safety protection with technical innovation and business considerations. I knew him best during the time he worked with me at Los Alamos but had the privilege to interact with him professionally on many technical committees and association activities. His independence relative to technical quality, detail and excellence often led people to consider him a maverick. But his non-compromising technical approach reflected his primary concern for introducing new ideas/approaches to the field of respiratory protection as quickly as possible. This in turn would improve the level of health protection provided and at the same time enhance the business position of his employer.

When I worked with Bill at Los Alamos I was impressed by his work intensity and attention to detail. His written output was overwhelming, and a trip or meeting report provided all the information that transpired at the event. Reading his resume one is impressed by the continual reference to "research", "technical" and "training". These are clearly the foundation of Bill's career activities, and are emphasized by this highly technical publication completed after his untimely death.

All involved with respiratory protection and worker health protection will miss Bill. In a small way this publication provides a continuing reminder of his style, interests, and contributions.

Harry Ettinger
Los Alamos, New Mexico
March 1996

TABLE OF CONTENTS

1 ESSENTIAL ELEMENTS OF AN EFFECTIVE RESPIRATORY PROTECTION PROGRAM

I. INTRODUCTION

Improper use of a respirator can have devastating effects on the life or health of the wearer. For instance, one worker died because his employer failed to require the use of a respirator when the tank he was pumping a saline solution into was the same time evacuating hydrogen sulfide, a very toxic gas, in excess of the maximum peak concentration allowed [Occupational Safety & Health Review Commission (OSHRC) Docket No. 85-166, April 17, 1991]. In another case, a worker was exposed to impermissible levels of cotton dust because she was allowed to remove her respirator before the cotton dust hazard was adequately controlled by other means [OSHRC Docket No. 84-767, February 27, 1991]. At a foundry, a foreman was given a respirator to wear as protection from airborne lead, which was in excess of permissible exposure limit (PEL), without benefit of periodical fit testing. His breathing air, inside the respirator, contained lead in excess of the PEL [OSHRC Docket No. 89-3688, October 26, 1990]. The point of these examples is that uninformed or improper wearing of respirators can have serious consequences. The identity and characteristics of the respiratory hazards that may be encountered, the respirators available for use, and how to use the respirator selected are crucial information that the wearer needs to know. The Occupational Safety and Health Administration (OSHA) of the Department of Labor mandates that employers prepare a written respirator program and use it to train employees, if the employees will be expected to work in conditions that require respirators. For those with a need to know and understand how and why respirators work, then, the first order of business is to examine these background issues of general respirator information and the OSHA requirements.[1]

A. General

In order to protect oneself, or anyone else, from airborne respiratory hazards, one must understand the respiratory hazard and its primary characteristics that render it hazardous. One must also understand the means of respiratory protection available and how they work relative to the hazard for which protection is necessary. Finally,

[1] All OSHRC Docket cases reviewed are found in The Bureau of National Affairs, *Occupational Safety & Health Cases,* Volume 14. Washington, D.C., 1991.

it sounds simplistic, but one needs to know how to use respirators properly as our previous examples — lack of respirator, early removal of respirator, and improperly fitting respirator — all lead to serious health consequences and even death, in the one case. Let us turn our attention briefly to the general information one needs to know in order to use respirators wisely.

1. Respiratory Hazard

A respiratory hazard exists in an industrial workplace whenever a substance is present in the atmosphere at a concentration that will cause bodily harm or whenever a deficiency of oxygen occurs in the atmosphere. Respiratory hazards in the industrial workplace may consist of:

1. Oxygen deficiency
2. Air contaminants
   a. Particulate matter, including dust, fume, and mist
   b. Vapor and gas
   c. Combination of particulate matter, vapor, and gas

The nature of the respiratory hazard, relative to the selection and classification of respirators, depends upon the atmospheric oxygen concentration; each contaminant's physical state, toxicity, and concentration; the presence of other contaminants or stress factors in the working environment; and amount of time the worker will be exposed and his or her susceptibility to exposure. Respiratory hazards may be classified as immediately or not immediately dangerous to life or health, and, in addition to the classifications listed above, each classification requires a different type of respiratory protection. When are these hazards a concern? Situations where airborne hazards threaten life and health are numerous in manufacturing operations, mining, maintenance of plants and process equipment, construction, and when responding to accidental spills of hazardous substances and other emergencies, including fires.

2. Primary Means of Respiratory Protection

The primary means of protecting workers against respiratory hazards are engineering controls, administrative rules, and work practices. Engineering controls may eliminate the respiratory hazards or reduce the levels of respiratory hazards in workplace atmospheres to levels sufficiently low so as not to cause bodily harm. That is, engineering controls make sure that the breathing air is of good quality. Administrative rules limit the time of exposure of workers to respiratory hazards in workplace atmospheres so that the workers do not suffer bodily harm. Work practices prevent or decrease the exposure of workers to respiratory hazards in workplace atmospheres so that bodily injury does not occur by the way tasks and procedures are performed relative to the hazard. Installing a local exhaust hood to capture and remove the offensive air contamination is an example of engineering control. Limiting each employee to 2 h per day at the hazardous task is an example of administrative control. Rewriting the work procedure, so that the chance of exposure to the potentially toxic substance is reduced, is an example of work-practice control.

3. *Use of Respirators*

As a means of respiratory protection, the use of respirators is the last resort and, therefore, is not listed as one of the primary means. However, situations do exist where engineering controls, administrative rules, and work practices may be inapplicable, impractical, infeasible, impossible, or not adequately effective. In addition, situations may exist when engineering controls are being developed, installed, and evaluated and protection is required in the interim. In such situations, employers must provide proper respirators to employees in order to protect them against respiratory hazards. Employers must also make proper respirators available to employees for emergency escape and for rescue operations. Firefighters also must be given proper respirators.

B. Respirator Program

1. *General*

An employer cannot simply hand out respirators and expect employees to fend for themselves. An employer must develop and implement a comprehensive respirator program to ensure that employees who depend on the wearing of respirators for protection against respiratory hazards are given adequate protection. The failure of an employer to implement a proper respirator program may result in injury or even death to employees as experience clearly demonstrates.

2. *Employer Responsibilities*

The establishment and maintenance of a well-thought out, sensible respiratory protection program is vital to the life and health of employees who are exposed to respiratory hazards that are not adequately controlled by engineering, administrative, or work-practice means. The Occupational Safety and Health Administration (OSHA) respirator regulations (found at 29 CFR 1910.134) require employers who provide respirators to employees to institute and carry out suitable respirator programs.

Comprehensive Respirator Program Outline. A comprehensive respirator program, as defined by OSHA in 29 CFR1910.34(b)(1) to (b)(11) covers the following matters:

a. *Written Program*
A complete respirator program containing written procedures covering each of the following issues shall be established and implemented.
The written respirator program shall include procedures concerning routine and nonroutine uses of respirators. Emergency, rescue, and escape situations requiring the use of respirators shall be anticipated and plans covering such respirator applications shall be included in the written respirator program. If applicable, respirator use in firefighting shall be covered in the written respirator program.

b. *Program Administration*
It is recommended that an individual having knowledge of respiratory protection sufficient to set up and supervise a proper respirator program be assigned as the respirator program administrator and he/she shall have the responsibility and authority for the respirator program.

c. *Plant Survey and Hazard Assessment*
A plant survey shall be completed in accordance with 29CFR1910.132(d) to determine which workplaces have respiratory hazards and to determine the conditions encountered by employees exposed to respiratory hazards. Surveillance of work area conditions and degree of employee exposure of stress should be routinely conducted. The information obtained is needed to permit the selection of the proper types of respirators.

d. *Respirator Selection*
The proper type of respirator shall be selected for each workplace where a respiratory hazard exists. The selection of the proper type of respirator for any given situation is based on the following factors:
- Workplace features and conditions
- Nature of the operation or process
- Location of the hazardous area with respect to a safe area having respirable air
- Employee activities
- Period of time for which respiratory protection is needed
- Nature of respiratory hazard (including type, physical and chemical properties, warning characteristics, physiological effects on human body, established permissible exposure limit, established concentration immediately dangerous to life or health, actual concentration (time-weighted average and peak) of toxic substance or oxygen deficiency)
- Physical characteristics and functional capabilities and limitations of various types of respirators
- Assigned protection factors for various types of respirators

e. *Approved Respirators*
Only respirators approved jointly by the National Institute for Occupational Safety and Health (NIOSH) of the U.S. Department of Health and Human Services and the Mine Safety and Health Administration (MSHA) of the U.S. Department of Labor shall be selected and used.

f. *Physiological and Psychological Limitations of Regulator Wearers*
A respirator may produce undesirable effects on the wearer. Respirators are uncomfortable and may reduce field of vision, require the wearer to carry extra weight, place an additional burden on his or her respiratory system, cause a feeling of claustrophobia, and may also result in a general feeling of anxiety. Employees should not be assigned to perform tasks requiring use of respirators unless a licensed physician has determined that they are physically able to both perform the work and use the prescribed respiratory equipment and any other chemical protective clothing required for the task. The local physician has the discretion to determine what health and physical conditions are pertinent. The two areas of greatest physiological concern are the respiratory system and the cardiovascular system. The examining physician should be given information and characteristics about the respiratory equipment to be used. He or she should know whether the equipment produces additional inspiratory and expiratory stress, whether it represents a significant additional weight, such as self-contained breathers, and whether it may cause an increase in the metabolic heat load, such as when worn with chemical protective clothing. Some other issues the physician may take into consideration are the physical demands of the job and workplace conditions not already mentioned and any other physiological or psychological factors he or she believes is pertinent in making decisions to qualify or disqualify the employee for respirator use. The respirator user's medical status should be reviewed periodically. OSHA suggests that this medical review be made annually. The examining

physician shall provide a written opinion which describes the ability of the employee to wear the prescribed respirator and recommends limitations on the use of respirators, if any. The report and opinion shall be forwarded to the employer for filing.

g. *Respirator Procurement*
An adequate supply of each type of respirator needed should be procured and made available for use. For respirators that are equipped with facepiece type respiratory-inlet coverings, a supply of respirators having more than one size and style of facepiece should be made available for selection and use, since no one size or style of facepiece will fit all faces.

h. *Training Respirator Wearers*
Selecting the appropriate respirator for protection against a given hazard is important, but equally important is using the selected device properly. Proper use can be ensured by carefully training employees in selection, use, fit testing, and maintenance of respirators. Therefore, the employer is required by OSHA to provide each respirator wearer with training. Unless the reasons for the use of respiratory protective devices and instructions on proper use and maintenance are thoroughly understood and ongoing training is provided, the devices will not be used properly or may not work properly even if otherwise used properly. Training activities should include:
   - Instruction in the nature of the hazard and discussion of what the results may be if the respirator is not used
   - Discussion of why a certain type of respirator is used in a particular environment
   - Description of respirator capabilities and limitations
   - Demonstrate actual respirator use (donning and perform tasks); repeat periodically. Wearers of SCBAs and emergency escape devices must be retrained every year
   - Conduct fit testing
   - Discuss actual full-service time operations of the unit (SCBA and emergency escape devices supplying oxygen or breathing air)
   - Discuss end-of-service-life indicator recognition
   - Describe how to recognize emergency situations
   - Cover methods to deal with emergency situations
   - Demonstrate cleaning and maintenance of respirators
   - Discuss OSHA regulations concerning respirator use
   - Explain why engineering controls, administrative policies, and work practices are not being used or are not adequate and what efforts are being made to diminish or eliminate use of respirators
   - Discuss proper maintenance and storage practices for the respirator assigned to each trainee
   - Instruct each respirator wearer how to recognize and cope with emergency situations pertaining to the use of respirators
   - Provide instructions as necessary for use of respirator in atmospheres immediately dangerous to life and health (IDLH), confined spaces, low-or high-temperature environments, firefighting, victim rescue under hazardous conditions, and emergency escape

i. *Respirator Fit Testing*
A negative pressure respirator is a respirator in which the air pressure inside the respiratory-inlet covering is positive during exhalation and negative during inhalation relative to the air pressure of the atmosphere surrounding the respirator

wearer. A positive pressure respirator is a respirator in which the air pressure inside the respiratory-inlet covering during both exhalation and inhalation is positive relative to the air pressure of the atmosphere surrounding the respirator wearer. Each potential respirator wearer of a negative pressure respirator or of a positive pressure respirator equipped with a tight-fitting respiratory-inlet covering (facepiece) shall be given a respirator fitting test prior to the initial use of the respirator. Afterwards, a respirator wearer shall be given a respirator fitting test annually, and whenever a different make, model, or size respirator is to be used. A qualitative or quantitative respirator fitting test prescribed by OSHA shall be used. The results of the fitting test shall be used to select a specific make, model, and size of respirator for use by the employee. Only a negative pressure respirator or a positive pressure respirator having a tight-fitting respiratory-inlet covering (facepiece) that provides an adequate fit on an employee during the fit test shall be permitted to be used by the employee.

j. *Special Respirator Wearing Problems*
Certain conditions prevent a proper seal of a particular type of respirator. When such conditions are encountered, it shall be necessary to eliminate the condition, or if this cannot be accomplished, then it shall be necessary to select a different type of respirator for use by the employee or to assign the employee to a job that will not require the use of a respirator. Conditions that prevent a proper seal of a respirator to an employee include:

- Hair (stubble, moustache, sideburns, beards, low hairline, bangs) which passes between a respirator wearer's face and the sealing surface of the respiratory-inlet covering
- Hair (moustache or beard) which interferes with the operation of a respirator valve
- Scars, hollow cheeks, excessively protruding cheekbones, deep creases in facial skin, absence of teeth or dentures, or unusual facial configurations which prevent a seal of a respirator facepiece to the employee's face
- Missing teeth or dentures which prevent the respirator mouthpiece from sealing to the employee's mouth
- A nose that has a shape or size that prevents the closing of the nose by the nose clamp of a respirator equipped with a mouthpiece/nose clamp type respiratory-inlet covering
- Spectacles that have temple bars or straps that pass between the sealing surface of a respirator facepiece and the respirator wearer's face
- Spectacles, goggles, face shield, welding helmet, or other eye and face protective device that interferes with the seal of a respirator to the respirator wearer
- A head covering that passes between the sealing surface of a respirator facepiece and the respirator wearer's face
- Special corrective lenses which are made to be mounted inside a respirator full facepiece are available and should be used by a person who needs corrective lenses and who must wear a respirator equipped with a full facepiece. Contact lenses may be worn by respirator wearers who need corrective lenses.

k. *Respirator Issue*
The proper type of respirator shall be specified for each application and shall be listed in the written respirator program. Only persons trained to ensure that proper

* Or when the user has a significant physical change, which may effect the fit (e.g., gained weight, lost weight, new scars, facial hair, etc.).

respirators are issued should be permitted to issue respirators to employees who need them for protection against harmful atmospheres.

l. *Respirator Inspection*

Prior to donning a respirator for use in a hazardous area, the respirator wearer shall inspect the respirator to ensure that it is in good working condition. After each cleaning and sanitizing, a respirator shall be inspected to determine if it is in good operating condition, if it needs replacement of parts or repair, or if it needs to be replaced. Each respirator stored for emergency or rescue use shall be inspected at least monthly and an inspection record of dates, findings, and remedial actions shall be maintained.

m. *Respirator Use*

Prior to entering into a hazardous area, the wearer of a respirator equipped with a facepiece type respiratory-inlet covering shall check the seal of the facepiece to his/her face by means of a fitting test or a positive and/or negative air pressure check or by using instructions provided by the respirator manufacturer.

A respirator wearer should immediately leave the hazardous area and go to a safe area for any respiratory protection related cause including:

- Failure of a respirator to provide adequate protection
- Malfunction of a respirator
- Detection of leakage of an air contaminant into the respirator
- Increase in resistance offered by the respirator to breathing
- Severe discomfort caused by wearing of the respirator
- To minimize skin irritation by washing his/her face and the respirator facepiece
- To replace air-purifying elements in an air-purifying type respirator when necessary
- To take a break in a safe area due to stress imposed by the use of the respirator
- Illness of the respirator wearer

n. *Monitoring Respirator Use*

The supervisor of employees wearing respirators in hazardous workplaces or the supervisor's appointee should frequently inspect the wearing of respirators to ensure that the correct types of respirators are being used, that respirators are being worn properly, and that the respirators are in good working condition.

o. *Monitoring Respirator Hazard*

The monitoring of respiratory hazards in workplaces is the surest way to determine the concentrations (time-weighted average and peak) of toxic substances and oxygen deficiency.

Initial monitoring of workplaces where respiratory hazards exist is recommended in order to obtain information needed to permit the selection of the proper types of respirators.

Monitoring of workplaces where respirators are routinely used should be carried out periodically to determine if the correct types of respirators are being used.

Whenever modifications are made to a workplace or to an industrial process that may change the level of a respiratory hazard or that may result in the generation of a new respiratory hazard, monitoring of the respiratory hazard should be carried out to obtain information to ensure that the correct type of respirator will be used.

p. *Respirator Maintenance and Storage*

Cleaning, sanitizing, inspecting, and repairing of respirators shall be carried out as follows.

Type of Respirator	Schedule
Routinely used respirator issued for exclusive use of a single employee	Before and after each use
Routinely used respirator issued for use by more than one employee	Before and after each use
Respirators used for emergency, rescue, escape, or firefighting	After each use, but at least monthly

Disposable respirators which cannot be cleaned and sanitized shall be discarded after a single day of use by one employee or more often as is necessary.

Each respirator for emergency, rescue, escape, or firefighting application shall be inspected to ensure that it is functioning properly before being placed in a workplace, after each use, and at least monthly.

Repairs of respirators shall be made only by experienced persons, and parts utilized in respirator repair shall be those designed for the specific respirator.

Respirators which no longer can be maintained in good operating condition shall be immediately discarded and replaced.

Air and oxygen cylinders of self-contained breathing apparatus type respirators shall be kept fully charged and the cylinders shall be recharged when the pressure of the air or oxygen falls below the respirator manufacturer's recommended pressure level.

All respirators shall be stored in a manner that protects against physical damage caused by dust, sunlight, heat, extreme cold, excessive moisture, and damaging chemicals.

Respirators for emergency, rescue, escape, or firefighting application shall be kept accessible to workplaces inside clean containers designed to protect them against physical damage, contamination, and weathering. Such containers shall be clearly marked to designate the applications of the respirators.

q. *Respirable Air and Oxygen for Supplied-Air Respirators and Self-Contained Breathing Apparatus*

Compressed gaseous or liquid oxygen should meet the requirements of the United States Pharmacopoeia for medical or breathing oxygen.

Chemically generated oxygen should meet the requirements of the U.S. Department of Defense Military Specification MIL-E-83252 or Military Specification MIL-O-1563c.

Compressed gaseous air shall meet at least the requirements for Grade D Air and liquid air shall meet at least the requirements for Grade M Air as described in the ANSI/CGA G-7.1-1989 American National Standard Commodity Specification for Air.

r. *Entry into Atmospheres Immediately Dangerous to Life or Health*

Respirators for entry into and use in atmospheres that are immediately dangerous to life or health are the self-contained breathing apparatus equipped with a full facepiece and operated in the pressure-demand mode or other positive pressure mode with a service life of at least 30 min or the supplied-air respirator equipped with a full facepiece and operated in the pressure-demand mode or other positive pressure mode and which contains an auxiliary self-contained air supply.

A confined space with a hazardous atmosphere shall be considered to have an atmosphere that is immediately dangerous to life or health unless proven otherwise or continuous forced air ventilation is provided in accordance with OSHA regulations.

When respiratory protective devices are used in atmospheres that are immediately dangerous to life or health, at least one standby person shall be present in a safe area. The standby person shall have a self-contained breathing apparatus equipped with a full facepiece and designed for operation in the pressure-demand or other positive pressure mode available for immediate rescue use. Communications (visual, voice, signal line, telephone, radio, or other suitable means) shall be maintained between the standby person and the respirator wearers in the immediately dangerous to life or health atmosphere.

The respirator wearers in the immediately dangerous to life or health atmosphere shall be equipped with safety harnesses and safety lines or other suitable retrieval equipment for removing the respirator wearers from the dangerous atmosphere if necessary.

s. *Firefighting*
For respiratory protection during firefighting, a person shall use a self-contained breathing apparatus equipped with a full facepiece and operated in the pressure-demand mode or other positive pressure mode with a service life of at least 30 min.

t. *Respirator Program Evaluation*
In order to ensure that the respirator program is effective in providing adequate protection to employees against inhalation of harmful atmospheres, the respirator program shall be evaluated at least annually. When possible, medical and bioassay surveillance of respirator wearers should be conducted and the results used in the program evaluation as a way to indicate the effectiveness of the program. The program evaluation should include an investigation of wearer acceptance of the respirators covering matters such as comfort, fatigue, resistance to breathing, obstruction of vision, interference with communications, restriction of movement, decrease in job performance, and confidence in the effectiveness of the respirator to provide adequate protection. If the program evaluation shows that problems exist, immediate action shall be taken to eliminate these problems. The results of the program evaluation shall be recorded. See also paragraph 3 in the next section — Employee Responsibility.

3. Employee Responsibility

The employee shall use the respirator provided by the employer in accordance with the instructions and training received from the employer.

The employee should take precautions to prevent damage to the provided respirator and he/she shall not alter the respirator in any way.

If the respirator worn by an employee in a hazardous area malfunctions, permits leakage of an air contaminant into the respirator, fails to provide adequate protection, offers excessive resistance to breathing, or causes severe discomfort, he/she shall immediately leave the hazardous area and go to a safe area where he/she shall report the problem to the person designated in the written respirator program. This person, the respirator program administrator, or a person assigned by the respirator program administrator shall investigate the problem and take corrective action to implement the solution and thus eliminate the problem.

Since wearing a respirator imposes stress on the wearer, an employee who is required to wear a respirator in a hazardous area shall be allowed to leave the hazardous area and go to a safe area to take a break in the safe area a reasonable number of times during each work shift.

An employee who is required to wear an air-purifying type respirator in a hazardous area shall be allowed to leave the hazardous area and go to a safe area to replace the air-purifying elements in the respirator when such replacement is necessary.

An employee who is required to wear a respirator equipped with a facepiece type respiratory-inlet covering shall be permitted to periodically leave the hazardous area a reasonable number of times during each work shift and go to a safe area to wash his/her face and to wash the respirator facepiece to minimize skin irritation.

2 RESPIRATION

I. CELLS

Some forms of life, such as bacteria, may consist of a single cell. Larger, more complex plants and animals are built of many cells. An adult human body contains about 100 trillion cells of several hundred different types with each type having a specific function to perform.

The substance that comprises a cell is called protoplasm. Most cells, whether plant or animal, have the same fundamental form. A cell is a tiny body with a liquid interior and a thin protective cover enclosing it.

The protective covering of the cell is called a membrane. This membrane is semipermeable which means that some substances can pass into and out of the cell through the membrane, but other substances cannot pass through the membrane. The membrane regulates the life of the cell since all nourishment that the cell receives must pass into the cell through the membrane and waste products of the cell must pass out of the cell through the membrane. If the membrane is damaged or destroyed, the cell dies.

Suspended in the fluid of the cell is a smaller body, called the nucleus, enclosed by another membrane. The nucleus is the control center of the cell. It contains substances called genes which are responsible for the traits of the cell.

The material outside the nucleus in the cell is called cytoplasm. Cytoplasm is mostly liquid and suspended in this liquid are tiny structures that have different functions. The cytoplasm also contains such materials as fat droplets, starch grains, mineral crystals, and gas bubbles. Dissolved in the liquid cytoplasm are substances such as protein enzymes, fatty acids, sugars, and salts. Oxygen molecules which enter the cell through the membrane react with foodstuffs stored in the cytoplasm such as sugars to release energy required for life. Likewise, carbon dioxide molecules and water molecules formed in the reaction pass out of the cell through the membrane. Metabolism is the term given to the wide variety of chemical reactions that occur in living cells.

The overall functioning of the human body is the combination of the functioning of all of its 100 trillion cells. Although different types of cells have different functions, they all have certain common functions which are the ability to live, grow, and reproduce.

II. TISSUES

A group of cells having the same special function to perform and work together is called a tissue. The human body contains six major types of tissues:

A. Epithelial

Epithelial tissues cover the surface of the body and line the cavities of the body. They protect the external and internal surfaces of the body. Some of these tissues absorb substances while others secrete substances. Some physiologists classify tissues that secrete substances as glandular tissues.

B. Connective

Connective tissues support and hold the body together. They also give the body its structure.

C. Muscle

Muscle tissues permit movements of some parts of the body in relation with other parts. Muscle tissues perform mechanical work by contraction — getting shorter and thicker.

D. Nerve

Nerve tissues consist of cells that transmit messages in the form of impulses from one part of the body to another. Nerve tissues help coordinate body activities. Nerve cells have many sizes and shapes. Some nerve cells may be up to 3 ft in length but microscopic in cross-section.

E. Vascular

Vascular tissues include blood cells of various types which are specialized to perform several important duties.

F. Reproductive

Reproductive tissues are specialized to produce the offspring of the body. They are the egg cells in the female and the sperm cells in the male.

III. SYSTEMS

A system is a group of organs that work together to perform a principal life function. All systems of the body work together efficiently to sustain the body as a living entity. The 11 major systems of the body include:

A. Integumentary

The integumentary system covers and protects the body. It includes the skin which protects the body against external agents and helps maintain a constant internal environment.

B. Skeletal

The skeletal system includes the bones and connective tissues which hold the bones together. This system supports the body, but still allows movement of the bones for body locomotion.

C. Muscular

The muscular system consists of over 600 muscles which move parts of the bony skeletal system to give the body locomotion.

D. Digestive

The digestive system takes in food, breaks the food up physically and chemically, and absorbs these substances into the blood.

E. Respiratory

The respiratory system takes oxygen from the air, transfers it to the blood which delivers it to the cells where it reacts with food substances to release energy required for life, takes carbon dioxide (a product of reaction of the oxygen with the food substances) from the cells and transports it in the blood and then removes it from the blood and expels it from the body.

F. Circulatory

The circulatory system consists mainly of the heart, blood fluid, blood vessels, lymph fluid, and lymph vessels. It transports food substances, oxygen, carbon dioxide, and waste materials from one part of the body to another.

G. Excretory

The excretory system eliminates waste products from the body. The kidneys, an important part of the excretory system, remove wastes from the blood. Sweat glands excrete a small portion of body wastes. Wastes also are expelled from the body by the rectum.

H. Nervous

The nervous system consists of the brain, spine, and peripheral nerves that are spread throughout the body. The nervous system controls many body activities. It sends impulses throughout the body to coordinate the activities of the systems.

I. Sensory

The sensory system includes several organs which receive stimuli from the outer environment and from various regions of the body. Stimulation of an organ results in impulses being sent to the brain via the nerves then the brain interprets the impulses to produce various sensations.

J. Endocrine

The endocrine system consists of several different glands which secrete hormones which control the rates of chemical reactions in the body.

K. Reproductive

The reproductive system consists of organs which provide the egg cell in the female and the sperm cell in the male, and which permits the union of these two cells in the female to result in a single cell which then divides into two cells, into four cells, and finally into many cells gradually becoming a human body.

IV. RESPIRATION

Respiration consists of all mechanisms of the body that serve to take oxygen from the air and deliver it to the cells, whereby it reacts with food substances in the cells to release energy needed for life. Respiration also includes removal of the carbon dioxide, the product of reaction of the oxygen with the food substances, from the cells, and the expulsion of the carbon dioxide from the body.

A. Internal Respiration

Internal respiration involves utilization of oxygen and food substances inside cells to release energy which the cells use to perform their special functions. Internal respiration is complex and it involves many chemical reactions. The following is a very simplified explanation of internal respiration that occurs inside cells.

Food substances are in the form of carbohydrates (the chief food substance), fats, and proteins. Oxygen reacts with food substances to release energy. The reaction of oxygen with food substances which are composed mainly of carbon and hydrogen results in conversion of these substances into carbon dioxide and water with the release of energy.

Some of the energy released is in the form of heat and the remainder is utilized to cause another chemical reaction to occur. This reaction involves the joining of ADP (adenosine diphosphate) and phosphate to produce ATP (adenosine triphosphate). Whenever a cell needs energy, the ATP breaks apart instantly into ADP and phosphate. In doing so, energy is released that is utilized by the cell to perform the life functions of the cell and to carry out the special functions of the cell. Whenever a cell demands energy rapidly and in large amounts, ATP decomposes into ADP and phosphate to release energy. Then, another substance, PC (phosphocreatine) immediately breaks down into creatine and phosphate to release energy which is used to rapidly cause ADP and phosphate to react together for more ATP. The energy release from the breakdown of PC occurs much more rapidly than energy can be obtained by the reaction of oxygen and food substances. Some of the energy obtained from the breakdown of the ATP is used to reform PC from the creatine and phosphate.

B. External Respiration

External respiration includes all the mechanical operations of the body involved in taking oxygen from the air, transporting the oxygen to the cells of the body where it is used to release energy, removing the carbon dioxide created in the cells from the cells, and expelling the carbon dioxide from the body. Breathing is the part of external respiration that involves the movement of air into and out of the body.

C. Respiratory Tract

The respiratory tract includes those organs of the body involved in the breathing phase of external respiration. Figures 1 and 2 show the main parts of the respiratory tract which include:

1. Nasal Cavity

The nasal cavity allows air to enter and leave the respiratory tract. It is divided into two passageways by a centrally located wall called the nasal septum. Each passageway is open to the outer atmosphere through openings called the nares or nostrils. Within the nares are hairs that project outward into the passageways. Several curved projections of tissue called turbinates project from the outer walls into each passageway. These passageways are lined with a membrane called the mucous membrane which on its surface contains cells that secrete a watery fluid called mucous and cells that contain minute fiber-like structures called cilia.

2. Mouth (Buccal) Cavity

The mouth (buccal) cavity permits air to enter and leave the respiratory tract. It also permits the intake of foods into the digestive tract. The mouth's upper wall is called the soft palate and its lower wall is called the hard palate. The mouth contains the tongue and the teeth. Its walls are lined with cells that secrete mucous.

3. Pharynx

The pharynx is an upright and round-shaped passageway that connects the nasal cavity and the buccal cavity at its upper end, and at its lower end divides into two separate passageways: the esophagus of the digestive tract and the larynx of the respiratory tract. The esophagus is a large tube through which food passes from the

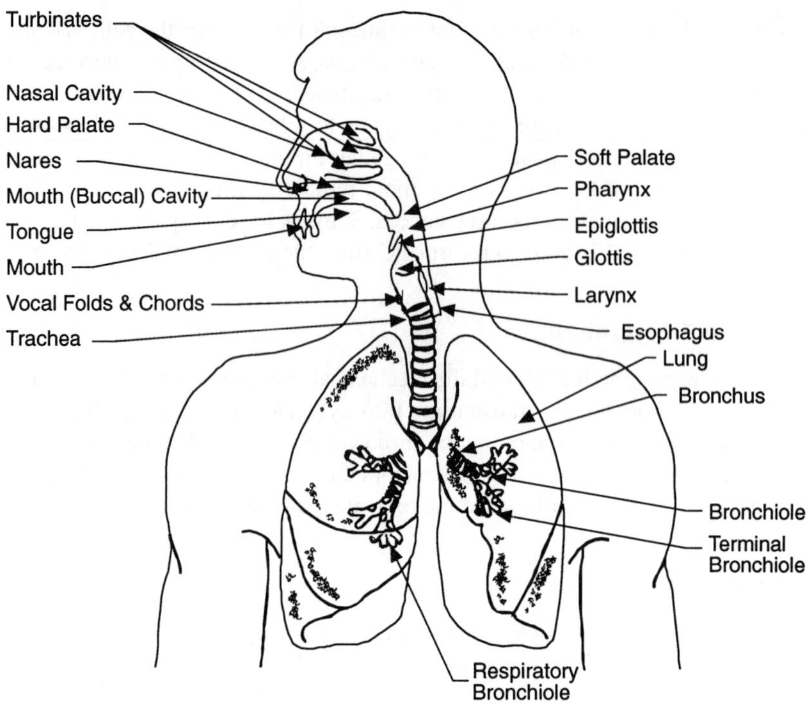

Figure 1 Respiratory tract.

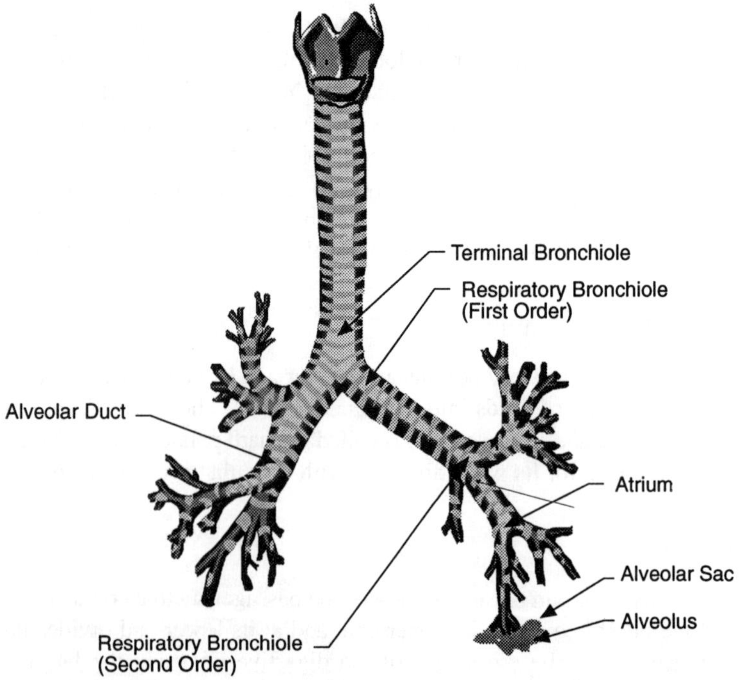

Figure 2 Terminal respiratory surfaces.

pharynx to the stomach of the digestive tract. The wall of the pharynx is lined with a mucous membrane that secretes mucous fluid. It also contains cells with cilia. Because the wall of the pharynx is supported by bone, it cannot be collapsed.

4. Larynx

The larynx is a tube-shaped organ which connects the pharynx to the trachea. The opening of the pharynx into the larynx is called the glottis and at this junction is a flap of tissue called the epiglottis which automatically moves to close the larynx whenever food is being swallowed to prevent food from entering the respiratory tract. Instead, the food slides into the esophagus. The epiglottis opens automatically during breathing to permit air to pass into and out of the larynx.

Each side of the larynx contains folds of tissue called vocal chords. Muscles can position these tissue folds so that their edges are either close together or far apart, and also so that they are stretched tightly or loosely relaxed. Air forced out of the lungs which moves upward through the larynx cause the positioned vocal chords to vibrate, and these vibrations produce sound waves in the air which are called the voice. The position and tension of the vocal chords determine the pitch of the sound produced. The position of the tongue, cheeks, palate, and lips affect the quality and complexity of the sound produced. The larynx also has a membrane lining with cells that secrete mucous fluid and cells that possess cilia.

5. Trachea

The trachea is a large diameter tube that sometimes is called the windpipe. The trachea divides into two tubes, each of which is called a bronchus. The wall of the trachea contains cartilage rings imbedded in connective tissue and muscle tissue which prevents collapse of the trachea. The inside lining of the trachea is a membrane which contains cells that secrete mucous and cells that contain cilia.

6. Bronchus

Bronchi are the plural of bronchus. One bronchus leads to the right side of the chest, and the other leads to the left side of the chest. Each bronchus is somewhat similar in structure to the trachea. The wall of a bronchus contains cartilage rings smaller than those in the trachea. As a bronchus extends into the chest, these cartilage rings change in form into isolated cartilage plates. The cartilage structures prevent collapse of the bronchi. The inner lining of a bronchus is a membrane which consists of both cells that secrete mucous and cells that possess cilia. Each bronchus divides into smaller tubes called bronchioles.

7. Bronchioles

The bronchioles are small diameter tube-shaped passageways. The bronchioles continue to divide again and again into ever smaller diameter tubes. The walls of the bronchioles do not contain cartilage supports but instead are composed of muscle tissue. The membrane which lines the walls of the bronchioles contains cells that secrete mucous and cells that contain cilia. As the bronchioles continue to branch out, their walls become thinner and thinner, but the inner membrane lining still contains cells that secrete mucous and cells that possess cilia.

8. Alveolar Ducts, Atria, Alveolar Sacs, Alveoli, and Pulmonary Membrane

The terminal bronchioles subdivide into very small tubes called alveolar ducts. The thin walls of the alveolar ducts and structures beyond them do not contain cells that possess cilia.

The alveolar ducts subdivide into tubes called atria (atria are the plural of atrium).

Extending from each atrium are three to six cavities called the alveolar sacs.

Projecting from the walls of the alveolar sacs are a several cup-like cavities called alveoli (alveolus is the singular of alveoli).

The walls of the alveolar ducts, atria, alveolar sacs, and alveoli comprise a membrane known as the pulmonary membrane. This membrane is very thin. Cells in this membrane secrete a fluid which covers the inner walls of the alveolar ducts, atria, alveolar sacs, and alveoli.

9. Cilia

Previous discussion indicated that the membrane lining of the nasal cavity, pharynx, larynx, trachea, bronchi, and bronchioles contain cells that possess cilia. A cilium is a minute fiber-like structure that projects from a cell's surface. A cilium is in constant motion. It moves in one direction very slowly with little force, and then it moves in the opposite direction very quickly and with much force. The cilia on the surface of a membrane are oriented so that the fluid (mucous) and objects immersed in the fluid are moved in the direction of the rapid and forceful motion of the cilia. The cilia located on the inner membrane walls of the pharynx, larynx, trachea, bronchi, and bronchioles move slowly with little force in the direction toward the terminal endings of the respiratory tract, and then move very suddenly and rapidly with great force in the opposite direction toward the mouth cavity. This motion of the cilia continuously moves the mucous on the inner walls of the mentioned tubular organs of the respiratory tract toward the mouth cavity where the mucous is either expectorated from the mouth and thus expelled from the body or is swallowed to go into the esophagus of the digestive tract. The mucous on the inner walls of the tubular passageways of the respiratory tract is never stagnant. It is always moving upward in the respiratory tract toward the mouth cavity by the continual ciliary action. In this way, the respiratory system is continuously ridding itself of contamination and debris.

10. Alveolar Phagocyte

The alveoli contain cells known as alveolar phagocyte which are capable of ingesting particles of matter that settle out of the air in the alveoli onto the surface of the pulmonary membrane. These alveolar phagocyte cells possess the ability to move independently although some may remain attached to the pulmonary membrane. It is theorized that these phagocyte cells are generated in the pulmonary membrane. Alveolar phagocyte tend to migrate to the respiratory bronchioles (the bronchioles connected to the alveolar ducts) where the movement of the mucous by the ciliary action will carry these phagocyte cells out of the respiratory tract via the bronchioles, bronchi, trachea, larynx, pharynx, and finally into the mouth cavity where the mucous is either expectorated and thus expelled from the body or is swallowed to enter the digestive tract via the esophagus.

11. Lungs

The cluster of connecting bronchioles, alveolar ducts, atria, alveolar sacs, and alveoli is called a lung. There are two lungs in the human body, one on the right side of the chest and one on the left side of the chest. A lubricated membrane called the visceral pleura covers the lungs. The chest cavity where the lungs are located, also called the thoracic cage, is lined with another lubricated membrane called the parietal pleura.

D. Other Organs Associated with Respiration

The thoracic cage is enclosed in the front by the sternum (breast plate), in the rear by the spinal column, encircled on the sides by the ribs, and bounded on the bottom by a dome- shaped muscle called the diaphragm.

The lungs are elastic and can slide freely within the thoracic cage. When the thoracic cage enlarges in volume during the inhalation phase of breathing, the lungs also enlarge in volume. Likewise, when the thoracic cage decreases in volume during the exhalation phase of breathing, the lungs also become smaller in volume.

Extremely fine elastic fibers form a continuous network in the lungs to encircle the vessels of the lungs and hold them in position to permit the lungs to expand and inflate during inhalation and to enable the lungs to diminish in volume and deflate during exhalation.

The alveolar ducts, atria, alveolar sacs, and alveoli are surrounded by extremely fine blood vessels called capillaries. The capillaries surrounding the alveoli are known as the pulmonary capillaries. The walls of the pulmonary capillaries are very thin.

The heart, a component of the circulatory system, is located behind and beneath the sternum and between the two lungs. It is a cone-shaped muscular organ that pumps blood through the body via the vessels of the circulatory system.

Large blood vessels called pulmonary arteries lead from the heart and subdivide repeatedly until they become the fine pulmonary capillaries that surround the alveoli. These capillaries join repeatedly into larger blood vessels called pulmonary veins which return back to the heart.

E. Details Concerning Respiration

Breathing is the result of successive enlargement and contraction of the thoracic cage. The inhalation phase (also called inspiration) involves taking air into the respiratory tract by an enlargement of the thoracic cage and the exhalation phase (also

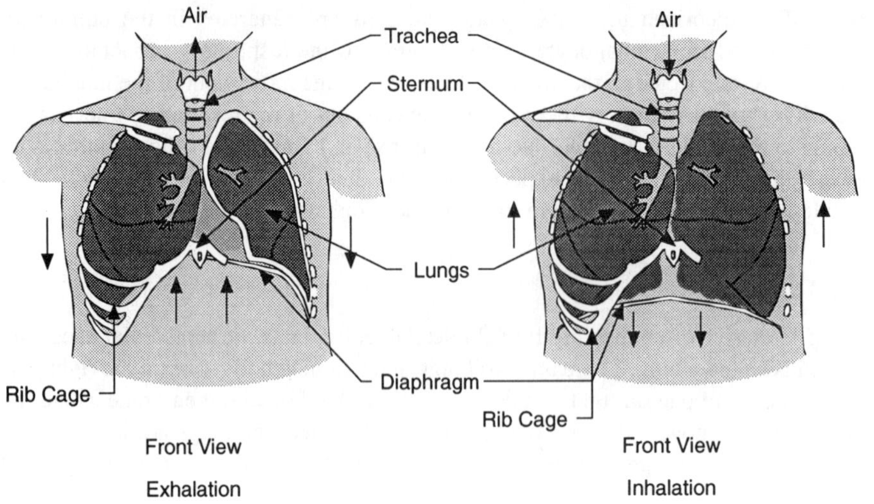

Figure 3 Sketch showing changes in sizes of lungs and position of ribs, sternum, and diaphragm during exhalation and inhalation.

called expiration) involves expelling air from the respiratory tract by a contraction of the thoracic cage. Expansion of the thoracic cage during inhalation requires muscular force that results from the contraction of certain muscles. Exhalation is more passive involving little muscular force, but instead it involves a relaxation of muscles (see Figure 3).

During inhalation, muscles connected to the ribs (external intercostal muscles) contract to cause the front ends of the ribs to move upward and outward, and small muscles located in the back of the neck also pull upward on the front of the thoracic cage. These muscular actions increase the size of the thoracic cage in cross-section. Simultaneously, the diaphragm contracts and moves downward which increases the elongation of the thoracic cage. When the thoracic cage enlarges in volume during inhalation, the elastic structure of the lungs permits them to fill the enlarged space of the thoracic cage. This increase results in a decrease in the pressure of the air in the lungs, and air from the atmosphere which is at a higher pressure will flow into the lungs via the nasal and buccal cavities, pharynx, larynx, trachea, bronchi, bronchioles, and so forth, and finally into the alveoli.

During exhalation, the external intercostal and neck muscles relax and thus lengthen to permit the front ends of the ribs to move downward and inward. Also, during exhalation, the diaphragm relaxes and moves upward. These actions decrease the volume of the thoracic cage to its original volume. This decrease causes the lungs to deflate. As the lungs deflate, the pressure of the air in the lungs increases. When the pressure of the air in the lungs increases above that of the atmospheric air, air is pushed out of the lungs via the various passageways of the respiratory tract and is expelled to the atmosphere. There is some muscular effort during exhalation which helps cause a contraction of the thoracic cage, but this muscular action is small in magnitude compared to that involved in inhalation. Abdominal muscles contract to help pull down on the front of the thoracic cage. These muscles also pull upward on the abdominal contents. Another set of muscles attached to the rib ends (internal intercostal muscles) help pull the ends of the ribs downward and inward.

It was stated previously that the pulmonary membrane which comprises the walls of the alveolar ducts, atria, alveolar sacs, and alveoli is very thin. It was also stated previously that the membrane wall of the pulmonary capillaries is only a few cells in thickness. These very thin membrane walls allow gases to readily pass through them. The total area of the pulmonary membrane of an adult human is about 30 m^2 (330 ft^2) during exhalation, and this area expands to about 100 m^2 (1080 ft^2) during the deepest inhalation. The pulmonary capillaries have a surface area of about 140 m^2 (1500 ft^2). This means that there is a very large area in the lungs through which gas molecules can pass from the air in the alveoli of the lungs to the blood in the pulmonary capillaries and through which gas molecules can pass from the blood in the pulmonary capillaries to the air in the alveoli of the lungs.

The pulmonary artery leaves the heart and subdivides repeatedly until the vessels become the very fine pulmonary capillaries that surround the alveoli of the lungs. These capillaries then join repeatedly into larger blood vessels called pulmonary veins which finally become a single large pulmonary vein that returns blood to the heart which the heart had pumped through the pulmonary capillaries. Another artery with this blood leaves the heart and divides repeatedly into smaller arteries which go to various parts of the body. These blood vessels become fine capillaries that surround cells in all parts of the body. These capillaries then repeatedly join to form larger blood vessels called veins which finally become a single large vein that returns the blood that had passed through the capillaries surrounding the cells to the heart.

When blood pumped by the heart passes through the capillaries which surround cells in various parts of the body, some of the oxygen in this blood leaves the blood and passes through the thin membrane wall of the capillaries and the very thin membrane wall of the cells where the oxygen reacts with the food substances in the cells. Thus, the blood which returns to the heart from these capillaries via veins contains less oxygen than the blood which had left the heart and had been pumped to these capillaries via arteries. The blood which had been pumped by the heart to cells in various parts of the body previously had come from the capillaries that surround the alveoli in the lungs. Therefore, blood pumped by the heart through the pulmonary artery, the pulmonary arteries, and into the pulmonary capillaries which surround the alveoli in the lungs will contain less oxygen than when it had left these pulmonary capillaries and had returned to the heart and then had been pumped by the heart to the capillaries surrounding the cells. This means that the concentration and pressure of the oxygen in the air in the alveoli are greater than the concentration and pressure of the oxygen in the blood entering the pulmonary capillaries. Therefore, oxygen will pass from the air in the alveoli through the thin membrane walls of the alveoli and the thin membrane walls of the pulmonary capillaries and enter the blood being pumped by the heart through the pulmonary capillaries.

Blood is composed of liquid plasma which contains large quantities of several different types of cells — the most numerous type being red blood corpuscles. A small portion of the oxygen entering the blood dissolves in the plasma, but most combines immediately with an iron-protein compound called hemoglobin which is contained in the red blood cells. The hemoglobin gives these red blood cells their red color. Each red blood cell contains about 265 million molecules of hemoglobin. Each molecule of hemoglobin can combine with one molecule of oxygen. There are about 25 billion red blood corpuscles in the adult human body. This means that the blood of an adult human can combine with about 6,625,000,000,000,000,000 molecules of oxygen. This blood returns from the pulmonary capillaries to the heart where it is pumped out through the arteries to various parts of the body where the arteries subdivide into tiny thin wall capillaries which surround the cells of the tissues throughout the body.

When the blood passes through the capillaries surrounding the cells, oxygen is released by the hemoglobin of the red blood cells, passes through the thin walls of the capillaries, diffuses through the fluid surrounding the tissue cells, enters the cells by passing through the very thin walls of the cells, and then the oxygen reacts with the food substances in the cells to form carbon dioxide and water and to release energy. Water molecules diffuse through the very thin walls of the cells and pass into the fluids surrounding the cells. The carbon dioxide molecules also pass through the very thin walls of the cells, diffuse through the fluid surrounding the cells, and passes through the thin membrane walls of the capillaries surrounding the tissue cells. A small portion of the carbon dioxide that enters the blood passing through these capillaries dissolves in the plasma of the blood and reacts with water in the plasma to become carbonic acid, and since the plasma is slightly alkaline, this carbonic acid becomes sodium carbonate, a salt. Most of the carbon dioxide enters the red blood corpuscles where it is carried in two different ways. Some of it becomes the salt, sodium bicarbonate, and some reacts directly with the hemoglobin and is chemically joined to the hemoglobin, but at points different than those where oxygen is held.

Air from the atmosphere enters the respiratory tract and passes into the alveoli when the lungs expand during the inhalation phase of breathing. The passage of oxygen from the air in the alveoli into the blood in the pulmonary capillaries occurs because the concentration and pressure of the oxygen in the alveoli is greater than the concentration and pressure of the oxygen in the blood in the pulmonary capillaries. The passage of oxygen from the blood in the capillaries surrounding the tissue cells into the tissue cells occurs because the concentration and the pressure of the oxygen in the blood in these capillaries is greater than the concentration and pressure of the oxygen in the tissue cells. The passage of carbon dioxide from the tissue cells into the blood in the capillaries surrounding the tissue cells takes place because the concentration and pressure of the carbon dioxide in the tissue cells is greater than the concentration and pressure of the carbon dioxide in the blood in the capillaries. The blood from the capillaries surrounding the tissue cells returns to the heart via the veins, and the heart then pumps it back to the pulmonary capillaries of the lungs via the pulmonary arteries.

When the blood passes through the pulmonary capillaries, carbon dioxide is released from its chemical combination with the hemoglobin of the red blood cells, from the sodium bicarbonate salt in the red blood cells, and from the sodium bicarbonate salt in the blood plasma. The released carbon dioxide molecules pass through the thin wall of the pulmonary capillaries and through the thin wall of the alveoli, and the carbon dioxide molecules enter into the air inside the alveoli. The passage of the carbon dioxide into the air inside the alveoli occurs because the concentration and pressure of the carbon dioxide in the blood in the pulmonary

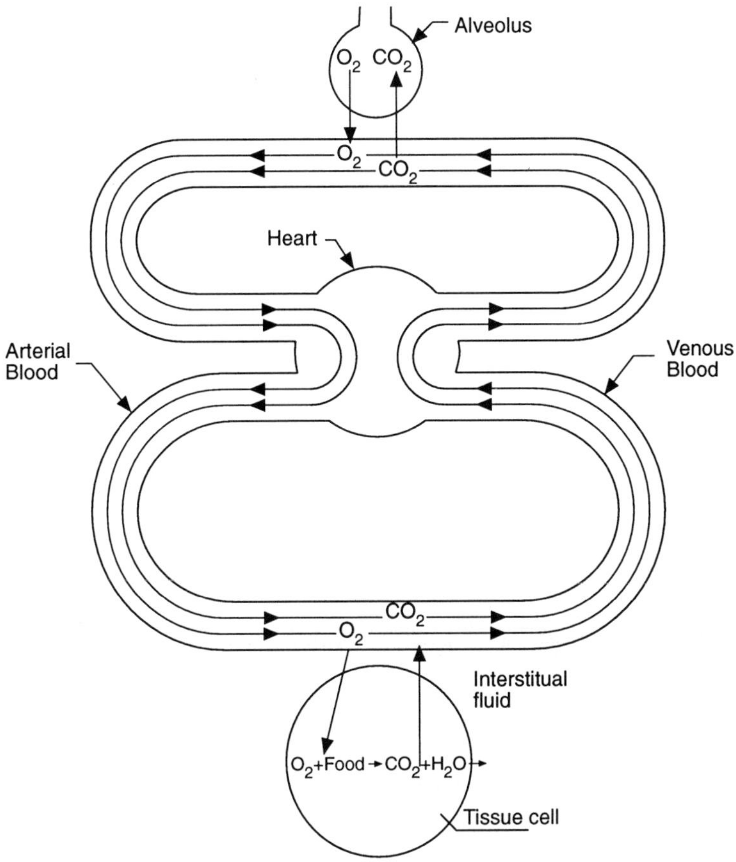

Figure 4 Transport of oxygen and carbon dioxide.

capillaries is greater than the concentration and pressure of the carbon dioxide in the air inside the alveoli. The carbon dioxide is expelled from the body when air is expelled from the lungs during the exhalation phase of breathing. The exchange of oxygen and carbon dioxide between the alveoli of the lungs and the tissue cells is shown in Figure 4.

In a previous discussion of the pulmonary membrane that forms the walls of the alveolar ducts, atria, alveolar sacs, and alveoli, it was stated that cells in this membrane secrete a fluid which covers the inner walls of these organs. This fluid does not impede the movement of gas molecules through the very thin pulmonary membrane so that gas molecules can pass readily from the air inside the alveoli into the blood flowing through the pulmonary capillaries which surround the alveoli and so that gas molecules can pass readily from the blood in the pulmonary capillaries into the air inside the alveoli. The film of fluid on the inner wall of the pulmonary membrane is kept to a minimum because this fluid has the tendency to diffuse through the pulmonary membrane wall and through the membrane wall of the pulmonary capillaries to pass into the plasma of the blood being pumped by the heart through the pulmonary capillaries.

Some of the excess water produced in the tissue cells when oxygen reacts with food substances in the tissue cells that passes into the fluid surrounding the tissue cells diffuses into the blood plasma. However, much of the excess water in the fluid surrounding the tissue cells is drained off by lymphatic ducts which empty into veins returning the blood to the heart. Excess water is removed from the body by perspiration, excretion, and exhalation. Removal of water by exhalation occurs when water on the moist surfaces of the respiratory tract vaporizes and enters the air in the respiratory tract and is eliminated from the body during the exhalation phase of breathing.

The lungs are never completely deflated during exhalation, and thus the lungs are not completely emptied of air during exhalation, nor do the lungs fill up only with fresh air during inhalation. Therefore, the air in the alveoli contains less oxygen and more carbon dioxide than is present in the atmospheric air.

Atmospheric air is composed mainly of nitrogen gas (about 79% by volume) and oxygen gas (about 21% by volume), and it contains only trace amounts of other gases. The amount of water vapor contained in atmospheric air varies widely, but is only a small percentage of the total gaseous constituents of the atmospheric air. Only a very small percentage of the atmospheric air is carbon dioxide gas (about 0.04% by volume).

Nitrogen gas which has a high concentration in the atmospheric air dissolves in the blood and in the fluids in the body including that in the tissue cells. Nitrogen does not react with any of the substances in the body, and thus can be considered to be inert under ordinary conditions. Its concentration in the body remains constant.

The air that passes into and out of the respiratory tract with each respiration is called the tidal air. The volume of the tidal air for an adult human is about 0.5 l. Since the normal breathing rate for an adult human at rest is about 12 respirations per minute, about 6 l of air passes into and out of the respiratory tract each minute.

A normal adult human can inhale with a maximum effort about 3 l of air more than is inhaled during a normal inhalation for rest conditions. This volume of air is called the inspiratory reserve volume. The sum of the tidal air and the inspiratory reserve volume is called the inspiratory capacity.

A normal adult human can exhale with a maximum effort about 1 l of air more than is exhaled during a normal exhalation for rest conditions. This volume of air is called the expiratory reserve volume.

After a maximum exhalation, the lungs still contain air that cannot be exhaled. The volume of this air is about 1.2 l for an adult human, and this volume of air is known as the residual volume. The sum of this residual volume and the expiratory reserve volume is called the functional residual capacity. It is the amount of air that permits oxygen and carbon dioxide to transfer into and out of the blood between periods of inspiration.

The total change in the volume of the respiratory tract resulting from a maximum inhalation and a maximum exhalation is the sum of the inspiratory reserve volume, the tidal volume, and the expiratory reserve volume. This is called the vital capacity. For a normal adult human, the vital capacity is about 4.5 l. A well-trained athletic adult male human may have a vital capacity of about 6 l.

Much of the air taken into the respiratory tract with each inhalation is exhaled without ever reaching the alveoli in the lungs. This air is useless in

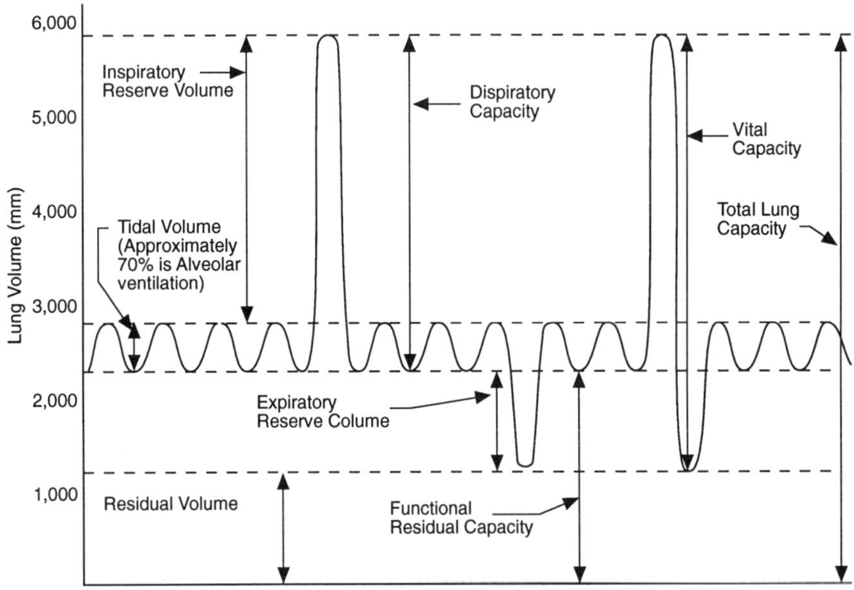

Figure 5 Spirogram showing divisions of respiratory air.

regard to oxygen entering the blood in the pulmonary capillaries and carbon dioxide leaving the blood in the pulmonary capillaries. The volume of the respiratory tract that does not function in regard to the oxygen and carbon dioxide exchange with the pulmonary capillary blood is called the dead air space. The volume of the dead air space for a normal adult human is about 0.15 l. This means that only about 0.35 l of the 0.5 l inhaled tidal air enters the alveoli and mixes with the air present in the alveoli. Figure 5 shows the lung volume change during inspiration and expiration.

F. Control of Respiration

The nervous system regulates the continuous rhythmic expansion and contraction of the thoracic cage which results in continuous rhythmic inhalation and exhalation. The respiratory control center is located in the medulla oblongata (the part of the brain connected to the spinal cord) and the pons (the part of the brain that connects the medulla oblongata to the upper sections of the brain). The respiratory control center sends impulses to the inspiratory muscles causing them to contract and thus expand the thoracic cage which results in an inhalation. A few moments later, the respiratory control center sends impulses to the inspiratory muscles causing them to stop further contraction and allowing them to relax; at the same time, impulses are received by the expiratory muscles causing them to contract. The relaxing of the inspiratory muscles and the contracting of the expiratory muscles causes the thoracic cage to decrease in volume resulting in an exhalation. This cycle repeats over and over. Reflex nerve actions tend to prevent overinflation and underinflation of the lungs and help maintain the basic rhythm of breathing.

Even though the basic rhythm of breathing is continuous, the rate of breathing and the volume of breathing vary tremendously in response to varying physiological conditions in the body. When human activity is increased, more oxygen is utilized by the cells to release energy and more carbon dioxide is produced. This means that the rate and volume of breathing must increase to bring more oxygen into the respiratory tract and to expel more carbon dioxide from the respiratory tract.

Carbon dioxide in the blood is the most powerful stimulus known to affect the respiratory control center. Increased body activity increases the concentration of the carbon dioxide in the blood, which will excite the respiratory control center to transmit more rapid and more intense impulses via nerves to the inspiratory and expiratory muscles which increase the rate and volume of breathing. This increases the rate that carbon dioxide in the air in the lungs is expelled from the body. When the concentration of carbon dioxide in the blood returns to normal when body activity decreases, the respiratory control center is no longer excited and the rate and volume of breathing then returns to normal.

A large increase in the concentration of carbon dioxide in the blood during vigorous body activity can increase the ventilation of the lungs by as much as 10 to 12 times that for normal rest conditions of the body. The concentration of carbon dioxide in the blood also affects a circulatory control center. This circulatory control center will send impulses via nerves to the heart to increase the rate of blood circulation in the body. The increase in blood circulation will increase the rate that carbon dioxide is carried from the tissue cells where it is produced to the pulmonary capillaries in the lungs where it can pass from the blood into the air inside the alveoli and then can be eliminated from the body by air expelled from the body during exhalation. A large increase in the concentration of carbon dioxide in the blood during vigorous body activity can increase the rate of blood circulation by as much as 6 to 8 times that for normal rest conditions of the body.

The described respiration control and circulatory control mechanisms can work in reverse to those discussed whenever there is a decrease in the concentration of carbon dioxide in the blood.

Under ordinary conditions, the hemoglobin of the blood in the pulmonary capillaries becomes almost saturated with oxygen, and extreme increase in ventilation of the lungs with air or moderate decrease in ventilation of the lungs with air has little effect on the amount of oxygen carried by the hemoglobin of the blood. Thus, there is no need for very acute and sensitive regulation of respiration by small variations in the concentration of oxygen in the blood to maintain a constant concentration of oxygen in the hemoglobin of the red blood cells that is necessary for normal body activity. However, there are occasions during special situations when the concentration of oxygen in the air in the alveoli falls too low to provide a sufficient amount of oxygen to the hemoglobin of the red blood cells present in the pulmonary capillaries to result in saturating the hemoglobin with oxygen. This can occur when a person ascends to a high altitude where the air has a low concentration of oxygen. This can also happen if a person enters an area where the air has an abnormally low concentration of oxygen. Certain respiratory diseases can cause a significant reduction in the ventilation of the lungs with air to cause insufficient oxygen to be present in the alveoli to provide the blood in the pulmonary capillaries with the needed quantity of oxygen for sustaining normal body activity.

In the arch of the aorta artery (the main artery of the human body extending upward from the heart in the chest), and in the left and right carotid arteries (arteries branching off the top of the aorta which pass upward in the neck into the head of the human body), small nerve endings are present that connect via nerves to the respiratory control center located in the medulla oblongata and pons. A decrease in the concentration of oxygen in the blood stimulates these nerve endings. When stimulated, these nerve endings send impulses via nerves to the respiratory control center to increase the rate and volume of breathing. This respiration control system is not as powerful a stimulator of respiration as the respiration control system which is stimulated by changes in the concentration of carbon dioxide in the blood. A very large decrease in the oxygen concentration in the blood will increase the ventilation of the lungs with air up to $1^2/_3$ times that for normal rest conditions of the body, whereas a very large increase in the concentration of carbon dioxide in the blood can increase the ventilation of the lungs with air up to 10 to 12 times that for normal rest conditions of the body.

The effect of a progressive decrease in the concentration of oxygen in the atmospheric air on breathing is not observable until the concentration of oxygen in the atmospheric air has decreased from its normal value of 20.8% by volume to about 13% by volume. At this concentration of oxygen in the atmospheric air, an increase in the rate and volume of breathing is observable. When the concentration of oxygen in the atmospheric air has decreased to about 11% by volume, a human will experience unconsciousness. A human will quickly die when the concentration of oxygen in the atmospheric air decreases to 8% by volume or less.

G. Characteristics of Breathing

Breathing is a rhythmic pulsating action. As an inspiration begins, the rate of flow of air into the respiratory tract starts from a value of zero and the air flow rate gradually increases as the lungs expand in volume until a maximum flow rate is attained, and then the rate of flow of air into the respiratory tract gradually decreases to a value of zero when the lungs stop expanding. Expiration now begins. When an expiration starts, the rate of flow of air out of the respiratory tract starts from a value of zero and the air flow rate gradually increases as the lungs decrease in volume until a maximum flow rate is reached, and then the rate of flow of air out of the respiratory tract gradually decreases to a value of zero when the lungs terminate deflation. The described breathing cycle repeats itself again and again as respiration continues. A single breathing cycle consists of a single inspiration and a single expiration.

The number of inspirations per minute or the number of expirations per minute is called the respiration rate or the breathing rate. It is the number of breathing cycles that occur in 1 min. The volume of air breathed per minute is called the minute volume. The minute volume is either the amount of air taken into the respiratory tract in 1 min or the amount of air expelled from the respiratory tract in 1 min.

An instantaneous breathing air flow rate is the value of the air flow at any single instant during the breathing cycle. The "average" instantaneous air flow rate has a value about twice that of the minute volume. The "maximum" instantaneous air flow rate has a value about four times that of the minute volume.

Table 1 Characteristics of Breathing by Healthy Adult Male Humans at Various Body Activities

Body activity	Work rate (kg-m/min)	Respiration rate (breath/min)	Minute volume (l/min)
Sedentary	0	14.6	10.3
Sitting	0	19.6	14.2
Light work	208–415	21.2–22.7	20.8–29.9
Moderate work	622–830	23.0–30.4	37.3–54.7
Heavy work	1107–1384	34.8–40.7	75.3–104.0
Maximum	1660	47.6	113.8

From Silverman, L., Plotkin, T., Sawyers, L. A., and Yancy, A., Air flow measurements on human subjects with and without respiratory resistance at several work rates, *Arch. Ind. Hyg. Occ. Med.*, 3, 461, 1951.

Table 1 lists various characteristics of breathing by healthy adult humans for various body activities. This information was obtained by the late Dr. Leslie Silverman and his associates at the School of Public Health of Harvard University.

3 AIR CONTAMINANTS

I. ATMOSPHERIC AIR

The gaseous atmosphere surrounding the earth has the following fixed composition at sea level.

Gas	Composition of atmospheric air by volume (%)
Nitrogen	78.09
Oxygen	20.95
Argon	0.92
Carbon dioxide	0.04
Neon, helium, krypton, hydrogen, xenon, nitrogen oxides, ozone	Trace amounts
Water vapor	Varies (up to 5% which dilutes other gases)

II. AIR CONTAMINANTS

Any of the normal constituents of the atmospheric air in greater than normal concentrations or any other substance present in the atmospheric air may be regarded as an air contaminant or air pollutant. Air contaminants vary widely in form (gas, vapor, liquid, solid) and in composition (elementary atomic particles, ions, simple molecules, complex molecules).

Air contaminants generally are classified according to their physical characteristics, chemical composition and properties, or physiological effects on humans.

III. PARTICULATE CONTAMINANTS

Systems that involve airborne particles consisting of discrete particles of either solid or liquid matter suspended in air may be classified according to their physical state and properties or they may be classified according to their physiological effects on the human body.

A. Aerosol

The term aerosol is often applied to a system consisting of particles suspended in air. An aerosol is a dispersal system in which air is the continuous phase or dispersing medium, and the particulate matter, in the form of discrete solid or liquid particles suspended in air, is the dispersed phase or the dispersoid.

1. Physical Classification of Aerosols

Sometimes aerosols are referred to according to the two major processes involved in their formation.

Mechanical Dispersoid. A mechanical dispersoid is defined as the particles of matter, solid or liquid, which are formed by and which are dispersed into air by mechanical means such as the disintegration processes of grinding, crushing, drilling, blasting, and spraying.

Condensation Dispersoid. A condensation dispersoid is defined as the particles of matter, solid or liquid, which are formed by and which are dispersed into air by physiochemical reactions such as combustion, vaporization, condensation, sublimation, calcination, and distillation. Generally the particles are formed by condensation from the vapor or gas state of the substance.

2. Physical Types

Dust. A dust is an aerosol in which the dispersed phase is a solid mechanical dispersoid. The particles of a dust may vary in size from submicroscopic to visible or macroscopic.

Spray. A spray is an aerosol in which the dispersed phase is a liquid mechanical dispersoid. The particles of a spray generally are of a visible or macroscopic size.

Fume. A fume is an aerosol in which the dispersed phase is a solid condensation dispersoid. The particles of a fume are generally less than 1 μm in size.

Mist. A mist is an aerosol in which the dispersed phase is a liquid condensation dispersoid. The particles of a mist vary considerably in size from submicroscopic to visible or macroscopic.

Fog. A fog is a mist that has an optical density of sufficient magnitude to perceptively intercept or obscure vision.

Smoke. Smoke usually is defined as a system which includes the products of incomplete combustion of organic substances in the form of solid and liquid particles suspended in air and gaseous products mixed with the air. The optical density of a smoke generally is sufficiently great to render it visible and to perceptively intercept or obscure vision.

Smog. A smog is a complex system which may consist of any combination of dispersoids, solid or liquid, suspended in air, gas, or vapor contaminants dispersed in the air. Smog sometimes is referred to as being a mixture of fog and smoke. The optical density of a smog usually is sufficiently great to make it visible and to perceptively intercept or obscure vision.

2. Physiological Classification of Aerosols

Nuisance and/or Relatively Inert. This type of aerosol may cause discomfort and minor irritation without causing injury. However, high concentrations of these aerosol particles may overwhelm the ability of the respiratory system to eliminate them from the respiratory tract, and large deposits of the particulate matter which remain in the respiratory tract may produce injury. Examples are marble and gypsum particles suspended as dust particles in air.

Pulmonary Fibrosis Producing. This type of aerosol produces nodulation and fibrosis in the lungs. Examples are suspensions of quartz (crystalline silica) dust particles and asbestos fibers suspended in air.

Carcinogens. This aerosol results in cancer in some persons generally occurring after a long latent period. Examples include chromate dust particles suspended in air, asbestos fibers suspended in air, and radioactive particles suspended in air.

Chemical Irritants. These aerosols produce irritation, inflammation, and ulceration generally in the upper portion of the respiratory tract. Examples include dusts, sprays, fumes, and mists which contain particles composed of acids, alkalies, or peroxides.

Systemic Poisons. These aerosols produce toxic pathological reactions in various parts of the body. Examples include dust, spray, fume, or mist that contains particles composed of lead or lead compounds which damage red blood cells to cause anemia; which injure the bowel to result in colic, constipation, and pain, which may damage the nerves to cause paralysis of certain muscles; which may damage the brain; and dust and fume containing particles of cadmium which cause lung damage and injury to the liver and kidneys.

Allergy Producing. This type of aerosol produces allergic, hypersensitive type reactions in the body such as itching of the nasal cavity, sneezing, and labored breathing. Examples include dust containing particles of pollen, plastic resins, gums, spices, fur fibers, and tobacco.

Febrile Reaction Producing. These aerosols produce chills followed by fever. Examples include fumes of zinc and copper, and dusts containing particles of certain textile fibers (cotton, hemp, jute, bagasse).

Pneumoconiosis. Pneumoconiosis literally means dust particulate matter deposited in the lungs whether the deposited particulate matter is injurious or harmless. In more common usage, pneumoconiosis has become a general term for any of the diseases of the lungs resulting from deposits of dust particulate matter in the lungs.

IV. GASEOUS AIR CONTAMINANTS

Air contaminants in the gaseous state, gases or vapors mixed with air, may be classified according to their chemical properties and composition, or they may be classified according to their physiological effects on the human body.

A. Chemical Classification of Gaseous Air Contaminants

Since there are many classes of chemical compounds, inorganic and organic, and because the chemical properties can vary widely within a class of compounds, the establishment of a detailed and complex chemical classification of gaseous air contaminants could cause confusion. Therefore, the following is a simple and broad system of chemically classifying gaseous air contaminants. Some of the classes depend on chemical properties and some classes depend on chemical composition.

1. Inert

Inert gaseous air contaminants are substances that do not react chemically with other substances under most conditions. However, these inert gaseous air contaminants may create a respiratory hazard by displacing air and thus producing an oxygen deficient atmosphere. Examples include helium, neon, argon, krypton, and xenon.

2. Acidic

Acidic gaseous air contaminants are substances such as acids or substances that react with water to produce acids. In water, they produce positively charged hydrogen ions and a pH <7. Examples of strong acidic gaseous air contaminants include hydrogen bromide, hydrogen chloride, hydrogen fluoride, sulfur dioxide, sulfur trioxide, bromine, chlorine, fluorine, nitrogen dioxide, and acetic acid. Examples of weak acidic gaseous contaminants include carbon dioxide, hydrogen sulfide, and hydrogen cyanide. The toxicity of an acidic gaseous air contaminant does not depend on the strength of the substance as an acid. Some of the most toxic acidic gaseous air contaminants are weak acids.

3. Alkaline

Alkaline gaseous air contaminants are alkalies or substances which react with water to produce an alkali. In water, they produce negatively charged hydroxyl ions and a pH >7. There are not really strong alkaline substances that exist in the gaseous state. Examples of gaseous air contaminants which are considered to be moderate to weak alkalies are ammonia and amines. Examples of gaseous air contaminants which are very weak alkaline substances include phosphine, arsine, and stibine. The toxicity of a gaseous alkaline air contaminant does not depend on the strength of the substance as an alkali. Some of the most toxic alkaline gaseous air contaminants are very weak alkalies.

4. *Organic*

Organic air contaminants are classified as vapors or gases. Organic compounds contain carbon atoms, which have four electrons each that can be shared with other atoms, including other carbon atoms, and therefore have the ability to form hundreds of thousands of compounds. The molecular structures of organic compounds are used to classify them. Some of the more common and important groups of organic compounds are gaseous air contaminants: saturated hydrocarbons (methane, ethane, propane, butane), unsaturated hydrocarbons (ethylene, acetylene), alcohols (methyl alcohol, ethyl alcohol, propyl alcohol), ethers (methyl ether, ethyl ether), aldehydes (formaldehyde, acetaldehyde), ketones (dimethyl ketone, methyl ethyl ketone), organic acids (formic acid, acetic acid), halides (chloroform, carbon tetrachloride, trichloroethylene, chlorobromoethane), amides (formamide, acetamide), nitriles (acetonitrile, acrylonitrile), isocyanates (toluene diisocyanate), amines (methylamine, ethylamine), epoxies (epoxyethane, epoxybutane, epichlorohydrin), and aromatics (benzene, toluene, xylene).

5. *Organometallic*

Organometallic compounds are those compounds in which metals are chemically bonded to organic groups. Some of the organometallic compounds are volatile and thus can act as gaseous air contaminants. Examples of organometallic air contaminants include ethyl silicate, tetraethyl lead, and organic phosphates.

6. *Hydrides*

Hydrides are compounds in which hydrogen is chemically bonded to metals and certain other elements such as metalloids. Examples of air contaminants which are gaseous hydrides include boron hydrides (diborane, pentaborane, decaborane).

B. Physiological Classification of Gaseous Air Contaminants

Air contaminants that are gases and vapors may also be classified according to their physiological effects on the human body. The physiological classification is not perfect because the physiological effects produced by many gases and vapors depend on their concentrations in air and some contaminants exert more than one type of physiological effect on the human body although one effect may predominate over the others.

1. *Asphyxiants*

Gaseous asphyxiants are substances that interfere with the supply or the utilization of oxygen in the body.

Simple asphyxiants are physiologically inert substances that reduce the oxygen supply to the body by diluting the oxygen in the atmospheric air below the concentration necessary to sustain internal respiration. Simple asphyxiants must be present in air in considerable partial pressure before they have an appreciable effect on respiration. Examples of simple asphyxiants include nitrogen, hydrogen, helium, methane, and ethane.

Chemical asphyxiants prevent either (1) the blood from transporting oxygen from the lungs to body tissue cells or (2) the tissue cells from utilizing the oxygen to release energy needed for life. Chemical asphyxiants may be dangerous even when their concentrations in air are low. Examples of chemical asphyxiants include carbon monoxide (which combines with the hemoglobin of the red blood cells and displaces the oxygen combined with the hemoglobin thus interfering with the oxygen-carrying capacity of the blood), hydrogen cyanide, cyanogen, and nitriles (which inhibit the utilization of oxygen in the tissue cells by interfering with the catalytic actions of enzymes that normally regulate the reactions of oxygen with food substances in the cells to release energy).

2. *Irritants*

Gaseous irritants are corrosive, may cause irritation and inflammation of the surfaces of the respiratory tract, and also may cause irritation and injury of the eyes and skin. Inflammation of the respiratory tract may result in pulmonary edema. In severe cases, this may effectively close the respiratory tract, filling the alveoli with fluid, severely interfering with the exchange of gases between the air in the lungs and the blood in the pulmonary capillaries. Pulmonary edema may cause death either by suffocation or by heart failure. Examples of irritant gaseous air contaminants that affect the upper portions of the respiratory tract include ammonia, hydrogen chloride, hydrogen fluoride, sulfur trioxide, formaldehyde, and acetic acid. Gaseous irritants that affect both the upper and lower parts of the respiratory tract include sulfur dioxide, iodine, bromine, chlorine, fluorine, ozone, and phosphorus trichloride. Gaseous irritants which affect mainly the lower portions of the respiratory tract are arsenic trichloride, nitrogen dioxide, and phosgene.

3. *Anesthetics*

Anesthesia is a partial or complete loss of feeling and sensation. Local anesthesia is a loss of sensation limited to a particular body area. General anesthesia is a loss of sensation over the whole body and is accompanied by a loss of consciousness. Gaseous anesthetics are substances that when inhaled exert a general anesthetic action on the body by depressing the activity of the central nervous system. Mild intoxication with anesthetics results in dizziness and loss of coordination. Intense intoxication with anesthetics results in unconsciousness, and can lead to respiratory paralysis and death. Some anesthetics, in addition to their anesthetic action, may also injure certain parts of the body. Anesthetics sometimes are called narcotics.

Examples of gaseous anesthetics which produce primarily anesthetic effects without other serious effects are nitrous oxide, hydrocarbons (such as propane, butane, ethylene, acetylene), and ethers (ethyl ether, isopropyl ether). Examples of gaseous anesthetics which injure various organs of the body in addition to their anesthetic action include carbon tetrachloride which severely damages the liver and the kidneys; chloroform which severely damages the liver and the heart; methyl chloride which severely injures the liver, kidneys, heart, and nervous system; and methyl alcohol which severely damages the nervous system especially the optic nerve.

4. *Systemic Poisons*

Gaseous systemic poisons produce injury to specific organs or specific systems of the body. Examples of gaseous systemic poisons include mercury vapor which is a protoplasmic poison that destroys the vitality of any living tissue it contacts and which chiefly injures the nervous system, kidneys, and certain glands, and which undermines general health; phosphorus which causes bone damage; hydrogen sulfide which paralyzes the respiratory control center; and arsine which results in the destruction of red blood cells and causes liver injury. Other gaseous systemic poisons include carbon tetrachloride which damages the liver and kidneys; hydrogen selenide which injures the liver and spleen, and methyl chloride which injures the nervous system (including the brain) and injures the liver and kidneys.

5. *Carcinogens*

Gaseous carcinogens cause cancer. Examples of gaseous carcinogens include vinyl chloride, benzene, and hydrazine.

V. CONCENTRATIONS OF AIR CONTAMINANTS

A. Concentrations of Airborne Particles

The concentrations of particles, solid or liquid, suspended in air generally are expressed by two different methods: (1) the mass of particulate matter per unit volume of air, and (2) the number of particles per unit volume of air. The former method is the most widely used method.

The expression of the concentration of airborne particles on the basis of mass of particulate matter per unit volume of air usually involves giving the number of milligrams of particulate matter per cubic meter of air (mg/m^3) or giving the number of micrograms of particulate matter per cubic meter of air ($\mu g/m^3$).

Stating the concentration of particles suspended in air on the basis of the number of particles per unit volume of air generally involves giving the number of millions of particles per cubic foot of air (mppcf) although sometimes the concentration is expressed as the number of particles per milliliter of air (ppml) or the number of particles per cubic centimeter of air (ppcc).

B. Concentrations of Gases and Vapors

Concentrations of gases and vapors in air usually are expressed by two different methods: (1) the mass of the gas or vapor per unit volume of air, and (2) the volume of the gas or vapor in a particular volume of air.

The expression of the concentration of a gas or vapor on the basis of the volume of the gas or vapor in a specific volume of air generally involves giving the number of parts (volumes) of the vapor or gas per million parts (volumes) of air (ppm) or the number of parts (volumes) of the gas or vapor per billion parts of air (ppb).

Stating the concentration of a gas or vapor in air on the basis of the mass of the gas or vapor per unit volume of air generally involves giving the number of milligrams of the gas or vapor per cubic meter of air (mg/m^3) or the number of micrograms of the gas or vapor per cubic meter of air ($\mu g/m^3$).

4 EFFECTS OF AIR CONTAMINANTS ON THE HUMAN BODY

I. TERMINOLOGY

An air contaminant which does not cause any adverse physiological effect on the human body can be considered to be inert.

A nuisance type air contaminant may cause discomfort and minor irritation without causing injury to the human body.

An air contaminant that can impair any vital function of the human body, temporary or permanent, or which adversely affects the health of the human body, from a slight degree to a maximum degree (death), is recognized as being toxic. Toxicity is the ability of a substance to produce an adverse physiological effect on the human body. Toxicity depends on the amount of the substance absorbed by the human body, the rate of absorption of the substance by the human body, and the site where the substance is absorbed by the human body. Toxicology is the study of the conditions under which and the mechanisms by which a substance may adversely affect the function of living tissue and so impair health.

A poison is a substance which results in an injury to tissue cells when applied to the tissue in a relatively small amount.

Dose is the specific quantity of a substance absorbed by the body.

Symptom is a sign or any functional evidence that living tissue has been adversely affected by a substance.

Acute is a term meaning short duration. Acute toxicity means that adverse effects on the body occur from a single dose of the substance or resulting from a very short time of exposure to the substance with the symptoms developing rapidly.

Chronic is a term meaning long duration. Chronic toxicity means that adverse effects on the body occur from repeated exposures to the substance over a relatively long period of time with the symptoms developing slowly.

II. ROUTES OF ENTRY OF TOXIC SUBSTANCES INTO THE HUMAN BODY

There are three routes for toxic substances to enter the human body:

1. Penetration of the skin.
2. Ingestion by the digestive tract.
3. Deposition in the respiratory tract.

III. ENTRY OF TOXIC SUBSTANCES INTO HUMAN BODY VIA SKIN

Four actions that are possible when a toxic substance contacts the skin include:

1. The skin acts as a barrier and the substance cannot injure or penetrate into the body.
2. The substance acts directly on the skin surface to produce injury.
3. The substance penetrates the surface of the skin and injures skin tissues beneath the surface.
4. The substance penetrates the skin, enters the blood stream, and is dispersed throughout the body. It may produce injury in various parts of the body.

Generally the skin is an effective barrier for the protection of the body and relatively few substances are taken into the body via the skin.

IV. ENTRY OF TOXIC SUBSTANCES INTO HUMAN BODY VIA DIGESTIVE TRACT

Accidental swallowing of a toxic substance can occur. In addition, particles in inspired air which are deposited in the respiratory tract and which are not soluble in the mucous on the walls of the respiratory tract and particles or gaseous substances which are soluble in the mucous lining the walls of the respiratory tract may be carried to the mouth cavity by the constant motion of cilia projecting from cells of the membrane lining the walls of the respiratory tract. Constant motion of the cilia tends to move the mucous toward the mouth. When the mucous containing the deposited contaminants reaches the mouth cavity, it will either be expectorated and thus be expelled from the body, or it will be swallowed. If the mucous is swallowed, the contaminants enter the digestive tract.

Because many substances taken into the digestive tract do not readily pass through the walls of the respiratory tract, they are not absorbed by the blood circulating through the body. These substances will be eliminated from the body either in feces or in urine.

Food and liquid in the digestive tract will dilute a toxic substance present in the digestive tract, and this will decrease the toxicity of the substance. Food and liquid in the digestive tract may react with certain toxic substances to produce a different substance. This new substance may be nontoxic. The new substance may be insoluble in the fluids present in the digestive tract, and thus it will be unable to penetrate through the walls of the digestive tract to be absorbed by blood circulating through the body.

A toxic substance which passes through the walls of the digestive tract and is absorbed by the blood circulating through the body will pass to the liver which may alter it and detoxify it. However, toxic substances which penetrate the walls of the digestive tract and are absorbed by the circulating blood may be carried to various parts of the body and cause injury.

V. ENTRY OF TOXIC SUBSTANCES INTO HUMAN BODY VIA RESPIRATORY TRACT

The respiratory tract is the most important route of entry of toxic substances into the body.

The surface area of the respiratory tract of an adult human during a deep inhalation is about 100 m^2 (1080 ft^2) which is very large compared to the 2 m^2 (21.5 ft^2) surface area of the skin, and is large compared to the 10 m^2 (108 ft^2) surface area of the digestive tract. The surface area of the pulmonary capillaries that surround the alveoli in the lungs of the respiratory tract also is very large, being about 140 m^2 (1500 ft^2). Thus, the area of the respiratory tract through which air contaminants can enter the body of an adult human is about 50 times that of the skin and about 10 times that of the digestive tract.

The membrane walls of the alveoli, the terminal air sacs in the lungs, and the membrane walls of the pulmonary capillaries, containing blood pumped by the heart, are very thin and allow gases and certain other materials to pass through them.

A huge quantity of atmospheric air is taken into the respiratory tract of an adult human during an 8-h work shift. A person working at a moderate rate may inhale between 10 m^3 (353 ft^3) and 25 m^3 (883 ft^3) of air in an 8-h work shift.

Thus, the large surface area of the respiratory tract, the large surface area of the pulmonary capillaries, the very thin walls of the alveoli and the pulmonary capillaries, and the huge quantity of air taken into the respiratory tract during an 8-h work shift make the respiratory tract the most important route by which toxic substances can enter the human body.

VI. ABSORPTION OF AIRBORNE PARTICLES BY THE HUMAN BODY VIA THE RESPIRATORY TRACT

Deposition of airborne particles, solid or liquid, in the respiratory tract depends on the rate and volume of breathing, on the size and density of airborne particles, the solubility of the substance in water, and the affinity of the substance for water.

In the following discussion, numerical particle sizes will be for aerodynamic particle diameters assuming spherical-shaped particles of unit density.

Inspired air passing into the two chambers of the nasal cavity is warmed and moistened by the warm and wet surface of the mucous membrane walls of the two chambers. The hairs projecting outward into these chambers just within the external openings of these passageways intercept and retain large size particles suspended in the inspired air. The turbinates projecting into these two passageways cause air passing through these chambers to change direction suddenly several times. These rapid changes in air flow direction result in the centrifugal forces acting on the particles carried in the inspired air to cause the larger and heavier particles to impinge on the mucous covering the surfaces of the passageways and thus become immersed in the mucous. The motion of the cilia of the membrane walls of the nasal cavity chambers moves this mucous containing the particles toward the pharynx.

The constant motion of the cilia moves the mucous containing the particles toward the buccal (mouth) cavity where the mucous is either expectorated (spit out) and thus expelled from the body or is swallowed and goes into the digestive tract. All particles that enter the digestive tract, if not dissolved in the digestive juices in the digestive tract, are expelled from the body in feces.

Large and dense particles (most particles larger than 10 μm) are removed from the inspired air in the nasal cavity and are carried in the mucous to the pharynx and then to the mouth cavity by ciliary action where they are either expectorated or swallowed.

Smaller airborne particles ranging in size from 10 to 3 μm, which pass through the nasal cavity, are deposited in the pharynx, larynx, trachea, bronchi, and bronchioles. Due to sharp changes in the direction of the flow of the inspired air in the nasal cavity, pharynx, larynx, trachea, bronchi, and bronchioles, centrifugal force (sometimes called inertial force) acts on airborne particles larger than 1 μm to cause these particles to impinge on the mucous film on the surfaces of these passageways of the upper respiratory tract. This process of deposition of airborne particles by centrifugal force is called impaction. As the particle size decreases, the deposition of airborne particles in the upper parts of the respiratory tract decreases, and few airborne particles smaller than 1 μm are deposited in the nasal cavity, pharynx, larynx, trachea, bronchi, and bronchioles.

Air molecules are in constant random motion. The bombardment of air molecules on low density and small size particles (usually particles less than 0.1 μm), causes these particles to move in a random, zig-zag pattern. This random, zig-zag motion increases with decreasing particle size and with decreasing particle density. The random motion of small size and low density airborne particles due to air molecule bombardment is often called Brownian motion. An English botanist named Brown discovered in 1827 that small particles, suspended either in a liquid or a gaseous fluid, would be in continual random motion due to bombardment of the particles by the liquid or the gas molecules.

In the alveoli, airborne particles have to move only a very short distance to contact with the moist pulmonary membrane. Deposition of airborne particles due to Brownian motion onto the moist pulmonary membrane of the alveoli increases as the size of the particles decreases below 0.1 μm.

The minimum deposition of airborne particles in the respiratory tract due to either impaction or Brownian motion is for particles ranging from 0.1 to 0.4 μm. Airborne particles of this range tend to remain suspended in air and if taken into the respiratory tract during inhalation, they will be expelled from the respiratory tract during exhalation.

Some airborne particles will settle out of the air and will be deposited on the surfaces of the respiratory tract due to the force of gravity. However, deposition of airborne particles in the respiratory tract resulting from settling due to the force of gravity occurs only with large size particles of high density. Deposition in the respiratory tract of airborne particles due to gravity settling is called sedimentation.

As the rate and volume of breathing increases, the centrifugal force acting on airborne particles increases and the deposition of airborne particles in the nasal cavity, pharynx, larynx, trachea, bronchi, and bronchioles increases.

As the rate and volume of breathing decreases, the Brownian motion becomes more important for deposition of small size airborne particles in the alveoli. This occurs because the movement of air into and out of the alveoli decreases which means that the time that the particles remain in the alveoli is increased to increase the opportunity for the random, zig-zag movement of small particles to contact and be deposited on the moist surface of the alveoli.

Breathing at a low rate and volume also increases the chance that airborne particles will be deposited in the respiratory tract by gravity settling.

A hygroscopic substance readily absorbs water. Airborne particles composed of a hygroscopic substance will increase in size due to absorption of water from the moisture in the air. Thus, airborne particles composed of a hygroscopic material will increase in size after it enters the humid atmosphere of the respiratory tract. This increase in particle size will increase the probability of deposition of hygroscopic particles in the respiratory tract due to impaction and sedimentation.

Airborne particles deposited on the moist walls of the nasal cavity, pharynx, larynx, trachea, bronchi, and bronchioles which are composed of substances that are insoluble in the mucous covering the walls of these organs are carried by the movement of the mucous resulting from ciliary action to the mouth cavity where they are either expelled from the body by expectoration or are swallowed to go into the digestive tract.

Airborne particles deposited in the alveoli which are insoluble in the fluid covering the walls of the alveoli may be ingested by phagocyte cells present in the alveoli. These phagocyte cells may move out of the alveoli and enter the bronchioles and then be carried in the mucous being moved toward the mouth cavity by the continual ciliary action. When they reach the mouth cavity, the phagocyte cells containing the ingested particles are either expectorated and thus expelled from the body or are swallowed to enter the digestive tract.

Some types of particles deposited in the alveoli, when ingested by the phagocyte cells present in the alveoli, destroy the phagocyte cells or immobilize the phagocyte cells. These actions can result in the accumulation of particulate matter in the alveoli. These particulate deposits in the alveoli will include a buildup of some tissue to form nodules within the alveoli.

Some particles deposited in the alveoli which are insoluble in the fluid present on the wall of the alveoli may penetrate through the membrane wall of the alveoli and enter into the fluid present in the interstitial space between the alveoli. These particles may be carried by this fluid into lymph vessels in the lungs and then will be deposited in lymph nodes located in the region in the lungs where the bronchi, blood vessels, and lymph vessels enter the lungs (the hilus). These particles deposited in the lymph nodes may result in the buildup with tissue of nodules retained within the lymph nodes. Particles penetrating into the interstitial space between the alveoli also may cause the formation of fibrous tissue in the interstitial space in the lungs.

Some insoluble particles may penetrate the walls of the alveoli and the pulmonary capillaries and enter into the blood where they are circulated to various parts of the body. Also, some insoluble particles deposited in the lymph nodes may penetrate membrane walls and enter blood vessels and the circulating blood will disseminate these particles to various parts of the body. The materials composing these particles may result in injury to tissue cells in various parts of the body. The buildup of nodules in the alveoli, the lymph nodes, and the fibrous tissue in the interstitial space in the lungs decreases the elasticity of the lungs which impedes breathing. This results in a decrease in the exchange of gases between the air in the alveoli and the blood passing through the pulmonary capillaries. The buildup of nodules in the alveoli and the buildup of fibrous tissue in the interstitial space between the alveoli reduces the surface area of the pulmonary membrane available for gas exchange between the air in the alveoli and the blood in the pulmonary capillaries.

Particles deposited in the respiratory tract which irritate mucous-secreting cells in the respiratory tract and gaseous substances in the inspired air which dissolve in the mucous in the respiratory tract and which irritate the mucous-secreting cells in the respiratory tract will increase the rate of secretion of mucous in the respiratory tract. Also, there are certain respiratory diseases which increase the rate of secretion of mucous in the respiratory tract. The increased rate of secretion of mucous in the respiratory tract will result in the buildup of a very thick mucous film on the walls of the organs of the respiratory tract which may be several times thicker than the length of cilia. This reduces the upward movement of the mucous in the respiratory tract by ciliary action to a negligible rate. This significant reduction means that particles deposited in the respiratory tract will not be effectively transported to the mouth cavity. Thus, clearance of particles from the respiratory tract becomes negligible.

Particles deposited in the alveoli that dissolve in the fluid on the membrane wall of the alveoli may cause an increase in secretion of fluid in the alveoli. The solution of particles composed of certain substances in the fluid in the alveoli may cause some disintegration of the membrane wall of alveoli to permit fluid from the interstitial space between alveoli to diffuse into the alveoli. An increased amount of fluid in the alveoli will impede the diffusion of oxygen and carbon dioxide between the air in the alveoli and the blood flowing through the pulmonary capillaries. An increase in the fluid inside the alveoli sufficient to fill some alveoli will render these alveoli unavailable for gas exchange between the alveoli and the blood in the pulmonary capillaries.

Substances from particles deposited in the alveoli which dissolve in the fluid in the alveoli may pass through the membrane wall of the alveoli and through the membrane wall of the pulmonary capillaries surrounding the alveoli to enter into the blood being pumped through the capillaries by the heart. Because the heart pumps this blood to various parts of the body, these substances may be absorbed by the tissue cells of certain organs in the body and cause injury to these organs.

Particles deposited in the mucous of the upper region of the respiratory tract which dissolve in the mucous and which do not have any effect on the cells of the membrane which line the organs of the upper portion of the respiratory tract will be carried to the mouth cavity by the upward movement of mucous caused by ciliary action, and either expelled from the body by expectoration or swallowed into the digestive tract.

Particles of some substances deposited in the upper portion of the respiratory tract which are soluble in the mucous on the membrane of the walls of the organs of the upper region of the respiratory tract may injure the cells of the membrane. The injury of the membrane may cause pain. Also, the injury of the membrane may result in increased secretion of mucous by cells in the membrane. An increase in the amount of mucous on the walls of the organs in the upper region of the respiratory tract may result in a reduction in the effective size of the opening of the passageways in the respiratory tract. In addition, injury of the membrane may cause a swelling of the membrane which will result in a decrease in the effective size of the opening of the respiratory tract passageways. A reduction in the effective size of the opening in any organ of the respiratory tract will increase the resistance to the flow of air through the organ which will make breathing more difficult.

Particles composed of certain substances deposited in the passageways of the upper organs of the respiratory tract, which may be either insoluble or soluble in the mucous on the membrane walls of these organs, may cause a constriction of some of these organs. This passageway constriction increases the resistance to the flow of air through the affected organs which impedes breathing.

Particles deposited in the mucous on the membrane walls of the respiratory tract, composed of substances which are soluble in the mucous, may penetrate the membrane walls of the respiratory tract organs, diffuse through interstitial fluid in the lungs, and pass through the walls of blood vessels in the lungs to enter the blood flowing through the vessels. The heart will circulate this blood to various parts of the body. Certain substances may cause injury to the tissue cells of particular organs in the body.

VII. ABSORPTION OF GASEOUS SUBSTANCES BY THE BODY VIA THE RESPIRATORY TRACT

A vapor is the gaseous state of a substance that is a liquid or solid at ordinary temperature and pressure. To simplify the discussion, a gas or vapor will be referred to simply as a gaseous substance.

Molecules of a gaseous substance are in constant random motion. When a gaseous substance is in contact with a liquid, some of its molecules bounce against the surface of the liquid and remain in the gaseous phase, and some of its molecules penetrate the liquid surface and become dissolved in the liquid. The dissolved molecules bounce among the moving molecules of the liquid in all directions, and some eventually reach the surface of the liquid again and bounce out of the liquid into the gaseous phase. When the number of molecules of the gaseous substance passing out of solution from the liquid equals the number of molecules of the gaseous substance passing into solution in the liquid, the solution is saturated with the dissolved molecules of the gaseous substance, and the gas molecules in the gaseous phase are in equilibrium with the dissolved gas molecules in the liquid. The amount of a gaseous substance that will dissolve in a liquid varies with the nature of the gaseous substance, the nature of the liquid substance, the pressure of the gas, and the temperature of the liquid. The quantity of a gaseous substance dissolved in a liquid increases as the pressure of the gaseous substance increases and decreases as the temperature of the liquid increases.

A gaseous substance may be nonreactive with body tissues which means that it will not be chemically altered by substances in the body. A gaseous substance that is nonreactive with body tissues is eliminated from the body mainly by the respiratory tract during breathing in its original composition.

A toxicological action in the body may be caused by a gaseous substance absorbed by the body in its original composition. A gaseous substance which is reactive with body tissues is chemically changed by the reaction into a product(s) of different chemical composition. The reaction product or products may or may not be toxic. A reactive gaseous substance absorbed by the body is eliminated by expired air or by urine or feces from the body in a different form than the original.

A gaseous substance in the inspired air that is highly soluble in aqueous solutions will be readily absorbed by the mucous on the surfaces of the passageways in the upper part of the respiratory tract, and little of this gaseous substance will reach the alveoli deep in the lungs. A gaseous substance in the inspired air having a moderate solubility in aqueous solutions may dissolve in the mucous on the membrane walls of the respiratory tract more or less uniformly throughout the respiratory tract. A gaseous substance in the inspired air having a low solubility in aqueous solutions will reach the alveoli deep in the lungs.

Some gaseous substances may severely irritate and inflame the respiratory tract. Gaseous substances which are very soluble in aqueous solution will attack the upper region of the respiratory tract. The attack by gaseous substances having moderate solubility in aqueous solution will be more or less uniform throughout the respiratory tract. However, with gaseous materials having low solubility in aqueous solution, the upper portion of the respiratory tract will suffer little, and the main damage will be deep in the lungs in the alveoli.

The severity of the effects of an irritant gaseous substance depends on its concentration in the inspired air. Exposure to a high concentration of an irritant gaseous material in the inspired air for even a short time has an intense effect on the respiratory tract.

A moderate exposure of the upper portion of the respiratory tract to an irritant gaseous substance will result in pain, swelling of the walls of the respiratory tract, and increased secretion of mucous. The swelling of the walls of the respiratory tract and the increased secretion of mucous effectively reduces the size of the opening of the passageways of the upper region of the respiratory tract which makes breathing difficult due to increased resistance to the flow of air through the passageways.

Severe irritation of the upper part of the respiratory tract not only causes inflammation of the walls of passageways and increased exudation of mucous but also results in plasma from blood vessels being passed into the respiratory tract passageways and the sloughing off of tissue from the walls of the respiratory tract organs. These reactions effectively reduce the size of the openings of the passageways of the upper region of the respiratory tract which results in increased resistance to breathing. In exceptionally severe cases of attack on the upper region of the respiratory tract by an irritant gas, death may occur from suffocation due to occlusion of the passageways of the respiratory tract.

The most serious, but least painful effects of inspiration of an irritant gaseous substance, occur when the irritant substance acts on the pulmonary membrane. This results in acute edema and may lead to suffocation. Fluid exudes from the pulmonary membranes and accumulates both between alveoli and passes into alveoli which seriously interferes with the exchange of oxygen and carbon dioxide between the air in the alveoli and the blood in the pulmonary capillaries. This also obstructs the flow of blood through the pulmonary capillaries which severely strains the heart. The loss of fluid from the blood may be sufficient to deplete the body of water which results in a decrease in the volume of blood and an increase in the viscosity of the blood. The actions of the irritant gaseous substance in the alveolar region of the lungs does not cause severe pain as does the actions of irritant substances on the upper regions of the respiratory tract. The major symptom of irritation in the alveolar region of the lungs is asphyxia in the form of cyanosis (bluish discoloration of the skin) without dyspnea (labored breathing), but slight exertion by the victim may result in immediate death. If severe edema in the alveolar region does not result in death, symptoms of bacterial infection occur because normal barriers to bacterial invasion are removed. This infection sometimes persists as a chronic condition and results in a protracted period of poor health.

A nonirritating gaseous substance which reaches the alveoli in the lungs may diffuse through the thin film of fluid lining the alveolar membrane wall, pass through the very thin alveolar membrane wall, pass through the very thin membrane wall of the pulmonary capillaries which surround the alveoli, and then enter the blood being pumped through the pulmonary capillaries by the heart. This blood returns to the heart which then pumps it to various parts of the body. The gaseous substances may diffuse from the blood into tissue cells in various parts of the body where they may react with constituents of the tissue cells to cause injuries.

Inspiration of air with a reduced quantity of oxygen caused by dilution of the oxygen in air by a gaseous substance which does not react with body tissues results in a deficiency of oxygen in the blood. This leads to physical asphyxia. An inadequate amount of oxygen in the blood means that the tissue cells of the body will receive insufficient oxygen for normal cell functions which results in a diminished amount of carbon dioxide in the blood which results in excessive breathing. In acute physical asphyxia, death may occur in a few minutes. When physical asphyxia occurs gradually, behavioral impairment and poor muscular coordination occur, followed by increased breathing and the feeling of fatigue, vomiting and nausea then occur, and finally convulsions, loss of consciousness, and eventually death.

Gaseous substances that are chemical asphyxiants include carbon monoxide and the cyanide compounds. Carbon monoxide passes from the air in the alveoli into the blood in the pulmonary capillaries and combines with the hemoglobin in the red blood cells to displace oxygen attached to the hemoglobin. This results in asphyxia with the symptoms depending on the concentration of the carbon monoxide in the air and the time of exposure of the body to the carbon monoxide. Symptoms include shortness of breath, headache, confusion, unconsciousness, respiratory failure, and death. The cyanide compounds pass from the alveoli into the blood in the pulmonary capillaries which is then pumped by the heart to various parts of the body. The cyanide compounds enter tissue cells in the body where they combine with certain enzymes that act as catalysts to control oxidation of food substances in the tissue cells to release energy needed for life. The combination of the cyanides with these enzymes inhibits the action of the enzymes to promote the oxidation of the food substances and thus the vital life functions of the tissue cells are suspended, resulting in asphyxia. Acute cyanide asphyxia results in rapid death after only a few breaths. In less than acute cases of asphyxia, headache, nausea, and a feeling of suffocation occur.

Gaseous substances which are primary anesthetics induce a general anesthetic action on the body without causing systemic effects on the body. On inhalation, a gaseous primary anesthetic passes from the alveoli into the blood passing through the pulmonary capillaries, and to the heart where it is pumped throughout the body. When the anesthetic gaseous substance is carried to the brain it causes a depression of the central nervous system. A low concentration of the anesthetic substance in the blood causes a slight impairment of judgment and a decrease in the coordination of body movements. A slightly higher concentration of the anesthetic substance in the blood results in confusion and a severe loss of muscular coordination. A somewhat higher concentration of the gaseous anesthetic in the blood produces excitement, loss of inhibitions, and convulsive muscular movements. A high concentration of the primary anesthetic material in the blood results in immobility and unconsciousness. Death may occur due to paralysis of the respiratory control center.

Gaseous substances which produce not only anesthesia but also result in organic injury to the body when inspired are known as secondary anesthetics. Generally, these substances cause damage to various organs in the body when they are present in air in concentrations too low to produce the anesthetic symptoms. Some result in irritation of the respiratory tract, and some produce injury to various body organs such as the liver, kidneys, spleen, and the central nervous system.

A number of gaseous substances act as systemic poisons when inhaled and absorbed by the body. On inhalation, the substance passes from the air in the alveoli to the blood flowing through the pulmonary capillaries and is then pumped by the heart to tissues throughout the body. Some of these substances produce degenerative changes in the liver, kidneys, spleen, and the nervous system. Certain gaseous substances transported in the blood to bone marrow where red blood cells, white blood cells, and blood platelets are generated injure the marrow which can result in anemia due to a reduction of red blood cells, a reduced resistance to infection caused by a reduction of white blood cells, and a delay in clotting time of the blood due to a decrease in platelets. Some gaseous substances carried in the blood induce dilation of blood vessels which results in a decrease in blood pressure. When certain gaseous materials pass from the air in the alveoli into the blood in the pulmonary capillaries, they alter the hemoglobin of the red blood cells to a substance known as methemoglobin. This methemoglobin holds oxygen much more firmly than hemoglobin and does not allow it to pass into tissue cells to react with food substances to produce energy needed for life, which results in anemia. Some gaseous substances depress the respiratory control center to cause respiratory failure. Certain gaseous substances are protoplasmic poisons which destroy the vitality of any form of living tissue which they come in contact with.

5 OXYGEN DEFICIENCY

I. CONCENTRATION AND PARTIAL PRESSURE OF OXYGEN IN ATMOSPHERIC AIR AND ALVEOLAR AIR

In a mixture of different gases such as air, the total pressure of the gas mixture is the sum of the partial pressures of all the gases. Gas pressure usually is measured in units of millimeters of the height of a column of mercury (mm Hg) which can be supported by the gas pressure.

The concentration of oxygen gas and the concentration of carbon dioxide gas in atmospheric air at sea level are 20.95% and 0.04% by volume, respectively. The pressure of the atmospheric air at sea level is 760 mm Hg. The partial pressure of oxygen gas and the partial pressure of carbon dioxide gas of atmospheric air at sea level are about 159 mm Hg and 0.3 mm Hg, respectively.

Inspired atmospheric air mixes with air left in the respiratory tract after expiration, which reduces the concentration of the oxygen and increases the concentration of the carbon dioxide in the mixed air in the respiratory tract. In the mixed air in the alveoli resulting from an inspiration, the partial pressure of oxygen and the partial pressure of carbon dioxide are about 110 mm Hg and about 40 mm Hg, respectively. The partial pressure of the oxygen and the partial pressure of the carbon dioxide in the blood entering the pulmonary capillaries coming from the heart after returning to the heart from tissues throughout the body are about 40 mm Hg and about 46 mm Hg, respectively. Since the partial pressure of the oxygen in the air inside the alveoli is greater than the partial pressure of oxygen in the blood entering the pulmonary capillaries, oxygen will pass through the thin membrane wall of the alveoli and through the thin wall of the pulmonary capillaries to pass into the blood in the pulmonary capillaries. For these conditions, the hemoglobin of the red blood cells which carry most of the oxygen pumped by the heart to tissue cells throughout the body becomes about 95% saturated with oxygen. Since the partial pressure of the carbon dioxide in the blood entering the pulmonary capillaries is greater than the partial pressure of carbon dioxide in the air inside the alveoli, carbon dioxide will pass through the thin membrane wall of the pulmonary capillaries, through the thin membrane wall of the alveoli, and into the air inside the alveoli.

As the altitude increases and the pressure of the atmospheric air decreases, the concentrations of oxygen and other gases in the atmospheric air do not change greatly. However, as the altitude increases, the partial pressures of oxygen and other gases in the atmospheric air are significantly reduced. Thus, it is not the concentration

Table 1 Observable Effects of Oxygen Deficiency at Sea Level Conditions

Oxygen concentration of atmospheric air at sea level (% by volume)	Observable physiological effects
16 to 12	Loss of peripheral vision Increased breathing volume Accelerated heartbeat Impaired thinking and attention and muscular coordination
12 to 10	Intermittent breathing Very faulty judgment Very poor muscular coordination Muscular exertion results in rapid fatigue which may damage the heart
10 to 6	Nausea Vomiting Inability to perform vigorous muscular exertion Unconsciousness followed by death
6 and less	Spasmatic breathing Convulsions Unconsciousness Death occurs within few minutes

measured in units of percentage by volume of the oxygen in the atmospheric air which is important in regard to respiration, but it is the partial pressure of oxygen measured in units of millimeters of mercury column that is important.

II. OXYGEN DEFICIENCY

As the concentration of oxygen in the atmospheric air is reduced, the partial pressure of oxygen in the atmospheric air is reduced which results in a reduced partial pressure of oxygen in the air inside the alveoli in the lungs. When the partial pressure of oxygen in the air inside the alveoli is reduced, the degree of saturation of the hemoglobin in the blood in the pulmonary capillaries also is reduced. Reducing the partial pressure of oxygen in the air inside the alveoli to about 60 mm Hg results in a reduction of the saturation of the hemoglobin in the blood in the pulmonary capillaries with oxygen to about 90%. When the heart pumps blood containing hemoglobin with a 90% saturation with oxygen to tissue cells throughout the body, symptoms of oxygen deficiency become observable.

As the altitude increases, the partial pressure of the oxygen in the atmospheric air decreases. This means that a higher concentration of oxygen in the atmospheric air is needed as the altitude is increased in order to increase the partial pressure of oxygen in the atmospheric air to obtain a high enough partial pressure of oxygen in the air inside the alveoli to ensure that the hemoglobin in the blood inside the pulmonary capillaries obtains adequate oxygen to prevent adverse effects on the body due to oxygen deficiency.

It is easier and more convenient to measure and use the concentration of oxygen in the ambient atmospheric air instead of the partial pressure of oxygen in the air inside the alveoli in the lungs to define oxygen deficient conditions. Therefore, values of the concentration of oxygen in the ambient atmospheric air in units of percentage of oxygen by volume are utilized to define oxygen deficient atmospheres.

Table 1 lists some observable effects of oxygen deficiency for concentrations of oxygen in the ambient atmospheric air at sea level conditions. Some adverse effects

Table 2 Oxygen Deficient Atmospheres

Altitude above sea level (feet)	Oxygen concentration below which oxygen deficiency, which is not IDLH, exists (% by volume)	Oxygen concentration below which oxygen deficiency, which is IDLH, exists (% by volume)
0 to 3000	19.5	16.0
3001 to 4000	19.5	16.4
4001 to 5000	19.5	17.1
5001 to 6000	19.5	17.8
6001 to 7000	19.5	18.5
7001 to 8000	19.5	19.3
8001 to 14000	see note	19.5

Note: Above 8000 feet altitude, an oxygen deficient atmosphere which is considered to be immediately dangerous to life or health exists when the oxygen concentration is less than 19.5% by volume.

of reduced oxygen concentration in the ambient atmosphere occur for oxygen concentrations less than 19.5% by volume but above 16% by volume; these effects are so slight that they are not readily noticed. Table 2 lists recommendations for oxygen concentrations in atmospheric air considered to be oxygen deficient but not immediately dangerous to life or health (not IDLH), and those that are oxygen deficient and immediately dangerous to life and health (IDLH).

6 REMOVAL OF AIRBORNE PARTICLES FROM AIR BY FIBROUS MEDIUM

I. INTRODUCTION

The term filter is sometimes applied to any device used to remove contaminants from air, such as particles in the solid or liquid state, gases, and vapors. However, filter more often refers to a device that utilizes fibrous material to remove particles from air, and filtration usually refers to the process of removal and retention of airborne particulate matter by a fibrous medium.

A small particle adheres to the surface of a fiber because at the point of intimate contact between the particle and the fiber there is a force of attraction. This adhesive force, which is intermolecular in nature, is known as Van der Waals force, named after a Dutch physicist who lived from 1837 to 1923 and who studied intermolecular forces.

II. MECHANISMS OF FILTRATION

A fibrous medium having characteristics desirable for removing particles from an air stream passing through it does not behave merely as a sieve removing and retaining particles because the spaces between the fibers of the medium are smaller in dimensions than the particles. A fibrous sieve could be used to cleanse the air of particulate matter, but the spaces between the fibers would be quickly blocked by retained particles. The plugged medium would offer so much resistance to air flow that it would be useless.

A fibrous medium ideal for removing particles suspended in air would have a network of fibers sufficiently open to offer a low resistance to air flow yet permitting particles to pass between the fibers. The fibrous network would be constructed so that the particles collide with the fibrous maze. A particle which collides with a fiber will adhere to the fiber because of Van der Waals force. If the collisions of particles and fibers are widely dispersed in the fibrous maze, the retained particles will be widely dispersed in the fibrous medium and thus will not block the spaces between the fibers. Consequently, the resistance offered to the flow of air will not increase excessively as the amount of particulate matter removed from the air stream and retained by the fibrous web is increased. In practice, it is possible to approach the ideal fibrous medium. An efficient fibrous medium allows only a very small penetration of particles but at the same time offers a low resistance to air flow which does not increase greatly as the amount of particulate matter removed from the air increases. For a particular fibrous medium, particle penetration and resistance to air flow are inversely interdependent.

The removal of particles from an air stream passing through a fibrous medium depends on the physical characteristics of the particles and the fibrous medium. The important characteristics of particles which influence their deposition in fibrous medium are size, shape, density, and electrical charge. Particle size and density are the most important characteristics. The significant factors of the fibrous medium which determine its effectiveness include fiber diameter, density of the medium, thickness of the medium, and the electrical nature of the fibers.

A fibrous medium removes particles by several mechanisms. The chief filtration mechanisms are: direct interception, gravity settling, impaction, diffusion, sieving, and electrostatic attraction. In the following discussion, numerical particle sizes will be used for aerodynamic particle diameters assuming spherical-shaped particles of unit density.

A. Gravity Settling

Gravity settling occurs when a particle suspended in air passing through the fibrous medium deviates from the pattern of the air flowing through the fibrous maze and settles onto the surface of a fiber to be retained there by Van der Waals force. Randomly moving air molecules restrict the fall of small particles due to gravity. As the particle size decreases, the effect of air molecular bombardment on particles increases to reduce the rate of gravity settling of the particles. For particles smaller than 1 μm, the gravity settling of particles is of minor importance.

B. Impaction

When a particle moves in a curved path, it is subjected to a centrifugal force which tends to cause it to move outward and away from the curved path. Thus, when a particle suspended in air passing through a fibrous medium begins to move in a curved path around a fiber, it is acted on by centrifugal force causing this particle to move in a less curved path. If this makes the particle collide with a fiber of the fibrous maze, the particle will be retained by the fiber because of the Van der Waals force. This mechanism of deposition is known as impaction. The centrifugal force acting on the airborne particle increases with increasing particle size and density and with the velocity of air flow past the fibers of the medium. This centrifugal force also increases with decreasing diameter of the fiber of the fibrous web since the curvature of the path of air flow around a fiber increases as the fiber diameter decreases. Impaction is one of the most important mechanisms of particle filtration by a fibrous medium. It is especially important for airborne particles larger than 1 μm.

C. Diffusion

Extremely small size particles suspended in air are driven about in random motion by the buffeting action of bounding air molecules. This random motion is called diffusion or Brownian motion (English botanist Robert Brown discovered in 1827 that small size particles suspended in either a liquid or gaseous fluid are bombarded by bounding molecules of the fluid to result in a random, zig-zag motion of the particles). This random, zig-zag motion may cause the particle to come into contact with a the surface of a fiber of the medium and be retained on the surface of the fiber due to the Van der Waals force. This mechanism of deposition is known as diffusion. It is one of the most important mechanisms of removal of particles from air passing through a fibrous maze. It is especially important for airborne particles smaller than 0.1 μm, but its effectiveness diminishes greatly as the particle size increases above 0.1 μm. The removal of particles by diffusion increases as particle size decreases, as the density of the particles decreases, as the diameter of the fibers decreases, and as the velocity of the air flowing past the fibers decreases.

D. Sieving

When particles suspended in air flowing into a fibrous medium are larger than the spaces between the fibers of the medium, the particles will be unable to penetrate between the fibers. As a result, the particles will be caught and retained by the fibrous network. Sieving is the name given to this mechanism. This process is undesirable since a relatively small quantity of retained particles can block the spaces between the fibers to result in a large increase in resistance offered by the fibrous medium to the flow of air. Most fibrous materials used to remove particles from air are designed to avoid sieving.

E. Electrostatic Attraction

Electrostatic charges set up a force field that spreads out into the space surrounding the charges. Negatively charged objects have excess electrons while positively charged objects have a deficiency of electrons. Objects with unlike charges attract each other, while objects with like charges repel each other.

Fibrous medium containing both negative and positive electrostatic charges may be produced in any of several ways. Felt is composed of wool fibers, synthetic organic fibers, or a combination of these fibers, and it has granules composed of an electrically insulting resin. An excess of negatively charged electrons exist on the surfaces of these granules. The surface areas on the fibers adjacent to these granules are positively charged. This fibrous material with the described electrostatic charges is known as "electrostatic felt." The negatively charged resin granules attract and retain positively charged particles suspended in air flowing through the fibrous maze. The positively charged surface areas attract and retain negatively charged particles suspended in air flowing through the fibrous network. The electrostatic charges in the fibrous material electrically polarize uncharged airborne particles in an air stream passing through the fibrous material. Polarization means that the electron clouds of the molecules composing the airborne particles are deformed so that one side of the particle becomes negatively charged relative to the other side of the particle which becomes positively charged. An electrically polarized airborne particle will be attracted toward the nearest accumulation of electrical charges, negative or positive, in the fibrous medium. The electrostatic field is more effective for removing large size particles than small size particles and is quite effective for airborne particles larger than 0.1 μm. Unfortunately, exposure of the electrostatic felt to very highly humid air, very hot air that softens the resin, or oily airborne particles that "wet" the surface of the resin greatly reduce the effectiveness of this fibrous material.

A fibrous medium composed of fibers made from plastic substances manufactured to have electric charges embedded inside the plastic fibers is known as an "electret" fibrous medium. The electric charges embedded in the fibers set up electrostatic fields within the fibrous web are effective for removing particles from air passing through the fibrous network. The effectiveness of the electret fibrous medium is less adversely affected by highly humid air, hot air, and oily particles than the electrostatic felt.

III. FILTERING EFFICIENCY AND RESISTANCE TO AIR FLOW

Low air flow velocities through the fibrous filtering medium enhance the filtration of small size airborne particles because more time is allowed for the diffusion mechanism of particle deposition (effective for small size particles), to occur. High air flow velocities enhance the filtration of large and dense airborne particles because high air flow velocity increases the centrifugal force that is effective for deposition of large and dense particles.

Small diameter fibers are more effective for removing particles than large diameter fibers. However, a fibrous network made only of small diameter fibers would have an excessively high resistance to the flow of air through it because the fibers would be closely packed and thus would have only very small spaces between them for the flow of air and would be rapidly plugged. Therefore, in order to utilize small diameter fibers to obtain a high filtering efficiency for airborne particles without simultaneously having an excessively high resistance to air flow, it is necessary to use an open lattice network of large diameter fibers as a means of dispersion for small diameter fibers.

Both theory and testing show that even for the most efficient fibrous filtering materials, a certain particle size range for airborne particles is the most difficult to remove from an air stream. This size range is from 0.1 to 0.4 μm. Decreasing the rate of flow of air increases the size of the airborne particle that has a maximum penetration through the fibrous medium. Increasing the rate of flow of air decreases the size of the airborne particle that has a maximum penetration through the fibrous medium. Thus, airflow can be adjusted to decrease the range of difficult to remove particle sizes.

If the flow of air is streamline, then the resistance to air flow offered by the fibrous medium increases proportionally with increasing air flow velocity. If the flow of air through a fibrous medium is turbulent, then the resistance to air flow offered by the fibrous medium increases at a continually faster rate as the air flow velocity is increased.

If the air flow velocity through a fibrous network remains constant at a specific value while the fibrous network continues to remove and retain particulate matter, the resistance to air flow offered by the fibrous network will differ markedly with variation in the pattern of distribution of particulate matter retained by the fibrous network.

If the fibrous medium employs the sieving filtration mechanism, then the openings between the fibers of the medium become rapidly plugged as filtration occurs and the resistance to air flow offered by the medium increases at a continually rapid pace. As the amount of retained particulate matter continues to increase and layers of retained particles accumulate on the surface of the fibrous medium, the resistance to air flow offered by the fibrous medium increases even more rapidly. The distribution of retained particles by a fibrous material in the form of layers of particles is called layer distribution. Since less space exists between packed aggregates of small size particles than between aggregates of large size particles, less space is available for air to flow through when layers of small size particles are present than when layers of large size particles are present. Thus, layers of retained small size particles will result in a more rapid increase in resistance to air flow offered by a fibrous medium than layers of retained large size particles.

The ideal fibrous medium for filtrating airborne particles has the following: (1) a sufficiently open latticework of fibers that allow particles to pass between the fibers without blocking the spaces between the fibers; (2) a structure that ensures eventual collision of the airborne particles with the fibers of the medium as the particles are carried by the air stream into the fibrous network; and (3) a structure that results in a wide dispersion of retained particles so that the spaces between the fibers are not rapidly reduced in size as the quantity of particulate matter retained by the fibrous medium increases. The distribution of retained particles by a fibrous material in a widely dispersed pattern is called deep bed distribution. As this type of fibrous network continues to filter an increasing amount of particulate matter, the sizes of the spaces between the fibers are reduced at a gradual pace and the resistance to air flow offered by the fibrous medium increases slowly. As filtration continues, sufficient particulate matter is eventually retained to reduce the size of the spaces between the fibers to that of the size of the particles. At this time, filtration of airborne particles by the sieving process occurs, and now the resistance to air flow offered by the fibrous medium starts to increase at an increasingly rapid rate. As filtration continues, layers of retained particles build up either within the fibrous network or on the surface of the fibrous network, and the resistance offered to the flow of air increases even more rapidly.

The quantity of airborne particulate matter carried by an air stream into a fibrous medium in a given period is called the particle load which is the product of the concentration of the particles in the air and the volume of air that flowed through the fibrous medium in the given period. The quantity of particulate matter that is filtered and retained by the fibrous medium in a given period is called the retained particle load or the particle deposit.

A fibrous medium that has a structure such that retained particles have a deep bed distribution offers a much lower resistance to air flow for a given retained particle load than the resistance to air flow offered by a fibrous medium that has a structure such that the same given particle load is in the form of layer distribution.

Liquid particles removed from air flowing through a fibrous network will be retained on the surface of the fibers in the form of globules if the viscosity of the liquid is high and will be retained on the surface of the fibers in the form of a film if the viscosity of the liquid is low.

IV. FILTER ELEMENTS

A filter element is a fabricated device containing fibrous material for filtration of airborne particles which may be either simple or complex in form and construction. A simple filter element may consist merely of a sheet of fibrous medium cut to some particular shape such as a circular disc or a rectangle. A complex filter element may be comprised of several different types of fibrous materials in series; a fibrous material may be formed into a variety of shapes; and the fibrous units may be housed in a rigid container such as a cylindrical-shaped cartridge.

Often, a relatively thick fibrous medium of low density with a somewhat open fiber latticework is employed upstream of a relatively thin fibrous medium of high density with a closely packed fibrous structure. The first medium with low to moderate efficiency for removing particles has a large storage capacity for retained particles and its resistance to air flow does not increase rapidly with increased quantity of retained particulate matter. This first fibrous medium serves as a prefilter to reduce the amount of airborne particulate matter that reaches the second fibrous medium. The second medium filters the small size particles that penetrate the first fibrous medium. This second medium serves as a final filter. Without the first medium, all the particulate matter suspended in the air stream would pass into the thin fibrous medium with the high density and closely packed fiber structure. Removal and retention of all the particulate matter by the thin, high density fibrous medium would mean a high filtering efficiency, but it would result in a rapidly increasing resistance to air flow as particulate matter is removed from the air stream and retained. The combination of the two fibrous filter materials as described results in a high filtering efficiency and a low resistance to air flow with increasing particle load.

Fibrous materials may be formed in various shapes to obtain large areas of filter material in small volumes. A large area of fibrous medium in a filter element is desirable since a large area results in a low air flow velocity which means that a low resistance to air flow offered by the fibrous network and a high efficiency for removal of small size particles is obtained. A filter element having a small volume is desirable since it is easier to handle than a large unit and requires less space for storage than a large unit. Typical shapes often utilized to obtain large areas of fibrous material in small volumes include pleats, convolutions, bellows, cones, and hemispherical domes.

7 REMOVAL OF GASEOUS SUBSTANCES FROM AIR BY SOLIDS

I. INTRODUCTION

Sorption is the process in which a solid takes up and retains gas, vapor, and liquid molecules. Sorption includes both absorption and adsorption.

II. ABSORPTION

Absorption occurs when a gas, vapor, or liquid penetrates the solid structure producing a solid solution in which the molecules of the gas, vapor, or liquid diffuse into the interior of the solid structure whereby they penetrate into the fields of the attractive force that exist between the constituent molecules, atoms, or ions composing the solid structure.

III. ADSORPTION

Adsorption is a surface phenomenon in which the molecules of a gas, vapor, or liquid are retained on the surface of a solid. Diffusion of the molecules of a gas, vapor, or liquid into extremely small size pores or capillaries of a solid, and retention of these molecules on the surface of the pores or capillaries is an adsorptive process. A pore or capillary in a solid is a long but extremely narrow channel or passageway. Solid materials used for adsorption of gases, vapors, or liquids are usually granular in nature which contain an abundance of pores or capillaries. The total surface area of the internal channels in a granule of a substance used for adsorption is much greater than the external surface area of the granule.

Adsorption occurs when the molecules of a gas, vapor, or liquid are held on the surface of a solid object, usually a granule, including the surface of the external walls of the solid object and the surface of channels in the interior of the solid object. Adsorption should not be confused with absorption in which the molecules of a gas, vapor, or liquid actually penetrate between the constituent molecules, atoms, or ions composing a solid object, such as a granule, and are retained internally in the solid object between the molecular, atomic, or ionic constituents of the solid object.

Being a surface phenomenon, adsorption usually is a very rapid process, while absorption being an internal diffusion phenomenon is usually a very slow process.

Adsorption of gases and vapors by solids is one of the chief ways to remove gaseous contaminants, gases, or vapors, from air. The solid substance that takes up and retains the gas or vapor is known as the *adsorbent.* The gas or vapor retained on the surface of the solid is called the *adsorbate.*

Whatever type of force that binds the constituent particles (molecules, atoms, or ions) of a solid together, a constituent particle located in the interior of the solid is subjected to equal forces extended in all directions; whereas a constituent particle located at the surface of the solid is subjected to unbalanced forces — the inward forces being greater than the outward forces. This results in a tendency to decrease the surface of the solid. The unbalanced forces located on the surface of the solid reach out into space beyond the solid. The molecules of a gas or vapor also are surrounded by fields that reach out into space. Thus, adsorption can be considered a process in which the fields of the adsorbed gas or vapor molecules and the fields of the constituent molecules, atoms, or ions of the solid, at the surface of the solid, are mutually satisfied.

Adsorption decreases the surface tension of the solid adsorbent which results in a decrease in surface energy. The molecules of the gas or vapor move freely in three dimensions. However, when the molecules are adsorbed onto the surface of a solid, they may be held at specific points on the solid surface, or they can move along the solid surface in either two dimensions. This means that the gas or vapor molecules retained on the surface of a solid are at a lower energy level than when these gas or vapor molecules were in the gaseous state. Since the energy of the system of the solid adsorbent and the retained gas or vapor adsorbate decreased when adsorption occurred, energy had to be released when adsorption took place. This energy is in the form of heat, called the heat of adsorption.

IV. PHYSICAL OR VAN DER WAALS ADSORPTION

The surfaces of many solids are chemically inert since the outer electron shells of their constituent particles (molecules, atoms, or ions) are satisfied by adjacent constituent particles. Gas or vapor molecules may be retained on these surfaces by mutual satisfaction of intermolecular forces of attraction of the gas or vapor molecules and the constituent particles (molecules, atoms, or ions) of the solid surface. The physical force of attraction between the retained molecules of the gas or vapor and the molecular, atomic, or ionic particles on the surface of the solid is known as Van der Waals force which is much weaker than the binding force between the molecular, atomic, or ionic constituent particles of the solid. The Van der Waals force is of the same order of magnitude as the physical force of attraction that causes molecules of a gas or vapor to condense from the gaseous state to the liquid state. This type of adsorption is called either physical adsorption or Van der Waals adsorption.

V. CHEMISORPTION OR ACTIVATED ADSORPTION

The surfaces of some solids are chemically reactive since the outer electron shells of the constituent particles (molecules, atoms, or ions) of the solid may not be fully satisfied by adjacent constituent particles. With this type of solid surface, gas or vapor molecules may be retained by chemical forces of attraction between the gas or vapor molecules and the constituent particles of the solid at the solid surface either by a transfer of electrons or by a sharing of electron pairs. The chemical type of force of attraction between the adsorbed gas or vapor and the solid is much stronger than the physical force of attraction between the adsorbed gas or vapor and the solid present when physical or Van der Waals adsorption occurs. Sometimes the chemical force of attraction between the adsorbed gas or vapor and the solid is much stronger than the force of attraction between the constituent molecular, atomic, or ionic particles of the solid. This type of adsorption is known as chemisorption or activated adsorption.

VI. VAN DER WAALS ADSORPTION AND CHEMISORPTION

When Van der Waals adsorption occurs, the heat of adsorption released is of the same order-of-magnitude as the heat of condensation, which is released when a gas or vapor is condensed from the gaseous state to the liquid state. When chemisorption occurs, the heat of adsorption released is of the order-of-magnitude of the heat of chemical reaction, which is released when two substances chemically react. The heat of adsorption released during chemisorption is much greater in magnitude than the heat released when Van der Waals adsorption occurs.

Removal of the gas or vapor adsorbate from the surface of the solid adsorbate is called *desorption.*

A gas or vapor retained by a solid via Van der Waals adsorption may be removed from the surface of the solid chemically unchanged by (1) greatly reducing the pressure of any vapor or gas, including air in contact with the solid, (2) by increasing the temperature of the solid, or by (3) a combination of both these conditions.

The force holding a gas or vapor on the surface of a solid, as a result of chemisorption, is extremely large. Because of this breaking apart the bond between the gas or vapor retained on the surface of a solid due to chemisorption requires an unusually high energy input in the form of heat (high temperature). Frequently, chemisorption is not reversible, which means that often it is not possible to remove a gas or vapor from a solid that is retained by the solid as a result of chemisorption. The bonds between the adsorbed substance and the adsorbent are so strong in this case that the adsorbed substance is not desorbed in its original chemical composition but as either a chemical compound consisting of a union between the adsorbate, a component of the adsorbate and the adsorbent, or a component of the adsorbent.

The more porous a solid is the more surface area it has for adsorption. The cross-section of the area of the pores or capillaries in a porous solid is important. These pores or capillaries may be so narrow that the molecules of some gases and vapors may not be able to enter into them. If gas or vapor molecules cannot enter the narrow channels in the solid, they cannot be adsorbed on the surfaces of the walls of these channels. Therefore, very fine pores or capillaries may act as sieves allowing penetration and adsorption of only small molecules, while keeping out and preventing adsorption of large molecules.

Sometimes, a porous solid having a large internal surface that is inert to a specific gaseous substance or a specific group of chemically similar gaseous substances is impregnated with a material that will form a chemical bond with the specific substances. The inert porous solid merely acts as a carrier for the chemically active material. The impregnated porous solid is used for chemisorption of gaseous substances. Sometimes a mixture of 2 or more solid adsorbents results in an adsorbent mass that retains a larger quantity of a gaseous substance than would the same mass of any of the solid adsorbents alone. Two or more solid adsorbents, either in an intimate mixture of the adsorbents or in a series of layers of the individual adsorbents, may be utilized in order to remove a specific gaseous substance, or a group of chemically similar gaseous substances, from air.

VII. CATALYSIS

A catalyst is a substance that influences the rate of a chemical reaction between other substances without undergoing permanent chemical change itself. Catalysis is a process in which the rate of a chemical reaction between substances is altered by a catalyst.

Catalysts can either accelerate or retard the rate of a chemical reaction. A catalyst that retards the rate of a chemical reaction is called an inhibitor. A substance which is incorporated into a catalyst to greatly increase the activity of the catalyst, even though this substance itself seems to be inactive, is known as a promotor.

Often, when a catalyst substance cannot be readily made into a porous solid with a large internal surface area, the catalyst material can be given a large surface area by incorporating it into a porous solid which is inert as a catalyst by itself. The inert porous solid merely serves as a carrier for the catalyst material which is spread out as a coating on the large surface area of the porous solid.

Catalysts are sometimes employed to accelerate the chemical reaction between a toxic gaseous substance in air and another substance to result either in a nontoxic substance or a substance which can be readily removed from air and retained by an adsorbent.

VIII. ADSORBENTS AND CATALYSTS

Solid adsorbents and solid catalysts used to remove gaseous contaminants from air generally are in the form of porous granules or porous pellets. The granule is the most common form. Granules or pellets are packed in beds of various shapes and sizes, and contaminated air is passed through these beds.

Different adsorbents and catalysts are utilized to remove toxic gaseous contaminants from air. A brief discussion of some important widely used adsorbents and catalysts employed to eliminate dangerous gases and vapors are presented below.

A. Activated Carbon

Activated carbon is an extremely porous solid composed of carbon which is available either in the form of granules or pellets. Some materials used for the manufacture of activated carbon are nut shells, wood, coal, petroleum, peat, and bones. Coconut shells consist of a very hard and porous carbon structure in the form of granules, which is a very efficient form for adsorption of gases and vapors. Pellet type activated carbons from coal and petroleum also are very effective for removing gases and vapors from air.

Several years ago, activated carbon made from coconut shells was the best type of activated carbon for adsorption of vapors and gases from air. Activated carbons made from coal and petroleum were next best in performance as adsorbents. However, modern technology improved activated carbons made from coal and petroleum and they are now about as good as activated carbon made from coconut shells.

The external surface area of a granule or pellet of activated carbon is infinitesimal compared to the internal surface area of the granule or pellet in the form of very fine pores or capillaries. Some commercial grades of activated carbon have surface areas ranging from 800 m^2/g (244,000 ft^2/oz) to 2000 m^2/g (610,000 ft^2/oz).

Activated carbon adsorbs many types of gases and vapors mainly through Van der Waals adsorption, however, some chemisorption adsorption by activated carbon is known to occur. Activated carbon is employed primarily for Van der Waals adsorption of organic gases and vapors from air since it has a great affinity for organic substances.

Activated carbon can be impregnated with selected substances which are useful for chemisorption of certain vapors and gases and particular groups of chemically similar gaseous substances. Activated carbon impregnated with salts of heavy metals such as copper sulfate, cobalt chloride, and nickel chloride is effective for chemisorption of ammonia from air. Activated carbon impregnated with oxides of chromium and manganese is used for chemisorption of acid gases such as hydrogen chloride and sulfur dioxide. Activated carbon impregnated with iodine is employed as a chemisorbent to remove mercury vapor from air by converting it to nonvolatile mercuric iodide which is retained by the carbon. Impregnating activated carbon with certain interhalogen compounds results in an impregnated activated carbon that is a superior chemisorbent for removing mercury vapor from air by converting the mercury vapor to a nonvolatile mercuric halide which is more readily held by the carbon.

B. Soda Lime

Soda lime is a porous alkaline solid available in the form of granules or pellets. Its active constituents are calcium hydroxide and sodium hydroxide. Since these substances are not porous, they are mixed with an inert porous solid such as diatomaceous earth and a small amount of Portland cement to make the product hard and strong. Soda lime needs moisture to insure its activity as a chemisorbent, and thus it is produced with a moisture content of 10 to 25% by mass.

Soda lime is used for the chemisorption of acidic gases and vapors such as hydrogen chloride, hydrogen fluoride, sulfur dioxide, and carbon dioxide.

C. Silica Gel

Silica gel is a porous solid composed of dehydrated silica (the anhydride of silicic acid). It is available in the form of granules. It is somewhat less porous than activated carbon. Its internal surface area, the surface area of pores within the solid, may vary from 500 m^2/g (153,000 ft^2/oz) to 900 m^2/g (374,500 ft^2/oz).

Silica gel is used chiefly for Van der Waals adsorption of ammonia. Silica gel has a high affinity for water vapor and is frequently used to remove moisture from air. Silica gel also adsorbs organic gases and vapors from air through Van der Waals adsorption, but it is not as effective as activated carbon for this application.

D. Activated Alumina

Activated alumina is a porous solid composed of aluminum oxide in the form of granules and pellets. It is much less porous than activated carbon and silica gel. Its internal surface area may vary from 200 m^2/g (61,000 ft^2/oz) to 350 m^2/g (122,500 ft^2/oz).

Activated alumina is used for chemisorption of acid gases and vapors from air. Activated alumina also has a great affinity for water vapor. Activated alumina impregnated with salts of heavy metals such as copper sulfate, cobalt chloride, and nickel chloride, is used as a chemisorbent in the removal of ammonia from air.

E. Molecular Sieves

Molecular sieves are porous solids composed of complex silicate compounds that contain aluminum. Molecular sieves are available in the form of both pellets and spherical beads. The surface area of pores within the molecular sieve pellets or spherical beads ranges from 600 m^2/g (183,000 ft^2/oz) to 800 m^2/g (244,000 ft^2/oz). Molecular sieves are available commercially with pores of different size cross-sections.

Molecular sieves have been used to remove organic gases and vapors from air through Van der Waals adsorption, but molecular sieve materials are not as good for adsorption of organic gases and vapors as activated carbon. Molecular sieves have a very high affinity for water vapor making them widely used for removal of water vapor from air.

F. Hopcalite

Hopcalite is a porous solid composed of a specially prepared mixture of activated manganese and cupric oxides available in the form of granules.

The chief use of hopcalite has been as a catalyst to enhance the rate of oxidation of carbon monoxide gas in air. Hopcalite also has been utilized to remove mercury vapor from air by acting as a catalyst to increase the rate of the oxidization of mercury vapor in air by the oxygen gas present in air to form nonvolatile mercuric oxide which is retained by the hopcalite.

G. Mixture of Activated Carbon and Soda Lime

An intimate mixture of activated carbon and soda lime, or a layer of activated carbon and soda lime followed by a layer of soda lime has been used to remove chlorine and phosgene from air.

8 RESPIRATORS

I. INTRODUCTION

A respirator is defined as a device worn by a person for protection against inhalation of a harmful atmosphere. The respirator provides protection to the wearer either by removing contaminants from the air before it is breathed or by supplying an independent source of respirable air.

A respirator which removes contaminants from the ambient air before the air is breathed by the wearer is called an air-purifying respirator. A respirator which supplies respirable air from a source other than the ambient atmosphere to the wearer is called an air-supplying respirator.

II. RESPIRATORY-INLET COVERINGS

All respirators are equipped with respiratory-inlet coverings which serves as a barrier against the harmful atmosphere and provides a way to connect the respirator wearer's respiratory system to an air-purifying device or source of respirable air. The two general types of respiratory-inlet coverings: "tight-fitting" and "loose-fitting".

A. Tight-Fitting Respiratory-Inlet Coverings

A tight-fitting respiratory-inlet covering consists of a facepiece that either covers the respirator wearer's nose and mouth or the respirator wearer's nose, mouth, and eyes. The covering is called a facepiece and is designed to make a gas-tight or particle-tight seal on the respirator wearer's face. Usually it is made of a molded flexible elastomer. A facepiece assembly includes the elastomeric face-covering, headbands, optional valves, and connections for air-purifying elements or sources of supplied respirable air. A facepiece that covers the nose and mouth with the lower sealing surface resting between the mouth and the chin is called a quarter-mask facepiece (see Figure 1). A facepiece that covers the nose and mouth with the lower sealing surface resting under the chin is known as a half-mask facepiece. A facepiece that covers the eyes, nose, and mouth, and usually extends from the hairline to below the chin, is called a full facepiece (see Figure 2). The full facepiece, which contains a lens for vision, theoretically provides the best and most reliable seal on the respirator wearer's face. The half-mask facepiece theoretically gives a better and more stable seal on the respirator wearer's face than the quarter-mask facepiece.

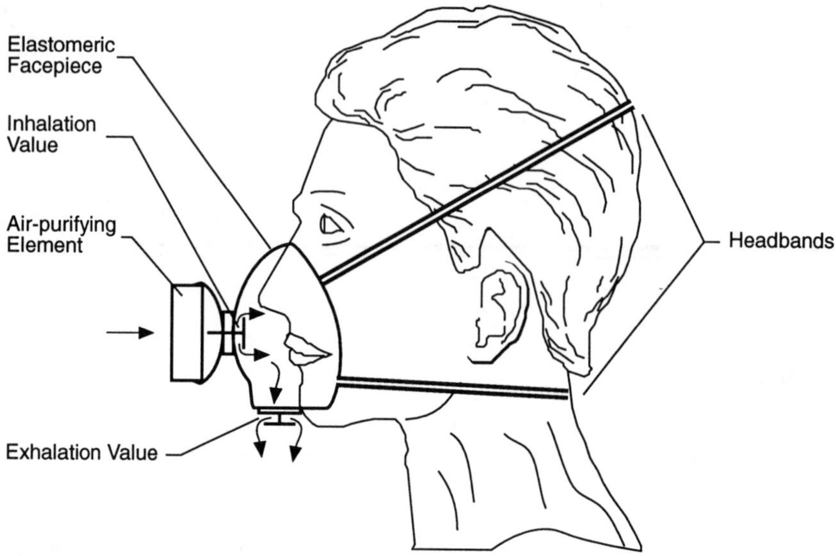

Figure 1 Reusable air-purifying respirator, nonpowered, equipped with quarter-mask type elastomeric facepiece.

There are disposable types of air-purifying respirators available. They do not contain elastomeric facepieces but instead have all fabric constructions which fit over the noses and mouths of respirator wearers. They generally are designed to protect against airborne particulate matter although some are designed to offer protection against both airborne particulate matter and low concentrations of certain gases and vapors.

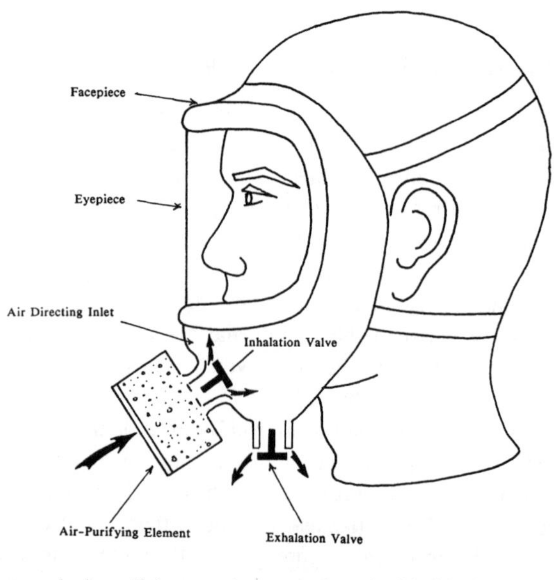

Figure 2 Nonpowered air-purifying respirator equipped with full facepiece respiratory-inlet covering and canister type air-purifying element.

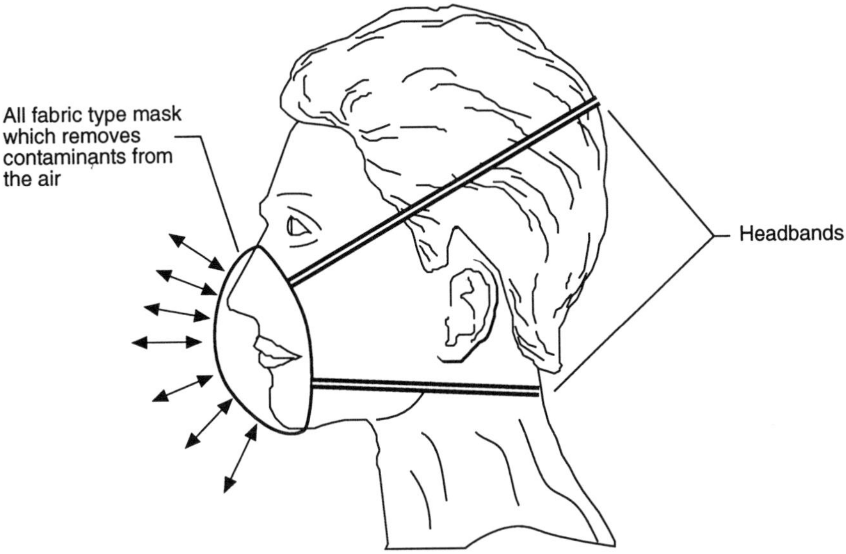

Figure 3 Disposable nonpowered air-purifying respirator with all fabric mask (valveless), half-mask type.

A fabric type nonpowered half-mask disposable respirator is shown in Figure 3.

Another type of tight-fitting respiratory-inlet covering consists of a mouthpiece and nose clamp which generally is used in respirators that are employed for escape from dangerous atmospheres. The mouthpiece is held between the teeth with the lips sealed around it and the nose clamp closes the nostrils (see Figure 4).

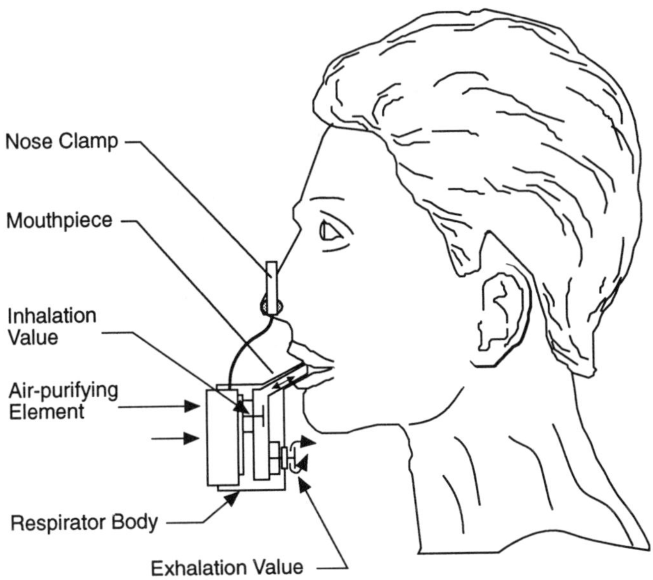

Figure 4 Nonpowered air-purifying respirator equipped with a mouthpiece and nose-clamp type respiratory-inlet covering.

B. Loose-Fitting Respiratory-Inlet Coverings

Loose-fitting respiratory-inlet coverings may cover a portion of the head, the entire head, the head and neck, the head, neck, and shoulders, the head and the upper trunk of the body (with or without the arms), or the entire body. A flexible tube or hose generally is used to supply respirable air to the loose-fitting respiratory-inlet covering. The quantity of respirable air provided to the loose-fitting respiratory-inlet covering must be sufficient to ensure that the air inside the covering has a pressure greater than that of the ambient atmosphere so that air passes out of the covering to the ambient atmosphere and so that ambient air does not leak into the covering. A loose-fitting respiratory-inlet covering can be used only with a powered type of air-purifying respirator and with an atmosphere-supplying respirator.

A loose-fitting respiratory-inlet covering which covers only a portion of the face, the nose and mouth, or the nose, mouth, and the eyes, is called a loose-fitting facepiece.

A loose-fitting respiratory-inlet covering which contains a rigid headgear designed to protect the head against injury is classified as a helmet.

A hood is a loose-fitting respiratory-inlet covering which encloses the wearer's head and neck or which encloses the wearer's head, neck, and shoulders. It usually is made of a flexible material that is impervious to the ambient atmosphere. A powered air-purifying respirator equipped with a hood type inlet covering is shown in Figure 5.

A blouse is a loose-fitting respiratory inlet-covering which envelopes the wearer's head and the trunk down to the waist. A blouse may not cover the wearer's arms, but

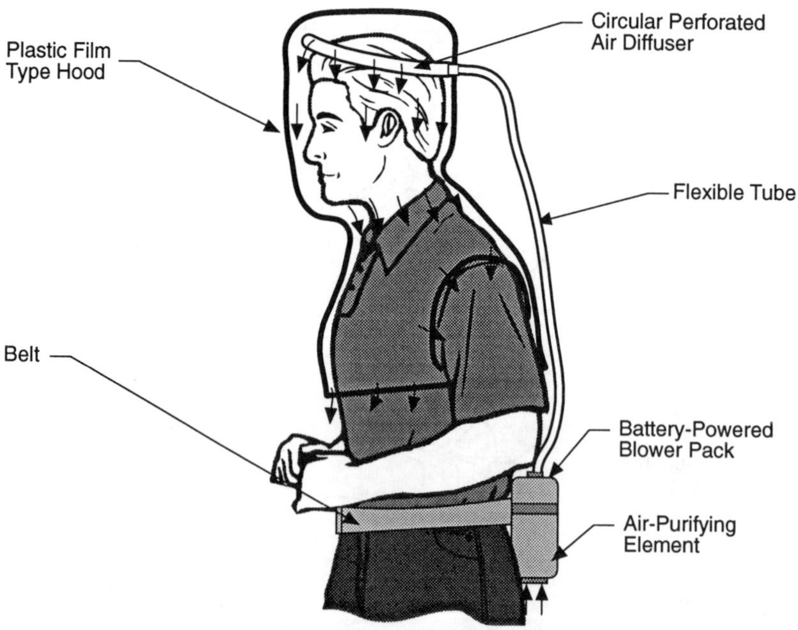

Figure 5 Powered air-purifying respirator equipped with hood type loose-fitting respiratory-inlet covering.

it sometimes is designed to include sleeves that cover the arms up to the wrists. It generally is constructed of a flexible material that is not permeated by the ambient atmosphere.

A suit type of respiratory-inlet covering encloses the wearer's entire body, and is made of a flexible material that is impermeable to the ambient atmosphere.

A hood, blouse, and suit may incorporate a rigid helmet to protect the head against injury or a facepiece to which respirable air is supplied.

A special type of loose-fitting respiratory-inlet covering is designed for use in respirators employed to protect a person engaged in abrasive-blasting. This respiratory-inlet covering uses materials that can withstand rebounding particles of the abrasive material and rebounding particles of the material being blasted.

The permeability of flexible materials used in loose-fitting respiratory-inlet coverings employed to protect persons against toxic gases and vapors including radioactive gases must be considered with utmost care.

III. CLASSIFICATION OF RESPIRATORS

Respirators generally are classified according to their mode of operation as follows:

1. Air-purifying respirators
   a. Nonpowered
      i. Particle-removing
      ii. Gas-vapor-removing
      iii. Combination particle-gas-vapor-removing
   b. Powered
      i. Particle-removing
      ii. Gas-vapor-removing
      iii. Combination particle-gas-vapor-removing
2. Atmosphere-supplying respirators
   a. Supplied-air
      i. Airline
         - Continuous-flow
         - Demand
         - Pressure-demand
      ii. Hose-mask
         - With blower
         - Without blower
   b. Self-contained
      i. Open-circuit
         - Continuous-flow
         - Demand
         - Pressure-demand
      ii. Closed-circuit
         - Rebreather with cylinder of oxygen
         - Recirculator with oxygen self-generator
   c. Combination supplied-air and self-contained
3. Combination atmosphere-supplying and air-purifying
   a. Combination supplied-air and air-purifying

IV. AIR-PURIFYING RESPIRATORS

In the operation of an air-purifying respirator, the ambient air, prior to being inhaled by the respirator wearer, is passed through air-purifying material or a device that removes the contaminant(s) from the air. Since air-purifying respirators only remove contaminants from air, they must never be used in oxygen-deficient atmospheres.

The breathing action of the wearer operates the nonpowered air-purifying respirator. When the respirator wearer inhales, ambient air is pulled through air-purifying material or through an air-purifying element(s) that removes the contaminant(s) from the air. When the respirator wearer exhales, the exhaled breath is expelled from the respirator by passing out of the respirator through an exhalation valve or by passing through the air-purifying material or element(s).

The powered air-purifying respirator is equipped with a blower which may be carried by the respirator wearer or which may be stationary. The blower passes ambient air through an air-purifying element(s) which removes the contaminant(s) from the air, and then the blower supplies the purified air to the respiratory-inlet covering of the respirator.

The respiratory-inlet covering of a nonpowered air-purifying respirator may be a quarter-mask, half-mask, full facepiece, or a mouthpiece with a nose clamp. The respiratory-inlet covering of a powered air-purifying respirator may be any of the types of facepieces or may be any of the types of loose-fitting respiratory-inlet coverings.

An air-purifying respirator may be equipped with air-purifying material or air-purifying device(s) to remove particles, a particular gaseous contaminant, a group of chemically similar gaseous contaminants, more than one group of chemically similar gaseous contaminants, or a combination of particulate contaminants and gaseous contaminants from air.

Most nonpowered and all powered air-purifying respirators are equipped with air-purifying elements that can be replaced when their useful service lives have been depleted. An air-purifying element must be replaced whenever its resistance to breathing becomes excessive or whenever penetration of an air contaminant is detected. Some nonpowered air-purifying respirators known as disposable air-purifying respirators are constructed so that it is impossible to replace the air-purifying material or element. This type of respirator must be discarded whenever its useful service life has been depleted.

Some gaseous air contaminants cannot be detected at concentrations in air equal to or below their permissible exposure limits by sensing odor, irritation, or taste. Some air-purifying elements are equipped with indicators known as "end-of-service-life indicators" that warn respirator wearers by visual or audible means when the useful service lives of the elements have been depleted in regard to penetration of particular gaseous air contaminants. An air-purifying respirator should not be used for protection against a gaseous air contaminant which cannot be detected at or below its permissible exposure limit by sensing odor, irritation, or taste unless it contains an effective end-of-service-life indicator, or unless the U.S. Department of Labor has issued a standard pertaining to the specific gaseous air contaminant that lists conditions permissible for the use of an air-purifying respirator. An air-purifying respirator should not be employed for protection against a gaseous air contaminant if there is no information available concerning warning properties of odor, taste, or irritation for the contaminant.

V. ATMOSPHERE-SUPPLYING RESPIRATORS

The atmosphere-supplying respirator provides the wearer with a respirable atmosphere from a source that is completely independent from the ambient atmosphere surrounding the wearer. This type of respirator provides wearers with protection against inhalation of harmful atmospheres that may be contaminated with toxic particulate matter, gases, and vapors, and atmospheres that may be oxygen deficient.

A. Supplied-Air Respirators

The supplied-air respirator (SAR) uses a source of respirable air that is stationary and is remote from the wearer. This respirable air is supplied to the respiratory-inlet covering of the respirator through a flexible hose.

There are two types of supplied-air respirators: the airline respirator and hose mask.

Respirable air from an air pump, air compressor, or a compressed air cylinder is supplied to the respiratory-inlet covering of an airline respirator by means of a small diameter flexible hose. The air-supply hose generally is attached to the respirator wearer by a belt or other suitable means. At this attachment, a quick connect/disconnect coupling is used to allow the wearer to connect and disconnect the air-supply hose.

Airline respirators may be operated in the following modes: continuous-flow, demand, and pressure-demand.

The hose mask supplies respirable air from a noncontaminated area through a large diameter flexible hose to a respiratory-inlet covering. The respirator may or may not be equipped with a blower.

1. Continuous-Flow Airline Respirator

A continuous-flow airline respirator maintains a flow of the respirable air at all times. The flow of respirable air to the respiratory-inlet covering is determined mainly by the pressure of the compressed air and the length of the flexible hose that connects the compressed respirable air to the respiratory-inlet covering. The flow of respirable air to the respiratory-inlet covering also may be partially controlled by an adjustable valve or orifice which is a component of the respirator located just downstream of the quick connect/disconnect coupling. A continuous flow airline supplied air respirator with air supply system is shown in Figure 6.

The respiratory-inlet covering of a continuous-flow airline respirator may be any of the tight-fitting types of facepieces or any of the various types of loose-fitting respiratory-inlet coverings. A continuous-flow airline respirator approved by NIOSH and MSHA under the provisions of the present Federal respirator test and approval regulations, 30 CFR Part 11, must maintain a flow of respirable air of at least 115

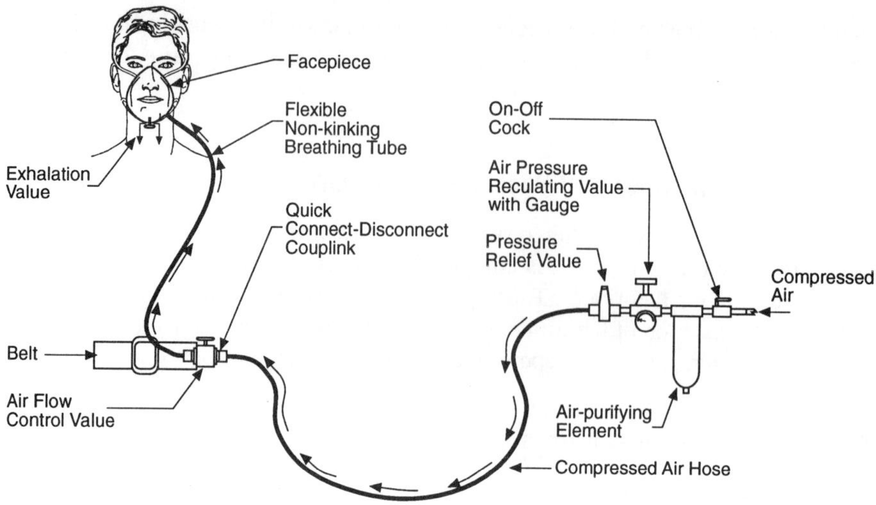

Figure 6 Continuous-flow airline type supplied-air respirator with half-mask facepiece.

liters per minute (4 ft^3/min or cfm) for a respirator equipped with a tight-fitting respiratory-inlet covering and at least 170 lpm (6 ft^3/min) for a respirator equipped with a loose-fitting respiratory-inlet covering. Two abrasive blasting continuous flow airline respirators equipped with facepiece-blouse or hood-blouse combination are shown in Figure 7.

2. *Demand Airline Respirator*

A demand airline respirator is equipped with a tight-fitting facepiece as the respiratory-inlet covering and an air flow regulating valve. This valve permits respirable air to flow to the facepiece only when the respirator wearer inhales and creates a negative air pressure inside the facepiece. The valve also stops the flow of the respirable air to the facepiece when the wearer exhales and creates a positive air pressure inside the facepiece. A negative air pressure is an air pressure lower than that of the ambient atmosphere outside the respirator; a positive air pressure is an air pressure greater than that of the ambient atmosphere outside the respirator.

The demand type of air flow regulating valve has a single stage and utilizes a flexible diaphragm in contact with the ambient atmosphere and levers for operation. This valve is attached to the respirator facepiece or to a belt secured to the waist of the respirator wearer. Sketches of the valve operation during inhalation and exhalation are shown in Figure 8.

3. *Pressure-Demand Airline Respirator*

A pressure-demand airline respirator is very similar to the demand airline respirator except that its air flow regulating valve maintains a positive air pressure inside the facepiece during both exhalation and inhalation.

The pressure-demand valve is similar to the demand valve except that it usually has a spring between the flexible diaphragm and the valve case to hold the valve

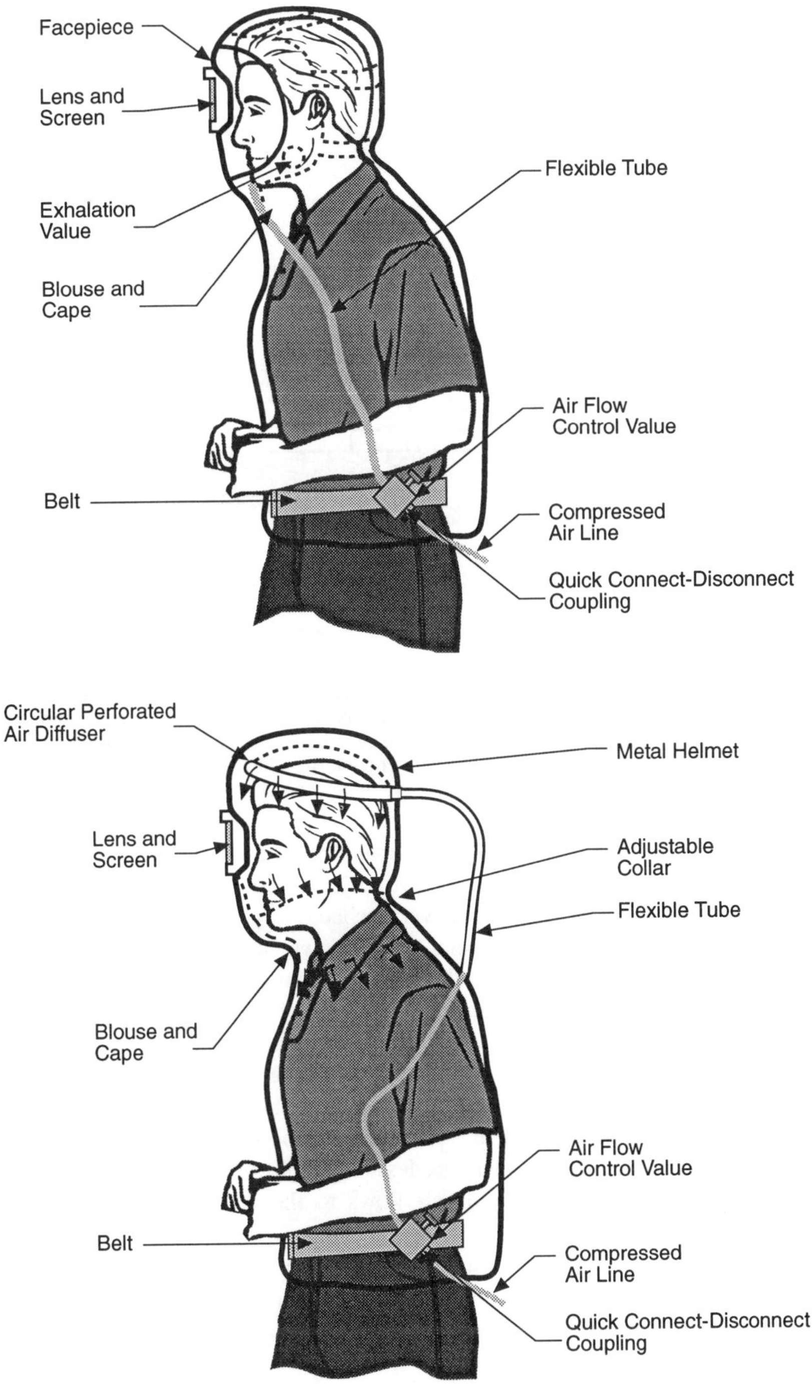

Figure 7 Abrasive-blasting continuous-flow airline respirators, one equipped with a facepiece and blouse combination and the other with a helmet and blouse combination.

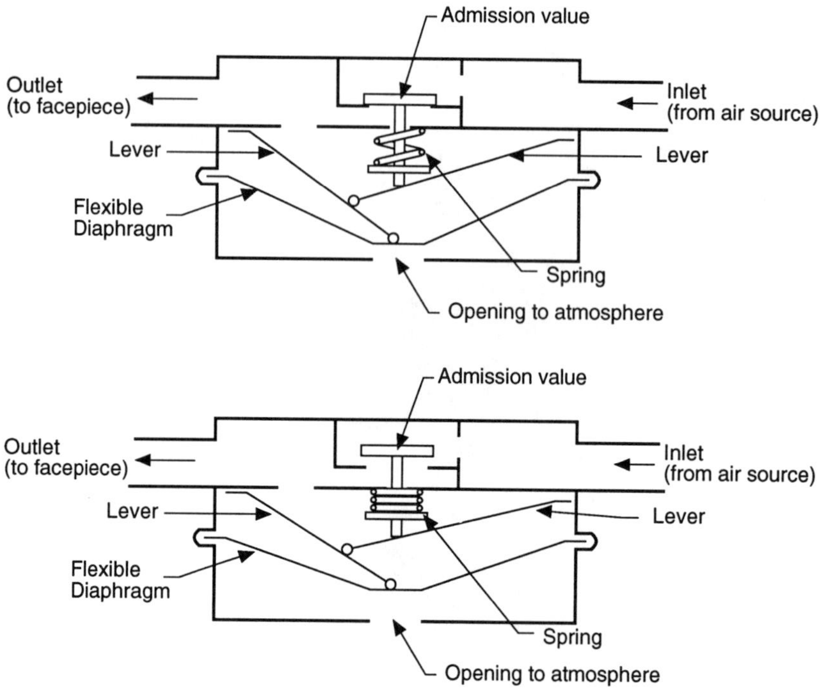

Figure 8 Sketches illustrating action of the simple single-stage demand valve. High pressure of exhaled air stretches diaphragm and the resulting movement of levers and action of spring causes admission valve to close and air flow ceases. Low pressure created by inhalation pulls diaphragm inward and the resulting movement of levers compress the spring and open the admission valve and air flows through the demand valve.

slightly open at all times. The pressure-demand airline respirator is equipped with a special exhalation valve which opens during exhalation but only when a certain positive air pressure inside the facepiece is reached. The combination of the modified air flow regulating valve and the special exhalation valve ensures that the air pressure inside the facepiece is positive at all times and that the additional respirable air needed by the respirator wearer during inhalation is provided during inhalation.

Since the air pressure inside the facepiece is always positive, any leakage between the facepiece and the respirator wearer's face is outward. This means that the pressure-demand airline respirator theoretically should provide a much higher level of respiratory protection than the demand airline respirator.

A demand airline respirator will use less respirable air than the continuous-flow airline respirator because respirable air flows to the demand airline respirator's facepiece only during inhalation. If the facepiece of a pressure-demand airline respirator seals well to the respirator wearer's face, the pressure-demand airline respirator will use only slightly more respirable air than the demand airline respirator. However, a poor seal will increase the consumption of respirable air since respirable air at a positive pressure inside the facepiece will leak outward from the facepiece to the ambient atmosphere air, which will reduce the service time if the source of respirable air is a cylinder of compressed respirable air.

4. *Hose-Mask without Blower*

The respiratory-inlet covering of a hose-mask without blower, a tight-fitting full facepiece, is connected to an area having respirable air by a large diameter flexible hose. The inlet of the hose should be anchored in the area having the respirable air. The federal respirator test and approval regulations (30 CFR Part 11) limit the length of the air-supply hose of the NIOSH/MSHA approved hose-mask without blower to a maximum of 75 ft. The air-supply hose is secured to a harness worn by the respirator wearer.

The respirator wearer draws air from the area having respirable air through the large diameter hose to the full facepiece by the wearer's own inspiratory effort. The facepiece contains both an inhalation and an exhalation valve. The inhalation valve closes during exhalation and the exhalation valve opens during exhalation to ensure that the expired air which is warm, moist, and rich in carbon dioxide is expelled from the facepiece to the ambient atmosphere.

Very few hose-masks without blowers are used at the present time. The most recent issue of the NIOSH Certified Equipment List shows that only one hose-mask without blower retains a NIOSH/MSHA approval.

5. *Hose-Mask with Blower*

A hose-mask with blower is equipped with a blower, either hand-operated or motor-operated, which takes air from an area having respirable air and pushes it through a large diameter flexible hose to a respiratory-inlet covering worn by the respirator wearer. The large diameter flexible hose is secured to a harness worn by the respirator wearer. The federal respirator test and approval regulations (30 CFR Part 11) limits the length of the large diameter flexible hose to 300 ft. The respiratory-inlet covering is a full facepiece equipped with inhalation and exhalation valves.

The blower is designed to permit free passage of air to the respirator wearer in the event that the blower ceases to operate. The blower is constructed so that if it should stop operating the respirator wearer can inhale through the blower and the large diameter hose. If the blower ceases operation and the respirator wearer is forced to inhale through the blower and hose, the inhalation valve closes during exhalation and the exhalation valve opens during exhalation to ensure that the exhaled air is expelled to the outside ambient atmosphere. Presently, very few hose-masks with blowers are used for respiratory protection. The most recently published NIOSH Certified Equipment List shows that only two hose-masks with blowers retain NIOSH/MSHA approvals.

B. Self-Contained Breathing Apparatus (SCBA)

The wearer of a self-contained breathing apparatus is not connected to a stationary source of respirable air such as an air compressor or a cylinder of compressed air. Instead, a supply of respirable air, oxygen, or oxygen-generating material is carried on the body of the wearer. The self-contained breathing respirator wearer is completely independent of the ambient atmosphere and thus the wearer's mobility is not restricted.

Most self-contained breathing respirators are equipped with a tight-fitting full facepiece. However, some self-contained breathing respirators are equipped with a tight-fitting quarter-mask or half-mask facepiece or a mouthpiece with nose clamp. Also, some self-contained breathing respirators used for escape from dangerous atmospheres utilize a loose-fitting hood or helmet.

The self-contained breathing apparatus protects wearers against inhalation of harmful atmospheres.

There are two major classes of self-contained breathing respirators: closed-circuit and open-circuit.

1. Closed-Circuit Self-Contained Breathing Apparatus

In the operation of a closed-circuit SCBA, air is rebreathed by the wearer after carbon dioxide exhaled by the wearer is removed from the exhaled air by material in a canister which is part of the apparatus and after the oxygen content of the air is restored from an oxygen source which is part of the apparatus. A closed-circuit SCBA sometimes is called a "rebreathing" device.

There are two basic types of closed-circuit SCBA. One type uses a cylinder of compressed gaseous or liquid oxygen while the other type utilizes a solid oxygen-generating material.

Closed-Circuit SCBA with Cylinder of Gaseous or Liquid Oxygen. In the device that employs a cylinder of compressed gaseous or liquid oxygen, air exhaled by the wearer passes through a one-way check valve in a breathing tube into a canister containing a granular material which removes the carbon dioxide from the exhaled air. This material also removes moisture from the exhaled air. Heat is released when the carbon dioxide and moisture are removed by the granular material. This causes the temperature of the air to rise. The air then passes into a cooling chamber which has a large wall surface area exposed to the ambient atmosphere. Heat is conducted through this wall to be absorbed by the ambient atmosphere which cools the air in the chamber. The air then passes into an inflatable breathing bag. Since the exhaled air has been depleted of carbon dioxide, the pressure of the air entering the bag will not be sufficient to keep the bag inflated. The bag collapses and in doing so, a pressure pad or plate in the bag presses on the stem of a low pressure admission valve which then opens to allow oxygen to flow into the bag. The oxygen which flows into the bag comes from either a cylinder of compressed gaseous oxygen or a cylinder of liquid air. Gaseous oxygen at high pressure from the cylinder passes through a pressure reducing valve or liquid oxygen from a cylinder is converted to low pressure gaseous oxygen. The oxygen only passes into the inflatable breathing bag if the admission valve has been opened. The oxygen added to the air in the breathing bag tends to inflate the bag. If the bag is sufficiently expanded, the pressure pad or plate no longer will press against the stem of the admission valve. This valve then closes and no more oxygen is admitted to the bag. The oxygen added to the air inside the breathing bag replenishes the air with oxygen to make up for the oxygen that had been consumed by the respirator wearer. As the respirator wearer inhales, air flows out of the bag through a tube into another one-way check valve and then into the wearer's respiratory tract.

The apparatus contains a saliva trap which has a valve to release accumulated saliva. The oxygen in the cylinder may contain a very small amount of nitrogen, and as oxygen is consumed from the cylinder during operation of the apparatus, the nitrogen in the air of the closed system of the apparatus increases. This increase reduces the concentration of oxygen in the air of the closed system. Unless the closed system is periodically purged, the concentration of oxygen in the air in the closed system may be reduced to a concentration below a safe level. The valve in the saliva trap is periodically opened and the respirator wearer exhales at that time to purge the system. The apparatus includes a manual bypass valve connecting the oxygen cylinder to the cooler or to the breathing bag that may be used to admit oxygen to the system in case the pressure reducing valve or the admission valve fails. A closed-circuit SCBA with compressed oxygen cylinder is shown in Figure 9.

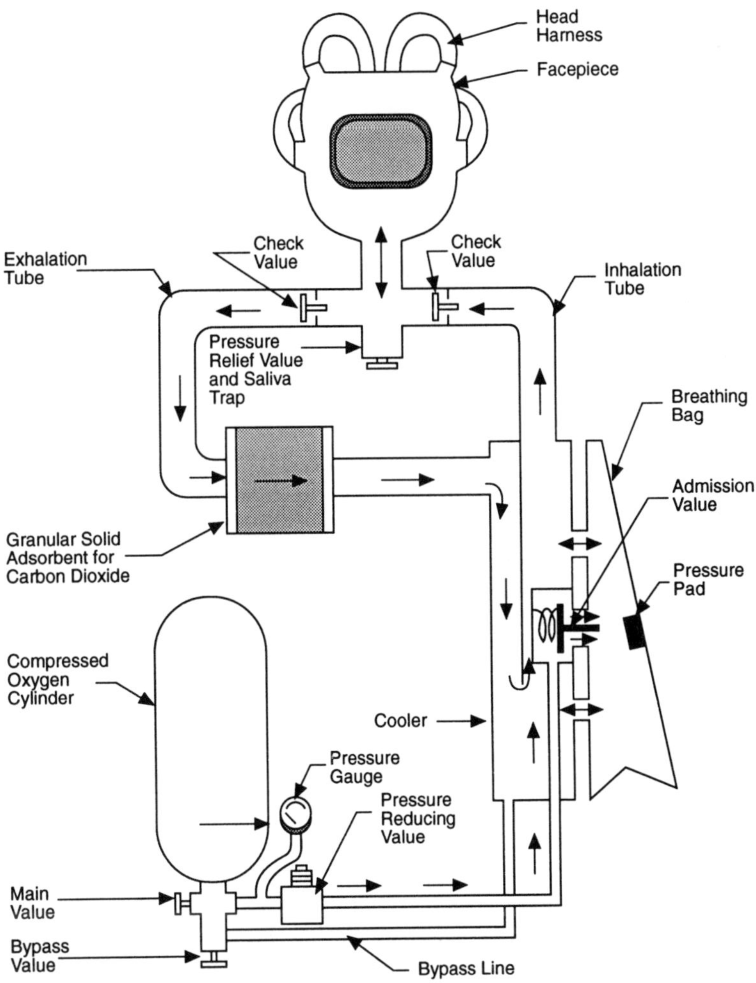

Figure 9 Closed-circuit self-contained breathing apparatus with cylinder of gaseous oxygen.

The apparatus is carried either on the wearer's chest or back. These devices are generally designed to provide 1 to 4 h of use.

Until recently, most devices were designed to operate so that a negative air pressure would be created inside the facepieces of the devices during inhalation. This could result in leakage of the hazardous ambient atmosphere into the air of the closed system of the apparatus if a poor seal between the facepiece and the wearer's face was achieved. Recently, devices which always maintain a positive air pressure inside the facepieces have been made available to provide 1 to 4 h of protection.

Closed-Circuit SCBA with Oxygen Self-Generator. In the operation of the closed-circuit SCBA with an oxygen self-generator, carbon dioxide and water vapor in the wearer's exhaled breath chemically react with a granular substance held inside a container to release oxygen to the air in the closed system so as to replenish the oxygen in the air of the closed system and make up for oxygen consumed by the wearer.

When the wearer of the apparatus exhales, the breath passes through a one-way check valve in a breathing tube into a canister containing a solid, granular substance (usually potassium superoxide), that reacts chemically with the carbon dioxide and the water vapor in the exhaled air. This chemical reaction results in retention of the carbon dioxide and the water vapor by the solid substance in the canister. This reaction produces oxygen which is added to the air passing through the canister. When the chemical reaction occurs, heat is released which increases the temperature of the air exiting from the canister. The heated air with a replenished supply of oxygen then passes into a breathing bag which serves as a reservoir for the air and which has a large wall surface area exposed to the ambient atmosphere. Heat is conducted through the wall of the breathing bag and is dissipated to the ambient atmosphere to cool the air inside the breathing bag. When the wearer of the device inhales, air is drawn from the breathing bag through another one-way check valve in a breathing tube and then into the wearer's respiratory tract. The assembly which connects the breathing tubes to the facepiece of the apparatus contains a manually operated valve which the wearer can open periodically to relieve any excess pressure that may accumulate in the closed system and to release accumulated saliva. This valve is designed so that the wearer can only exhale through it. The wearer cannot inhale so that he or she does not contaminate the air in the closed system. A closed circuit SCBA with oxygen generator is shown in Figure 10.

The apparatus usually is designed to provide 1 h of respiratory protection. Once the apparatus is started, it cannot be readily turned off. The apparatus contains a mechanical timer that warns the wearer that the oxygen-generating capacity of the granular substance that removes the carbon dioxide and water vapor from the exhaled air is almost depleted.

The apparatus is carried either on the chest or the back. A negative air pressure is created inside the facepiece of the apparatus whenever the wearer inhales. This can cause leakage of the hazardous ambient atmosphere into the air of the closed system of the apparatus if a gas-tight seal between the facepiece and the wearer's face is not achieved. Recently, an oxygen-generating SCBA which always maintains a positive air pressure inside a respiratory-inlet covering has been made available to provide 1 h of protection for use in emergency situations.

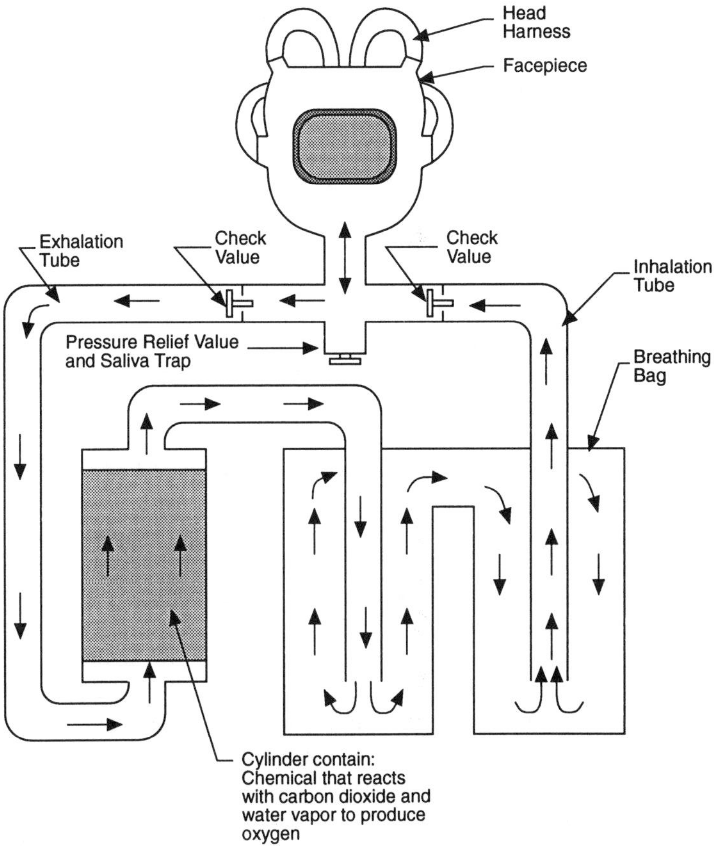

Figure 10 Closed-circuit self-contained breathing apparatus with oxygen self-generator.

2. Open-Circuit SCBA

An open-circuit SCBA usually utilizes a cylinder of compressed gaseous air or oxygen as a source of a respirable atmosphere and exhausts the exhaled breath to the ambient atmosphere. Generally, an open-circuit SCBA is designed for use with compressed gaseous air.

Although compressed gaseous oxygen could be used in an open-circuit SCBA designed for compressed gaseous air, it must not be done. Compressed gaseous oxygen should never be used in an open-circuit SCBA designed for compressed gaseous air because any minute amount of oil or other foreign oxidizable substance in the apparatus can cause an explosion.

Because there is no recirculation of the breathing atmosphere in the open-circuit SCBA and because the container of compressed air or oxygen of the open-circuit SCBA must provide the total volume of the breathing atmosphere, the useful service period of protection provided by the open-circuit SCBA is usually shorter than that provided by the closed-circuit SCBA.

Most open-circuit SCBA devices used for entry into a hazardous atmosphere provide either a 30-min or 60-min useful service period of protection. Open-circuit

SCBA units which offer less than a 30-min useful service period of protection are approved by NIOSH and MSHA only for use in emergency situations.

There are three types of open-circuit SCBA devices: demand, pressure-demand, and continuous-flow.

Demand Self-Contained Breathing Apparatus. A cylinder of compressed gaseous air at high pressure, 2000 to 4500 pounds per square inch gage (psig), supplies air through a main shutoff valve to a two-stage regulating valve that reduces the air pressure to permit it to be delivered to the tight-fitting respiratory inlet covering (usually a full facepiece) of the SCBA. This two-stage regulating valve also passes air to the facepiece on demand. A bypass valve is available that can be used to pass air directly from the cylinder to the facepiece without the air passing through the main shutoff valve and the two-stage regulating valve in case the regulating apparatus fails. This bypass valve also reduces the pressure of the compressed air to an appropriate value. The first stage of the two-stage regulator reduces the air pressure to about 50 to 100 psig at the entrance of the admission valve to the second stage. The second stage utilizes a flexible diaphragm in contact with the ambient atmosphere and levers for operation. When inhalation by the wearer of the SCBA produces a negative pressure inside the facepiece, the diaphragm in the second stage of the two-stage regulator is depressed and, by means of the levers associated with the diaphragm, the admission valve is opened and air then flows to the facepiece until the wearer exhales. The air flowing out of the second stage of the regulator is at a low pressure of about 25 to 50 mm water column (mm H_2O). When the wearer exhales, the pressure inside the facepiece becomes positive. This causes the diaphragm to move outward and, by means of the levers associated with the diaphragm, the admission valve is closed which stops the flow of air to the facepiece until the wearer inhales. An exhalation valve in the facepiece opens during exhalation by the wearer to ensure that exhaled air is expelled to the ambient atmosphere. A two-stage regulating valve for demand SCBA is shown in Figure 11.

Since the air pressure inside the facepiece becomes negative when the SCBA wearer inhales, this can result in leakage of the hazardous ambient atmosphere into the facepiece if a gas-tight seal between the facepiece and the wearer's face has not been achieved.

The regulating valve apparatus of the SCBA is mounted either on a harness that carries and supports the cylinder of compressed air or it is mounted directly on the facepiece. The use of demand open-circuit SCBA devices is declining, and the use of pressure-demand SCBA devices which always maintain a positive air pressure inside the facepiece during both exhalation and inhalation is increasing. A open-circuit SCBA is shown in Figure 12.

Pressure-Demand Self-Contained Breathing Apparatus. The pressure-demand open-circuit SCBA is very similar to the demand open-circuit SCBA except that its two-stage regulating valve maintains a positive air pressure inside the facepiece during both exhalation and inhalation.

The two-stage regulating valve of the pressure-demand open-circuit SCBA has a design similar to the two stage-regulating valve of the demand open-circuit SCBA except that it usually has a spring located between the flexible diaphragm and the

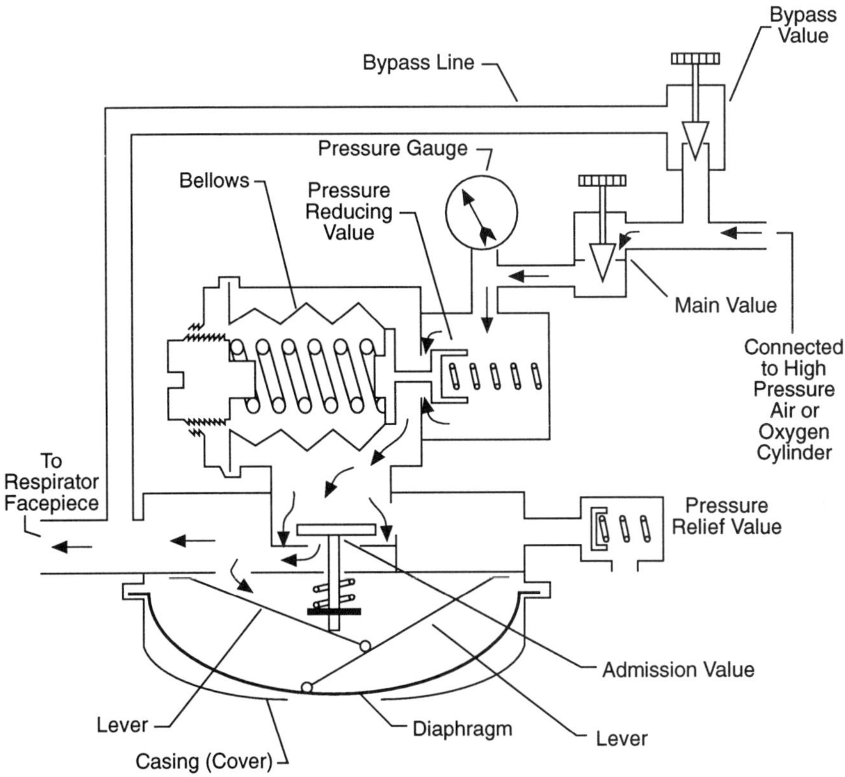

Figure 11 Typical two-stage (pressure-reducing and demand) valve and bypass for demand open-circuit self-contained breathing apparatus.

regulator casing which holds the admission valve slightly open at all times. Theoretically this permits air to flow continuously into the SCBA facepiece. The pressure-demand open-circuit SCBA also has a special exhalation valve that requires more than the normal positive air pressure to open it. This special exhalation valve (usually a spring-loaded valve) maintains a minimum positive air pressure of about 75 to 150 mm H_2O inside the SCBA facepiece. The exhalation valve opens only when the air inside the facepiece exceeds value. Thus, the combination of the modified two-stage regulating valve and the special exhalation valve ensure that the air pressure inside the SCBA facepiece is positive at all times and that the additional air needed by the wearer during inhalation is provided.

Because the air pressure inside the SCBA facepiece is always positive, any leakage between the facepiece and the SCBA wearer's face is outward. This means that the pressure-demand open-circuit SCBA theoretically should provide a much higher level of protection than the demand open-circuit SCBA.

Theoretically, the demand open-circuit SCBA should use less respirable air than the pressure-demand open-circuit SCBA since respirable air will flow to the facepiece of the demand SCBA only when the wearer inhales. However, if the facepiece of the pressure-demand SCBA seals very well on the face of the wearer, the pressure-demand SCBA will use only slightly more respirable air than that used by the demand

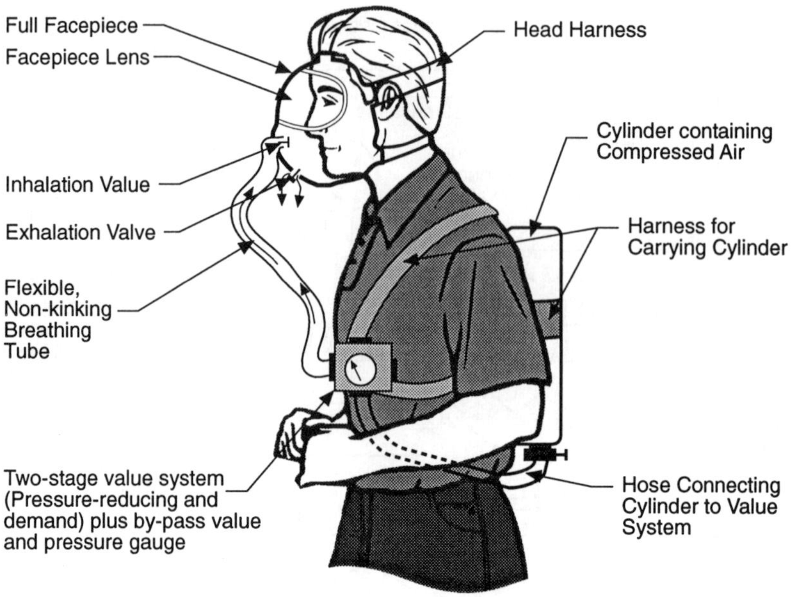

Figure 12 Open-circuit self-contained breathing apparatus.

SCBA. A poor seal of the facepiece on the face will increase the amount of respirable air used and also will reduce the period for depletion of respirable air in the cylinder.

Continuous-Flow Open-Circuit Self-Contained Breathing Apparatus. Several simple continuous-flow SCBAs that provide useful service time periods ranging from 5 to 10 min are available. These respirators have been approved by NIOSH and MSHA for use as emergency escape devices.

The continuous-flow SCBA device uses a small size cylinder of compressed gaseous respirable air. The pressure of the compressed gaseous air inside a cylinder ranges from 2000 to 3000 psig.

Usually, the continuous-flow SCBA used for emergency escape is equipped with a simple hood made of a flexible, transparent plastic film which covers the wearer's head and neck. The hood generally contains an elastic band that draws the lower edge of the hood close to the wearer's neck.

A simple, single-stage valve is attached to the cylinder outlet and is connected to the plastic film hood by means of a small diameter flexible hose. When this valve is opened, it reduces the pressure of the compressed air to a low pressure and allows the air to flow continuously to the hood at a constant flow rate. The flow rates for various makes and models of these SCBA devices range from 35 lpm to slightly more than 70 lpm. Air exists from the hood at its lower edge at the wearer's neck.

The cylinder is carried on SCBA wearer at the neck, chest, or waist level with the location depending on the make and model of the SCBA device. A pouch or harness is used to secure the SCBA device to the wearer's body.

The weight of these SCBA units range from about 4 1/2 to about 10 1/4 lb.

Performance evaluation of continuous-flow SCBA devices have indicated that for safe use of such a device for emergency escape, the rate of flow of respirable air to the hood should not be less than 60 lpm.

C. Combination Supplied-Air Respirator with Auxiliary Self-Contained Breathing Apparatus

Loss of the source of the respirable air supply to an airline respirator worn by a person in a hazardous atmosphere leaves the wearer without respiratory protection. This situation may occur for reasons such as damage to the air-supply hose, failure of an air compressor to operate, and depletion of the air in a cylinder of compressed respirable air. Because of the possibility that the described situation may occur, the airline respirator is not approved by NIOSH and MSHA for use in atmospheres immediately dangerous to life or health (IDLH).

However, an airline respirator with an auxiliary self-contained breathing air supply is approved by NIOSH and MSHA for use in an IDLH atmosphere. The service time for the auxiliary self-contained breathing air supply ranges from 5 to 15 min or even longer. The combination respirator generally is used for emergency escape from an IDLH atmosphere.

The wearer of a combination airline respirator with an auxiliary self-contained breathing air supply may use it when the wearer needs to temporarily disconnect the airline respirator section from a source of compressed respirable air when changing locations in a hazardous area.

A combination airline respirator with an auxiliary self-contained breathing air supply may be used for emergency entry into an IDLH atmosphere to connect the airline section of the device to a source of compressed respirable air if the auxiliary self-contained breathing air supply has a service period of 15 min or longer and if not more than 20% of the supply of the air of the auxiliary self-contained breathing section is used during entry into the IDLH atmosphere.

VI. COMBINATION ATMOSPHERE-SUPPLYING AND AIR-PURIFYING RESPIRATORS

NIOSH and MSHA have approved combination airline type atmosphere-supplying and air-purifying respirators. This type of respirator is approved under the provisions for the air-purifying section of the combination device since this combination provides the least protection to the wearer.

This combination provides respiratory protection to the wearer when entering and/or leaving a hazardous area without being connected to a source of compressed respirable air. When entering and leaving the hazardous area, the respirator wearer uses the air-purifying section of the combination respirator for protection.

When the wearer reaches his or her location the wearer connects the airline section of the combination respirator to a source of compressed respirable air. The wearer then uses the airline section while performing the job.

The wearer can use the air-purifying section of the combination respirator when moving from one area to another where sources of compressed respirable air are located. The wearer can use the air-purifying section of the combination respirator to escape from a hazardous area if the source of compressed respirable air for the airline section of the combination respirator fails.

VII. REQUIREMENTS FOR RESPIRABLE AIR AND OXYGEN FOR SUPPLIED-AIR RESPIRATORS AND SELF-CONTAINED BREATHING APPARATUS

1. Compressed gaseous or liquid oxygen shall meet the requirements of the U.S. Pharmacopoeia for medical or breathing oxygen.
2. Chemically generated oxygen shall meet the requirements of U.S. Department of Defense Military Specifications MIL-E-83252 or MIL-O-1563c.
3. Compressed gaseous air shall meet at least the requirements for Grade D Air; liquid air shall meet at least the requirements for Grade M Air as described in the ANSI/CGA G-7.1-1989 American National Standard Commodity Specification for Air.
4. Compressed gaseous or liquid oxygen shall not be used in open-circuit self-contained breathing apparatus that have previously used compressed gaseous or liquid air.
5. Compressed gaseous or liquid oxygen shall not be used with supplied-air respirators.
6. Containers for compressed breathing gas and liquified breathing gas shall be tested and maintained in accordance with the Shipping Container Specification Regulations of the U.S. Department of Transportation, 49 CFR Part 100-178. These containers shall be permanently marked to identify their contents (e.g., compressed breathing air, compressed breathing oxygen, liquified breathing air) and the markings on the containers shall be in accordance with the requirements of the ANSI/CGA C-4-1978 American National Standard Method of Marking Portable Compressed Gas Containers to Identify the Material Contained.
7. An air pump or an air compressor used to provide respirable air to a supplied-air respirator and a high pressure air compressor for use in filling compressed air cylinders for a SCBA device shall be constructed and located so as to avoid the entry of contaminated ambient air into the air supply system.
8. It is recommended that an air pump or an air compressor to be used to provide respirable air to a supplied-air respirator and a high-pressure air compressor for use in filling compressed air cylinders for a SCBA device be a type that is not lubricated with oil.
9. The air pump or the air compressor for use to supply respirable air to a supplied-air respirator and the high pressure air compressor for use in filling compressed air cylinders for a SCBA device shall be equipped with an appropriate air purification element to ensure that the compressor will supply acceptable breathing air that meets the requirements for Grade D Air as listed in the ANSI/CGA G-7.1-1989 American National Standard Specification for Air.
10. The air pump or the air compressor for use to provide respirable air to a supplied-air respirator to be used in a very cold atmosphere shall contain equipment to reduce the moisture content of the breathing air so that its dew point is at least 10°C (18°F) lower than the coldest temperature expected in the atmosphere where the respiratory protective device is to be used.
11. The high pressure air compressor for use to provide respirable air for filling compressed air cylinders of SCBA devices shall contain equipment to reduce the

moisture content of the air so that its dew point is at –55°C (–67°F) or lower. If a SCBA device is to be used in an atmosphere at a temperature less than –32°C (–25.6°F), then the moisture content of the respirable air in the compressed air cylinder of the SCBA unit should have a dew point of –75°C (–103°F) or lower.

12. The air pump or the air compressor for use to provide breathing air to a supplied-air respirator shall be equipped with an alarm, visual and audible, which shall be actuated during use of the air pump or compressor whenever the pump or compressor is unable to provide breathing air at a present pressure to the respirator. The alarm shall be located so that the respirator wearer will be alerted that it has been actuated, or a person shall be assigned to observe the alarm and to immediately notify the respirator wearer if the alarm is actuated and to instruct the respirator wearer to immediately leave the hazardous area.
13. If an oil-lubricated air pump or air compressor is used to provide respirable air to a supplied-air respirator, it shall be equipped with a high temperature alarm, visual and audible, and a carbon monoxide alarm, visual and audible. The high temperature alarm shall be preset to be actuated at 50°C (122°F) by increasing temperature of the air being provided by the pump or compressor. The carbon monoxide alarm shall be preset to actuate whenever the air being provided by the pump or compressor reaches a concentration of 10 ppm carbon monoxide gas. Each alarm shall be located so that the respirator wearer will be alerted that it has been actuated, or a person shall be assigned to observe the alarms and to immediately notify the respirator wearer if an alarm has been actuated and to instruct the respirator wearer to immediately leave the hazardous atmosphere.
14. Couplings (quick connect/disconnect) used in a plant for connecting the air-supply hoses of supplied-air respirators to a source of respirable air shall be incompatible with couplings for nonrespirable plant air or other plant gas systems to prevent accidental connections of supplied-air respirators to plant systems containing nonrespirable air, oxygen, nitrogen, acetylene, and any other gases.

VIII. ADVANTAGES AND DISADVANTAGES OF VARIOUS TYPES OF RESPIRATORS

1. Nonpowered air-purifying respirators
   a. Advantages
      i. Small and compact
      ii. Lightweight
      iii. Simple construction for most
      iv. No restriction of wearer's mobility
      v. Low initial cost
   b. Disadvantages
      i. Cannot be used in oxygen-deficient atmosphere
      ii. Cannot be used in IDLH atmosphere
      iii. Gas-/vapor-removing type cannot be used for gases and vapors with poor warning properties unless for emergency escape, or unless equipped with end-of-service-life indicator, or permitted by U.S. Department of Labor standard for particular hazardous substances
      iv. Negative air pressure inside facepiece during inhalation by wearer may result in leakage of hazardous atmosphere into facepiece which is inhaled by wearer
      v. Cannot be used by person having a beard or mustache because it passes between facepiece sealing surface and wearer's face

- vi. Great care must be taken to select proper type for specific air contaminant(s)
- vii. Must use fitting test to select particular make and model for each wearer
- viii. Fit of all fabric type disposable respirator cannot be checked adequately by wearer prior to entry of wearer into hazardous atmosphere, which and this may result in excessive leakage of air contaminant into respirator and its inhalation by respirator wearer
- ix. Wearing discomfort may limit use period
- x. Resistance to breathing may cause discomfort and fatigue
- xi. Eye irritation caused by air contaminant requires use of full facepiece
- xii. Many types are limited for use against only low concentrations of air contaminants
- xiii. High cost for frequent replacement of disposable type or frequent replacement of air-purifying element in reusable type
- xiv. High cost for maintenance of reusable type

2. Powered air-purifying respirators
   a. Advantages
      - i. No restriction on mobility of wearer
      - ii. Minimal resistance to breathing
      - iii. Cooling effect on wearer in warm atmospheres
      - iv. Less wearing discomfort than for nonpowered air-purifying type
      - v. May be able to wear for extended time as long as air flow to respiratory-inlet covering is not reduced below required minimum value and as long as air contaminant does not penetrate air-purifying element to result in contaminant concentration above the permissible exposure limit in air supplied to respiratory-inlet covering
      - vi. Fit testing not required for devices equipped with loose-fitting respiratory-inlet covering
      - vii. Many persons who are unable to get satisfactory fit with facepiece may be able to use powered air-purifying devices equipped with loose-fitting respiratory-inlet coverings
      - viii. Devices equipped with loose-fitting respiratory-inlet coverings can be used by wearer's with beards and moustaches
   b. Disadvantages
      - i. Cannot be used in oxygen-deficient atmosphere
      - ii. Cannot be used in IDLH atmosphere
      - iii. Gas- or vapor-removing type cannot be used for gases and vapors with poor warning properties unless for emergency escape, or unless equipped with end-of-service-life indicator, or permitted by U.S. Department of Labor standard for particular hazardous substance
      - iv. Great care must be taken to select proper type for specific air contaminant(s)
      - v. A fitting test of the facepiece of device equipped with facepiece type respiratory-inlet covering is required for all wearers
      - vi. Eye irritation caused by air contaminant requires use of full facepiece or loose-fitting respiratory-inlet covering that protects the entire face
      - vii. Functional limitations of many types of air-purifying elements restrict use of respirator only to low concentrations of several air contaminants
      - viii. Poor design of respiratory-inlet coverings of some devices, poor construction of some devices, and/or air flow rates which are too low for some devices have resulted in low values for assigned protection factors (APF) for powered air-purifying respirators

   ix. Failure of blower to operate while wearer is in hazardous atmosphere results in loss of protection provided to wearer and requires wearer to immediately leave the hazardous area and go to safe area
   x. Use in cold atmosphere may result in severe discomfort to wearer
   xi. Battery for operating blower requires frequent recharging
   xii. Wearer generally cannot detect when inadequate quantity of respirable air is supplied to respiratory-inlet covering due to low battery power or plugging of particle-filtering element by retained particulate matter
   xiii. High initial cost
   xiv. High cost for frequent battery replacement
   xv. High speed motor of blower requires frequent replacement which is costly
   xvi. High cost of frequent replacement of air-purifying elements
   xvii. Complex design and construction makes maintenance difficult and costly

3. Airline type of supplied-air form of atmosphere-supplying respirators
   a. Advantages
      i. May be used for long periods
      ii. Minimal resistance to breathing
      iii. Low weight
      iv. Low bulk
      v. Minimal discomfort
      vi. Many persons who are unable to obtain satisfactory fit with facepieces may be able to use airline devices equipped with loose-fitting respiratory-inlet coverings
      vii. Cooling effect on wearer in warm atmospheres
      viii. Fit testing not required for devices equipped with loose-fitting respiratory-inlet covering
      ix. Devices equipped with loose-fitting respiratory-inlet coverings can be used by wearers with facial hair such as beards and mustaches
      x. Continuous-flow devices and especially pressure-demand devices offer high levels of protection
      xi. Can be used for oxygen-deficient atmospheres not IDLH
      xii. Moderate initial cost
      xiii. Low operating cost
      xiv. Low maintenance cost
   b. Disadvantages
      i. Cannot be used in IDLH atmosphere unless it is a combination pressure-demand airline respirator equipped with a full facepiece and an auxiliary self-contained breathing supply of respirable air
      ii. Trailing air-supply hose restricts mobility of wearer unless it is a combination airline respirator with an auxiliary self-contained breathing supply of respirable air, or a combination airline respirator with a provision for air-purifying protection
      iii. Air-supply hose is vulnerable to damage which may result in loss of respirable air being provided to respiratory-inlet of respirator
      iv. Sudden loss of source of supply of respirable air due to damage to air-supply hose or failure of air pump or air compressor to provide respirable air leaves wearer without respiratory protection unless the respirator is a combination airline respirator with an auxiliary self-contained breathing supply of respirable air, or a combination airline respirator with a provision for air-purifying protection

v. Quality of air provided to respirator must be checked to ensure it meets pertinent requirements
vi. Inlet of air pump or air compressor used to provide respirable air to respirator must be located in area having respirable atmosphere
vii. Oil-lubricated compressor used to provide respirable air to respirator must be equipped with both a high temperature alarm and a carbon monoxide gas alarm
viii. Couplings in plant used to connect air-supply hoses of airline respirators to source of respirable air must be incompatible with couplings for nonrespirable plant air and any other gases
ix. Demand type device does not provide wearer with high level of protection. Negative air pressure inside facepiece of demand type device during inhalation may result in leakage of hazardous atmosphere into facepiece which will be inhaled by wearer
x. Complexity of air flow control valve of pressure-demand unit requires that maintenance be performed only by adequately trained person
xi. A fitting test of the facepiece of device equipped with facepiece type respiratory-inlet covering is required for all wearers

4. Hose-mask type of supplied-air form of atmosphere-supplying respirators
   a. Advantages
      i. May be used for long periods
      ii. Minimal resistance to breathing by hose-mask with blower
      iii. Can be used in oxygen-deficient atmospheres not IDLH
      iv. Simple construction
      v. Low bulk
      vi. Easy maintenance
      vii. Low operating cost
   b. Disadvantages
      i. Cannot be used in IDLH atmosphere
      ii. Cannot be used by person having hair such as beard or mustache because it passes between facepiece sealing surface and wearer's face
      iii. Negative air pressure inside facepiece of hose-mask without blower during inhalation by wearer may result in leakage of hazardous atmosphere into facepiece which will be inhaled by wearer
      iv. Hose-mask without blower provides wearer with low level of protection
      v. A low air flow rate by hose mask with blower to facepiece results in providing wearer with low level of protection
      vi. A fitting test of the facepiece is required for any potential wearer
      vii. Inlet of hose of hose-mask without blower and inlet of blower for hose-mask with blower must be located in area having respirable air
      viii. Air-supply hose is vulnerable to damage which may result in loss of respirable air being provided to facepiece of respirator
      ix. Sudden loss of source of respirable air due to damage of air-supply hose or failure of blower of hose-mask with blower leaves wearer without respiratory protection
      x. Resistance to breathing by hose-mask without blower causes discomfort and fatigue
      xi. Availability of only two NIOSH/MSHA approved hose-masks with blower and only one NIOSH/MSHA approved hose-mask without blower; all are made and marketed by the same company which severely restricts the selection of product

5. Self-contained breathing apparatus type atmosphere-supplying respirators
   a. Advantages
      i. SCBA wearer carries own source of respirable air, and thus is independent of ambient atmosphere
      ii. Minimal restriction of mobility of wearer
      iii. Provides protection against air contaminants of all types, particulate and gaseous, and oxygen deficiency
      iv. Pressure-demand open-circuit type with full facepiece provides very high level of protection, and can be used for IDLH contaminated atmosphere, IDLH oxygen-deficient atmosphere, entry into atmosphere with unknown level of contamination or oxygen deficiency, and for firefighting
      v. Positive pressure type closed-circuit type can be used for escape in mining operations.
   b. Disadvantages
      i. Complex construction
      ii. Most are heavy and bulky making them unsuitable for wearers engaged in strenuous work or for wearers who must enter and work in a confined space
      iii. Limited service period makes them unsuitable for use for long periods
      iv. Short service period of open-circuit type device limits its use to where wearer can travel to and from easily in a short period and perform a task in a short period
      v. A fitting test of the facepiece of a device equipped with facepiece type respiratory-inlet covering is required for any person who will use the device
      vi. Extensive training of potential wearers is required
      vii. Negative pressure inside facepiece of demand open-circuit device and nonpositive pressure closed-circuit device during inhalation by wearer may result in leakage of hazardous atmosphere into facepiece which will be inhaled by wearer
      viii. High initial cost
      ix. High operating cost
      x. Complexity of device makes maintenance time-consuming and costly
      xi. Complexity of two-stage flow control regulating valve employed in open-circuit device that requires maintenance of valve be performed only by adequately trained individuals
      xii. Respirable air and oxygen must meet stringent requirements

9 RESPIRATOR FIT TESTING AND PROTECTION FACTORS

I. INTRODUCTION

In order for a respirator equipped with a tight-fitting facepiece to provide adequate respiratory protection to its wearer while in a hazardous atmosphere, the facepiece must form a satisfactory seal that minimizes leakage of the hazardous atmosphere into the facepiece where it would be inhaled by the wearer. Since each individual's face varies in size and shape, the ability of the facepiece of a respirator to achieve a satisfactory fit on a particular worker's face must be determined by a fit test.

There are two basic types of respirator facepiece fitting tests: qualitative and quantitative.

II. QUALITATIVE FIT TESTING

A qualitative respirator facepiece fitting test assesses the adequacy of the facepiece of a respirator to seal on a particular person's face by determining whether or not the person detects penetration of an airborne test agent into the facepiece by sensing odor, irritation, or taste of the test agent. The qualitative fit test is a pass/fail type test. Any detection of the airborne test agent means that the facepiece has failed to achieve a satisfactory seal.

III. QUANTITATIVE FIT TESTING AND FIT FACTOR (FF)

A quantitative respirator fit test assesses the adequacy of the facepiece of a respirator to seal on a particular person's face by using instrumentation to measure the ratio of the concentration of an airborne test agent in the air of the ambient atmosphere and in the air inside the facepiece in the breathing zone of the wearer. The ratio is called the "fit factor."

IV. WORKPLACE PROTECTION FACTOR (PF)

Predicting the performance of a respirator worn by a person in a workplace on the basis of respirator facepiece fit test data is uncertain. The uncertainty is caused by differences in the conditions of the fit test and the conditions in the workplace, variations in the position of the facepiece on a person's face after different donnings (unlike headband strap tensions during different wearings of the respirator), variations in equipment and methods of measuring respirator performance during a fit test and during work on a job site, and different face, head, and body movements by the respirator wearer during a fit test and during actual work.

These results are expressed as a "protection factor" (PF). The PF is the ratio of the measured average concentration of the air contaminant present in the workplace atmosphere to the measured average concentration of the contaminant in the air inside the facepiece of a respirator on the wearer. In practice, this number is rarely determined in the field but is, rather, determined by NIOSH at the Los Alamos National Laboratory.

The protection factor determined by NIOSH expresses the minimum expected level of performance that a class of respirators would provide in the workplace provided that the respirators worn by persons in workplaces are in good operating condition and that the respirator wearers have been properly fit tested with the respirators and have been trained to wear the respirators in a suitable manner. The APF is considered to be an achievable ratio of the average concentration of the air contaminant measured in the ambient workplace atmosphere to the average concentration of the air contaminant inside the respirator facepiece in the breathing zone of the respirator wearer. PF values listed for the various classes of respirators are based on the results of laboratory respirator fitting tests. Therefore, they should be applied with caution by someone with the experience to consider all the potential reasons the PF might not be achieved.

V. MAXIMUM USE CONCENTRATION (MUC)

The product of the PF for a specific class of respirators and the permissible exposure limit (PEL) listed by OSHA for a particular hazardous substance is called the maximum use concentration (MUC) for the specific class of respirators for the particular hazardous substance. In mathematical terms,

$$\text{MUC} = \text{PF} \times \text{PEL}$$

The MUC for which a specific class of respirators can be used cannot exceed the use limitations for the specific class of respirators prescribed by the NIOSH/MSHA approval for this class of respirators.

VI. REQUIRED RESPIRATOR FIT TESTING

An employer shall carry out respirator fit testing for employees who will be required to wear tight-fitting air-purifying respirators and tight-fitting atmosphere-supplying respirators in workplaces. The employer may use either qualitative fit testing or quantitative fit testing.

If respirators equipped with elastomeric facepieces are to be used, the employer should provide an assortment of respirators for selection by persons who will wear respirators in hazardous workplaces. A good practice is to make at least three sizes for each type of facepiece from at least two different respirator manufacturers available for selection by the employees. However, when respiratory protection is necessary, the requirement is placed on the employer, not the employee, to select the equipment necessary to protect the health of the employee. Therefore, it is typically the better practice to merely provide the proper equipment in sizes that allow the employees to pass the fit testing procedure rather than to give them a smorgasbord menu from which to select. The central issue in the selection and use of respiratory protective devices is that health and safety factors must be considered, such as nature of the hazard, intended uses and limitations of respiratory protective devices, movement and work rate limitations, emergency escape time and distance requirements, and training requirements. Additional general considerations in selecting the appropriate respirator are sorbet efficiencies, odor warning properties, eye irritation potential, protection factors, lower flammability limit (LFL), and conditions that are immediately dangerous to life or health (IDLH — see 29 CFR 1910.120).

Employees who wear negative pressure respirators are required to be fit tested at least every six months. Additionally, because the sealing of the respirator on the face may be affected, a qualitative fit testing must be repeated immediately and a quantitative fit testing should be repeated as soon as possible anytime the test subject (the respirator wearer) has a weight change of 20 lb or more, receives significant facial scarring in the area of the facepiece seal, undergoes significant dental changes such as multiple extractions without prosthesis, undergoes reconstructive or cosmetic facial surgery, or endures any other condition that may interfere with facepiece sealing.

Prior to conducting the fit test, the person who is to wear a respirator shall be trained in how to don and wear it by a competent person. The worker who will wear the respirator and his or her supervisor (both) must have an opportunity to handle the respirator, properly fit it to their face, test its facepiece-to-face seal, and wear it long enough to get familiar with it before wearing it in a fit test atmosphere. Every respirator wearer must receive fitting instructions that include a demonstration on how to don and wear the respirator as well as a chance to practice wearing it before the respirator fit test, unless he or she is already experienced at wearing that particular type of respirator. The instructor should demonstrate how to adjust the respirator to improve the fit as well as how to determine, in the field, if the respirator is fitting properly. Whenever the respirator is not fitting properly, the wearer must clearly understand that it must not be worn further except to immediately evacuate the contaminated area. The wearer must also review those conditions that can prevent a good face seal, such as facial hair, skullcaps, eyeglass temple pieces, and certain hearing protectors. The worker should thoroughly understand the importance of diligence in avoiding the loss of a facial seal and encouraged to check the facepiece fit each time he or she dons the respirator by following the manufacturer's instructions.

If a qualitative fit test is to be performed, a determination shall be made in accordance with the appropriate protocol to determine if the person can detect the presence of the test agent. If the person is unable to detect it, the person shall not be permitted to perform the fit test.

The person who is to wear a respirator shall be permitted to select the respirator that is the most comfortable.

Qualitative and quantitative respirator fitting tests shall be carried out in accordance with protocols specified by the U.S. Department of Labor in respiratory protection regulations contained in 29 CFR 1910.134 or specified in standards for specific hazardous substances or particular work operations promulgated by the U.S. Department of Labor in subpart Z of 29 CFR 1910.

At the present time, airborne test agents included in protocols for qualitative respirator fitting tests accepted by the U.S. Department of Labor include isoamyl acetate vapor, sodium saccharin aerosol, and stannic oxychloride irritant smoke. Currently, airborne test agents for use in quantitative respirator fitting tests accepted by the U.S. Department of Labor include sebecate aerosol, corn oil aerosol, and sodium chloride aerosol. Also, the U.S. Department of Labor has sanctioned the use of airborne particulate matter present in "room air" as a test agent when using a condensation nuclei counter (CNC) for conducting quantitative respirator fitting tests.

When performing a respirator fitting test, the respirator wearer is required to carry out a series of various exercises such as normal breathing, deep breathing, turning head from side to side, nodding head up and down, and talking; bending down to touch the toes, reaching upward with the arms, running in place, and normal breathing again is not presently required in OSHA 1910.134, although it may be prudent to do. Prior to entering the test chamber, the fit test subject must be given complete instructions as to his or her part in the test procedures. The fit test subject performs the listed exercises in the order given. In the normal standing position, without talking, the subject breathes normally for at least 1 min. Remaining in the normal standing position, the subject does deep breathing for at least 1 min, pausing as necessary so as not to hyperventilate. Standing in place the subject slowly turns his or her head from side to side between the extreme positions. The head is held at each extreme position for at least 5 s. This part of the test is performed for at least 1 min. Still standing in place, the subject slowly moves his or her head up and down between the extreme position straight up and the extreme position straight down. The head is held at each extreme position for at least 5 s and the complete test exercise is performed for at least 1 min. The subject is then required to read the "rainbow passage" out slowly and loud, so as to be heard clearly by the test conductor or monitor. The Rainbow Passage is:

> When the sunlight strikes raindrops in the air, they act like a prism and form a rainbow. The rainbow is a division of white light into many beautiful colors. These take the shape of a long round arch, with its path high above, and its two ends apparently beyond the horizon. There is, according to legend, a boiling pot of gold at one end. People look but no one ever finds it. When a man looks for something beyond reach, his friends say he is looking for the pot of gold at the end of the rainbow.

Next the test subject is made to grimace, smile, frown, and generally contort the face using the facial muscles. This phase is continued for at least 15 s in order to check the reseal of the respirator after seal is broken. Changing pace, the test subject is required to bend at the waist and touch his or her toes and return to the upright position. This exercise is repeated for at least 1 min. Next, the test subject jobs in place for at least 1 min. Finally, the test subject returns to normal breathing, as above. The fit test is terminated whenever any single peak penetration exceeds 5% for half-masks and 1% for full facepieces. The test subject may be refitted and retested if these penetration values are exceeded. If two of the three required tests are terminated, the fit is considered inadequate.

Qualitative Respirator Fit Tests are performed to supplement quantitative fit testing or when quantitative testing is impractical or infeasible for specific contaminants such as asbestos or acrylonitrile. The test is performed using the respirator that was determined to be the most effective during the last quantitative fit test or selected by a competent person. Isoamyl acetate, called banana oil, is the agent used for qualitative testing. Individuals must be tested to ensure that they can detect the odor of isoamyl acetate, else it may be necessary to use the irritant smoke test. The fitting process is conducted in a room separate from the fit-test room to prevent odor fatigue. But both rooms must be well-ventilated and separated far enough to avoid cross-contamination. Prior to the selection process, the test subject is shown how to put on a respirator, how it should be positioned on the face, how to set strap tension, and how to assess comfort. A mirror must be available in order to assist the subject in evaluating the fit and positioning of the respirator. This review does not constitute formal training on respirator use. Assessment of comfort includes:

- proper placement of the chin
- positioning of mask on nose
- strap tension
- fit across nose bridge
- room for safety glasses
- distance from nose to chin
- room to talk
- tendency to slip
- cheeks filled out
- self-observation in mirror.

Also, adequate time must be allowed for this assessment.

The test subject then conducts the conventional negative and positive-pressure fit checks (see American National Standards Practices for Respiratory Protection, ANSI Z88.2-1980). Before conducting the negative or positive-pressure checks, the subject seats the mask by rapidly moving the head side-to-side and up and down, taking a few deep breaths. The test subject is now ready for qualitative fit testing. After passing the fit test, the test subject is questioned again regarding the comfort of the respirator. If it has become uncomfortable, another model of respirator shall be tried.

Each respirator used for fitting and fit testing is equipped with organic vapor cartridges. After selecting, donning, and properly adjusting a respirator, the test subject wears it to the fit testing room. Each test subject wears his or her respirator for at least 5 min before starting the fit test. Once again, an exercise regiment including normal breathing, deep breathing, turning the head from side to side, moving the head up and down, reading the Rainbow Passage, bending over and touching the toes, and jogging in place followed by normal breathing as at the start is used for the test. If at anytime during the test the subject detects the odor of the testing agent, he or she must quickly exit from the test chamber and leave the test area. If the entire test is completed without the test subject detecting the odor of the testing agent, the qualitative test is passed and the respirator selected is judged adequate.

VII. APPLYING THE RESULTS OF FIT TESTING

Respirators used in fit tests shall be equipped with appropriate air-purifying elements that do not permit the airborne test agent to penetrate. Facepieces used in powered air-purifying respirators and in atmosphere-supplying respirators shall be tested without air-supplying equipment or attachments. This may be accomplished by testing a particular facepiece that is available for use in an atmosphere-supplying respirator or powered air-purifying respirator and also in a nonpowered air-purifying respirator with an appropriate air-purifying element.

The fit factor determined by the quantitative fit test is expressed as the ratio of the challenge concentration outside the respirator to the concentration inside the respirator. The challenge agent is the testing agent introduced into a test chamber for photometric based systems. Ambient air is the challenge agent when condensation nuclei counter (CNC) devices are used. The average test chamber concentration is the arithmetic average of the concentration of the challenge agent in the test chamber at the beginning and end of the test. The average peak concentration of the challenge agent inside the respirator is the arithmetic average of the peak concentrations for each of the eight exercises of the test, which are computed as the arithmetic average of the peak concentrations found for each breath during the exercise. The average peak concentration for an exercise may be determined graphically if the peak concentrations do not vary greatly during a single exercise. When fit factors are calculated by a computer, the average concentration, instead of average peak concentration, may be used.

OSHA analyzed some fit test records and found that the median fit factor was 3000 for a half-mask. Therefore, fit test subjects should not be permitted to wear a half-mask if a minimum fit factor of 500 cannot be obtained. For a full facepiece respirator, the minimum desirable fit factor is 3000. If hair growth or apparel interferes with a satisfactory fit, then the hair or clothing may be altered or removed so as to eliminate the interference and allow a satisfactory fit to be obtained.

Generally, if a qualitative fit test is passed for a particular quarter-mask or half-mask facepiece, the employee who wore it during the test may wear any nonpowered air-purifying respirator with the same make and model facepiece so long as the concentration of the contaminant to be encountered does not exceed ten times the PEL for the contaminant ($10 \times$ PEL). The same employee may wear any powered air-purifying respirator or an air-supplying respirator with the same make and model facepiece so long as the concentration of air contaminant to be encountered is less than the respirators PF times the PEL (PF $\times$ PEL). This is called the Maximum Use Concentration (MUC) and is summarized in Table 1.

Table 1 Maximum Use Concentration (MUC)

Half-mask, nonpowered air-purifying	10 × PEL
Full facepiece, nonpowered air-purifying	50 × PEL
Half-mask, powered air-purifying	250 × PEL
Full facepiece, powered air-purifying	500 × PEL
Half-mask, pressure demand supplied air	250 × PEL
Full facepiece, pressure demand supplied air	500 × PEL
Full facepiece, pressure demand supplied air, escape respirator	1000 × PEL
Full facepiece, pressure demand SCBA, entry and escape	May be used in IDLH and unknown concentrations
Full facepiece, continuous flow SCBA, escape only	IDLH concentration
Mouthpiece only, continuous flow SCBA, escape only	IDLH concentration

VIII. RESPIRATOR FIT CHECK

An important requirement of an effective respirator program is that prior to each entry into a hazardous area, the wearer must check the seal of the facepiece. This could be accomplished by performing a qualitative or quantitative respirator fit test. However, this often is not practical. Another way to check the seal of the respirator involves having the respirator wearer perform both a negative air pressure respirator fit check and a positive air pressure respirator fit check.

In a negative air pressure respirator fit check, the respirator wearer closes the inlet of a respirator air-purifying element or squeezes a flexible breathing tube to close or block its inlet. The wearer inhales and holds his/her breath for several seconds. If the facepiece collapses and no inward leakage of air into the facepiece between the wearer's face and the sealing surface of the facepiece is detected, it is assumed that the seal of the facepiece on the wearer is satisfactory.

In a positive air pressure respirator fit check, the respirator wearer closes an exhalation valve or breathing tube, or both. The wearer then exhales. If air pressure increases inside the facepiece and air is not detected to leak outward between the wearer's face and the sealing surface of the facepiece, it is assumed that the seal of the facepiece on the wearer is satisfactory.

Disposable air-purifying respirator models typically cannot be given a satisfactory negative or positive air pressure fit check because their design makes it impossible to effectively block passage of air into the respirator and/or to effectively block passage of air out of the respirator.

The described negative and positive air pressure respirator fit checks are not qualitative fit tests. These fit checks cannot be used as a way to determine whether or not a specific make and model respirator should be used by a particular person for respiratory protection in a hazardous workplace.

IX. TABLE OF ASSIGNED PROTECTION FACTORS

Table 2 lists recommended assigned protection factors (APF) for various classes of respirators. Since these assigned protection factors are based mainly on the results

Table 2 Recommended Assigned Protection Factors for Classes of Respirators

Class of respirator	Assigned protection factor
Air-purifying respirators[a]	
Nonpowered type	
Disposable dust/mist (all fabric mask body)	5
Disposable[b] and reusable with quarter- or half-mask facepiece[c]	10
Reusable with full facepiece	50[d]
Powered type	
With quarter- or half-mask facepiece[c]	50
With full facepiece	250
With loose-fitting full facepiece, helmet, blouse, or hood	25
Atmosphere-supplying respirators[e]	
Airline (supplied-air) type	
Demand type	
With quarter- or half-mask facepiece[c]	10
With full facepiece	50
Continuous-flow type	
With quarter- or half-mask[c]	50
With full facepiece	250
With loose-fitting full facepiece, helmet, hood, or blouse	25
Pressure-demand type	
With quarter- or half-mask facepiece[c]	1000
With full facepiece	1000
Self-contained breathing apparatus	
Demand open-circuit and nonpositive pressure closed-circuit	
With quarter- or half-mask facepiece[c]	10
With full facepiece	50
Pressure-demand open-circuit and positive pressure closed circuit	
With full facepiece[f-g]	>1000
Combination airline (supplied-air) with auxiliary self-contained breathing air supply	
Demand with quarter- or half-mask facepiece[c]	10
Demand with full facepiece	50
Pressure-demand with full facepiece[f]	>1000
Combination atmosphere-supplying and air-purifying respirators[f]	
Combination airline respirator with air-purifying provision	
Demand with quarter- or half-mask facepiece[c]	10
Demand with full facepiece	50
Pressure-demand with full facepiece	50

[a] Air-purifying respirators may not be used in oxygen-deficient or IDLH atmospheres.

[b] Disposable air-purifying respirator with elastomeric quarter- or half-mask facepiece only.

[c] Quarter- or half-mask facepieces cannot be used in atmospheres with air contaminants that produce eye irritation. If an air contaminant produces eye irritation, the following types of respirators must be used: a nonpowered air-purifying respirator must be equipped with a tight-fitting full facepiece; a powered air-purifying respirator must be equipped with a tight-fitting full facepiece, a loose-fitting full facepiece, helmet, hood, or blouse; a demand airline respirator must be equipped with a tight-fitting full facepiece; a continuous-flow airline respirator must be equipped with a tight-fitting full facepiece, a loose-fitting full facepiece, helmet, hood, or blouse; a continuous-flow airline respirator must be equipped with a tight-fitting full facepiece, a loose-fitting full facepiece, helmet, hood, or blouse; a pressure-demand airline respirator must be equipped with a tight-fitting full facepiece; a demand open-circuit SCBA and a nonpositive pressure closed-circuit SCBA must be equipped with a tight-fitting full facepiece; a combination demand airline respirator with an auxiliary self-contained breathing air supply must be equipped with a tight-fitting full facepiece; a combination demand airline respirator with air-purifying provision must be equipped with a tight-fitting full facepiece.

[d] An APF of 50 is permitted for a nonpowered air-purifying respirator equipped with this type of facepiece only when a quantitative respirator fitting test has been performed and the respirator wearer has achieved a FF of 500 or greater. If the wearer obtains a FF less than 500 in the quantitative fitting test, the person shall not be allowed to use a nonpowered air-purifying respirator equipped with this type of facepiece. If a qualitative respirator fitting test is conducted, and the respirator wearer passes the test, the person shall be permitted to use an APF of only 10 for a nonpowered air-purifying respirator equipped with this type of facepiece.

[e] Any atmosphere-supplying respirator may be used in an oxygen-deficient atmosphere where the concentration of the oxygen in the atmosphere is above that for an IDLH oxygen-deficient condition.

[f] Only a positive pressure type SCBA equipped with a full facepiece (full facepiece pressure-demand open-circuit SCBA and full facepiece positive pressure closed-circuit SCBA) and a combination positive pressure type supplied-air respirator equipped with a full facepiece with an auxiliary self-contained breathing air supply (full facepiece combination pressure-demand airline respirator with an auxiliary self-contained breathing air supply) may be used for entry into an IDLH contaminated atmosphere, an IDLH oxygen-deficient atmosphere, and a hazardous atmosphere with unknown conditions.

[g] Only a positive pressure type SCBA equipped with a full facepiece (full facepiece pressure-demand open-circuit SCBA and full facepiece positive pressure closed-circuit SCBA) may be used for firefighting.

of laboratory respirator fitting tests, care must be taken in applying them to use of respirators in workplaces.

The product of the APF for a particular class of respirators and the established PEL for a specific hazardous substance listed by the U.S. Department of Labor is the MUC for the particular class of respirators for that substance. This MUC however cannot exceed any use limitation for the particular class of respirators prescribed by the NIOSH/MSHA approval for this class of respirators.

The recommended APF values for the various classes of respirators listed in Table 2 should be used to select the types of respirators that should be used to protect persons who are required to wear respirators in hazardous atmospheres in workplaces. Only classes of respirators with MUCs (MUC = APF × PEL) greater than the TWA (time-weighted average) concentration of a particular airborne substance in the hazardous atmosphere of the workplace may be considered for selection for use by persons who must wear respirators for respiratory protection in the workplace.

10 TEST, CERTIFICATION, AND APPROVAL OF RESPIRATORS

Until the summer of 1995, provisions of the federal respiratory protection regulations given in 29 CFR 1910.134 require that respirators provided by employers to employees for their protection against inhalation of hazardous atmospheres in workplaces be those which have been tested, certified, and approved by the National Institute for Occupational Safety and Health (NIOSH) of the U.S. Department of Health and Human Services and the Minc Safety and Health Administration (MSHA) of the U.S. Department of Labor. The program of testing, certifying, respirators by NIOSH and MSHA is carried out under the authority of Part 11 of Title 30 of the Code of Federal Regulations (30 CFR 11). Since 1995 this task has been relegated to NIOSH alone.

The regulations at 30 CFR 11 do the following: (a) establish procedures and requirements for filing applications to the Federal Government for test, certification, and approval of respirators or for modification of certified and approved respirators; (b) establish a schedule of fees to be charged to applicants for work done by the Federal Government in processing the application, conducting examinations, and performing tests; (c) provide for the issuance of certifications of approval for respirators or for modifications of certified and approved respirators; and (d) describe criteria for a quality assurance program for manufactured respirators, describing minimum performance requirements for respirators, and prescribing methods for examining and testing the performance of respirators.

Although the program was always operated by NIOSH, the certifications were issued jointly by NIOSH and MSHA. The work of the program is still conducted by the Certification Branch (CB) of the Division of Safety Research (DSR) of NIOSH which is located in Morgantown, West Virginia. Manufacturers of respirators submit specimen respirators, drawings, parts and materials lists, quality assurance plans, and performance test data to the NIOSH facility at Morgantown, West Virginia, and they request that NIOSH test the respirators to determine if these devices meet performance criteria given in 30 CFR 11, certify that the devices do meet the performance requirements, and approve the devices for use in protecting persons against certain types of hazardous atmospheres in workplaces. Prior to submitting a respirator to NIOSH for testing, a respirator manufacturer will have conducted performance tests on the respirator to determine that it meets the performance criteria. The manufacturer submits copies of test data that indicates that the respirator meets the performance requirements to NIOSH. If the results of testing by NIOSH confirm that the respirator does perform adequately and NIOSH approves the quality assurance program for the respirator submitted by the manufacturer, NIOSH then issues a certification to the manufacturer which states that the respirator meets the required performance criteria and is approved for the protection of persons against inhalation of particular hazardous atmospheres which occur in workplaces. NIOSH provides the manufacturer with an approval label which states that NIOSH has approved the respirator for particular respiratory protection applications. The manufacturer is required to either affix replicas of the approval label on either production respirators to be sold or to the packaging for the production respirators.

The manufacturer of a respirator that has been certified and approved by NIOSH is required to perform a quality assurance program, to keep records of quality assurance inspections and to test specimens of production respirators. This program is conducted to ensure that only production lots of respirators found to be satisfactory are accepted for sale and shipment to customers. The manufacturer is permitted to ship to customers only production lots of respirators that have been found to meet the requirements of the quality assurance program which NIOSH has found to be satisfactory.

NIOSH periodically purchases respirators on the open market and tests these devices to determine whether they meet the performance criteria. If NIOSH finds that a certified and approved respirator in the marketplace fails to meet the performance requirements, NIOSH immediately notifies the manufacturer of the problem and directs the manufacturer to take immediate action to eliminate the problem. Depending on the severity of the problem, NIOSH may require the manufacturer to stop the manufacture of the respirator until the problem has been eliminated, may require the manufacturer to recall respirators in the marketplace, and may require the manufacturer to terminate shipping of the respirator to customers until the problem has been solved. If the manufacturer fails to take action as directed by NIOSH to eliminate the performance problem, or if it is found that the performance defect in the respirator cannot be eliminated, then NIOSH may withdraw the certification and approval for the respirator.

NIOSH annually publishes a document entitled "NIOSH Certified Equipment List." This document lists the model numbers of certified and approved respirators, the names of the respirator manufacturers, the NIOSH approval numbers, and the approved applications for the respirators.

The current respirator test, certification, and approval program was established under authorization of the Coal Mine Health and Safety Act of 1969 and the Federal Mine Safety and Health Amendments Act of 1977. The goal of the program is to ensure the protection of workers against inhalation of harmful atmospheres by certifying that respirators meet minimum performance criteria listed in 30 CFR 11 and by approving respirators for use in harmful atmospheres.

Testing, certifying, and approving respirators by the Federal Government began in 1919 when the Bureau of Mines (BM) of the U.S. Department of Interior was authorized to test, certify, and approve mine rescue type respirators. The BM tested, certified, and approved respirators from 1919 until 1972 under the provisions of Schedules of the BM. On May 25, 1972, Title 30, Code of Federal Regulations, Part 11 (30 CFR 11), "Respiratory Protective Devices; Tests for Permissibility; Fees" superseded and revoked the previous Bureau of Mines Schedules for respirators. 30

CFR 11 granted jurisdiction for joint test, certification, and approval of respirators to both NIOSH and BM.

A reorganization of the U.S. Department of Interior occurred in 1974 which resulted in the creation of the Mining Enforcement and Safety Administration (MESA) which was given the health and safety activities of the BM, including the respirator test, certification, and approval program. Subsequent respirator certifications and approvals were jointly issued by NIOSH and MESA.

The Federal Mine Safety and Health Amendments Act of 1977 transferred the authority for the enforcement of mining safety and health from the U.S. Department of Interior to the U.S. Department of Labor in 1978. This Act created the Mine Safety and Health Administration (MSHA) in the U.S. Department of Labor which replaced MESA of the U.S. Department of Interior. MSHA acquired the respirator test, certification, and approval functions of MESA. Certifications and approvals for respirators now are jointly issued by NIOSH and MSHA but 1995 changes aimed at reducing government duplication placed the responsibility for the program solely on NIOSH.

11 SPECIAL PROBLEMS

I. FACIAL HAIR

Facial hair such as a beard, mustache, sideburns, low hairline, bangs, and a slight growth of stubble can permit leakage of a hazardous atmosphere into the interior of the respiratory-inlet covering. Such leakage may occur with tight-fitting facepieces of nonpowered air-purifying respirators and even with tight-fitting facepieces and loose-fitting facepieces of powered air-purifying respirators and atmosphere-supplying respirators. Also, facial hair between the sealing surface of a tight-fitting facepiece of an atmosphere-supplying respirator and the respirator wearer's face may permit loss of respirable air supplied to the facepiece which is a waste of respirable air and may result in a reduced use period of the respirator if the respirable air comes from a container having a fixed quantity of respirable air. Facial hair may also interfere with the operation of a valve located in a respiratory-inlet covering. Therefore, a person having facial hair should be prohibited from wearing a respirator with such a respiratory-inlet covering in a hazardous area.

A powered air-purifying respirator and a continuous-flow airline respirator with a loose-fitting respiratory-inlet covering that completely encloses facial hair can be permitted for use in a hazardous atmosphere. However, conditions such as an IDLH atmosphere, an oxygen-deficient atmosphere in the case of the powered air-purifying respirator, firefighting, an APF which is too low, etc. increase the risk of use of these respirators by a person with facial hair. An escape type SCBA which contains a hood type respiratory-inlet covering that would completely enclose facial hair may be used for emergency escape from a hazardous atmosphere.

II. HEAD COVERINGS

Passage of part of a head covering between the sealing surface of a respiratory-inlet covering of a respirator and the wearer's face will cause the same problems as discussed under facial hair. Therefore, a head covering which passes between the sealing surface of a respiratory-inlet covering of a respirator and the face of a respirator wearer shall not be allowed to be worn by a person who is required to use the respirator in a hazardous atmosphere.

III. FACIAL DEFORMITIES

Facial deformities such as scars, severe acne, deep skin creases, hollow cheeks, protruding cheekbones, absence of teeth or dentures, or unusual facial configurations may prevent a good seal of a respiratory-inlet covering to a person's face. Therefore, a person with a facial deformity should be prohibited from wearing a respirator with a respiratory-inlet covering in a hazardous atmosphere if a facial deformity may adversely affect the seal of the respiratory-inlet covering to person's face.

Missing teeth or dentures may prevent an adequate seal of a respirator mouthpiece in a person's mouth. A nose may have a shape, structure, or size that prevents closing of the nose by the nose clamp of a mouthpiece/nose clamp type respiratory-inlet covering. A person having these types of facial deformities should not be permitted to use a respirator with a mouthpiece/nose clamp type respiratory-inlet covering.

IV. VISION

Spectacles that have temple bars or straps that pass between the sealing surface of a respiratory-inlet covering of a respirator and the respirator wearer's face prevents a good seal of the respiratory-inlet covering to the face, and therefore is prohibited from use with a respirator having a respiratory-inlet covering with a sealing surface affected by the spectacle. Special corrective lenses which can be mounted inside a respiratory-inlet covering are available from respirator manufacturers. Care must be taken to ensure that the special corrective lenses are properly installed inside a respiratory-inlet covering.

Spectacles and protective goggles may also interfere with the seal of a quarter-mask or half-mask facepiece, or a mouthpiece/nose clamp of a respirator to the respirator's face. If this occurs, a full facepiece with installed special corrective lenses, if necessary, should be used.

Soft, gas-permeable contact lenses may be used with respirators, including respirators equipped with full facepieces and atmosphere-supplying respirators with respiratory-inlet coverings that cover the eyes. Care must be taken to ensure that air entering the respiratory-inlet covering does not contain matter that may penetrate between the contact lenses and the respirator wearer's eyes to cause eye discomfort or eye injury.

V. COMMUNICATIONS

Talking while wearing a respirator equipped with a tight-fitting facepiece may break the seal between the facepiece and the respirator wearer's face and may cause leakage between the sealing surface of the facepiece and the wearer's face. This may result in the leakage problems addressed previously in the discussion about facial hair. When communication by a respirator wearer in a hazardous atmosphere is necessary, hand signals, use of a signal line, or other means of communication that do not interfere with the seal of a respiratory-inlet covering may be used.

Special communicating equipment which may alleviate the communications problem may be obtained from some respirator manufacturers. Installed special communicating equipment in powered air-purifying respirators and continuous-flow airline respirators employing loose-fitting respiratory-inlet coverings that enclose the respirator wearer's head may be a solution to the communications problem provided that these respirators are acceptable for use in the particular hazardous areas where the respirators must be worn.

VI. PLANS AND PROCEDURES FOR USE OF RESPIRATORS IN DANGEROUS ATMOSPHERES

If an IDLH atmosphere may occur in normal operations or during anticipated periodic maintenance operations or other anticipated special operations, then the written respirator program shall include procedures for the safe use of respirators in the IDLH atmosphere(s).

Emergency, rescue, and escape situations requiring the use of respirators shall be anticipated, and if the occurrence of such situations are possible, then the written respirator program shall include procedures for the safe use of respirators whenever such situations occur.

Instructions and training on the use of respirators in dangerous atmospheres shall be provided to employees.

VII. RESPIRATOR USE IN IDLH ATMOSPHERES IN NORMAL OPERATIONS, ANTICIPATED MAINTENANCE OPERATIONS, AND OTHER ANTICIPATED SPECIAL OPERATIONS

During the time that a person wearing a respirator is in an IDLH atmosphere performing routine work, conducting a maintenance operation, or carrying out a special task, at least one standby person shall be present in a safe area for emergency rescue. If more than one person wearing a respirator is to be present simultaneously in an IDLH atmosphere, the employer shall determine if more than one standby person in a safe area is necessary. If more than one standby person is necessary, the employer shall ensure that they are present in a safe area while the respirator wearers are in the IDLH atmosphere.

Communications between the standby person(s) and the respirator wearer(s) in the IDLH atmosphere by suitable means shall be maintained. Each standby person shall have a self-contained breathing apparatus equipped with a full facepiece and designed for operation in the pressure-demand mode or other positive pressure mode available for immediate use. The respirator wearers in the IDLH atmosphere shall be equipped with safety harnesses and safety lines or other suitable retrieval equipment for their removal from the IDLH atmosphere to a safe area if necessary.

VIII. RESPIRATOR USE FOR EMERGENCY ESCAPE

When emergency escape situations requiring the use of respirators are anticipated, suitable respirators shall be stored in areas where dangerous atmospheres may suddenly occur. The respirators shall be stored in protective containers marked to show that they contain respirators for use in emergencies, that are located so that the respirators can be quickly obtained, and can be opened readily to remove the respirators for immediate donning. The employer shall make sure that these employees know where the respirators for emergency escape are located. The employer also shall instruct and train the employees in the safe use of the respirators for emergency escape.

Each respirator stored for emergency escape shall be inspected at least monthly to determine if it is in good operating condition, or after each use. If inspection shows that the respirator is in unacceptable condition, it shall be replaced immediately. A record of inspection dates, findings, and remedial actions shall be maintained for each respirator stored for use in emergency escape.

Air and oxygen cylinders of self-contained breathing apparatus stored for use in emergency escape shall be kept fully charged, and the cylinders shall be recharged when the pressure of the air or oxygen falls to 90% of the respirator manufacturer's recommended pressure level.

IX. RESPIRATOR USE IN CONFINED SPACES

A confined space is an enclosure such as a storage tank, process vessel, boiler, silo, tank car, duct, tube, pipeline, vault, tunnel, pit, or narrow and deep trench having limited means of egress and poor ventilation. Such a confined space may have an atmosphere which contains toxic air contaminants, is oxygen deficient, or is flammable or explosive. A confined space should be considered to have an IDLH atmosphere unless proved otherwise.

Before a person is permitted to enter a confined space, testing of the atmosphere of the confined space shall be performed to determine the concentration of any known air contaminant and any anticipated air contaminant that may be toxic, flammable, or explosive, and to determine the concentration of oxygen.

The confined space should be force-ventilated to keep the concentrations of any toxic air contaminants to levels below established permissible exposure concentrations, to keep the concentrations of any airborne flammable or explosive substances at safe levels, and to maintain an adequate oxygen concentration so that the atmosphere is not oxygen deficient.

No person shall be permitted to enter a confined space that contains a flammable or explosive atmosphere.

Any person entering a confined space which has an atmosphere that contains a toxic air contaminant at a level greater than the established permissible exposure concentration or which has an atmosphere that is oxygen deficient, must wear the proper type of respirator.

Even if the atmosphere inside a confined space is not flammable or explosive and has a toxic air contaminant at a level below the established permissible exposure limit and the atmosphere is not deficient in oxygen, the safest procedure is to force-ventilate the space continuously and to continuously monitor the contaminant and oxygen concentrations.

Air-purifying respirators and airline respirators may be worn in a confined space only if testing of the atmosphere in the confined space show that the atmosphere contains adequate oxygen and that the concentration of any toxic air contaminant is below the IDLH concentration. The atmosphere inside the confined space shall be monitored continuously while persons wearing the mentioned types of respirators are inside the confined space.

If the atmosphere inside a confined space is IDLH because of a high concentration of an air contaminant or because of the level of oxygen deficiency, a person who must enter this confined space shall be required to wear a pressure-demand open-circuit SCBA equipped with a full facepiece, a positive pressure type closed-circuit SCBA equipped with a full facepiece, or a combination full facepiece pressure-demand airline respirator with a self-contained breathing air supply.

While any person is in a confined space, the requirements pertaining to standby personnel given in the previous discussion concerning "respirator use in IDLH atmospheres in normal operations, anticipated maintenance operations, and other anticipated special operations" shall be complied with. These standby requirements shall be complied with for any conditions where a person is in a confined space which includes situations where a person may not be required to wear a respirator in a confined space.

X. LOW TEMPERATURE ATMOSPHERES

A low temperature atmosphere may cause fogging of a lens in a respiratory inlet covering. Coating the inner surface of a lens with an antifogging agent available from a respirator manufacturer may prevent fogging of the lens at a temperature as low as 0°C (32°F), but severe fogging of the coated lens may occur at –18°C (0°F). A full facepiece with an inlet air deflector inside the facepiece that deflects cold inlet air away from the lens. A nose cup inside the facepiece prevents the warm and moist expired air from contacting the lens and directs this exhaled air to pass out of the facepiece through an exhalation valve preventing lens fogging to a temperature as low as –34.4°C (–30°F).

Air at a very low temperature may cause the exhalation valve in the facepiece of a respirator to freeze either in the open or closed position.

Dry respirable air should be used with airline respirators and with the types of SCBA devices which contain air cylinders when these respiratory protective devices are used in low temperature atmospheres. It is recommended that the respirable air for an airline respirator intended for use in a cold atmosphere have a reduced moisture content such that its dew point is at least 10°C (18°F) lower than the coldest temperature expected in the atmosphere where the respirator is to be used. The respirable air in the compressed air cylinder of a SCBA device should have a reduced moisture content such that its dew point is –55°C (–67°F) or lower. If the SCBA device is intended for use in a cold atmosphere having a temperature of –32°C (–25.6°F) or lower, then the moisture content of the respirable air in the compressed air cylinder of the SCBA device should have a dew point of –75°C (–103°F) or lower. The approval label of a NIOSH/MSHA approved SCBA device lists the minimum use temperature for the SCBA device.

High pressure connections made of metal on airline respirators and on SCBA units which use cylinders of high pressure compressed air or oxygen may leak at low temperatures because of metal contraction. Care must be taken to ensure that these metal connections are not overtightened to prevent leakage at low temperatures, since they may crack or rupture when the respiratory protective equipment is brought to a higher temperature atmosphere.

Some airline respirators are equipped with a vortex tube which can be adjusted to warm the respirable air supplied to the respiratory-inlet covering of the respirator. (The vortex tube also can be used to cool the respirable air supplied to an airline respirator and this will be mentioned in the discussion concerning high temperature atmospheres.)

An elastomeric facepiece may become stiff and distorted at a low temperature which will prevent a good seal of the facepiece to a respirator wearer's face. Elastomeric components such as valve flaps, diaphragms, gaskets, and breathing tubes used in various respirators may stiffen at low temperatures which may prevent proper operation of these parts.

XI. HIGH TEMPERATURE ATMOSPHERES

Wearing a respirator places a physical stress on the wearer. Likewise, a person wearing a respirator in a high temperature atmosphere must endure an additional heat stress.

Light respirators and respirators that offer low resistance to breathing will minimize the stress placed on a respirator wearer.

Airline respirators generally are light; continuous-flow and pressure-demand airline respirators offer minimal resistance to breathing. Therefore, the airline respirator should be considered for persons who must wear respirators and work in high temperature atmospheres. A vortex tube is available as a component of some airline respirators. It can be adjusted to cool the respirable air supplied to the respiratory-inlet covering of the airline respirator.

A powered air-purifying respirator offers minimal resistance to breathing. Many modern powered air-purifying respirators are light. Thus, consideration should be given to the use of a powered air-purifying respirator in an atmosphere at an elevated temperature.

Components of respirators made of certain elastomeric materials deteriorate at accelerated rates when exposed to high temperatures, and this deterioration may cause these parts to malfunction. The elastomeric materials used in some respirator facepieces soften at even moderately elevated temperatures. This distortion may permit leakage of a hazardous atmosphere between the sealing surface of the facepiece and the respirator wearer's face. Respirators used in high temperature atmospheres should be inspected more often than respirators used in normal room temperature atmospheres to ensure that respirators with elastomeric components that have deteriorated because of exposure to high temperature are found so that they can be discarded and replaced with respirators in good operating condition.

XII. PHYSIOLOGICAL STRESS ON RESPIRATOR WEARERS

Wearing any respirator, alone or in conjunction with other types of protective equipment, places some physiological stress on the wearer.

The weight of equipment worn by a person will increase the amount of energy used by the person to perform a task. Resistance to breathing offered by a respirator also results in an increase in energy expenditure to perform a task. If a respirator wearer has an impaired cardiovascular system or impaired respiratory system, wearing a respirator may constitute an unacceptable risk.

It is very important that the physician, who is responsible to determine whether an employee is to be permitted to be assigned to a task that requires the employee to wear a respirator consider the following factors:

- The characteristics of the type of respirator the employee would wear
- The characteristics of any additional protective equipment to be worn by the employee
- The physical demands of tasks to be performed by the employee while wearing the respirator
- The conditions of the workplace where the employee would wear the respirator and perform the work tasks
- The results of pertinent medical tests on the employee and the medical history of the employee
- Any relevant psychological factors concerning the employee

Selection of a type of respirator for use in protecting a person against inhalation of a hazardous atmosphere in a workplace should take into account the weights of various types of respirators, the breathing resistances offered by different types of respirators, the level of respiratory protection provided by the various kinds of respirators, and the person's tolerance of wearing respirators.

Some hazardous atmospheres require that an employee wear protective clothing to protect the employee against contact with a harmful material in addition to requiring that the employee wear a respirator to prevent inhalation of a harmful atmosphere by the employee. If the protective clothing is impervious to gases and liquids it will prevent dissipation of body heat by evaporation of perspiration and by convection. This will result in imposing a severe heat stress on the wearer even in a moderate temperature atmosphere. Any of the following will provide the wearer with both respiratory and body protection against hazardous substances and will prevent heat stress even in an atmosphere at an elevated temperature:

- Suit made of impervious material but provided with compressed air for both breathing and body ventilation
- Suit made of impervious fabric but provided with cool air from a vortex tube for both breathing and body ventilation
- Suit made of impervious material containing a circulating cooling water system and an airline respirator
- Suit made of impervious material combined with an ice-packet cooling vest and an airline respirator
- Suit made of impervious material combined with a thermoelectric cooler which provides cool air for both breathing and body ventilation

ADDITIONAL READING

1. AIHA/ACGIH Joint Respirator Committee: Respiratory Protective Devices Manual, American Industrial Hygiene Association and American Conference of Governmental Industrial Hygienist, 1963.

2. Ballantyne, B and Schwabe, PH (editors): Respiratory Protection: Principles and Applications, Year Book Medical Publishers, Inc., 1981.
3. National Fire Protection Association: SCBA - A Fire Service Guide to the Selection, Use, Care and Maintenance of Self-Contained Breathing Apparatus National Fire Protection Association, Quincy, MA, 1981.
4. Hannan, DG and Merkel, CRE: Sorbent and Catalysts for Respirator Cartridges and Gas Mask Canisters. Am Ind Hyg Assoc J 29:136-139 (1968).

12 DEVELOPMENTS IN RESPIRATORY PROTECTION

The enactment of the Occupational Safety Health Act in 1970 has had a significant impact on respiratory protection. Congress, in creating OSHA to enforce regulations on safety and health, also created NIOSH. One of its functions is to test and certify respirators. An existing consensus standard for respiratory protection, ANSI Z88.2-1969, was adopted and enforced by OSHA as its standard on respiratory protection. OSHA also mandated the use of MSHA/NIOSH approved respirators. There had been significant developments in respiratory protection in the past two decades. Details of some of these developments will be addressed in subsequent chapters. A brief review of major developments follow.

I. RESPIRATORS

Respirator manufacturing has become more international. Many foreign companies have entered the U.S. market such as Dräger of Germany, Interspiro of Sweden (now a subsidiary of Comsec), and Racal Airstream of U.K. Some U.S. manufacturers were acquired by foreign companies such as Survivair (Comsec of France), Norton (Siebe of U.K., now Siebe North), Willson (INCO of Canada, and later Dalloz of France), and Glendale (Bilsom of Sweden). There are also affiliations with foreign manufacturers among domestic manufacturers such as International Safety Instruments (Sabre of U.K.), and 3M (Shigematsu of Japan). Acquisitions also occurred among domestic manufacturers such as Norton's purchase of Welsh from Textron. Cabot purchased American Optical from Warner-Lambert. Willson purchased Pulmosan, and Racal acquired Globe, Glendale was purchased by Bilsom, and UVEX acquired Protech.

A. Open-Circuit Self-Contained Breathing Apparatus:

The first open-circuit positive-pressure (pressure demand) self-contained breathing apparatus (SCBA) was developed in the mid 1960s by Scott Aviation. The pressure demand SCBA maintains a positive-pressure inside the facepiece during inhalation. This action significantly reduces the leakage of high concentrations of toxic air contaminants into the facepiece of the SCBA during use.[1] This device received wide acceptance among firefighters because it provided a higher level protection than the negative-pressure demand type. In 1979, OSHA published a standard on fire brigade[2] which required that positive-pressure SCBA must be used for structural firefighting after July 1, 1983.

OSHA regulations are not applicable to employees of states, counties, and municipalities in those states without a state OSHA program. The issue of mandating the use of pressure demand SCBAs for public employees such as firefighters was left to the American National Standards Institute (ANSI) Z 88.5 Subcommittee on respiratory protection for fire services. Due to the higher breathing resistance and the higher cost of switching from demand to pressure demand SCBAs, there was considerable resistance for the ANSI Z 88.5 Subcommittee members to agree on the issue of mandating the use of positive-pressure SCBA for firefighting. Finally, an agreement was reached to permit a 6-year phase-in period for pressure demand SCBAs. The final phase-in date for compliance was January 1987. There are virtually no demand type open-circuit SCBAs used today for firefighting and there is almost no market demand for demand type SCBAs.[3]

Another problem encountered by the fire service is emergency rescue. If the SCBA of a firefighter fails, he/she could only escape by sharing the facepiece with a fellow fire fighter. This is a dangerous practice because most fire environments are immediately dangerous to life or health. However, a device called "buddy breather" was developed. It is a connector which allows a connection of two SCBAs. Two firefighters can then escape to safety by sharing the air from one breather. This practice is accepted by OSHA. The buddy breathing connector is approved by MSHA/NIOSH as a component of the SCBA assembly but NIOSH specifically prohibits its use. A more recent development is a connector device called "quick fill" which permits the transfer of air between two cylinders. This appears to be safer than the buddy breather, and it is approved by MSHA and NIOSH (Figure 1).

A major problem associated with SCBA use is that the rated service life of the air cylinder is based on a moderate work rate. An approved 30-min air cylinder may last only 18 min under a heavy work rate. There is also a weight limitation of 35 lbs for all approved SCBAs, which may prevent the development of SCBAs with longer service life.

The advent of the aluminum cylinder and composite materials has reduced the weight of the air cylinders appreciably as compared to the steel cylinder equipped SCBAs. This development was initiated by the National Aeronautics and Space Administration (NASA) whose interest in space technology prompted the development of lightweight materials. Two contracts were awarded for the development of lightweight SCBA and air cylinders in 1974. A lightweight breather was developed by Scott Aviation (Scott 4.5), and the composite cylinder was developed by Martin-Marietta. Scott marketed a SCBA equipped with a glass fiber reinforced hoop wrapped aluminum cylinder in 1976. The first SCBA equipped with a 60-min service life composite cylinder with a weight less than 35 lbs was marketed by Scott in 1980 (Figure 2). An SCBA with a fully wrapped aluminum composite cylinder can weigh even less. A comparison of the weight of various types of air cylinders is shown in Table 1.

The fully wound composite cylinders cost more than the hoop wrapped cylinders which do not have fiber glass resin reinforcement at the top and bottom of the cylinder. Because the Department of Transportation, which has jurisdiction over the performance of air cylinders, requires that burst test be performed only for the

Figure 1 Quick fill connector for the MSA SCBA.

sidewall of the cylinder, many SCBA manufacturers selected the hoop wrapped design as a cost saving measure. There were numerous failures of the hoop wrapped cylinders in the early 1980s. It was found that the failures were due to an excessive quantity of impurities in the aluminum alloy. The problem was later corrected.

While SCBA have been used widely for firefighting, they are not specifically designed to provide such service. The facepiece pressure may go negative under heavy work load, the harness material may soften under high radiant heat, and the regulator may freeze up under low temperature. The National Fire Protection Association developed a standard (NFPA 1981) which prescribes the performance requirements for SCBAs used in the fire service.[4] This is the first time the respirator users dictated the development of a device to serve their needs. The main reason this occurred is that firefighters are more organized than other user groups, and many States recognize the standards promulgated by the NFPA (Figure 3). At this date, most of the performance requirements developed in NFPA 1981 have not been considered by NIOSH in their proposed revision of 30 CFR 11 (now 42 CFR 84). The major performance criteria prescribed in the NFPA 1981 are:

- High Peak Air Flow
- High Temperature Resistance
- Low Temperature Resistance
- Thermal Resistance

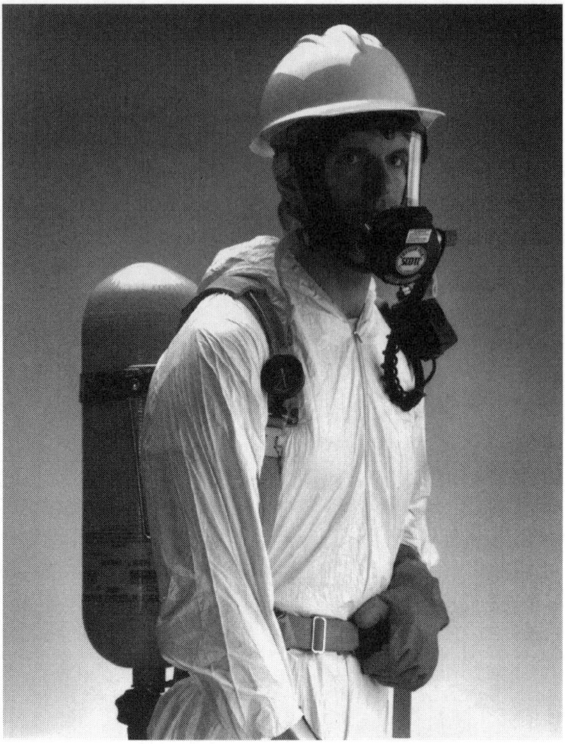

Figure 2 Scott 4.5. The first SCBA equipped with a 60-min service life cylinder.

- Flame Resistance
- Corrosion Resistance
- Shock and Vibration Resistance
- Lens Abrasion Resistance
- Communication

Because none of these performance requirements are contained in the existing 30 CFR 11, these breathers are essentially certified by their manufacturers at this time. The Lawrence Livermore National Laboratory was asked by OSHA to evaluate the conformance of a commercial lot of SCBAs procured by the agency to the NFPA standard. This is the first time a third party has evaluated the compliance of an SCBA manufactured to self-certification. The test results confirmed compliance.[5]

Table 1 Weight of SCBAs with Various Types of Air Cylinders

Cylinder material	Service life (in min)	Cylinder weight (in lb)	Whole assembly weight (in lb)
Steel	30	19	32
Aluminum (Al)	30	17	30
Hoop Wound	30	12	25
Fully Wound	30	9.5	22.5
Fully Wound	60	19	34

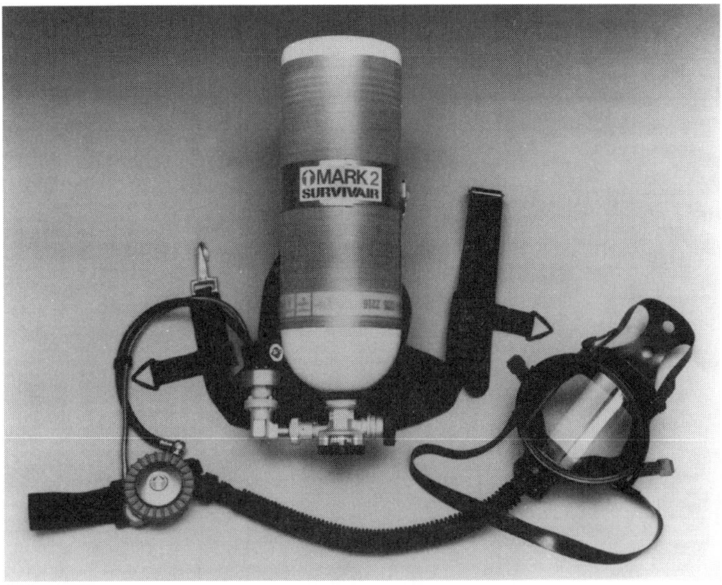

Figure 3 SCBA meets NFPA-1981 requirements — Survivair Mark II.

With the entry of foreign SCBAs into the U.S. market, some new concepts have been introduced to this country. For example, the face mounted regulator is much smaller than the belt mounted regulator. This design, very common in Europe, is now available in many SCBAs sold in the U.S.

B. Positive-Pressure Closed-Circuit SCBA

The first positive-pressure closed-circuit SCBA (PPCCSCBA) with a 1 h service life was approved in 1978 (Figure 4). The advantages of such device are that it provides a longer service life than the open-circuit SCBA at the same work rate and it also weighs less. Since there is no approval schedule in the 30 CFR 11 for such a device, it can only be certified as a negative-pressure device. The device has been accepted by OSHA as a positive-pressure SCBA. Recently, MSHA/NIOSH has approved these devices as positive-pressure SCBAs.

A great deal of controversy was generated when the Biomarine 60 P, the first PPCCSCBA, was introduced. Because it uses compressed oxygen for breathing, concerns were expressed regarding its safety in a fire environment. Hearings were conducted by OSHA on this issue during the promulgation of the fire brigade standard.[6] However, no conclusive evidence was found to prove that such a device is unsafe for firefighting. Furthermore, the U.S. Navy has used an oxygen generating closed-circuit SCBA in submarines, the Chemox, for years. The fire safety issue of using PPCCSCBA has since faded out. More manufacturers have produced this type of breathers and their service life has been increased to between 2 and 4 h (Figure 5). In spite of the advantage of providing a positive-pressure during use, MSHA has not mandated the use of PPCCSCBA for mine rescue operations; therefore, the demand for this device is not great.

Figure 4 Biomarine 60 P, PPCCSCBA with a 1-h service life.

Figure 5 Biomarine BioPak 240P, 4-h PPCCSCBA.

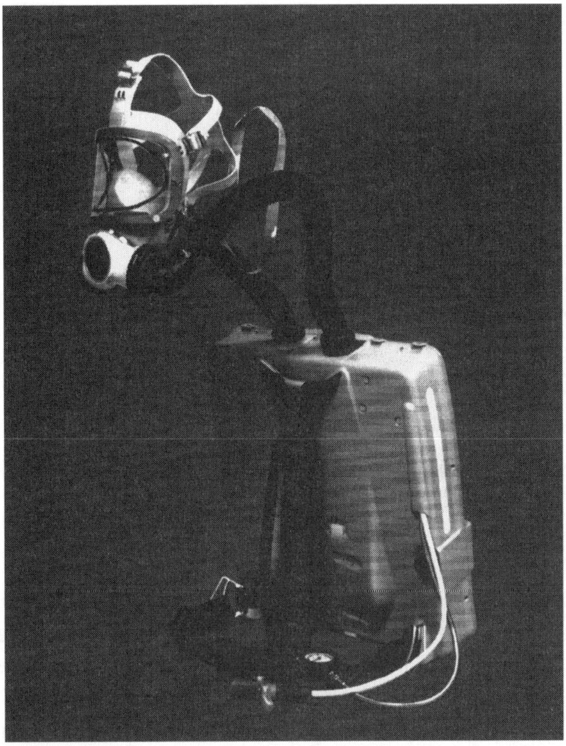

Figure 6 Hybrid design Litton PPCCSCBA.

There are inherent disadvantages to closed-circuit SCBAs as compared with the open-circuit devices, such as higher temperature of the breathing air and not being able to remove contaminants when they are introduced into the facepiece. These obstacles may be overcome by exhausting a certain percentage of air inside the facepiece. A hybrid design SCBA has been developed using such a principle (Figure 6). Because about 20% of the air is flushed from the facepiece, in order to maintain adequate level of oxygen the compressed air has a higher oxygen content of 30%.

Humans have an ability to adopt to hostile environments. At a meeting on the use of closed-circuit SCBAs for firefighting held at the Lawrence Livermore National Laboratory, one speaker stated that all the new hired firefighters in his department were given training on the use of PPCCSCBA. Because these trainees were never trained in the use the open-circuit SCBA, they had developed more tolerance to the higher operating temperature of the breather.[7]

C. Continuous Flow Escape SCBA

The hooded type escape SCBA is a new development. The device, made by Robertshaw, was first approved in 1977 (Figure 7). It has an open-circuit continuous flow design and utilizes a high pressure coil as an air reservoir and a hood instead of facepiece as an inlet covering. The air flow of this device is about 28 liters per minute (l/m) and it has a service life of 5 min. Donning the device is relatively simple. The wearer pulls the hood out of the storage case, activates the air flow, places the

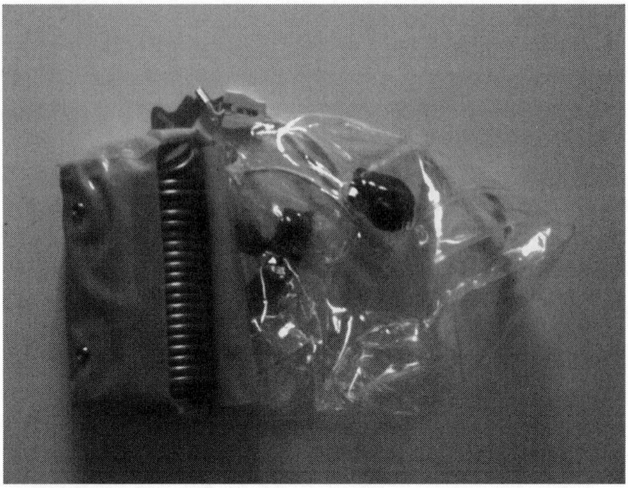

Figure 7 Robertshaw Air Capsule ESCBA.

hood over the head, and then tightens the draw string. Another Survivair design consisted of two air cylinders, a pressure regulator and a hood (Figure 8). In order to maintain an acceptable concentration of oxygen and carbon dioxide inside the hood, the compressed air has a higher oxygen content of between 26 and 30%. A longer service life version was developed by Scott. This device, using a chemically generated oxygen, provides a service life of 15 min (Figure 9).

In 1982, a design with a single rechargeable air cylinder was developed by Sabre of the United Kingdom and manufactured by its affiliate, the International Safety Instruments (ISI), in the U.S. (Figure 10). The cylinder air pressure is about 2000 lb and it can be recharged without special equipment. The air flow is about 40 l/m. The weight was reduced by the use of an aluminum cylinder. This design became very popular and replaced all other designs.

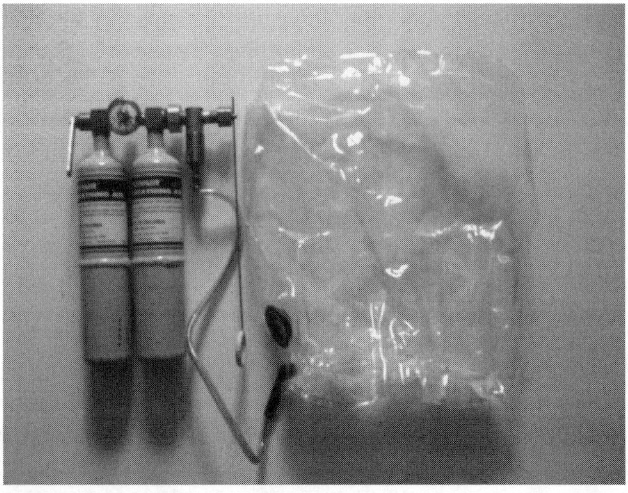

Figure 8 Survivair-Emergency Air ESCBA.

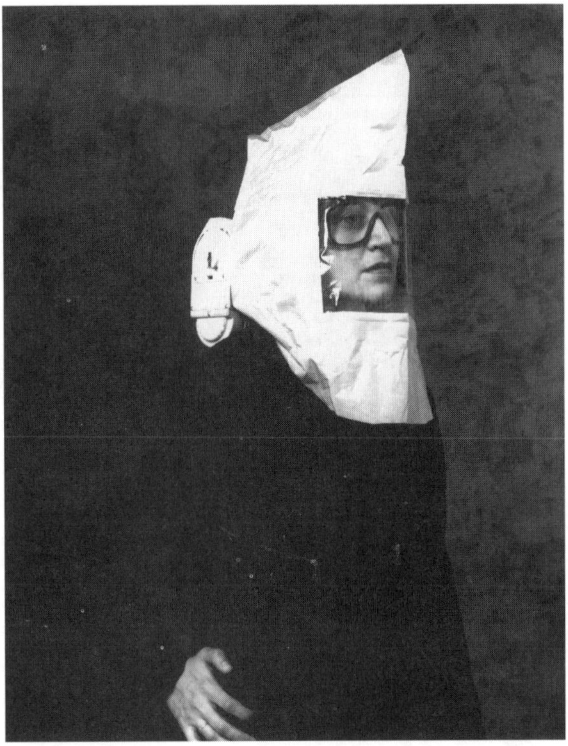

Figure 9 Scott-Scram ESCBA.

Figure 10 ISI-ELSA ESCBA.

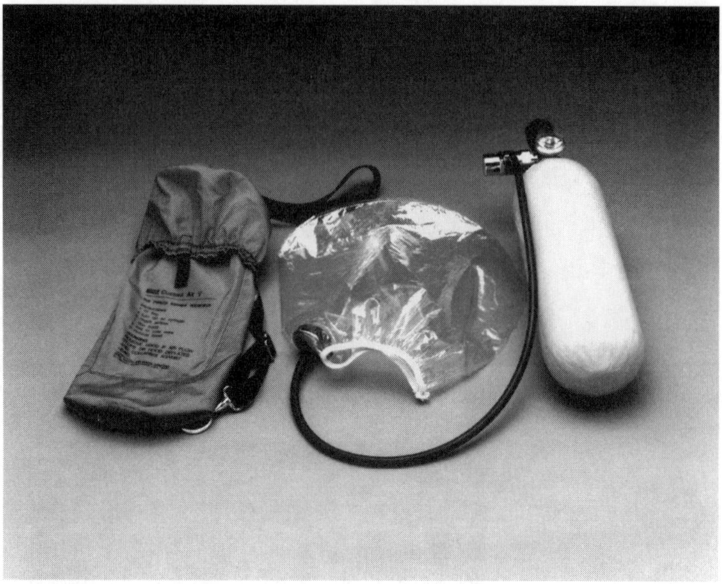

Figure 11 MSA-Custom Air V, ESCBA with a 70 l/m air flow.

Because all the air inside the hood is not flushed out, the oxygen and carbon dioxide concentrations can reach an unsafe level if the wearer egresses at a fast pace. However, the continuous flow escape ESCBA is a class of SCBA for which there is no performance requirement in the 30 CFR 11 concerning the levels of oxygen and CO_2 (Figure 11). Some devices having a higher air flow, as much as 70 l/m, were developed which reduce the level of CO_2 and increase the level of oxygen when the user exits to safety at a moderate pace. To compensate for the weight increase of the extra air, several manufacturers use a fully wound composite cylinder. In order to provide a longer egress time, several devices with a service life of 10 min at flow rates between 40 and 70 l/m are available (Figure 12). However, there is only one unit with an air flow of 70 l/m.[8]

D. Powered Air-Purifying Respirators

The concept of developing a powered air-purifying respirator is not new. Patents for PAPRs had been issued in the 1930s. Silverman and Burgess in the U.S.,[9] and Sherwood and Greenhalgh in the U.K.[10] both developed prototype PAPRs in the mid 1960s. It appears that Americans are interested in the tight-fitting design and the British are more interested in the loose-fitting design. Commercial production of a personal size PAPR was not begun until the 1970s. A PAPR made by White Cap may be the first PAPR approved in the U.S. The White Cap was very bulky and was not practical for personal use.

The first MSHA/NIOSH approved personal size PAPR was made by a British manufacturer, Racal (Figure 13). The device is a helmet with a blower and the filter element located inside the helmet. A visor is integrated to the helmet to provide eye and face protection. The Racal AH-1 PAPR is equipped with a dust/mist filter. Because the U.S. and U.K. have major differences in filter certification requirements,

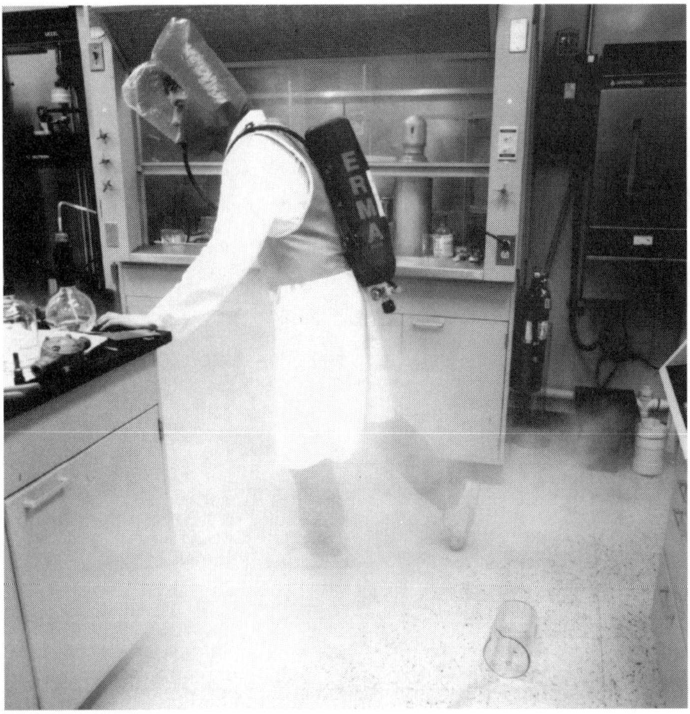

Figure 12 Draeger-ERMA ESCBA, with a 10-min service life at 70 l/m.

Figure 13 Racal AH-1 PAPR.

Figure 14 MSA PAPR.

the U.S. test is a loading method with the filter being challenged with a high dose of a relatively large silica dust aerosol; the British method uses submicrometer NaCl aerosol particles as a challenge agent to the filter. In order to meet the U.S. certification requirements, Racal had to develop a less efficient filter to pass the silica dust test.

Shortly, after the approval of the Racal PAPR, Mine Safety Appliance Company (MSA) received an approval for the first U.S. made personal size PAPR in 1977 (Figure 14). The MSA PAPR is equipped with either a half-mask or a full facepiece, with two high-efficiency particulate air (HEPA) filter elements located on either side of the blower assembly. A breathing hose connects the facepiece and the blower. Another PAPR developed by the 3M company also received approval. The 3M PAPR, the Air Hat, is similar to the Racal AH-1 in design (Figure 15). A Racal PAPR equipped with HEPA filters at a belt mounted position was approved in 1979. Racal later developed a PAPR, model AH-3, which comes equipped with a pair of HEPA filter elements which were mounted on a belt mounted blower (Figure 16).

A properly designed PAPR would provide the same level of protection as a continuous flow supplied air respirator (SAR). The PAPR has the additional advantage of no restriction on mobility. The design of the PAPRs is very versatile. The inlet covering of the PAPRs can be classified as tight-fitting, loose-fitting, helmet, and hood type. The blower has either a face, helmet, or belt mounted position. The filter element can be placed inside the helmet or mounted with the blower and carried on the belt. These configurations are summarized in Table 2.

The current air flow certification requirements for the tight and loose-fitting PAPRs are 115 and 170 l/m respectively which are the same flow rates for the supplied air respirators. The flow rate is based on moderate work rate. OSHA

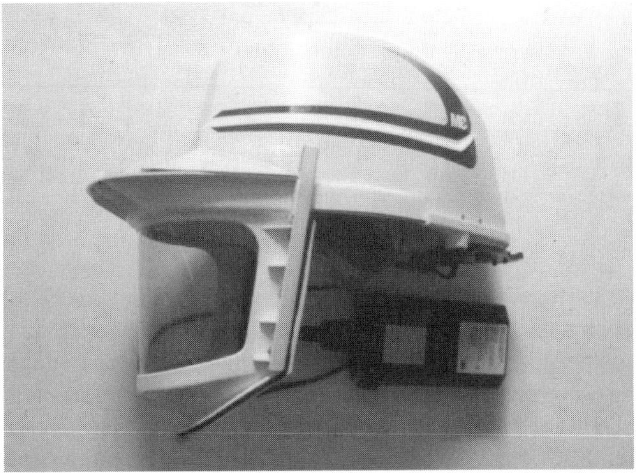

Figure 15 3M Air Hat PAPR with helmet mounted HEPA filter and blower.

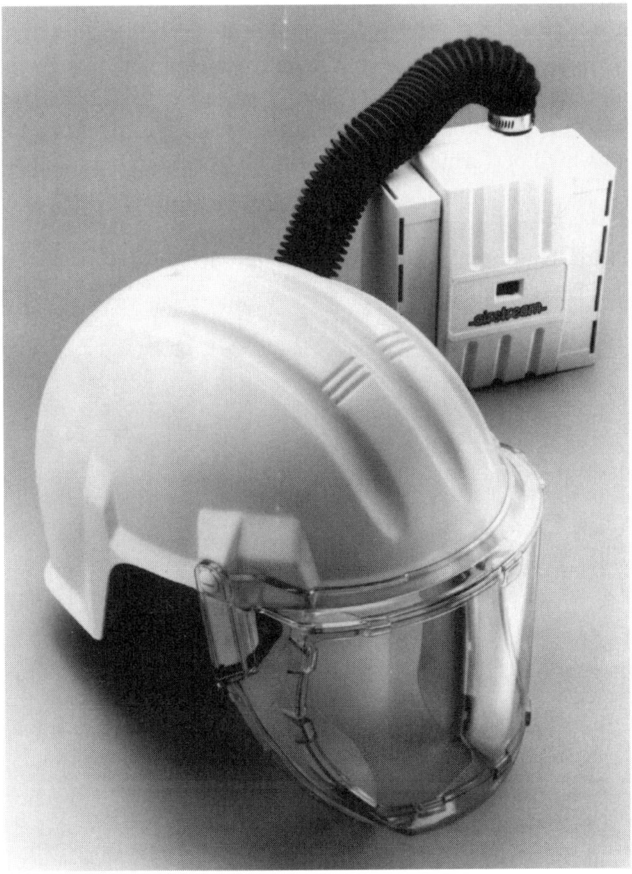

Figure 16 Racal AH-3 PAPR.

Table 2 Classification of PAPRs

By inlet covering	
	Tight-fitting
	Loose-fitting (helmet)
	Hood
By blower location	
	Face mounted
	Belt mounted
	Helmet mounted

requested that the Lawrence Livermore National Laboratory (LLNL) to conduct a study to determine the performance of PAPRs under heavy work load at various air flow rates. The results indicated that the tight-fitting PAPR can maintain a positive-pressure under heavy work load when run at a minimum flow rate of 170 l/m. However, none of the helmet type PAPRs could maintain a positive-pressure, even at a flow rate of 280 l/m.[11] In response to the result of this study, one tight-fitting PAPR manufacturer has redesigned their device to increase the flow rate to 170 l/m (6 cfm) (Figure 17).

The hood type PAPR is a relatively new design. The first model designed by Bullard was approved in 1987 (Figure 18). This device has two design features that are first in the industry: (1) a maximum flow of 9 cfm which provides a very high level of protection; and (2) the use of a more reliable gel cell battery instead of the nickel cadmium rechargeable battery. This battery also provides a service life of up to 16 h. Because users are not well informed about the advantages of a higher air flow rate PAPR, this high performance PAPR which costs more to produce does not sell too well and was discontinued.

The first PAPR which provides protection for gases or vapors was developed in 1983. The Racal Breathe-Easy PAPR has a unique design which uses three cartridges

Figure 17 MSA OptimAir 6 PAPR with high air flow.

Figure 18 Bullard Quantum high air flow hood type PAPR.

to reduce the flow resistance and meets the certification requirement (Figure 19). The cartridges are mounted on a blower called the "turbo unit" with a separate battery pack. Both tight, loose, and hood configurations are available. In 1990 MSA modified its 6 cfm model PAPR. The new design permits the use of gas and vapor

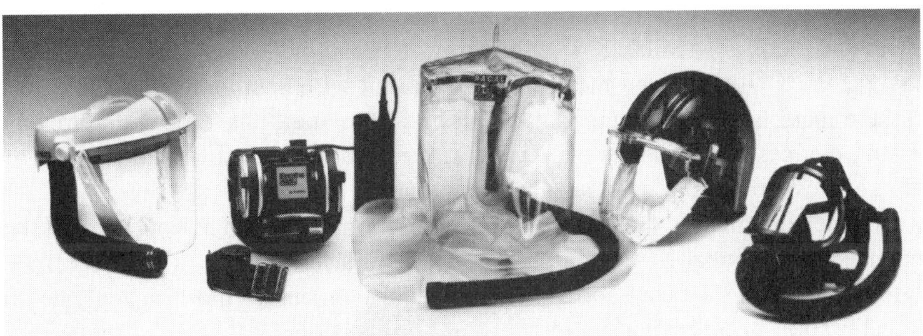

Figure 19 Racal Breathe-Easy with different inlet coverings.

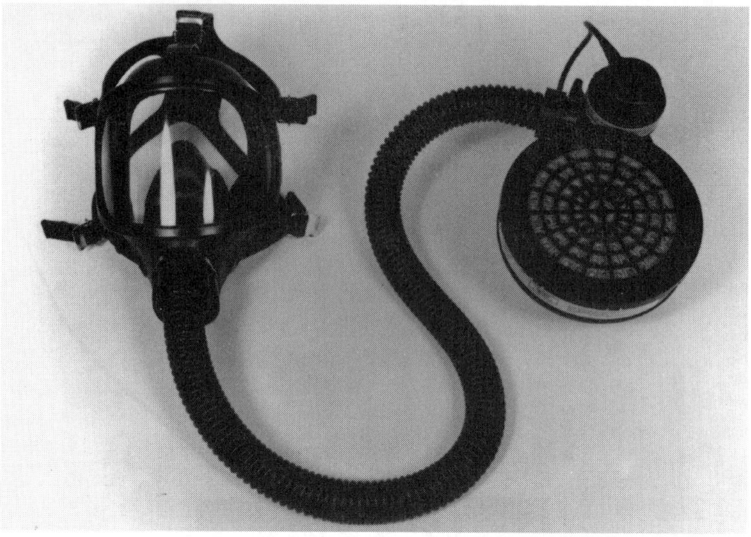

Figure 20 Racal Breathe-Easy with a single filter element.

cartridges. Racal later developed a large single cartridge that replaces the three cartridges used in the Breath-Easy line PAPRs (Figure 20).

The first PAPR with a face mounted blower and a single HEPA filter was developed by Racal and approved in 1985 (Figure 21). This device is strictly designed for the asbestos abatement industry. Since this is a billion dollar industry, almost every manufacturer has an approved device available.

The major design disadvantage with the constant flow PAPRs is that pressure inside the inlet covering of the device may go negative under high work rates and it may provide more air than the wearer actually needs under lower work rates. The excess air to be filtered can shorten the service life of the sorbent. A possible solution is the development of a pressure demand PAPR. The power source for most PAPRs is a nickel-cadmium battery pack which is not a very reliable device. Since PAPRs are used in work environments with high levels of air contaminants, a reduction in air flow could cause overexposure which is especially critical for the loose-fitting and the hood configurations.

A recent development in the PAPRs is a warning device for the presence of inadequate air flow or negative-pressure inside the respirator inlet covering. Three sensing methods are available. The flow measurement method is used most often by manufacturers. An air flow meter such as the hot wire anemometer, a differential pressure meter such as a venturi meter, or an orifice meter have been employed as sensing devices for inadequate flow rates. These devices would indicate the true value of flow irregardless of outside conditions.

The second method measures the motor performance, such as voltage or current which is proportional to the output air flow of the blower unit. Because motor performance varies for each motor, a set of calibration curves must be generated for the calibration of individual motors.

The third type measures the facepiece pressure by a sensor. The alarm activates when it detects a negative-pressure inside the facepiece. This is the best system for

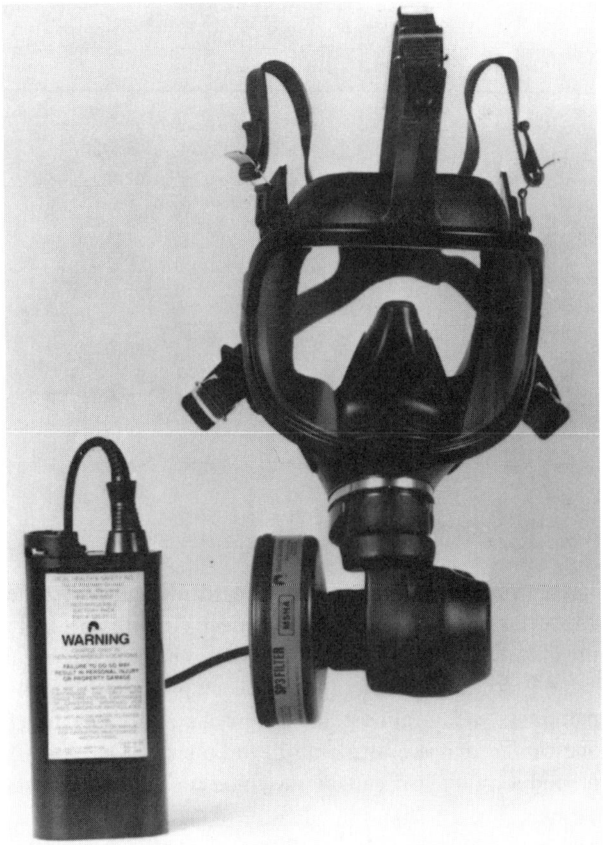

Figure 21 Racal Powerflow PAPR with face mounted blower.

direct measurement of the protection provided by the respirator. One manufacturer is developing a PAPR based on this concept.

At least three approved PAPRs have a low air flow warning device available. This desirable feature should become a requirement in the revised 42 CFR 84.

E. Evolution of Respirator Facepiece Design

In the early 1970s, most respirator manufacturers had only one size of facepiece available in each model. Multisize facepieces were available only after Edwin Hyatt of the Los Alamos Scientific Laboratory published the results of quantitative fit testing (QNFT).[12] The Hyatt study was conducted on an anthropometrically selected test panel which represents the adult work population in the U.S. The Hyatt study concluded that no single facepiece size could fit all of the panel members and that some facepieces tested could only fit a very small portion of the test panel.

The first half-mask respirator with three sizes was developed by MSA and approved by MSHA/NIOSH in the mid 1970s (Figure 22). Most facepieces were made of rubber or plastic such as PVC. Full facepieces are usually available in only one size. However, many manufacturers now have at least three sizes available.

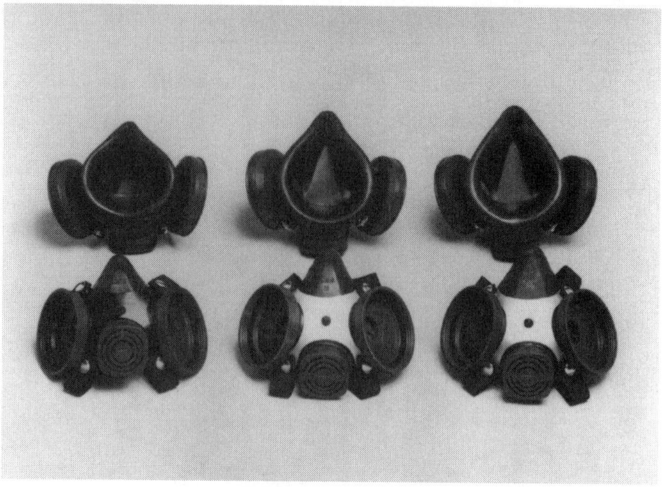

Figure 22 MSA multisize Comfo II half-mask.

MSA, which has an in-house sculptor available for design facepieces, is the only manufacturer that offers three sizes of full facepiece (Figure 23). In 1980, Survivair introduced a line of multisize facepieces which were made of silicone rubber (Figure 24). Since silicone is soft and pliable, the facepiece provides a much better fit than other stiffer materials. Today, almost all half-masks and many full facepieces are made of silicone rubber and are available in three sizes. One manufacturer, Scott Aviation, developed a multisized half-mask which uses a common shell with interchangeable faceseals (Figure 25).

Another design that improves the stability of the half-mask on the face replaces the upper strap with a cradle. Norton (North) was the first one to introduce the head cradle (Figure 26). This design is available in almost all makes and models. In the atmosphere supplying respirators, a net is used to replace the head straps. This design can be found in the Scott 4.5 SCBA and most Robertshaw SARs and SCBAs.

To further improve the facepiece seal, newly developed facepieces have wider profiles with deeper crease of the sealing area (Figure 27). One caution for the use

Figure 23 MSA multisize full facepieces.

Figure 24 Survivair multisize half-masks.

Figure 25 Scott 66 half-mask respirators with interchangeable face seal.

Figure 26 North half-mask with head cradle.

Figure 27 MSA Comfo Elite half-mask with wide profile.

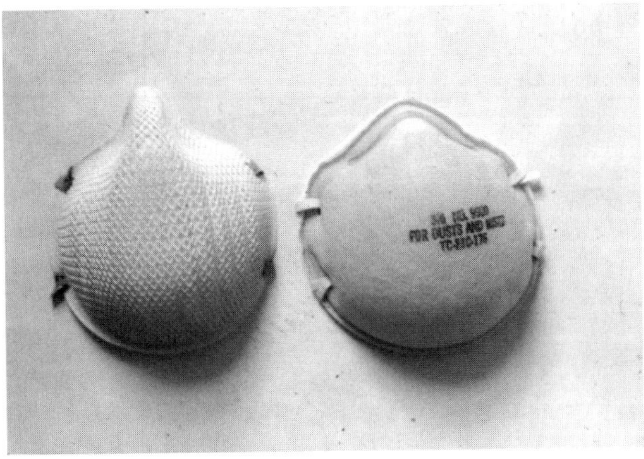

Figure 28 3M and Moldex disposable respirators.

of silicone facepiece is that silicone has very low permeation resistance against a variety of chemicals.

F. Disposable Respirators

The first disposable respirator (single use, maintenance free, or filtering facepiece) was approved by the Bureau of Mines in 1970. The 3M 8710 was the first disposable respirator approved by MSHA/NIOSH. This is the only class of respirators which fit test is not an approval requirement. There was a QNFT using 1.2 μm coal dust in the Bureau of Mines (BM) test schedule 21B; however, it was deleted in the 1972 revision of 30 CFR 11.

Since fit testing is not a requirement for certification, there is no restriction in respirator design. The commonly used 3M disposable respirator was based on women's bra cup, and the mesh type coating on the Moldex disposable respirator is based on the women's swim wear (Figures 28 and 29).

The name of the disposable respirators keeps changing. They were originally called "single use" respirators in the 1970s. The names changed to "disposable" in the 1980s. The name has now been changed to "maintenance free." However, the most appropriate term for this class of respirators should be "*filtering facepiece*" because this term reflects its design. In the last decade, there was some improvement in the efficiency of the filter media and straps. However, basic problems concerning how to perform fit testing and an effective facepiece seal test on the disposable respirators were never resolved. The fit test is not a requirement for certification. In 1982, OSHA accepted a saccharin mist qualitative fit testing method developed by the 3M company specially for disposable respirators. All validation data submitted by 3M were based on testing with elastomeric facepieces and a "Q" respirator which has different filter media and is stiffer than the regular 3M disposable respirators. There are no published data to indicate whether this method ever worked on disposable respirators. The particle size of saccharin mist is considered too large to penetrate through the facepiece seal.[13-16] Yet some brands of disposable respirators failed the saccharin fit test.

Figure 29 Various types of disposable respirators.

The only satisfactory method to check the facepiece seal prior to entry into the hazardous place is to perform a negative-pressure and positive-pressure fit check. Because most disposable respirators have no exhalation valve or have inaccessible exhalation valve, many disposable respirator manufacturers recommend a "positive-pressure fit check" which requires the respirator wearer to cup both hands over the filter area of the disposable respirator and inhale. The purpose of a negative-pressure fit check test is to block the inhalation areas which is easily performed by blocking the air vents of the filter cartridges on an elastomeric facepiece. The disposable respirator manufacturers' recommended method cannot block all the filter area of the disposable respirator, so it is not clear what this test accomplishes. Unfortunately, both the OSHA and ANSI respirator standards accept the manufacturer's recommended fit check method.

The OSHA lead standard prohibited the use of single use (disposable) respirators. Since 1979, most disposable respirators have been reapproved by NIOSH as respirators with replaceable dust/mist filters. Since OSHA recognizes NIOSH's approval, disposable dust/mist respirators become acceptable to OSHA. Recently, a disposable respirator equipped with an elastomeric facepiece and permanently attached or replaceable filter element was approved. This respirator looks like an elastomeric facepiece respirator. This approval further confused users regarding what is a "disposable" respirator.

G. Limited Use Respirators

In the late 1980s, a new class of respirator was approved. This respirator has an elastomeric facepiece with nonremoval cartridges. This respirator was called "maintenance free" by the manufacturer. It has been advertised as disposable with no-maintenance. A replaceable cartridge version was introduced later. Since respirator sales is a very competitive market, many manufacturers followed the suit. Some manufacturers presented this line of respirators as "limited use." This line of respirators sells less than the manufacturer's regular line of elastomeric facepiece respirators. In order to reduce the cost, the facepiece is made of cheaper and thinner material, mainly plastic, and with

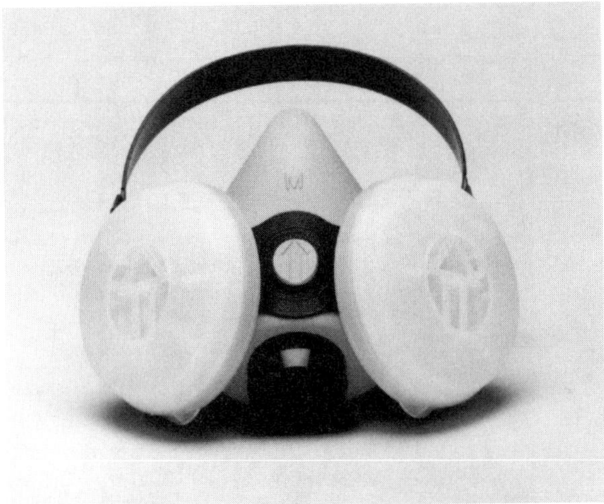

Figure 30 Survivair limited use respirators.

smaller size cartridges to reduce weight. Only one manufacturer, Survivair, who was the first manufacturer to introduce three sizes of silicone facepiece, has introduced the facepiece made of silicone rubber (Figure 30). One manufacturer, MSA, has introduced three sizes of "limited use" full facepieces (Figure 31). Due to weight limitation, the facepieces of these "limited use" respirators are not as sturdy as the regular facepiece. The cartridge contact point to the facepiece has been reduced from "area" to "point," which may provide a less firm contact to the facepiece. The smaller cartridge may also

Figure 31 MSA advantage full facepiece respirators.

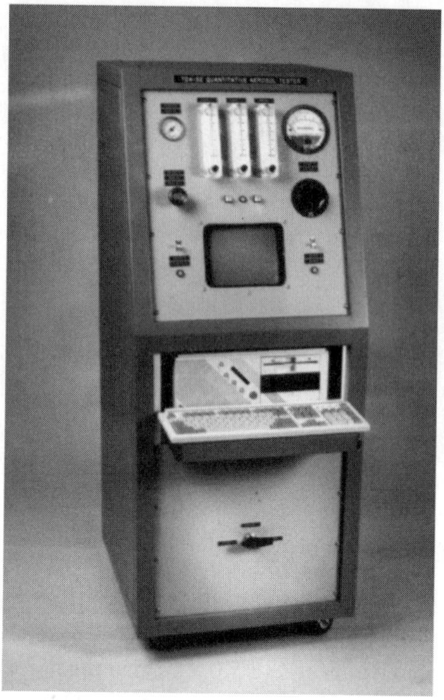

Figure 32 ATI QNFT instrument.

reduce the reserve capacity provided by the regular size cartridges. One ecological consideration is that these nondegradable "maintenance free" facepieces may aggravate the solid waste disposal problem.

II. RESPIRATOR QUANTITATIVE FIT TESTING

Quantitative fit testing equipment was not commercially available until the mid 1970s. These instruments use oil mist as the challenge and a light scattering photometer as a detector. A computerized version was developed later (Figure 32).

After OSHA amended the lead standard which accepted QLFT as an alternative to QNFT, the sale of the QNFT equipment declined and one manufacturer even ceased operation. Without the introduction of the PortaCount® (Figure 33), a portable QNFT, and the implementation of the OSHA asbestos standards which required QNFT, the whole QNFT equipment industry would have been out of business. PortaCount® was developed under a contract with the U.S. Army. It is essentially a condensation nuclei counter which uses ambient air as a challenge agent. The PortaCount® counts all particles inside and outside the facepiece. It is a lightweight, push button, computer programmed operation.

Since the late 1980s, several nonaerosol based quantitative fit test methods have been developed. Carpenter and Willeke[17] proposed a method which measured pressure decay rate caused by respirator leakage. Another respirator fit test method using air pressure was proposed by Crutchfield.[18,19] The controlled negative-pressure (CNP) fit test method measures leak rates through the facepiece as a way to determine the fit factor.

Figure 33 TSI PORTACOUNT® QNFT instrument.

One important aspect for conducting any respirator fit testing is that the exercises performed by the test subject must be able to simulate actual movements occurring in the workplace. However, the test protocol of these two methods does not allow the test subject to perform any dynamic exercise during the test.

Both the Willeke and Crutchfield studies proposed a promising alternative method to perform respirator fit testing. However, more studies are needed to prove the virtue of these methods.

III. PROBE ERROR

NIOSH reported in 1980 that there was a significant sampling bias (up to 300%) due to the location of the probe on the facepiece. The International Society of Respiratory Protection (ISRP) organized a conference to address this issue. Opponents of QNFT seized this opportunity to declare that QNFT was invalid and that respirators should be fit tested by QLFT. However, they failed to mention that all the QLFT validation data were derived from QNFT. Further studies reduced the magnitude of probe bias from 300 to 40%.[20-22]

IV. STANDARDS AND REGULATIONS DEVELOPMENT IN RESPIRATORY PROTECTION

The respiratory protection standards and regulations developed over this time period can be separated into use regulations (OSHA), testing and certification regulations (MSHA/NIOSH), and consensus standards (ANSI or ASTM).

A. Testing and Certification Regulations (NIOSH)

Prior to NIOSH, respirator testing and certification was administrated by the U.S. Bureau of Mines under the provisions of 30 CFR 11 which was strictly a voluntary program. Later, Bureau of Mines transferred the program to NIOSH and the respirators are now certified jointly by MSHA and NIOSH under authority of the Coal Mine Safety and Health Act. OSHA also requires the use of MSHA/NIOSH certified respirators. Many individuals incorrectly believe that NIOSH is the only certifying body, which is not true.

30 CFR 11 was revised in 1972 and was essentially the same as the Bureau of Mines testing schedules except that a quality assurance program was added. Since most testing schedules listed in 30 CFR 11 were outdated, NIOSH called a public meeting in 1979 to solicit comments for revision. A NIOSH proposal was published in 1987[23] and a revision is now in progress. MSHA has transferred the certification function to NIOSH and now it is under 42 CFR 84.

Listed below are several problem areas with the proposed NIOSH revision.

1. Particulate Air-Purifying Respirators

A Challenge Concentration of 200 mg/m³ for Aerosol Particles for Use in Evaluating the Aerosol Particle Filtering Efficiency for Particulate Air-Purifying Respirators. This very high challenge concentration will plug respirator filers so rapidly that it will be impossible to measure and record changes in instantaneous aerosol particle penetrations of the respirator filters. Tests using low concentrations of airborne particles have shown that some respirator filters first show a decreased particle filtering efficiency until plugging of the fibrous filtering material by retained particulate matter becomes sufficient to improve particle filtering efficiency.

The use of a very high concentration of airborne particles like 200 mg/m^3 will result in very rapid plugging of the respirator, increasing the filtering efficiencies of these respirator filters so quickly that this will falsely make them appear to be better than they really are. An aerosol particle concentration of 200 mg/m^3 is far above actual concentration of aerosol particles that occur in industry today. An aerosol particle concentration of 10 to 30 mg/m^3 would better represent aerosol particle concentrations found in industry today rather than 200 mg/m^3. This concentration range of 10 to 30 mg/m^3 would be suitable for use in evaluating the effectiveness of particulate filtering respirators.

Minimum of 90 to 95% Efficiency for Filtering Aerosol Particles by Nonpowered Particulate Air-Purifying Respirators. A minimum efficiency of 90 to 95% for removing aerosol particles from air by a filter used in a nonpowered particulate air-purifying respirator is much too low. This is especially true when considering that quarter-mask and half-mask facepiece respirators are assigned a facepiece-fit factor of 10 which means that these respirators are allowed to have a combination leakage of 10% of a contaminant in the ambient air between the facepiece and the faceseal of the respirator wearer and through the filter. Any filter leakage in excess of 1% would not be acceptable.

No Final Maximum Resistance to Air Flow Requirement by a Particulate Air-Purifying Respirator after Completion of an Aerosol Test. The proposed requirements call for a maximum *initial* inhalation resistance and a maximum *initial* exhalation resistance offered by a particulate air-purifying respirator. Eliminating a maximum *final* inhalation resistance and a maximum *final* exhalation resistance after completion of an aerosol test of the respirator means that the filter(s) of a particulate air-purifying respirator will be allowed to plug rapidly by retained particulate matter. This will shorten the useful life of the respirator filter(s). The elimination of final resistance to air flow requirements for particulate air-purifying respirators will greatly increase their sales because of the need for increased frequency of filter replacement by respirator users. A rapid increase in the resistance offered to breathing by particulate air-purifying respirators may increase the fatigue of respirator wearers if the wearers fail to replace respirator filters when they should.

Conditioning of Respirator Filters for Only 24 h in an Atmosphere at 80% Relative Humidity and at 38° ± 2.5°C. Previous work performed in the 1970s by the Los Alamos National Laboratory (LANL) showed that conditioning respirator filters *passively* by merely keeping them in an atmosphere at an elevated temperature and humidity for even a week (7 days) was inadequate to have much effect on the performance of dust/mist filters composed of electrostatic felt on removing aerosol particles from air. These tests also showed that a relative humidity of 80% was too low for conditioning respirator filters prior to testing the performance of the filters for removing particulate matter from air.[24] The conditioning of respirator filters at an elevated temperature and humidity is supposed to replicate storage of respirator filters under summertime conditions in the sunbelt area. Conditioning of the respirator filters for 30 days at a temperature of 38° ± 2°C and a relative humidity for 90% would be more satisfactory.

Low Air Flow Warning for Powered Air-Purifying Respirators. The high protection provided by the PAPR is maintained by the constant flow of clean air to the inlet covering of the PAPR. A reduction in air flow could severely reduce the effectiveness of the device. This is especially critical for the loose-fitting and the hood configurations. PAPRs should provide an audio or visual warning for a reduction in air flow or fail to maintain a positive-pressure inside the respirator inlet covering. High air flow would also reduce the service life of the sorbent for protection against gases or vapors. Since the air consumption for the wearer with normal activities is only 28.5 l/m (1 cfm), a pressure demand PAPR design should be developed in which the speed of the air blower varies with breathing rate of the wearer. The pressure demand design would increase the service life for sorbent as well as the battery service life for protection against particulate exposures. A test method for the pressure demand design should be developed.

Minimum Air Flow Rates for PAPR and Continuous Flow SAR of 115 l/m (4 cfm) for Respirators Equipped with Tight-Fitting Inlet Covering and 170 l/m (6 cfm) for Respirators Equipped with Loose-Fitting Respiratory-Inlet Coverings. The 115 l/m minimum air flow rate for PAPRs has been shown by testing to be too low. These low air flow rates result in excessive inward leakage of contaminants in the ambient atmosphere into the interior of the respiratory-inlet coverings when respirator wearers work at high work rates and/or have extensive body movements. Test data for commercially available PAPRs resulted in the recognition by NIOSH and OSHA of low assigned protection factors (APF) for PAPRs. Minimum air flow rates for PAPRs of 170 l/m (6 cfm) for respirators equipped with tight-fitting respiratory-inlet coverings and 285 l/m (10 cfm) for respirators equipped with loose-fitting respiratory-inlet coverings would greatly improve the performance of commercially available PAPRs.

2. *Approval of Air Supplied Suits*

Air supplied suits provide whole body protection against gaseous contaminants. The cool air flow through the suit would reduce the body temperature when the respirator wearer works in a hot environment. The Department of Energy is approving suits for its contractors. Their testing procedures should be incorporated into the 42 CFR 84.

3. *SCBA for Firefighting*

The major use of the SCBA in this country is for firefighting. However, the requirements such as high peak air flow, high and low temperature resistance, and high performance standards for SCBA are not addressed in the proposal. The National Fire Protection Association (NFPA) developed a standard on SCBA for firefighting (NFPA, 1981) which has been discussed above. SCBA performance testing which meets the requirements prescribed in the NFPA should be adopted and incorporated into the final standard.

4. *Continuous Flow Escape SCBA*

Because the continuous flow escape SCBA (CFESCBA) was developed after the 1972 revision of the 30 CFR 11, there is no test requirement concerning the minimum required air flow. Because not all the air inside the hood is flushed out, the oxygen and carbon dioxide concentrations can reach an unsafe level if the wearer egresses at a fast pace. The LLNL study[8] demonstrated that at an air flow of 40 l/m, the levels of CO_2 and oxygen inside the hood reach the immediately dangerous to life or health (IDLH) concentration when the user exits to safety at a moderate pace. Devices with an air flow as low as 28 l/m have been approved. Performance requirements for CFESCBA should be developed.

5. *Respiratory Protection Against Biological Hazards*

Most of the OSHA health standards address chemical hazards with little emphasis on biological hazards. Bloodborne pathogens are the first substance specific OSHA standard regulating biological hazards; however, no respiratory protection is required to minimize these hazards. The Mycobacterium tuberculosis (TB) was the first biological hazard to generate the health care industry's attention to OSHA's regulations on respiratory protection. TB is a communicable disease. The incidence of disease has been increased since 1985. The outbreak of multidrug-resistant TB has heightened concern since the transmission of the disease has been reported for persons working in health care facilities, homeless shelters, and correction institutions. The case is further complicated to the high risk of TB among persons infected with the human immunodeficiency virus (HIV). Transmission is most likely to occur via the respiratory route when infected individuals cough, sneeze, expectorate, or talk.

In 1990, the Center for Disease Control and Prevention (CDC) issued a guideline for TB control which requires engineering controls and respiratory protection to reduce or minimize the transmission of TB to health-care workers.[25] Based on the reported TB bacteria size range of 1 to 5 μm, CDC recommended the use of a respirator equipped with a dust/mist filter for protection against TB, with the assumption that the size of the silica dust is 0.6 μm. However, CDC has not recognized the fact that the aerodynamic diameter for the silica dust is around 2 μm, rather than 0.6 μm, when measured as a projected diameter by optical microscopy.

On July 7 1992, the OSHA Regional Office in New York issued an enforcement guideline which required that employers must use a respirator equipped with a dust/fume/mist filter as a minimum protection against exposure to TB under the following conditions:[26]

- When entering a pulmonary isolation room occupied by a known or suspected infectious TB patient.
- While performing certain high hazard medical procedures such as aerosol administration of medication (pentamidine), bronchoscopy, and diagnostic sputum induction.
- When transporting TB disease patients.

In September 1992, NIOSH issued recommended guidelines for personal respiratory protection of workers in health-care facilities potentially exposed to tuberculosis.[27] Because there is no threshold of exposure of TB, it would be treated as a potential carcinogen. Based on the selection criteria listed in the NIOSH Respirator Decision Logic,[28] NIOSH usually recommends a positive-pressure self-contained breathing apparatus (SCBA) or a pressure demand supplied air respirator (PDSAR) for protection against TB. Instead, NIOSH made the following recommendations:

1. A half-mask powered air purifying respirator (PAPR) equipped with HEPA filter(s) is required when confirmed or potential TB transmitters are present, or potentially present, at hazardous locations and procedures involve "medium" risk. An air flow of 170 l/m (6 cfm) or higher than the approval flow rate of 115 l/m is recommended.
2. A half-mask positive-pressure supplied air respirator (PDSAR) is required when confirmed or potential TB transmitters are present at certain other hazardous locations and when procedures involve "high" risk.

This is a more practical consideration since the bulk, weight, and short service life of the SCBA prevents its application for the health care facilities.

In the guidelines, NIOSH also discussed the deficiencies of the filtering facepieces (disposable respirators) used by health providers because they resemble the surgical mask which is commonly used to control the transmission of liquids. *"What is more relevant, the face-seal leakage for cup-shaped, disposable masks can be considerably higher than 10% to 20% if these masks are not properly fitted to each wearer's face, fit tested by a qualified individual, and then fit checked by each wearer before respirator use. Both fit testing and fit checking are essential elements in any effective and reliable personal respiratory protection program... Cup-shaped, disposable masks cannot be reliably fit checked by wearers. Therefore, the efficacy and reliability of the face seals on cup-shaped, disposable masks are undependable because there are no proven reliable fit tests nor reliable fit checks. Such devices cannot be relied upon to assure protection of workers against exposure to aerosolized droplet nuclei containing tubercle bacilli".*

On October 8 1993, OSHA issued a TB compliance policy which requires the use of a respirator under the following conditions cited in the CDC 1990 guidelines:[29]

- When employees enter rooms housing individuals with suspected or confirmed infectious TB diseases
- When employees perform high hazard procedures on individuals who have suspected or confirmed TB diseases
- When emergency medical-response personnel or others must transport, in a closed vehicle, an individual with suspected or confirmed TB diseases.

In the above circumstances, the employer must provide and ensure the use of NIOSH-approved HEPA filter respirators as the minimum acceptable level of respiratory protection. Use of filtering facepieces is acceptable provided the respirator maintains its structural and functional integrity. In addition, the employer must establish a respiratory protection program in accordance with 29 CFR 1910, 134 (b).

OSHA compliance officers have conducted inspections in hospitals with patients with TB. Many citations have been issued for violations of OSHA respiratory protection standards. Citations include: no respiratory protection program, use of incorrect respirators, no fit testing performed, and no training. Because filtering facepieces equipped with a HEPA filter element costs about five times more than a filtering facepiece equipped with a dust/mist filter, many hospitals have contested this violation because these respirators are disposed of after a single wear. Many contested citation cases were settled and hospitals agreed to issue respirators equipped with HEPA filters. OSHA is also in the process of developing a standard for protection against exposure to TB.

On October 12 1993, CDC issued for public comment, a draft guideline regarding the prevention of TB transmission in health care facilities.[30] The guideline provides a thorough discussion of the advantages and disadvantages of various air-purifying particulate filtering respirators and also a discussion of the characteristics of the fit test, fit check, and faceseal leakage of each respirator. Based on these discussions, CDC proposed four criteria for certifying a class of respirator for TB protection:

1. The ability to filter particles 1 μm in size in the unloaded state with a filter efficiency of ≥ 95%, given flow rate of up to 50 l/m.
2. The ability to be qualitatively or quantitatively fit tested in a reliable way to obtain a faceseal leakage of no more than 10% for most workers.
3. The ability to fit health care workers with different facials sizes and characteristics, based on an availability of at least three sizes of respirators.
4. To ensure proper protection, the facepiece fit should be checked by the wearer each time he or she puts on the respirator, in accordance with OSHA's standard and good industrial hygiene practice.

CDC also states that among the currently approved respirators, only the respirator equipped with HEPA filters meets the proposed certification criteria. Since the proposed criteria is not a complete test protocol for certifying respirators, it leaves many unanswered questions. For example, the type of aerosol and the mean and standard deviation of the test aerosol must be specified. Test variables such as the air flow, challenge concentration and test time that would affect the efficiency of the filter are also lacking. The proposed guidelines allow a combined respirator leakage of 15% which accounts for a filter leakage of 5% and a faceseal leakage of 10%. Since OSHA and NIOSH regulations only allow a maximum leakage of 10% for a half-mask respirator, it is not clear how CDC has developed this respirator certification criteria.

Under urging from health-care providers and the CDC, NIOSH announced that it will use the modular concept to develop respirator certification regulations. The proposal for the particulate module was published on May 20, 1994.[31] The anticipated date for proposed rule of other modules has the following schedule:

Subject	Date
Assigned protection factors	Late 1994
Administrate program	Early 1995
Quality assurance requirements	Early 1995
Gas and vapor requirements	Mid 1995
Positive-pressure SCBA requirements	Early 1996
Simulated workplace protection factor test	Early 1997

This approach contrasts with the complete proposal published in 1987.[23] NIOSH states that the first module addressed the most widely used type of respirators: the air-purifying respirator with a particulate filter. This new proposal would replace the inadequate and outdated current procedures with better methods to assess a respirator's ability to filter out toxic substances. In addition to benefitting industrial workers, the improved testing requirement will also benefit health-care settings implementing the current draft CDC Guidelines for Preventing the Transmission of Tuberculosis. While respirators with HEPA filters are the only respirators currently available which meet these criteria, the proposed regulation will enable manufacturers to produce a broader range of certified respirators which provide the necessary level of protection.

B. Use Regulations (OSHA)

The OSH Act authorized OSHA to adopt national consensus standards without rulemaking for the first 2 years of its existence. The American National Standards Institute (ANSI) Standard on the Practice of Respiratory Protection (ANSI Z 88.2-1969), was adopted with minor revisions in 1971. An advance notice for proposal rulemaking (ANPR) for revising 1910.134 was announced in 1985 and a proposal is in progress. Major features of the proposal are the assigned protection factor for each class of respirators, medical evaluation, and fit testing.

There are two types of OSHA standards which prescribe requirements on respiratory protection. The standard on respiratory protection, 29 CFR 1910.134, sets general requirements concerning respirator usage. Because the 1910.134 is more than 20 years old, it does not address some new requirements such as fit testing and maximum use limit. These requirements are specified in substance specific health standards, such as asbestos, benzene, or lead.

It is quite common that OSHA health standards are challenged in court by both management and labor. There have also been litigations on the respiratory protection requirements in health standards promulgated by OSHA.

1. Lead Standard

The final OSHA standard on lead requires that only HEPA filters be used on air-purifying respirators and QNFT be performed on negative-pressure respirators. The dust/mist or fume filter respirators were unacceptable because of low performance. Shortly after OSHA published the lead standard in 1979, 3M, a major disposable respirator manufacturer, filed a petition to OSHA which stated that the hearing records did not support the prohibition on the use of dust/mist or fume filters and that there was no published procedure or guidelines available for performing QNFT. The commonly used QNFT fit testing agent di-2-ethyl hexyl phthalate was a suspected carcinogen. The QNFT equipment was expensive and the supply was limited. 3M had no respirators equipped with HEPA filters and they could not be fit tested by the QNFT. If relief is not granted, 3M will lose $2 millon in sales.

Regulatory agencies cannot add requirements to the final standard if these issues were not addressed in the hearing or in the records. OSHA determined that the exclusion of dust/mist or fume respirators from the selection table is not supported by substantial evidence on the record as a whole and was made without adequate notice thereby denying 3M the right and opportunity to comment. Administrative reconsideration on this issue was warranted. OSHA permitted the continued use of dust/mist or fume filters pending a reconsideration. The stay on the QNFT requirement was denied. However, OSHA delayed the start date for one year from the effective date which permitted employers to obtain equipment or necessary services.[32]

In February 1980, 3M filed another petition with OSHA for administrative stay and reconsideration for QNFT requirement in the lead standard.[33] 3M's request was denied by OSHA. In March 1980, 3M filed a petition with the U.S. Court of Appeals in Washington, DC for administrative stay and reconsideration of the quantitative fit testing requirement of the OSHA lead standard.

OSHA held a hearing to determine whether qualitative fit testing could be accepted as an alternative to QNFT. Three proposed QLFT methods were introduced at the hearing for comments. These testing agents are isoamyl acetate developed by the Du Pont Company, the irritant fume developed by the American Iron and Steel Institute (AISI) and sodium saccharin developed by 3M. Both isoamyl acetate and irritant fume have been used widely as test agents for QLFT. Du Pont and AISI refined these methods. However, there was little information on the use of saccharin, an artificial sweetener and a suspected carcinogen, as a fit testing agent. Both of these methods claimed that the test subject who passes the QLFT will achieve a minimum fit factor of 100 as verified by QNFT. The QLFT method would reject inadequate fit (fit factor < 100).

Questions were raised at the hearing regarding whether the large size of saccharin would penetrate through the facepiece. After the hearing, OSHA requested that Los

Alamos National Laboratory (LANL) conduct a study to determine the performance of the saccharin QLFT method.[34] The test results concluded that as many as 8% of the test subjects who passed the saccharin QLFT could not achieve a fit factor of 100. The method is only effective in rejecting respirator fits with measured fit factors of less than 10 for half-mask negative-pressure respirators. The LANL also conducted a study for OSHA to determine the performance of the irritant fume method.[35] OSHA accepted all three QLFT methods and added these methods along with a QNFT protocol to the amended lead standard. These four methods become a "boiler plate" for fit testing in all health standards promulgated by OSHA including gaseous air contaminants such as benzene and formaldehyde. Because saccharin has a measured particle size of 3 µm, it is not clear whether a worker fit tested with saccharin would receive adequate protection when exposed to gaseous air contaminants.

In 1992, NIOSH made a determination regarding the use of sodium saccharin for performing fit testing "We do not recommend the use of any carcinogens as test agents for fit-testing. Because sodium saccharin is a potential carcinogen, we recommend that it *not* be used for respirator fit testing."[36]

2. *Cotton Dust Standard*

After the revised OSHA cotton dust standard was published, 3M filed a petition on February 10, 1986 to request an administrative stay for a provision in the cotton dust standard respirator selection table which assigns a protection factor of 5 for disposable respirators. 3M made the following claims:

Both NIOSH and the ANSI Standard, Practice for Respiratory Protection, Z-88.2, 1980, have recognized that disposable respirators are equivalent to nondisposable respirators with replaceable filters.

OSHA allows use of disposable respirators for protection against lead for 10 times the PEL.

There are available and accepted fit tests to assess the face fit of disposable respirators.

Limiting the use of disposable respirators to a PF of 5 would force the cotton dust industry to purchase more expensive and less comfortable respirators. If relief is not granted, irreparable harm to 3M will result regarding lost sales and profits.

OSHA's response can be summarized as the following:[37]

NIOSH does judge "disposable" respirators equal to respirators with replaceable filters, but NIOSH's certification system does not quantitatively account for leakage which may result from improper fit. The NIOSH certification does not indicate whether the disposable respirator provides as much protection as the replaceable-filter half-mask respirators when fit as well as filter efficiency are taken into account.

While ANSI did perform QLFT on high-efficiency respirators, it did not perform these tests on disposable respirators. Hence, the ANSI respirator protection factor table was developed without any fit testing for disposable respirators.

The OSHA lead standard allows the use of QLFT to establish a PF of 10 for disposable respirators. The blood lead test required under the OSHA lead standard provides a reasonably direct indication of whether lead is getting into the employee's breathing zone. In contrast, the pulmonary function testing required under the cotton dust standard cannot indicate how much cotton dust enters the breathing zone.

The facepiece fit check method can be performed effectively only on rigid elastomeric half-mask respirators, which have inhalation and exhalation valves. These valves can be blocked off easily by the employee's hand. The disposable respirators permitted for use under the cotton dust standard do not, as the preamble explains, have either inhalation or exhalation valves. Therefore, a simple effective fit check cannot be performed on these respirators.

OSHA had reviewed the data submitted by 3M which claimed that disposable respirators can be effectively fit checked to allow for a protection factor of 10. A total of 23 subjects were tested in several trials to see how many of those who failed a QNFT would also fail a positive-pressure fit check (PPFC) procedure performed by 3M. Using a QNFT screen level of 1008, as many as 37 per 100 improperly fitted wearers of 3M's 8710 respirator could be erroneously passed by 3M's PPFC procedures. If a QNFT screening level of 10 is selected, as many as 41 per 100 improperly fitted wearers of 3M's 8710 respirator could be erroneously passed by 3M's PPFC procedures.

A regulatory analysis indicated that virtually all workers in cotton textile manufacturing operations and few, if any, in waste processing operations are exposed to cotton dust at concentrations that exceed 5 times the PEL. Consequently, little weight can be given to 3M's speculative claims regarding lost sales and profits.

OSHA denied 3M's request for a stay of limiting the use of disposable respirators to situations where cotton dust concentration is no greater than 5 times the PEL. 3M filed a petition in the U.S. District Court and the Count denied 3M's petition.[38]

3. Asbestos Standard

In July 1986, 3M made a request for a stay and reconsideration of portions of the respirator selection table in the OSHA revised asbestos standard which prohibits the use of disposable respirators and other respirators without HEPA filters at asbestos concentrations up to 10 times the PEL. In addition to the technical issues raised by 3M such as there is not sufficient information to indicate that dust/mist disposable respirators are inadequate, 3M also stated that if 3M cannot sell its non-HEPA and disposable respirators it will have to give serious consideration to layoffs of up to 100 employees forever. 3M would lose $700,000 in product labels and recall products. 3M will never effectively regain non-HEPA filter or disposable respirator markets because of the doubts created about those products.

OSHA denied 3M's request.[39] OSHA's response can be summarized as the following:

The Agency's conclusion that HEPA filters provide greater protection against asbestos than non-HEPA filter is supported by the record.

The Agency's conclusion that disposable respirators with or without HEPA filters permit greater faceseal leakage than elastomeric facepiece respirators is supported by the record.

Contrary to 3M's assertion that workers find disposable respirators more comfortable and are therefore more likely to use them, the record indicated that workers and industrial hygienists unanimously opposed the use of disposable respirators.

The two workplace protection factor studies on disposable respirators in the record do not support 3M's contention that disposable respirators provide equivalent protection to elastomeric facepiece respirators.

3M has presented only bare allegations and unsupported speculations about some possible economic injury.

3M filed a lawsuit against OSHA in the U.S. Court of Appeals in the District of Columbia. 3M later withdrew the case.

In the late 1970s, one respirator manufacturer voluntarily withdrew their asbestos approval for the dust/mist filters and recommended that customers use HEPA filters. However, none of the other manufacturers followed the suit. This manufacturer lost substantial sales in the asbestos market.

4. *Method of Compliance*

One issue which draws much attention is whether respiratory protection should be accepted as equivalent to engineering controls. Current OSHA health standards require that feasible engineering controls be used to control air contaminants in the workplace. Respirators can only be used when engineering controls are not feasible or inadequate. This policy has been criticized as being too inflexible, not cost-effective, and often unnecessary for health protection.

On February 22 1983, OSHA issued an advanced notice of proposed rulemaking on methods of compliance. A hearing was conducted on May 30, 1990.[40] At the hearing, OSHA stated that there are five sets of circumstances that have been identified by OSHA from data in the record where engineering controls may be generally infeasible:

During the time necessary to install feasible engineering controls;

Where feasible engineering controls result in only a negligible reduction in exposure.

During emergencies, life saving, recovery operations, repair, shutdowns, and field situations where there is a lack of utilities for implementing engineering controls.

Operations requiring added protection where there is a failure of normal controls; and

Entries into unknown atmospheres.

OSHA was requesting information on whether respirators can provide reliable and predictable exposure control for the following:

As a control means for compliance with established short-term exposure limits (such as found in formaldehyde, ethylene oxide or benzene standards).

As an exposure control means during any activity that is intermittent and/or brief in duration.

As an exposure control means during any maintenance operations whether routine or nonroutine.

As a control means where the employer has "budgeted" the use of respirators (e.g., limited use to a set number of days per year or hours per day).

As an exposure control means after approval by OSHA of a comprehensive written respiratory protection program.

As an exposure control means based on cost-effectiveness.

Table 3 Comparison of Available Control Strategies

Strategy	Capital investment	Annual expense (yr)	Net present value (NPV)
Engineering	$262,640	$0	–$197,655
Combination	$144,820	$8,020	–$166,171
Respiratory	$0	$46,820	–$334,485

The data indicate that installing engineering controls is more cost-effective than using respirators.

The opening statement made by Charles Adkins, OSHA Director of Health Standards Programs, stated that existing data did not convince the Agency that in these areas, implementation of even a strong respiratory program would result in equivalency of protection afforded by respirators as compared to feasible engineering controls. The inherent limitations of respirators preclude their providing equivalent protection to engineering controls. The final standard on Methods of Compliance is being prepared.

If only the cost of procuring a respirator is considered, respirators are always cheaper than engineering controls. However, if the cost of maintaining an effective respiratory protection program is considered, then using respirators may not be as cost-effective as engineering controls. The comments submitted by the Organization Resources Counselors, Inc. at the OSHA hearing on Methods of Compliance provided an example to illustrate that use of a respirator alone is the least cost-effective way to control exposure in the workplace.[41]

By using an economic model, the least costly strategy from a range of strategies can be selected. Based on a life of 15 years, the analysis shows that to reduce the benzene exposure from 10 to 1 ppm in an ethylbenzene plant, the costs for using engineering control, respirators alone, or a combination of engineering controls and respirators are shown in Table 3.

C. Consensus Standards

The current OSHA standard on respiratory protection, 29 CFR 1910.134 was adopted from the 1968 version of the American National Standards Institute (ANSI) standard for the Practice of Respiratory Protection, ANSI Z-88.2, 1968. From the requirement prescribed in the OSH Act, OSHA cannot adopt any national consensus standards 2 years after the effective date of the Act. It must be promulgated by rulemaking. There were two revisions of the Z-88.2 — in 1980 and 1992. However, this standard was not adopted by OSHA because it was developing a revision of 29 CFR 1910.134. The ANSI standard is developed by a subcommittee to which a number is assigned. The developed standard bears the number of the subcommittee. ANSI requires that every standard must be revised or reaffirmed every 5 years. The revising of the Z-88.2 standard was begun in 1986 and was completed in 1992. ANSI assigns all standards on respiratory protection in the Z88 group. It has the following standards:

- Respiratory Protection Against Radon Daughters, Z88.1-1969.
- Practices for Respiratory Protection, Z88.2-1992.*

* Being revised.

- Respiratory Protection During Fumigation, Z88.3-1983.
- Practices for Respiratory Protection for the Fire Service, Z88.5-1981.*
- Respirator Use-Physical Qualification for Personnel, Z88.6-1984.*
- Testing Methods for Air-Purifying Respirators, Z-88.8.**
- Testing Methods for Atmosphere Supplying Respirators, Z-88.9.**
- Respirator Fit Testing Methods, Z-88.10.**
- Respiratory Protection Against Blood-borne Pathogens, Z-88.12.**

1. Use Standard

The revision of the ANSI Z-88.2 standard was started in 1984.[42] The Z88.2 Subcommittee was chaired by Thomas Nelson of du Pont Company. Several interesting provisions were proposed during the development of the standard:

The respirator equipped with dust/mist filters is acceptable for protection against particles having a size of 1 μm even when these filters are tested against a 2 μm silica dust during the certification test.

Supplied air respirators are not required to be used if a suitable air supply is not available.

The oxygen deficiency immediately dangerous to life or health concentration has been reduced from 16 to 14%. There may not be a sufficient margin of safety at this concentration.

The quantitative fit testing (QNFT) is not considered any better than the qualitative fit testing (QLFT) even when all QLFT methods are validated with QNFT.

The commonly used continuous flow escape SCBA which costs more than $300 is a disposable respirator by definition.

The half-mask negative-pressure respirator has an assigned protection factor (APF) of 10; the full facepiece negative-pressure respirator has an APF of 100. The half-mask positive-pressure respirator such as a PAPR or a SAR has an APF of 50; the full facepiece positive-pressure PAPR or SAR has an APF of 1000. The difference of protection factors between the negative-pressure half-mask and full facepiece respirators is by a factor of 10. For the more protective positive-pressure respirators, the difference between the half-mask and full facepiece should differ only by a factor of 2.[12] However, the subcommittee decided that the difference in APF between the half-mask and the full facepiece positive-pressure respirators should have a factor of 20.

Disposable respirators have been assigned the same PF as the half-mask elastomeric facepieces respirators because they all provide a workplace protection factor (WPF) of 10. Only field studies were considered in setting the APFs. Three WPF studies on disposable respirators were available for review.[43-45] Both of these studies were unpublished and both claimed a WPF of 10. The third published study, conducted by NIOSH,[46] showed a WPF of 3.5 but was declared a non-WPF study and was not considered in setting the WPF for disposable respirators.

One of the accepted studies was conducted in a pigment operation containing cadmium. The ambient contaminant concentration was very low and respiratory protection was not needed. No details on testing conditions were available for review. The second accepted study was performed in an asbestos removal operations. Half-mask deposable respirators, half-mask elastomeric facepiece respirators and SCBAs were selected for testing. It is not clear why PAPRs, which are commonly used in asbestos removal, were not included in this study. WPF values were determined by fiber count. The

* Being revised.
** Being developed.

OSHA or NIOSH method requires that counting stop after 100 fields. Since the ambient asbestos concentrations were very low, the number of fields counted had been increased to 500 in order to search for sufficient fibers. The study concluded that the 3M 8710 disposable respirator provided more protection than the Scott pressure demand SCBA. The average in-mask fiber count for the SCBA was only 2 fibers; however, the average in-mask fiber count for the 3M 8710 was 10 fibers.

2. *Fit Test Standard*

The American Society for Testing and Materials E-34 Committee had formed a Task Force, E-34.14, to develop a standard on respirator quantitative fit testing in 1978. A draft standard was developed, but was never finalized.

An ANSI Subcommittee, Z-88.10, was formed to develop fit testing methods. The subcommittee was chaired by Barry McNeil of Reynolds Electric Company. The subcommittee was responsible for developing quantitative and qualitative fit testing methods. The membership could not reach agreement on such issues as:

Whether to accept two QLFT methods using isoamyl acetate and sodium saccharin as validated methods. These two methods were submitted by the same organizations which submitted these methods to OSHA for adoption. However, both methods failed validation criteria proposed by the Z88.10 Subcommittee.

Whether the validation of the QLFT method should require that concentration of the challenge agent inside the test enclosure be reasonably stable and consistent.

To accept a test aerosol with a mass median aerodynamic diameter of 5 μm for performing QLFT.

The standard has been submitted to the chairman of the ANSI Z-88 Committee; however, no action was taken. New Z88.6 and Z 88.10 Subcommittees were organized in 1995.

D. Update of Respiratory Protection Standards Development Activities

1. *NIOSH*

In the summer of 1994, the National Institute for Occupational Safety and Health (NIOSH) published a proposed rule on the certification of particulate filters for air-purifying respirators under the provisions of 42 CFR 84.[31] A detailed discussion of the proposed rule appears elsewhere in this book. A final rule which was published in the summer of 1995,[47] can be summarized as the following:

1. NIOSH is the sole authority for approving respirators for non-mine use.
2. Except for the non-powered particulate filters, all other provisions in 30 CFR 11 have been incorporated into 42 CFR 84.
3. One additional class of filters has been added to the final standard. There are three classes of certified particulate filters for non-powered air-purifying respirators: N, R, and P.
4. Each class of filters has three efficiency level: 95%, 99% and 100 (99.97)%.
5. N-class filters are not acceptable for protection against oil based particulates.
6. R-class filters are acceptable for protection against oil based particulates for only one shift.

7. P-class filters are acceptable for protection against oil based particulates without use time limitation.
8. Worst case testing conditions (higher air flow and smaller challenge particle size) have been employed for filter testing.
9. Paint spray and pesticide respirators will be no longer be approved under 42 CFR 84.
10. A different test method will be developed at a later date for the powered air-purifying respirators (PAPR). NIOSH will accept only high-efficiency particulate air (HEPA) filters for new PAPR approval during the interim period.
11. Respirator manufacturers have a period of three years to sell respirators approved under the old 30 CFR 11.
12. Only P-100 series filters can carry a magenta color which is used for high-efficiency particulate air (HEPA) filters approved under 30 CFR 11.

In addition, NIOSH announced plans to develop a user's guide for workers and respirator program managers who are not safety or health professionals. The user's guide was released in February 1996. Over 50 filters have been approved under the provisions of 42 CFR 84 in 1995. With a few exceptions, most of these approved filters are either N-95 filtering facepieces or P-100 mechanical filters. There is no substantial cost increase of the new filters as compared to the old filters. NIOSH has not announced the release date for the next module.

2. OSHA

On November 15, 1994, the Occupational Safety and Healthy Administration (OSHA) published a proposed revision of the respirator use regulation, 29 CFR 1910.134.[48] This proposal is a complete update of the current regulation that was published in 1970. This is a continuation of the rulemaking process which was initiated in 1982 for which an advanced notice for the proposed rule (ANPR) was published to solicit information. In 1985, a draft proposal was transmitted to the OSHA Construction Safety and Health Advisory Committee for review and comments. Under the current regulations each of the three branches of industry under OSHA's jurisdiction has its own regulation on respiratory protection: the Construction industry (29 CFR 1926.103), the Maritime industry (1915.152), and the General Industry (1910.134). The proposed revision will be applicable to all three industries. Since the current 1910.134 does not contain detailed requirements on selection and fit testing, these requirements were developed for substance specific standards listed in Subpart Z of the OSHA General Industry Standards such as asbestos (1910.1001) or lead (1910.1025). These requirements in substance specific standards will be deleted and these requirements prescribed in 1910.134 will be incorporated to these standard by reference.

Each employer is required to set up a program which consists of the following elements:

- Written program
- Program administration
- Selection
- Medical evaluation

- Use
- Fit testing
- Training
- Maintenance
- Program evaluation
- Recordkeeping
- Air Quality

The proposal requires that one individual be responsible for program administration.

The selection section requires that respirators be provided to employees without cost. Multiple sizes of facepieces will be provided to facilitate selection. A workplace evaluation must be performed prior to selection. The current proposal requires the use of the assigned protection factor (APF) listed in the NIOSH Respirator Decision Logic (RDL). Most of the comments received at the hearing indicated that the RDL is not up to date. OSHA may develop the APF table for the 1910.134, since NIOSH has not set a definite date for revising the RDL. The APF table will replace the selection table listed in each of the substance specific standards, such as asbestos or lead. Only MSHA/NIOSH approved respirators are acceptable.

The proposal requires that a medical evaluation be performed if respirators are used for more than five hours per week. OSHA has listed three alternatives for medical evaluation: written physician's opinion, medical history and examination, and questionnaire. A majority of comments favored the third alternative.

The respirator use section requires that written standard operating procedures (SOP) be developed for routine and emergency use situations. Procedures are also required for respirator use under oxygen deficient and immediate dangerous to life or health (IDLH) atmospheres.

Fit testing must be performed before respirator assignment. Either qualitative fit testing (QLFT) or quantitative fit testing (QNFT) is acceptable. Annual fit testing is also proposed. A new provision is that fit testing is also required for positive-pressure respirators. However, it is performed under the negative-pressure mode. Criteria are also proposed for accepting new fit testing methods.

Annual training is required for respirator wearers. The training must include the hazards for which the workers are exposed; the capabilities and limitation of respirators in use; and the use, care, and maintenance of respirators.

The respirator maintenance section requires that respirators be cleaned and disinfected after each use. Inspection is performed for routine and non-routine use respirators. Requirements are also set for storing respirators.

Annual program evaluation shall be performed, including random inspection of the workplace, to ensure that the program is being implemented effectively.

The recordkeeping requirement states that medical and exposure records will be maintained in accordance with the requirements prescribed in the standard of access to medical records, 1910.20.

The air quality section requires the use of Grade D air, which meets the current version of specifications of the Compressed Gas Association (CGA). In-line filters and/or sorbent beds are required for oil lubricated air compressors. The air compressor must be located so as to avoid the entry of contaminated air into the intake of the air compressor.

The proposal also requires that an approval label that cannot be removed or defaced during use be attached to the respirator.

Other issues of interest are the use of contact lenses and the continued use of irritant smoke and sodium saccharin as fit test agents.

OSHA received over 450 comments on the proposed rule. A hearing was held between June 6 and June 20, 1995 where more than 30 individuals testified. OSHA also made a unprecedented move to hold a panel discussion on workplace protection factors (WPF) during the middle of the hearing. This action was due to the strong lobbying by the filtering facepiece (disposable respirator) manufacturers. These manufacturers have informed their users that filtering facepiece would provide the same level of protection as the elastomeric facepiece respirators. Both the NIOSH Respirator Decision Logic[28] and the OSHA Cotton Dust Standard[37,38] assigned the disposable respirator with a protection factor of 5 while the elastomeric facepiece has been assigned a protection factor of 10. To avoid potential liabilities, one filtering facepiece manufacturer has conducted workplace protection studies that demonstrated that under the chosen test conditions, the filtering facepiece would achieve an APF of 10.

Since there is no consensus regarding how the WPF data should be analyzed, OSHA requested Mark Nicas, a faculty member from the University of California at Berkeley, to develop a document to address this issue.[49] In November 1995, OSHA reopened the record to solicit comments on the Nicas report. The revision of the respiratory protection standard has been designated as a high priority project and is scheduled for completion in September 1996.

3. *ANSI*

The draft standards on respiratory protection against blood-borne pathogens (Z-88.12) was finalized in 1995. A new subcommittee on respirator fit testing methods (Z-88.10) was organized in and has held meetings since 1995. Two new subcommittees on practices for respiratory protection (Z88.2-1992) and respirator use-physical qualification for personnel (Z88.6-1984) were organized in and have met since 1996. Since many of the members of the practices for respiratory protection for the fire service subcommittee (Z88.5) belong to the fire service, and the National Fire Protection Association (NFPA) is developing a respiratory protection standard for the fire service. The function of developing an updated standard for Z88.5 has been transferred to NFPA.

REFERENCES

1. Hack, A., Trujillo, A., and Bradley, O. D., Respirator studies for the nuclear regulatory commission, October 1, 1977–September 30, 1978: Evaluation of open-circuit self-contained breathing apparatus. Los Alamos Scientific Laboratory, NUREG/CR-1235, LA-8188-PR, 1980.
2. OSHA., Title 29 Code of Federal Regulations, Part 1910.156. Occupational Safety and Health Administration, 1993.
3. ANSI., Practices for respiratory protection for the fire service, ANSI-Z88.5, 1981, American National Standards Institute (ANSI), New York, NY.

4. The National Fire Protection Association, Open-circuit self-contained breathing apparatus for fire fighting, NFPA 1981. The National Fire Protection Association, Quincy, MA, 1992.
5. Johnson, J. S., da Roza, R. A., and McCormack, C. E., Evaluation of a commercial SCBA's compliance to the NFPA 1981 standard for fire fighters and measurement of simulated workplace protection factors at high work rates, Lawrence Livermore National Laboratory, 1992.
6. Occupational Safety and Health Administration Hearing on Fire Brigade, OSHA Docket S-006.
7. Lawrence Livermore National Laboratory Symposium, The role of closed-circuit breathing apparatus in structural fire fighting, October 23–24, 1984, Lawrence Livermore National Laboratory, Livermore, CA.
8. Johnson, J. S., da Roza, R. A., Foote, K. L., and Held, K., An evaluation of emergency escape respirators for use in a space launch environment, Lawrence Livermore National Laboratory, Livermore, CA, 1991.
9. Silverman, L. and Burgess, W. A., A self-contained positive supply air filter respirator, *Am. Ind. Hyg. Assoc. J.*, 25, 329, 1964.
10. Sherwood, R. J. and Greenhalgh, D. M. S., A self-contained pressurized blouse, *Ann. Occup. Hyg.*, 8, 247, 1965.
11. da Roza, R. A., Cadena-Fix, C. A., and Kramer, J. E., Powered air-purifying respirator study final report. Lawrence Livermore National Laboratory, UCRL-53757, 1986.
12. Hyatt, E. C., Respirator protection factors. Los Alamos Scientific Laboratory, UC-41, 1976.
13. Holton, P. M., Tackett, D. Y., and Willeke, K., Particle size-dependent leakage and losses of aerosol in respirators, *Am. Ind. Hyg. Assoc. J.*, 48, 848, 1987.
14. Holton, P. M., Tackett, D. Y., and Willeke, K., The effect of aerosol size distribution and measurement method on respirator fit, *Am. Ind. Hyg. Assoc. J.*, 48, 838, 1987.
15. Hinds, W. C. and Kraske, G., Performance of dust respirators with facial seal leaks. I. Experimental, *Am. Ind. Hyg. Assoc. J.*, 48, 836, 1987.
16. Hinds, W. C. and Bellin, P., Performance of dust respirators with facial seal leaks. II. Predictive model, *Am. Ind. Hyg. Assoc. J.*, 48, 842, 1987.
17. Carpenter, D. R. and Willeke, K., Non-invasive, quantitative respirator fit testing through dynamic pressure measurement, *Am. Ind. Hyg. Assoc. J.*, 49, 485, 1988.
18. Crutchfield, C., Eroh, M., and Van Ert, M., A feasibility pressure, *Am. Ind. Hyg. Assoc. J.*, 52, 172, 1991.
19. Crutchfield, C., Murphy, R. M., and Van Ert, M., A comparison of controlled negative-pressure and aerosol quantitative respirator fit test system using fixed leaks, *Am. Ind. Hyg. Assoc. J.*, 52, 349, 1991.
20. Myers, W. R., Allender, J. A., Plummer, R., and Stobbe, T., Parameters that bias the measurement of airborne concentration within a respirator, *Am. Ind. Hyg. Assoc. J.*, 47, 106, 1986.
21. Myers, W. R., Allender, J. R., Iskander, W., and Stanley, D., Cause of in-facepiece sampling bias. I. Half-facepiece respirators, Ann. Occup. Hyg., 32, 345, 1988.
22. Myers, W. R., Allender, J. R., Iskander, W., and Stanley, D., Cause of in-facepiece sampling bias. II. Full-facepiece respirators, *Ann. Occup. Hyg.*, 32, 361, 1988.
23. NIOSH, Revision of tests and requirements for certification of permissibility of respiratory protective devices used in mines and mining: Notice of proposed rulemaking, 42 CFR Part 84. Federal Register, 52, 32402, 1987.
24. Douglas, D., Revoir, W. R., Davis, T. O., Pritchard, J. A., Lowry, P. L., Hack, A., Richards, C. P., Geoffrion, L. A., Bustos, J. M., and Hesch, P. R., Respirator studies for the National Institute for Occupational Safety and Health, July 1, 1974 through June 30, 1975, Los Alamos National Laboratory, LA-6386-PR, 1976.

25. CDC, Guidelines for preventing the transmission of tuberculosis in health-care settings, with special focus on HIV-related issues. Centers for Disease Control and Prevention, Mobility and Mortality Weekly report (MMWR). 39:No.RR17, December 7, 1990.
26. Stanley, J. W., OSHA Enforcement Guidelines for Occupational Exposure to Tuberculosis, OSHA Regional Office, New York, NY, July 7, 1992.
27. NIOSH, Guideline for personal respiratory protection of workers in health-care facilities potentially exposed to tuberculosis, September 14, 1992.
28. NIOSH, Respirator decision logic. National Institute for Occupational Safety and Health. DHHS (NIOSH) Publication, 87-108, May 1987.
29. Clark, R. A., Enforcement policy and procedures for occupational exposure to tuberculosis. OSHA Directorate of Compliance Programs, October 8, 1993.
30. CDC., Draft guideline for preventing the transmission of tuberculosis in health-care facilities. Centers for Disease Control and Prevention (CDC), *Federal Register*, 58 (195), 52810, October 12, 1993.
31. NIOSH., Respiratory protective devices: Proposed rule, *Federal Register*, 59, 26850, May 24, 1994.
32. Letter from Eula Bingham, Assistant Secretary for OSHA to Nelson E. Schmidt, counsel for 3M, Jan 16, 1979, OSHA Docket H-049A, 43 FR 52952 November 14, 1978, 43 FR 54354, November 21, 1978.
33. Letter from Nelson Schmidt, 3M, to Eula Bingham, OSHA. February 18, 1980, OSHA Docket H-049A.
34. Marsh, J. L., Evaluation of saccharin qualitative fitting test for respirators, Los Alamos Scientific Laboratory, LA-UR-82-2389, 1982.
35. Marsh, J. L., Evaluation of irritant smoke qualitative fitting test for respirators, Los Alamos Scientific Laboratory, LA-9778-MS, 1983.
36. Millar, J. D., Director, NIOSH, Letter to D. Bevis, January 31, 1992.
37. Letter from Patrick Tyson, Acting Assistant Secretary for OSHA to Peter G. Nash, Counsel for 3M, April 15, 1986, OSHA Docket H-033.
38. *3M, petitioner, v. Occupational Safety and Health Administration*, Respondents, U.S. Court of Appeals, District of Columbia Circuit. Nos. 78-2014, 86-1075 and 86-1157. Argued Jan. 16, 1987. Decided Aug. 7, 1987. 825 Federal Reporter, 2d Series, 482, 1987.
39. Letter from Frank White, Deputy Assistant Secretary for OSHA to Peter G. Nash, Counsel for 3M, September 5, 1986.
40. Adkins, C. E., Statement made at the informal public hearing on the proposed OSHA standard regarding method of compliance, May 30, 1990, OSHA Docket H-160.
41. Birkner, L. R. and Boggs, R. F., Comments submitted to notice of proposed rulemaking for methods of compliance, Appendix B. Organization Resources Counselors, Inc., Washington, DC, OSHA Docket H-160, May 4, 1990.
42. ANSI., Standard Practices for Respiratory Protection, ANSI Z88.2-1992. American National Standards Institute, New York, NY, 1992.
43. Chemical Manufacturers' Association, CMA/Cadmium pigments study. Unpublished. Undated. Washington D.C.
44. Nelson, T. J. and Dixon, S. W., Respirator protection factors for asbestos, Parts I & II. Presented at the American Industrial Hygiene Conference, Las Vegas, NV, May 1985.
45. Phillips, J. L., Attorney, Du Pont Company, Letter to OSHA Docket Officer on Proposal to Revise 29 CFR 1910.1001 — Asbestos Standard Post Hearing Brief, October 16, 1986.
46. Reed, L. D., Lenhart, S. W., Stephenson, R. L., and Allender, J. R., Workplace evaluation of a disposable respirator in a dusty environment, *Appl. Ind. Hyg.*, 2, 53, 1987.

47. NIOSH., Respiratory Protective Devices, Final Rule. National Institute for Occupational Safety and Health 31.NIOSH: Federal Register, 60: 30335–30398, June 8, 1995.
48. Occupational Safety and Health Administration: Respiratory Protection Proposed Rule, 29 CFR 1910.134 et al. Federal Register 59: 58884–58956, November 15, 1994.
49. Nicas, M., The Analysis of Workplace Protection Factor Data and the Deviation of Assigned Protection Factors. U.S. DOL Purchase Order No. B9F53504. September 19, 1995.

13 TESTING AND CERTIFICATION REGULATIONS

I. RESPIRATOR TEST REQUIREMENTS

A. Classification of Test Schedules

Respirators are approved according to the requirements prescribed in the Code of Federal Regulations, title 30 and part 11 (30 CFR 11),[1] which in turn corresponds to the Bureau of Mines (BM) approval schedules. The following subparts are listed in the 30 CFR 11 for approving each class of respirator:

Subpart H:	Self-contained breathing apparatus (SCBA) (Schedule 13)
Subpart I:	Gas Masks (Schedule 14)
Subpart J:	Supplied air respirators (Schedule 19)
Subpart K:	Dust, fume, and mist respirators (Schedule 21)
Subpart L:	Chemical cartridge respirators (Schedule 23)
Subpart M:	Special use respirators (Schedules 14 or 23)

Each respirator has an approval number which corresponds to the test schedule listed in the 30 CFR 11. For example, the self-contained breathing apparatus (SCBA) is numbered TC-13F-XXX. The TC stands for testing and certification, which is different from the approval number issued by the BM.

Each subpart has minimum requirements for components and performance. Although some components are common for each type of respirator, performance requirements are unique to each type. The common component or performance requirements are:

- Description
- Required components
- Breathing tubes
- Harness installation and construction
- Apparatus container
- Half-mask and full facepiece minimum fit
- Facepieces or eyepiece
- Inhalation and exhalation valves
- Head harness
- Breathing resistance, inhalation and exhalation
- Exhalation valve leakage
- Facepiece leakage test (isoamyl acetate)

B. SCBA Performance Tests

The SCBA has the following additional component or performance requirements:

- Breathing bags (closed-circuit only)
- Maximum weight (35 lbs)
- Timer/service life indicator
- Gas flow
- Service time
- Carbon dioxide
- Low temperature
- Man test

C. SCBA Man Test

The man test determines:

1. The performance of the apparatus under different orientations.
2. The operating and breathing characteristics of the apparatus during actual use.
3. Whether the device meets the service time requirement.
4. Whether or not the lens of the facepiece causes fogging under different temperature and humidity ranges, as determined by the required service time.

The test subjects perform activities such as walking on a treadmill, climbing or weight carrying during the tests. Inspired air samples are collected every 15 min to determine whether the sample meets the requirements for oxygen and carbon dioxide. The activities carried out by the test subject for a 30-min service life SCBA are listed as below.

Activities	Time (min)
Sampling	2
Walking (4.8 km)	3
Carrying (23 kg)	4
Walks (4.8 km)	3
Climbing	1
Sampling	2
Walking	2
Climbing	1
Carrying (23 kg)	6
Walking (4.8 km)	3
Climbing	1
Sampling	2

D. Supplied Air Respirators Performance Tests

The supplied air respirators (SAR) are classified into the following three categories:

1. Type "A" SAR — a hose mask without a blower.
2. Type "B" SAR — a hose mask with a blower.
3. Type "C" SAR — which operates in continuous flow, demand, or pressure demand mode.

Each type also has "AE," "BE," and "CE" approvals available. The "E" designates an approval for abrasive blasting in which the SARs are equipped with additional devices designed to protect the wearer's head and neck against impact and abrasion from rebounding abrasive material. A test schedule is prescribed to test these devices.

The use of these SARs operated in the demand mode is declining because they are operated in a negative pressure mode, which provides much less protection than the continuous flow and pressure demand SARs.

Subpart J has the following performance tests prescribed for the type "C" SARs:

- Air velocity and noise level
- Air flow
- Nonkinkability of the hose
- Strength of hoses and couplings
- Permeation of hose by gasoline

The maximum length of the air hose is 91 m (300 ft). The manufacturer must specify a pressure range at the point of attachment of the air supply hose to the air supply system. For example, the pressure range is from 280 to 550 kN/m^2 (40 to 80 psi) for a hose between 6 and 76 m (15 to 250 ft). The tight-fitting SARs must maintain a minimum air flow of 115 liters per minute (l/m) (4 ft^3), and the loose-fitting SAR must maintain a minimum air flow of 170 l/m (6 ft^3) at the maximum hose length. The maximum flow rate must not exceed 425 l/m (15 ft^3). The pressure inside the facepiece of the pressure demand SAR (PDSAR) must not exceed 50 mm (2 in.) of water column at the flow rate specified above. In addition, the PDSAR must maintain a static pressure of less than 38 mm (1.5 in.) of water pressure inside the facepiece. However, this requirement has no relationship to the work rate of the respirator wearer.

E. Dust, Fume, and Mist Respirators Performance Tests

The test schedule includes performance tests for dust, mist, fume, and high-efficiency particulate air (HEPA) filters. Usually, a respirator that receives the high-efficiency approval also includes the fume, dust and mist approvals. The fume approval also includes the approval for dust and mist. The dust approval also includes the mist. Single use or disposable dust respirators are the only respirators in the 30 CFR 11 for which an isoamyl acetate facepiece seal test is not required.

Silica dust having a projected particle size of 0.4 to 0.6 µm (measured by a microscope) is used for testing. The test is performed inside a dust chamber at a concentration of 50 to 60 mg/mm^3. The temperature is approximately 25°C and the relative humidity may vary from 20 to 80%. The test period is 90 min at a flow rate of 32 l/m for nonpowered respirators. For powered air-purifying respirators (PAPR), the test period is 4 h. Thc flow rate for tight and loose-fitting inlet covering PAPRs is 115 l/m (4 ft^3), and 170 l/m (6 ft^3), respectively. When the filter is placed in pairs or in triplets, the air flow to each filter element will be reduced accordingly. For example, if two filter elements are used on a twin cartridge respirator, the air flow to each filter will be half of 32 l/m or 16 l/m. The maximum allowable filter penetration is 1% for all filters listed in Subpart K except for the HEPA filter.

Table 1 Test Conditions for Particulate Filters

Filter type	Dust	Mist	Fume	HEPA
Test aerosol	SiO_2	SiO_2	Pb	DOP
Particle size (μm)	0.6	0.6		0.3
Concentration (mg/m³)	55	25	15	100
Maximum penetration (%)	1	1	1	0.03
Single Use	40 × 24[1]			
Flow rate (l/m)				
Test time (min)	90			
Maximum retained wt (gm)	1.8			
Reusable				
Flow rate (l/m)	32	32	32	32/85
Test time (min)	90	312	312	5–10"
Maximum retained wt (gm)	1.5	2.5	1.5	
PAPR, tight-fitting				
Flow rate (l/m)	115	115	115	32/85
Test time (min)	240	240	240	5–10"
Maximum retained wt (gm)	14.4	6.9	4.2	
PAPR, loose-fitting				
Flow rate (l/m)	170	170	170	32/85
Test time (min)	240	240	240	5–10"
Maximum retained wt (gm)	21.3	10.2	6.2	

[1] Test is performed on a breathing machine at 24 respirations per minute with a minute volume of 40 l/m.

The performance test requirements for the silica mist is similar to the test for silica dust, except that an aqueous suspension of silica at 20 to 25 mg/m³ replaces the silica dust, and the test period is 312 min and 4 h for the nonpowered and powered respirators, respectively.

The lead fume test is performed inside a test chamber with freshly generated lead oxide fumes at a concentration from 15 to 20 mg of Pb per cubic meter of air. Other test conditions, including air flow, are the same as the silica dust test except that the test period is 312 min for nonpowered respirators, and 4 h for powered respirators. The maximum allowable filter penetration is 1%.

The di-2-ethylhexyl phthalate (DEHP or DOP) test is performed on the high-efficiency particulate air (HEPA) filters. The filter is challenged with a heat generated monodisperse oil mist of DEHP having a size of 0.3 μm at a concentration of 100 mg/m³. The air flow is 32 and 85 l/m for a single filter, and 16 and 42.5 l/m when the filters are used in pairs. The maximum allowable penetration is 0.03 % after a test period of 5 to 10 s. The HEPA filter cartridge is the only respirator component listed in Subpart K of 30 CFR 11 for which the DOP test is performed on each cartridge manufactured. The test requirements for each type of particulate filter are listed in Table 1.

F. Gas Mask Performance Tests

The gas masks include the chin style, front-mounted or back-mounted, and the escape type. Each type may have approval for the following gases or vapors:

- Acid gas: Chlorine, chlorine dioxide, sulfur dioxide or hydrogen chloride
- Alkaline gas or vapor: Ammonia or methyl amine
- Organic vapor: Carbon tetrachloride
- Carbon monoxide
- Combination: Either two, three, or all of the above

Table 2 Bench Test Requirements for Front or Back Mounted Canisters

		Test Atmosphere				
Test condition	Gas or vapor	Concn. (ppm)	Flow rate (l/m)	No. of tests	Maximum penetration (ppm)	Minimum service life (min)
AR[a]	SO_2	20,000	64	3	5	12
AR	Cl_2	20,000	64	3	5	12
EQ[b]	SO_2	20,000	32	4	5	12
EQ	Cl_2	20,000	32	4	5	12
AR	CCl_4	20,000	64	3	5	12
EQ	CCl_4	20,000	32	4	5	12
AR	NH_3	20,000	64	3	50	12
EQ	NH_3	30,000	32	4	50	12
AR	CO	20,000	64[c]	2	([e])	60
AR	CO	5,000	32[d]	3	([e])	60
AR	CO	3,000	32[e]	3	([e])	60

[a] AR: tested as received.

[b] EQ: equilibrated at 25% and 85% RH for 6 h at room temperature and an air flow of 64 l/m.

[c] Tested at 95 ± 3% RH and 25 ± 2.5°C.

[d] Tested at 95 ± 3% RH and 0°C.

[e] Maximum CO penetration will be 385 cm^3 during the minimum life. The peak penetration shall not exceeed 500 ppm.

The performance test for gas masks set requirements for breakthrough concentration and minimum service life under the following bench test conditions:

- Temperature: 20 to 30°C
- Relative humidity: 45 to 55%
- Flow rate: 32 or 64 l/m
- Equilibration: 64 l/m for 6 h at room temperature, and at 25% and 85% RH
- Challenge concentration: 3000 to 30000 ppm
- Breakthrough concentration: 5 or 50 ppm
- Minimum service life: 12 to 60 min

Test requirements for front or back mounted canisters and the chin style canisters are listed in Tables 2 and 3.

G. Chemical Cartridge Respirator Performance Tests

The performance test requirements for chemical cartridge respirators are similar to those for gas masks; however, the challenge concentrations, breakthrough concentrations, and service life requirements are different. There is one additional approval for the paint spray protection which is not available for the gas masks. Examples of testing requirements are listed in Table 4.

The lacquer and enamel mist test is a very old test. The lacquer and enamel formulations used in these tests are rarely used today. It is rumored that the only customers for these testing materials are respirator manufacturers and NIOSH. The concentration of cellulose nitrate lacquer inside the test chamber is 95 to 125 mg/m^3. The air flow is 32 l/m for nonpowered respirators, and the air flow is 115 and 170 l/m for tight and loose-fitting PAPRs, respectively. The test time is 156 and 240 min

Table 3 Bench Test Requirements for Chin Style Canisters

		Test Atmosphere				
Test condition	Gas or vapor	Concn. (ppm)	Flow rate (l/m)	No. of tests	Maximum penetration (ppm)	Minimum service life (min)
AR[a]	SO_2	5,000	64	3	5	12
AR	Cl_2	5,000	64	3	5	12
EQ[b]	SO_2	5,000	32	4	5	12
EQ	Cl_2	5,000	32	4	5	12
AR	CCl_4	5,000	64	3	5	12
EQ	CCl_4	5,000	32	4	5	12
AR	NH_3	5,000	64	3	50	12
EQ	NH_3	5,000	32	4	50	12
AR	CO	20,000	64[c]	2	([e])	60
AR	CO	5,000	32[d]	3	([e])	60
AR	CO	3,000	32[e]	3	([e])	60

[a] AR: tested as received.

[b] EQ: equilibrated at 25% and 85% RH for 6 h at room temperature and an air flow of 64 l/m.

[c] Tested at 95 ± 3% RH and 25 ± 2.5°C.

[d] Tested at 95 ± 3% RH and 0°C.

[e] Maximum CO penetration will be 385 cm^3 during the minimum life. The peak penetration shall not exceeed 500 ppm.

for nonpowered and powered air-purifying respirators, respectively. The maximum allowable penetration is 5, 28, and 41 mg for nonpowered, tight, and loose-fitting PAPRs, respectively. The test conditions for the enamel test are similar except that the maximum allowable penetration for the pigment (weighed as ash) is 1.5, 8.3, and 12.3 mg for nonpowered, tight, and loose-fitting PAPRs, respectively.

H. Pesticide and Special Use Respirators Performance Tests

Pesticide respirators consist of canister and cartridges for powered and nonpowered air-purifying respirators. Three series of tests are required for certifying these respi-

Table 4 Bench Test Requirements for Chemical Cartridges

		Test Atmosphere				
Test condition	Gas or vapor	Concn. (ppm)	Flow rate (l/m)	No. of tests	Maximum penetration (ppm)	Minimum service life (min)
AR	SO_2	500	64	3	5	30
EQ	SO_2	500	32	4	5	30
AR	Cl_2	500	64	3	5	35
EQ	Cl_2	500	32	4	5	35
AR	HCl	500	64	3	5	50
EQ	HCl	500	32	4	5	50
AR	CCl_4	1,000	64	3	10	50
EQ	CCl_4	1,000	32	4	5	50
AR	NH_3	1,000	64	3	50	50
EQ	NH_3	1,000	32	4	50	50
AR	CH_3NH_2	1,000	64	3	10	25
EQ	CH_3NH_2	1,000	32	4	10	25

Note: AR: as received; EQ: equilibrated at 25% and 85% RH for 6 h at room temperature with an air flow of 25 l/m for nonpowered respirators and 115 l/m and 170 l/m for tight and loose-fitting PAPRs, respectively.

rators: silica dust, lead fume, and organic vapor. An additional DOP penetration test is for canisters approval only. The test methods are the same as described above.

The special use respirators prescribed in subpart N include approvals for substances with poor warning properties that are not acceptable for testing with the air-purifying respirators schedules listed in subparts I and L. These respirators are equipped with an end-of-service-life indicator. The testing schedule was specifically developed for approving vinyl chloride respirators. The performance requirements listed in subpart N are slightly different from the methods prescribed in subparts I and L for testing canisters and chemical cartridges. The basic difference is that canisters or cartridges are equilibrated at 25 ± 5°C and 85 ± 5% RH for 6 h. The air flow rate is 64 l/m and 32 l/m for canisters and cartridges, respectively. The respirator is then tested at the same temperature and humidity as during the equilibration. The required service life is 6 h for canisters and 144 min for cartridges. The maximum allowable penetration for the canister is 1 ppm at a challenge concentration of 25 ppm. The maximum allowable penetration for the cartridge is 1 ppm at a challenge concentration of 10 ppm. If the cartridge or canister is equipped with an end-of-service-life indicator, it must show a color change at 80 ± 10% of the total service life of 1 ppm breakthrough. Canisters for protection against vinyl chloride are tested under the requirements of this subpart. Since many employers had implemented engineering controls to reduce vinyl chloride exposure, there was little demand for vinyl chloride respirators; this class of respirators was discontinued.

I. End-of-Service-Life Indicators

The 30 CFR 11 specifies that cartridges or canisters will not be approved for a specific chemical if the chemical has poor warning properties. On July 19, 1984, NIOSH announced in the Federal Register[2] that NIOSH was accepting applications for certification cartridges or canisters for protection against chemicals with poor warning properties provided that the device was equipped with an end-of-service-life indicator (ESLI). This testing schedule prescribed in subpart N for testing vinyl chloride devices was adapted to approve cartridges or canisters equipped with an ESLI. At this time, respirators equipped with an end-of-service-life indicator for mercury vapor, hydrogen sulfide, and ethylene oxide have been approved.

The testing requirements for the mercury vapor cartridge are summarized at the following Table. The testing requirements for other chemicals are similar, except for the challenge concentration.

Testing Requirement for Mercury Cartridges with ESLI

Concn. (mg/m³)	Equilibration[a]	No. of cart.	Humidity (%)	Time (min)	Penetration (mg/m³)
PEL	No	1	50	480	PEL
21.5	No	3	50	480	PEL
21.5	Yes	2	85	480	PEL
21.5	Yes	2	85	480	PEL

Note: Air flow is 64 l/m; equilibrated at 25°C, 85% RH, and 64 l/m for 6 h.

The passive ESLI must change color at 90% or less of the actual service of the sorbent. However, the test requirement does not address the desorption problem during multiday use. The mercury cartridge is tested at the saturated concentration at 25°C, which is 210 times the OSHA PEL.

The testing condition for the ethylene oxide (EtO) canister is similar to the mercury cartridge except that in addition to tests at the PEL concentration, a concentration at 50 times PEL or 50 ppm, and a high concentration of 5000 ppm is performed for emergency escape. The allowable penetration is 1 ppm for a challenge of 50 ppm EtO after 240 min, and 1 ppm for a challenge of 5000 ppm EtO after 12 min. The hydrogen sulfide cartridge is approved for escape only because it has poor odor warning properties. The test conditions for the H_2S cartridge are similar to those for the mercury cartridge, except that the challenge concentration of H_2S is 1000 ppm. A maximum penetration of 10 ppm H_2S is permitted for a minimum service life of 30 min.

J. Fit Testing Requirement

A quantitative fit testing (QNFT) requirement for respirator certification was deleted from the final version of the revised 30 CFR 11 in 1972 because respirator manufacturers argued that the QNFT requirement was not addressed in the proposed revision. The only available method for checking the facepiece seal is the isoamyl acetate (IAA) qualitative fit testing method (QLFT). A QNFT test schedule using coal dust developed by the Bureau of Mines which is useful for fit testing single use dust respirators was also not included in the 1972 revision. The facepiece tightness test is a requirement for all respirators equipped with tight-fitting facepieces. This requirement made the single use and disposable dust/mist respirators the only class of respirators for which fit testing is not required because these devices cannot be tested by the isoamyl acetate challenge.

The isoamyl acetate QLFT method is similar to the QNFT method developed by the Bureau of Mines using dichlordifluoromethane (Freon 12) except that Freon 12 is replaced with isoamyl acetate. The test chamber has a concentration of 100 ppm for testing half-mask respirators, and has a concentration of 1000 ppm for testing full facepiece respirators. The exercises to be performed by the test subjects include 2 min walking, nodding, and shaking head in normal movements, and 3 min exercising and running in place. The same exercises will be performed when the respirator is equipped with a half-mask, full facepiece, or for Type C supplied air respirators operated in continuous flow, demand or pressure demand mode.

K. Breathing Resistance

The breathing resistance for each class of respirator is also specified in each subpart. These requirements are summarized in Table 5.

L. Exhalation Valve Leakage Test

Dry exhalation valves and valves seats will be subjected to a suction of 25 mm water pressure while in a normal operating position. The leakage between the valve and valve seat shall not exceed 30 mL/min.

Table 5 Maximum Permissible Breathing Resistances ($mm\ H_2O$)

Type of respirator	Inhalation: Initial	Inhalation: Final	Exhalation
Particulate filters			
Single use	12	15	15
Dust/mist	30	50	20
Fume	20	40	20
Radon daughters	18	25	15
Asbestos	18	25	15
Gas mask			
Front/back mount	60	75	20
Front/back mount-P[a]	70	65	20
Chin style	40	55	20
Chin style-P	65	80	20
Escape	60	75	20
Escape-P	70	85	20
Chemical cartridge			
Gases and vapors	40	45	20
Gases, vapors, and filter	50	70	20
Gases, vapors, and paint Spray	50	70	20
SCBA			
Open-circuit[b]			
Demand	32		25
Pressure demand	32		sp + 51[c]
Closed-circuit	er ± 100[d]		51
Supplied air respirator			
Negative pressure	50		25
Continuous flow	25		25
Positive pressure	38		51

[a] P: Equipped with a particulate filter.

[b] Tested at an air flow rate of 120 l/m.

[c] sp: Static pressure in the facepiece, and the static pressure at zero flow shall not exceed 38 mm H_2O.

[d] er: Exhalation resistance, the inhalation resistance shall not exceed the difference between er and 100 mm H_2O.

II. PROPOSED REVISION OF 42 CFR 84

A. Background

The current 30 CFR 11 was revised in 1972 to adopt the BM test schedules with little modification except that a quality assurance program was added to the revision. There has been widespread criticism of both the performance test procedures and criteria for respirators used to protect persons against inhalation of harmful atmospheres in workplace as delineated in the governmental respirator test and approval document — Part 11 of title 30 of the Code of Federal Regulations (30 CFR 11). This policy is administered by the Mine Safety and Health Administration (MSHA) and the National Institute for Occupational Safety and Health (NIOSH) in carrying out a respirator test and approval program in the U.S.

The criticism concerns the use of respirator performance test procedures and criteria developed by the BM as long ago as 50 years, e.g., the lack of physiological data to justify many respirator performance criteria, the failure to apply newly developed technology in evaluating the performance of respirators, and the inadequacy of fitting tests to insure that approved respirators satisfactorily fit the majority of the adult work population. There is a lack of correlation of the results of laboratory tests on respirators with the results of field testing, and inattention is given to test data and recommendations of recognized respirator researchers.

For example, the dust filter certification test requires a very high concentration (50 mg/m^3) of silica dust, and the integrated penetration is the only measurement at the end of a 90-min test. The integrated measurement method would permit the use of less efficient filters which allow the penetration of fine particles until a cake is built up on the respirator filter surface. This decreases the penetration of particles, and meets the integrated penetration requirement. Most current exposure standards for toxic particulates are less than 0.5 mg/m^3. The filter would permit the continued penetration of low concentrations of toxic air contaminants without the opportunity for building up a cake on the filter surface, therefore improving efficiency of the filter. The respirator wearer would be exposed to these toxic substances because the approved filter is ineffective against low concentrations of challenge.

On July 28, 1980, NIOSH called a public meeting at the National Bureau of Standards in Gaithersburg, MD to discuss deficiencies in the current respirator testing and certification regulations (30 CFR 11). In October 1980, the American National Standards Institute (ANSI) established a Subcommittee of its Z88 Committee on Respiratory Protection to assist MSHA and NIOSH in improving 30 CFR 11. This group was the ANSI Ad Hoc Respirator Test and Approval Subcommittee. The Subcommittee was chaired by William Revoir. The members of the Subcommittee were employed by various government agencies other than NIOSH and MSHA, which are involved in occupational health and safety, research institutions, private consulting firms, industrial companies, and manufacturers of respirators. Members of the Ad Hoc Subcommittee agreed to the following:

1. There is confidence in the performance of respirator high-efficiency particulate air (HEPA) filtering elements in NIOSH/MSHA approved air-purifying respirators for removal of particulate type carcinogens from air.
2. There is no confidence that the performance criteria of 30 CFR 11 for other types of respirator particulate-filtering elements and for respirator vapor-/gas-removing elements ensures that these devices in NIOSH/MSHA approved air-purifying respirators are effective in removing specific carcinogens from air.
3. The respirator fitting requirements prescribed in the 30 CFR 11 are not adequate to ensure that facepieces in NIOSH/MSHA approved air-purifying respirators adequately fit the majority of the adult U.S. work population, male and female.
4. The 30 CFR 11 should be revised to list a minimum protection factor (fit factor) for each type of respirator; and each make and model of respirator submitted to NIOSH and MSHA for approval should be quantitatively fit tested using an appropriate panel of human test subjects representative of the U.S. adult work population, male and female, to determine whether or not it meets the appropriate protection factor (fit factor) criterion.

During the meetings, the ANSI Ad Hoc Respirator Test and Approval Subcommittee established three working groups to develop recommendations for changes in the technical sections of 30 CFR 11 for submittal to NIOSH and MSHA. These three working groups are the Respirator Quantitative Respirator Fitting Test Working Group, the Atmosphere-Supplying Respirator Working Group, and the Air-Purifying Respirator Working Group.

The discussion pointed out that the most widely used type of air-purifying respirator is the particulate-filtering respirator. Therefore, the Air-Purifying Working Group decided that the first projects would concern particulate-filtering respirators. Under the leadership of the late William Revoir, the Air-Purifying Respirator Working Group conducted extensive testing to evaluate the performance of approved negative pressure air-purifying respirators and to investigate a new testing agent for certifying particulate filters.[3-4] A report with recommended changes for the 30 CFR 11 was submitted by the ANSI Ad Hoc Subcommittee for Respirator Test and Approval to NIOSH in 1982.[5] Many of these recommendations were adopted by NIOSH in the revised 30 CFR 11.

The first proposal for the revision of 30 CFR 11 was published on August 27, 1987[6] and a hearing was conducted in January 1988.[7] Since MSHA is only interested in the approval of respirators for mine applications, MSHA decided that the agency would only be involved in the approval of respirators used in the mine environment. Therefore, NIOSH became the sole approval agency for respirators used for nonmine operations, and 30 CFR 11 became 42 CFR 84.

The major changes in the 42 CFR 84 include the workplace respirator performance testing, the particulate filter testing, and the facepiece seal test. Because the performance test methods prescribed in the 30 CFR 11 are mainly bench tests performed in the laboratory, it may not have any relevancy to the actual performance of the respirators in the workplace. NIOSH proposed that all respirators be field tested before being certified. Since no detailed test protocol was prescribed in the proposed 42 CFR 84, respirator manufacturers and other groups objected that it was difficult to perform field studies without knowing the requirements. In a revised proposal prepared in 1988,[7] NIOSH withdrew the requirement of field testing from the 42 CFR 84 and also set another rulemaking on this subject. A requirement for testing a canister or cartridge at 85% relative humidity after the device was equilibrated at 85% relative humidity, was also withdrawn from the proposal.

B. Filter Test

The proposed test method for particulate filters would replace all current classifications for dust, fume, mist, and HEPA filters with three new classes: I, II and III. All filters would be conditioned at 38°C and 80% humidity for 24 h before testing. The challenge is either a solid aerosol or a liquid aerosol. The solid challenge is a monodisperse sodium chloride aerosol having a size of 0.2 μm. The challenge concentration is up to 200 mg/m^3 at a flow rate of 85 l/m if the filter is used singly, or tested at a flow rate of 42.5 l/m if the filters are used in pair. The test would be terminated when 200 mg of sodium chloride (or liquid aerosol) particles had been collected on the respirator filter surface. The penetration is measured on a real time basis. The filter is considered acceptable when it meets the following penetration requirements.

Class	Nonpowered (%)	Powered (%)
Class I	10 (5)	
Class II	1	1
Class III	0.03	0.03

Note: 1988 revision (1987 proposal).

Class I filters are not permitted for PAPRs. The maximum penetration for Class I filter was 5% in the 1987 proposal. The penetration requirement of the Class III is equivalent to the HEPA filter prescribed in 30 CFR 11. However, no information is available to indicate whether current dust, mist, or fume filters would meet the requirements of Classes I and II. A summary of test conditions is shown in Table 6. The allowable penetration has been increased to 10% for Class I filters in the September 1988 revised proposal.

C. Facepiece Seal Test

The proposed 42 CFR 84 prescribes that all respirators equipped with tight-fitting inlet coverings pass a quantitative fit testing (QNFT) performed on an anthropometrically selected human test panel of 25 test subjects that represents the adult work population of the U.S. (see Figure 1). The test aerosol is either a polydisperse NACl solid aerosol or a polydisperse oil mist aerosol. The MMAD for the testing aerosol is 0.6 μm with a standard deviation of 2.2 or less. The test subject is required to wear safety spectacles and the test subject is also required to wear the respirator for at least 15 min prior to beginning the faceseal test. The probe location on the facepiece is also specified. The maximum permissible facepiece leakage for the negative pressure respirators is listed in the following table.

Facepiece	Type I	Type II	Type III
Quarter or half	0.15	0.06	0.05
Full	0.10	0.02	0.01

The exercise regimen consists of the following:

1. Normal breathing.
2. Deep breathing.
3. Turning head from side to side.

Table 6 Test Conditions for Particulate Filters

Condition	Solid	Liquid
Test aerosol	NaCl	oil mist
Particle size (μm)	0.2–0.3	0.2–0.3
Concentration (mg/m³)	<200	<200
Nonpowered		
Flow rate (l/m)	85/42.5	85/42.5
Maximum retained wt (mg)	200	200
PAPR		
Flow rate (l/m)	40 × 24[a]	40 × 24[a]
Maximum retained wt (gm)	2000	2000

[a] Test is performed on a breathing machine at 24 respirations per minute with a minute volume of 40 l/m.

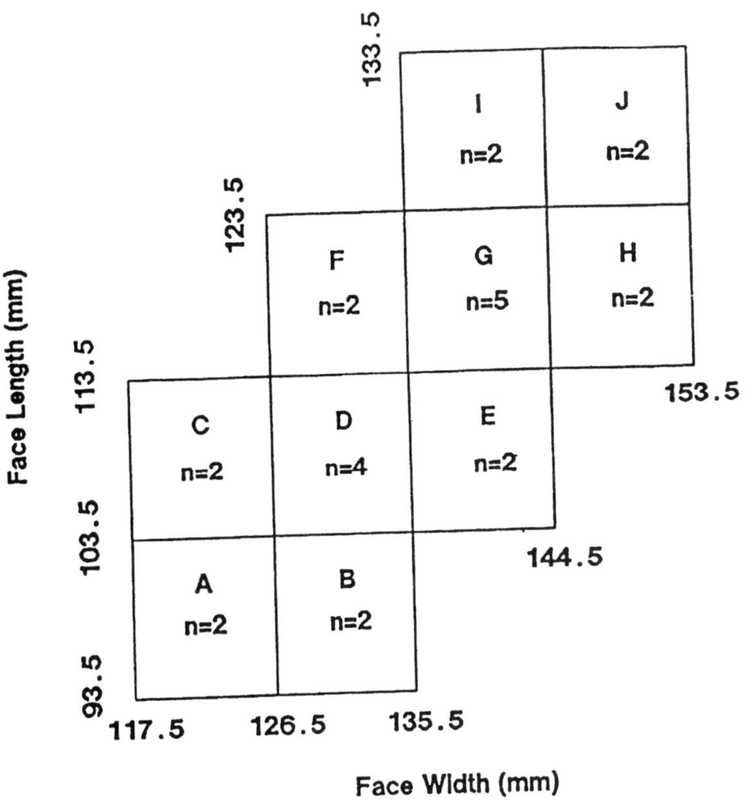

Figure 1 Anthropometric test panel specifications for face seal performance testing [42 CFR 84.232(b)]. (From Reference 7.)

4. Nodding head up and down.
5. Repeatedly raising arms upward and simultaneously looking upward.
6. Bending forward at waist and simultaneously extending arms downward toward shoes.
7. Talking.
8. Grimacing or frowning.
9. Normal breathing.

In the 1988 revised proposal,[7] exercises which simulate body movements, such as Exercises 5 and 6 above and the grimacing exercise which could induce leaks around the facepiece were deleted from the original proposal. However, the OSHA asbestos standard does have two exercises of bending over and touching toes and jogging in place that simulate body movements.

D. Cartridge and Canister Performance Test

There are no major changes in the testing requirements for chemical cartridges and canisters except that the front or back mounted canister is now called "high capacity" canister and the chin styled canister is called the "small capacity" canister. In addition, hydrogen sulfide has been added to the list. The test conditions for cartridges and canisters are summarized in Tables 7 through 9.

Table 7 Bench Test Requirements for Chemical Cartridges

	Test atmosphere				Approval requirements	
Gas or vapor	Concn. (ppm)	Flow rate nonpowered (l/m)	PAPR Tight	PAPR Loose	Maximum penetration (ppm)	Minimum service (min)
NH_3	1000	64	115	170	50	50
Cl_2	500	64	115	170	5	35
ClO_2	500	64	115	170	0.1	30
HCl	500	64	115	170	5	50
H_2S	1000	64	115	170	10	30
Hg	21.5[a]	64	115	170	0.05	480
CH_3NH_2	1000	64	115	170	10	25
SO_2	500	64	115	170	5	30
CCl_4	1000	64	115	170	5	50

[a] Test concentration in mg/m^3.

Table 8 Bench Test Requirements for High Capacity Canisters

	Test atmosphere				Approval requirements	
Gas or vapor	Concn. (ppm)	Flow rate nonpowered (l/m)	PAPR Tight	PAPR Loose	Maximum penetration (ppm)	Minimum service (min)
NH_3	30,000	64	115	170	50	12
CO	20,000	64[a]			[c]	60
CO	5000	32[b]			[c]	60
CO	3000	32[a]			[c]	60
Cl_2	20,000	64	115	170	5	12
H_2S	20,000	64	115	170	10	12
SO_2	20,000	64	115	170	5	12
CCl_4	20,000	64	115	170	5	12

[a] Tested at 95 + 3% RH and 25 ± 2.5°C.

[b] Tested at 95 ± 3% and 0°C.

[c] Maximum penetration < 385 ml.

Table 9 Bench Test Requirements for Low Capacity Canisters

	Test atmosphere				Approval requirements	
Gas or vapor	Concn. (ppm)	Flow rate nonpowered (l/m)	PAPR Tight	PAPR Loose	Maximum penetration (ppm)	Minimum service (min)
NH_3	5000	64	115	170	50	12
CO	20,000	64[a]			[c]	60
CO	10,000	32[b]			[c]	60
CO	5000	32[a]			[c]	60
Cl_2	3000	64	115	170	5	12
H_2S	5000	64	115	170	10	12
SO_2	5000	64	115	170	5	12
CCl_4	5000	64	115	170	5	12

[a] Tested at 95 ± 3% RH and 25 ± 2.5°C.

[b] Tested at 95 ± 3% and 0°C.

[c] Maximum penetration < 385 ml.

Table 10 Maximum Permissible Breathing Resistances (mm H_2O)

Type of respirator	Inhalation	Exhalation
SCBA, Open-Circuit		
Negative pressure	32	25
Pos pressure, >static	≥0	51
Pos press, inc. static	≥0	89[a]
Static press, no flow	NA	89
SCBA, Closed-Circuit		
Negative pressure	100-MER[b]	64
Positive pressure	≥0	89
Supplied Air Respirators		
Negative pressure	51	25
Continuous flow	25	25
Positive pressure	51	25
Gas Mask		
High capacity	60	20
High capacity — P[c]	70	20
Low capacity	50	20
Low capacity — P	65	20
Chemical Cartridge		
Gases and vapors	40	20
Gases, vapors, and filter	50	20
Particulate Filters		
All classes	30	20

[a] MER, measured exhalation resistance in mm H_2O.

[b] Including the pressure required to fully open the relief valve if applicable.

[c] P, particulate filter.

E. Breathing Resistance

The breathing resistance testing requirements for each class of respirator are generally the same as those in 30 CFR 11 except that the final inhalation resistance requirement is deleted. A summary of the breathing resistance requirements is shown is Table 10.

F. Atmospheric Supplying Respirators

There are few changes in the testing requirements for SCBAs in 42 CFR 84, except that testing requirements for positive pressure closed-circuit SCBAs, a facepiece flammability test, and a vibration test have been added and the maximum air flow has been upgraded. Because the main use for the SCBA in this country is for firefighting, it is not clear why the performance requirements prescribed in NFPA 1981 have not been adopted in 42 CFR 84. There is also no testing requirement for certifying escape self-contained breathing apparatus.

There are very few changes in performance requirements for supplied air respirators (SAR) in the 42 CFR 84 except that testing requirements for type A and type B airline respirators have been deleted and the gasoline permeation test for airline hoses has also been deleted. There is no performance testing requirement for certifying supplied air suits.

G. Respirators for Protection Against TB

On October 12, 1993, the Center for Disease Control and Prevention (CDC) issued a draft guideline for preventing the transmission of tuberculosis (TB) in health care facilities for public comments.[8] The guideline discusses the advantages and disadvantages of various classes of air-purifying particulate filtering respirators and the characteristics on fit test, fit check, and faceseal leakage of each class of respirators. Based on these discussions, CDC proposed four criteria for certifying a class of respirator for TB protection:

1. The ability to filter particles of 1 μm in the unloaded state with a filter efficiency of ≥ 95%, given flow rate of up to 50 l/m.
2. The ability to be qualitatively or quantitatively fit tested in a reliable way to obtain a faceseal leakage of no more than 10% for most workers.
3. The ability to fit health care workers with different facial sizes and characteristics; availability of at least three sizes of respirators.
4. To ensure proper protection, the facepiece fit should be checked by the wearer each time he or she puts on the respirators, in accordance with OSHA's standard and good industrial hygiene practice.

CDC also states that among the currently approved respirators, only the respirator equipped with HEPA filters would meet the proposed certification criteria. Because the proposed criteria is not a complete test protocol for certifying respirators, it leaves many unanswered questions. For example, the type of aerosol and the mean and standard deviation of the test aerosol must be specified. Test variables such as the air flow, challenge concentration, and test time that would affect the efficiency of the filter are also lacking. The proposed guidelines allow a combined respirator leakage of 15% which accounts for a filter leakage of 5% and a faceseal leakage of 10%. Since OSHA and NIOSH regulations only allow a maximum leakage of 10% for a half-mask respirator, it is not clear how the CDC developed this respirator certification criteria.

H. Modular Respirator Approval Concept

Under urging from health-care providers and the CDC, NIOSH announced that it will use the modular concept to develop respirator certification regulations.[9] The proposal for the particulate module was published on May 20, 1994. The anticipated date for proposed rule of other modules has the following schedule.

Subject	Date
Assigned protection factors	Late 1994
Administrate program	Early 1995
Quality assurance requirements	Early 1995
Gas and vapor requirements	Mid 1995
Positive pressure SCBA requirements	Early 1996
Simulated workplace protection factor test	Early 1997

This approach is in contrast to the complete proposal published in 1987.[6] NIOSH states that "the first module addressed the most widely used type of respirators — the air-purifying respirator with a particulate filter. This proposal would replace the inadequate and outdated current procedures with better methods to assess a respirator's ability to filter out toxic substances. In addition to benefitting industrial workers, the improved testing requirement will also benefit health-care settings implementing the current draft CDC Guidelines for Preventing the Transmission of Tuberculosis. While respirators with HEPA filters are the only respirators currently available which meet these criteria, the proposed regulation will enable manufacturers to produce a broader range of certified respirator which provide the necessary level of protection".

NIOSH made the following changes from the 1987 proposal:[6]

1. Three more classes of filters with solid aerosol approval only in addition to three classes of solid and liquid approval have been added.
2. Particle loading has been increased from 100 to 200 mg.
3. Instead of requiring that the filter penetration test be continued until filter penetration became stabilized, the proposed test terminates after the required loading level has been achieved.
4. The particle size for the challenge has been reduced from 0.2–0.3 μm to 0.06–0.11 μm for the solid NaCl aerosol and 0.17–0.22 μm for the polydisperse liquid DOP aerosol which replaces the monodisperse DOP.
5. The test sample size has been increased to 30. The mean maximum penetration, m, and standard deviation, s, for a given class of filter penetration, p, must meet the test statistic U, while

$$U = m + 2.22\,s \leq p \qquad (1)$$

I. Informal Hearing on the Filter Module

An informal hearing was held on June 23 and 24, 1994. Persons who provided comments were mainly representatives from respirator manufacturers and health care industry.

NIOSH's introductory remarks can be summarized as follows:

1. The OSH and MSH Acts require the use of respiratory protective equipment.
2. The first BM testing requirement was established in 1919. The last revision was made in 1969.
3. The first particulate filter penetration test was developed in 1934; this August will be the 60th anniversary of the silica dust test method. This testing method is ineffective against submicrometer particles. The proposed revision requires the use of most penetrating size range to provide significant improvement in protection provided to respirator wearer.
4. MSHA still retains authority for approving respirators for mine rescue and mine use.
5. The modular concept would address emerging health hazards. This approach would facilitate the transition into new requirements.
6. This proposed module would provide significant protection to wearers of respirators.
7. The proposed module would enable users to easily discern the level of protection that can be expected when using a respirator.
8. The proposed module would enable classification of the filters on their ability to inhibit penetration of particulates of the most penetrating size.

The representative from OSHA made the following statements.

1. Congratulates NIOSH's effort in updating the respirator test and certification regulations.
2. The modular format would expedite the revision process.
3. OSHA is in the process of revising its respiratory protection standard which relies on the respirator certification standard established by NIOSH.
4. The issue of protection factors has an important impact on the OSHA proposal, and OSHA encourages NIOSH to expedite its work in this area.
5. A respirator which cannot be fit checked after donning may not achieve the required protection factor. NIOSH should evaluate all the manufacturers use instructions before issuing approvals.
6. NIOSH needs to explain the difference between the current and proposed regulations in filter testing requirements, such as solid NaCl aerosol challenge, neutralized liquid DOP challenge at a concentration of 200 mg/m^3, and the testing methods for combination respirators.

The representatives from the health care industry supported NIOSH's intention to adopt the October 1993 CDC draft guidelines which set forth four criteria for certifying respirators for protection against TB. All of the presenters stated that using the HEPA filter for protection against TB exposure is based on a theoretical benefit, that it has no sound scientific ground, and that the data are not available to determine the effectiveness of currently approved respirators. The NIOSH proposed filter testing module would expedite the approval of respirators that specifically protect health care workers against TB. Other presenters commented that the OSHA requirements for a respirator program and fit testing have added considerable cost to the health care industry. In addition, the required respirator cannot accommodate bearded workers and eye glass wearers.

Comments from the respirator manufacturers association, the Industrial Safety Equipment Association (ISEA), and several manufacturers can be summarized as follows:

1. The manufacturers support NIOSH's modular approach since it is innovative, results oriented, and has measurable success. The stepwise incremental approach is feasible since complete rulemaking is overly burdensome.
2. The revision of 42 CFR 84 is costly since the current electrostatic D/M filter media cannot meet both the solid and liquid testing requirements for all classes. The price for currently approved filter media is between \$0.60 and \$1.0/yd^2; however, the filter media which would meet the 42 CFR 84 requirements would cost between \$12 and \$17/yd^2.
3. There are ambiguities in the module interrelations, such as the combination of gas or vapor/particulate cartridges, combination air-purifying and supplied air respirators, and PAPRs. The removal of ambiguous and unrealistic requirements would expedite advancement and minimize cost impact.
4. A user's guide should be developed which would inform the regulatory community and minimize the confusion of the new regulations. The guide should provide guidance of selection to avoid incorrect applications.
5. The proposed grandfather clause should be modified. The two year limitation on sale and distribution should be extended to 4 years. The 6 month limit on extensions of approvals which may affect filter media should be extended to 2 years.

6. There are some interlaboratory variations of test results on filter efficiency. There should be little inconsistency regardless of the testing location.
7. There are no specific requirements on preconditioning parameters such as container volume or the time between the conditioning and testing.
8. There is a lack of tolerance in air flow requirements. A tolerance of ± 2% is recommended.
9. The breathing resistance requirements are unnecessarily high. The inhalation resistance should be increased from 30 to 35 mm; the exhalation resistance should be increased from 20 to 25 mm.
10. The efficiency requirements are too stringent for Type B and C filters. The efficiency of Type B filter should be changed to 96% to allow an APF value of 25 and the efficiency of the Type C filters should be changed to 90% to allow for an APF value of 10.
11. PAPRs are not adequately addressed in this module. A new module should be added to address specific requirements on PAPR.
12. The proposal does not reflect worker's need for modern design and technology. The proposal is unclear in its intent and execution. The consequence is that the proposed design is restrictive and inconsistent.
13. There is interaction among filter penetration, air flow, and fit testing. The measurement of one cannot be made in isolation of others.
14. Preliminary test data show that with a NaCl aerosol loading of 2000 mg for a PAPR, there is a significant increase in pressure drop for four types of common filter media. There is a 2 to 1 increase in pressure drop for a mechanical filter media. For a filter containing 9% electrostatic filter media, the pressure drop increase is 7 to 1. For a filter containing 30% electrostatic filter media, the pressure drop increase is 10 to 1. For a filter containing 90% electrostatic filter media, the pressure drop increase is 30 to 1. The caking effect was also evident unless the tests were interrupted.
15. The test statistics for filter acceptance would raise costs and lead to overdesign. Not all filters types follow the normal distribution. The instrument resolution can distort the distribution. Even when the manufacturer follows the spirit of regulation by performing 100% testing, the respirator can still fail the statistical test.

    The AQL (acceptable quality level):

    $$U = m + k\,s \tag{2}$$

    when the sample size has been increased to 30 and *k* is getting high. Filters will be overdesigned with excessive sampling and testing regimens. This will lead to expensive and bulky filters. Types B and C filters will be significantly affected.

    The k factor should be reduced from 2.22 to 1.778. The reason is that due to the 95% confidence limit for 90% of the test samples requirement was proposed in the 1987 revision. The current requirement is a 95% confidence limit for 95% of test samples.
16. The proposed two tier solid and solid/liquid approvals can lead to misuse since the test method may overstate filter efficiency. For example, asbestos is a solid aerosol, but during demolition operations, the worker's exposure to asbestos is a mixture of fiber and water spray. The average worker does not know the difference between solid and liquid aerosol approvals.
17. Since the filter efficiency is the deciding factor for purchase and the filters with solid approval are always cheaper, misuse and misapplication may occur.

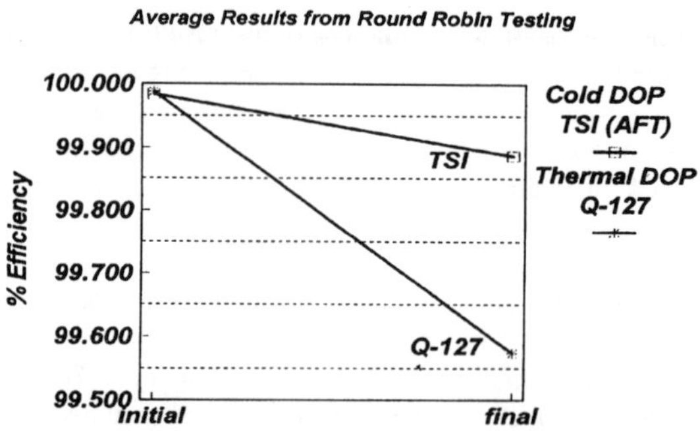

Figure 2 Average results from round robin testing for type A filters. (From Reference 10.)

18. Based on the ISEA round robin tests on the permanently charged electret filter media,[10] there is a significant difference in filter efficiency between the cold generated (polydisperse) and the heat generated (monodisperse) liquid DOP challenge. The cold generated polydisperse DOP aerosol is not the most penetrating "worse case" test aerosol as defined in the current proposal. Based on the round robin tests performed by respirator manufacturers, there is a significant difference in filter penetration between the cold and heat generated DOP aerosols when the test was performed on the electret filter media. The difference in average filter efficiency between the cold and hot DOP aerosol challenges for Types A, B, and C electret filter media are 99.87% and 99.52%; 99% and 94.5%; and 93.5% and 84.5%, respectively (see Figures 2 through 4). However, there is little difference in filter penetration between cold and heat generated DOP aerosols for different classes of mechanical filter media. The difference in efficiency between the cold and hot DOP aerosol challenges for Type A, B, and C mechanical filter media are 99.87% and 99.89%; 99.56% and 99.6%; and 97.47% and 97.53%, respectively. The ISEA round robin test results of the electret filter media challenged with hot monodisperse DOP aerosol (measured by the Q-127 instrument) and the cold polydisperse DOP aerosol (measured by the TSI aerosol fit tester (AFT)) are shown in Figure 5. The ISEA round robin test results of mechanical filter media by the same test aerosols are shown in Table 11.

The cold generated polydisperse DOP aerosol consistently overestimated filter efficiency with each type of electret filter media. The heat generated monodisperse DOP is a more penetrating aerosol; it should replace the cold generated DOP as the "worse case aerosol."

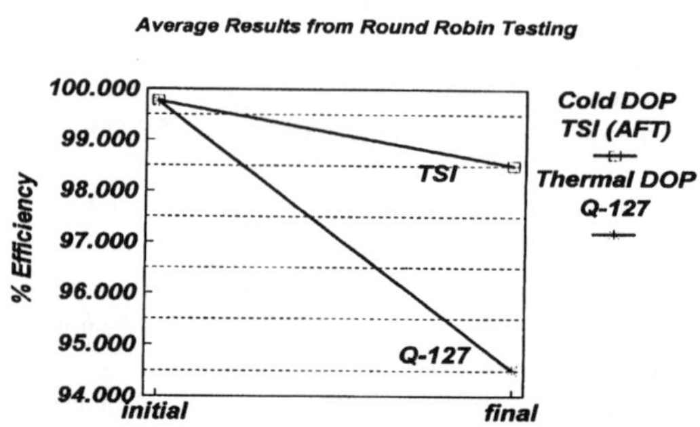

Figure 3 Average results from round robin testing to type B filters. (From Reference 10.)

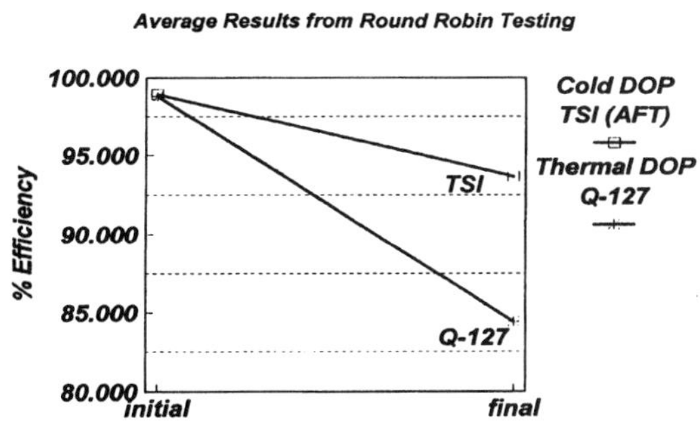

Figure 4 Average results from round robin testing for type C filters. (From Reference 10.)

19. Studies conducted by Brown,[11] Smith,[12] and Kennedy and Reed[13] indicated that electret filter media could be degraded by aerosols such as silkstone, foundry fettling, foundry burning, carbon brick, lead smelter, lead battery, and coal tar. Because users cannot detect, by taste or smell of filter penetration, there is no indication that the electret type filter is losing efficiency.
20. If the filter penetration increases when the 100 mg challenge point is reached, the test should be continued until the efficiency and penetration have been stabilized.

III. OTHER TESTING AUTHORITIES

For establishments under OSHA's jurisdictions, only MSHA/NIOSH approved respirators are acceptable. Other regulatory agencies such as the Department of Energy (DOE) or the Nuclear Regulatory Commission (NRC) may impose their own rules. Generally, they accept respirators approved by MSHA/NIOSH. However, they could set additional requirements for special use respirators or respirators which are not covered by 30 CFR 11. The supplied air suit is one example. It is not clear why NIOSH refrains from developing a testing schedule for this device.

A. National Fire Protection Association

The SCBAs certified under 30 CFR 11 are based on a moderate work rate. A device approved with a 30 min service life may last only 20 min when it is used under a heavy work load such as firefighting. Furthermore, the device is not designed to be used in the fire environment. The National Fire Protection Association (NFPA) has developed a standard (NFPA-1981) which sets performance requirements for SCBAs being used in the fire environments.[14] Generally, the SCBA will be tested on a breathing machine at a flow rate of 100 l/m, under the assumption that a positive pressure would be maintained inside the facepiece at such a high work rate. The SCBA testing determines whether the device maintains positive pressure when it is subjected to most of the performance criteria prescribed as the following.

1. Air Flow Performance

The SCBA is tested on a breathing machine with a minute volume of 103 l/m, a respiratory frequency of 30 breaths per minute and a tidal volume of 3.4 l.

Type A Filter Media

TSI (AFT)	Test Site 1 (Moldex)					Test Site 2 (3M)					Test Site 3 (North)					Test Site 4 (MSA)					Test Site 5 ** (Survivair)					% pen. avg.
	A1	A2	A3	A4	A5	A1	A2	A3	A4	A5	A1	A2	A3	A4	A5	A1	A2	A3	A4	A5	A1	A2	A3	A4	A5	
initial pen.	0.015	0.018	0.014	0.011	0.013	0.015	0.012	0.013	0.009	0.012	0.026	0.027	0.019	0.036	0.037	0.009	0.012	0.006	0.008	0.010	*0.587*	*0.424*	*0.454*	*0.362*	*0.254*	0.018 initial pen.
max penetr.	0.099	0.122	0.103	0.084	0.096	0.108	0.118	0.106	0.061	0.088	0.161	0.143	0.112	0.220	0.213	0.072	0.111	0.075	0.074	0.095	*1.270*	*1.000*	*1.050*	*0.904*	*0.714*	0.113 max penetr.

Q-127	Test Site 1 (Moldex)					Test Site 2 (3M)					Test Site 3 (North)					Test Site 4 (MSA)					Test Site 5 (Survivair)					
	A1	A2	A3	A4	A5	A1	A2	A3	A4	A5	A1	A2	A3	A4	A5	A1	A2	A3	A4	A5	A1	A2	A3	A4	A5	
initial pen.	0.002	0.003	0.001	0.002	0.001	0.024	0.023	0.017	0.017	0.021			0.016	0.017	0.020	0.027	0.025	0.019	0.008	0.021	0.001	0.001	0.001	0.001	0.002	0.012 initial pen.
max penetr.	0.105	0.098	0.038	0.033	0.035	0.983	0.795	0.695	0.786	0.911			0.490	0.480	0.400	0.720	0.750	0.570	0.330	0.740	0.260	0.150	0.170	0.094	0.160	0.426 max penetr.

Type B Filter Media

TSI (AFT)	Test Site 1 (Moldex)					Test Site 2 (3M)					Test Site 3 (North)					Test Site 4 (MSA)					Test Site 5 ** (Survivair)					avg.
	B1	B2	B3	B4	B5	B1	B2	B3	B4	B5	B1	B2	B3	B4	B5	B1	B2	B3	B4	B5	B1	B2	B3	B4	B5	
initial pen.	0.194	0.129	0.148	0.214	0.195	0.301	0.223	0.144	0.165	0.169	0.242	0.320	0.348	0.514	0.345	0.207	0.169	0.138	0.154	0.150	*1.460*	*1.330*	*1.350*	*1.550*	*2.000*	0.223 initial pen.
max penetr.	1.270	0.966	1.170	1.670	1.810	2.060	1.820	0.943	1.070	1.270	1.380	1.680	1.780	1.840	1.830	1.920	1.290	1.080	1.320	1.670	*4.080*	*3.910*	*4.180*	*4.600*	*5.630*	1.492 max penetr.

Q-127	Test Site 1 (Moldex)					Test Site 2 (3M)					Test Site 3 (North)					Test Site 4 (MSA)					Test Site 5 (Survivair)					
	B1	B2	B3	B4	B5	B1	B2	B3	B4	B5	B1	B2	B3	B4	B5	B1	B2	B3	B4	B5	B1	B2	B3	B4	B5	
initial pen.	0.052	0.091	0.066	0.064	0.044	0.215	0.299	0.249	0.329	0.263	0.380	0.460	0.430	0.450	0.370	0.290	0.380	0.510	0.200	0.330	0.120	0.080	0.110	0.100	0.170	0.242 initial pen.
max penetr.	1.950	2.950	2.350	2.300	1.500	6.310	7.260	7.730	9.340	8.080	5.900	8.900	7.300	6.600	6.800	6.100	7.900	7.900	4.900	6.400	3.600	3.200	4.700	3.800	5.300	5.483 max penetr.

Type C Filter Media

TSI (AFT)	Test Site 1 (Moldex)					Test Site 2 (3M)					Test Site 3 (North)					Test Site 4 (MSA)					Test Site 5 ** (Survivair)					avg.
	C1	C2	C3	C4	C5	C1	C2	C3	C4	C5	C1	C2	C3	C4	C5	C1	C2	C3	C4	C5	C1	C2	C3	C4	C5	
initial pen.	0.78	0.99	0.85	0.95	1.00	0.82	0.98	0.70	1.00	0.97	1.41	1.63	1.71	1.17	1.10	0.95	1.26	0.80	1.05	0.90	*4.97*	*4.65*	*4.93*	*5.68*	*6.34*	1.051 initial pen.
max penetr.	4.69	5.78	5.67	6.53	6.43	5.18	6.30	5.48	6.47	5.36	6.39	7.26	7.57	5.77	5.83	7.98	8.50	5.80	6.30	6.99	*14.30*	*12.60*	*13.10*	*13.80*	*16.70*	6.309 max penetr.

Q-127	Test Site 1 (Moldex)					Test Site 2 (3M)					Test Site 3 (North)					Test Site 4 (MSA)					Test Site 5 (Survivair)					
	C1	C2	C3	C4	C5	C1	C2	C3	C4	C5	C1	C2	C3	C4	C5	C1	C2	C3	C4	C5	C1	C2	C3	C4	C5	
initial pen.	0.53	0.65	0.45	0.53	0.48	1.37	1.48	1.42	1.45		2.40	2.20	1.60	1.40	1.90	1.50	1.50	0.80	1.10		0.83	0.80	0.74	1.10	1.10	1.19 initial pen.
max penetr.	10.20	13.00	8.40	9.90	9.80	22.50	21.90	21.20	23.10		19.00	18.50	16.00	16.00	17.50	14.00	19.00	11.00	16.00		12.00	14.00	12.00	15.00	18.00	15.57 max penetr.

Figure 5 ISEA round robin testing results. Test site 5 reported that filter preconditioning was at relative humidities much higher than that called for in 42CFR84. It is believed that this increase caused dramatically higher initial penetrations and final penetrations. Therefore, those values in italics were not used in the "average % penetration" calculation. ** Test Site 5 reported that filter preconditioning was at relative humidities much higher than that called for in 42CFR84. It is believed that this increase caused dramatically higher initial penetrations and final penetrations. Therefore, those values in italics were not used in the "average % penetration" calculation. (From Reference 10.)

Table 11 Mechanical Filter Test of Media

Type A filter media				Type B filter media				Type C filter media			
ID	Test site	AFT	Q-127	ID	Test site	AFT	Q-127	ID	Test site	AFT	Q-127
A1	1	0.012	0.009	B1	1	0.386	0.280	C1	1	2.270	2.300
A1	2	0.009	0.013	B1	2	0.353	0.382	C1	2	2.430	2.550
A1	3	0.018	0.016	B1	3	0.524	0.480	C1	3	2.870	3.200
A1	4	0.014	0.016	B1	4	0.486	0.480	C1	4	2.700	3.000
A1	5	0.013	0.008	B1	5	0.431	0.350	C1	5	2.500	2.000
A2	1	0.010	0.007	B2	1	0.405	0.290	C2	1	2.300	2.200
A2	2	0.010	0.016	B2	2	0.380	0.410	C2	2	2.450	2.430
A2	3	0.017	0.013	B2	3	0.539	0.550	C2	3	2.840	2.900
A2	4	0.014	0.017	B2	4	0.448	0.530	C2	4	2.590	2.700
A2	5	0.013	0.003	B2	5	0.408	0.320	C2	5	2.570	2.400
A3	1	0.012	0.006	B3	1	0.389	0.280	C3	1	2.350	2.100
A3	2	0.012	0.013	B3	2	0.441	0.411	C3	2	2.410	2.400
A3	3	0.018	0.017	B3	3	0.521	0.550	C3	3	2.850	3.000
A3	4	0.014	0.013	B3	4	0.454	0.530	C3	4	2.480	2.200
A3	5	0.014	0.007	B3	5	0.426	0.410	C3	5	2.330	2.200
A4	1	0.012	0.005	B4	1	0.370	0.280	C4	1	2.340	2.050
A4	2	0.010	0.017	B4	2	0.402	0.447	C4	2	2.340	2.430
A4	3	0.017	0.015	B4	3	0.520	0.530	C4	3	2.900	2.800
A4	4	0.012	0.015	B4	4	0.469	0.390	C4	4	2.580	2.500
A4	5	0.013	0.004	B4	5	0.427	0.310	C4	5	2.440	2.700
A5	1	0.011	0.007	B5	1	0.367	0.320	C5	1	2.340	1.950
A5	2	0.010	0.016	B5	2	0.381	0.422	C5	2	2.410	2.510
A5	3	0.016	0.016	B5	3	0.516	0.510	C5	3	2.890	2.800
A5	4	0.013	0.010	B5	4	0.470	0.350	C5	4	2.530	2.400
A5	5	0.013	0.004	B5	5	0.414	0.290	C5	5	2.460	2.100
	Average	0.013	0.011		Average	0.437	0.404		Average	2.527	2.473

From Reference 10.

The facepiece pressure will be more than 0 mm water column but less than 89 mm water column above ambient pressure throughout the usable service life of the device.

2. Environmental Temperature Performance

a. The device is cold soaked at –32°C for a minimum of 12 h and then tested for air flow as prescribed in test 1 above at –32°C.
b. The device is hot soaked at 71°C for 12 h and then tested for air flow as prescribed in test 1 above at 71°C.
c. The device is hot soaked at 71°C for 12 h and then transferred to a chamber with an air temperature of –32°C. The air flow performance will be tested within 3 min after the transfer.
d. The device is cold soaked at –32°C for 12 h and then transferred to a chamber with an air temperature of 71°C. The air flow performance will be tested within 3 min after the transfer.

3. Flame Resistance Performance

The test requirement is adopted from Method 5903, Flame Resistance of Cloth; Vertical, of Federal Test Method Standard No. 191, Textile Test Methods.

4. Corrosion Resistance Performance

The test requirement is adopted from Method 509.3, Salt Fog, Section II, of MIL-STD-810E, Environmental Test Methods. After being subjected to the test, the SCBA will be tested in accordance with test 1 prescribed above.

5. Lens Abrasion Performance

The lens of the facepiece is subject to a abrasion test with steel wool. Upon completion of the test, the specimen is evaluated in accordance with ASTM D-1003, Standard Test Method for Haze and Luminous Transmittance of Transparent Plastics.

6. Communication Performance

The method of measuring intelligibility is in accordance with ANSI S3.2, Method for Measurement of Monosyllabic Word Intelligibility.

7. Vibration Resistance Performance

The SCBA will be tested in accordance with Method 514.4, Vibration, of MIL-STD-810E, Environmental Test Methods. After being subjected to the test, the SCBA will be tested in accordance with test 1 prescribed above.

8. Particulate Resistance Performance

A fully charged SCBA will be subject to the testing requirements prescribed in Method 510.3, Sand and Dust, Section II-3, Procedure 1 of MIL-STD-810E, Environmental Test Methods. After being subjected to the test, the SCBA will be tested in accordance with test 1 prescribed above.

9. Fabric Heat Resistance Performance

The specimens of fabric components will be tested in a force air circulation oven with an air temperature of 260°C. The fabric will be observed for melting or ignition to determine pass or fail.

10. Thread Heat Resistance Performance

The thread utilized will be tested in accordance with Method 1534, Melting Point of Synthetic Fibers, of Federal Test Method Standard 191 A, Textile Test Methods, to a temperature of 260°C. The fabric will be observed for melting or ignition to determine pass or fail.

All SCBAs must be MSHA/NIOSH approved prior to being tested for NFPA 1981 compliance. Since there is no authority to certify compliance with the NFPA 1981, it is essentially a manufacturer's self-certification program. In order to determine compliance with the NFPA 1981, OSHA asked the Lawrence Livermore National Laboratory (LLNL) to evaluate the performance of the Agency procured SCBAs in 1989.[15] OSHA specifically required that the thermal resistance performance, lens abrasion resistance performance, vibration and shock resistance performance, and the communication performance will be conducted by the contractor and witnessed by a LLNL representative. The test results for other required tests prescribed in the NFPA 1981 will be reviewed by LLNL. The test results indicated that the device supplied by Survivair has met all performance requirements prescribed by NFPA 1981. Currently, almost all SCBA manufacturers sell SCBAs meeting the NFPA 1981 standard.

B. Department of Energy

The 30 CFR 11 does not have a test schedule for the supplied air suit which is needed to protect workers in the nuclear industry who handle radioactive gases, such as tritium. The Department of Energy (DOE) has set requirements to approve supplied air suits and has formed a Respirator Advisory Committee to oversee the approval of the supplied air suits.[16] The certification program is administrated by the Los Alamos National Laboratory (LANL). The DOE contractor submits a suit to LANL for testing. After testing, the LANL submits the test results to the Respirator Advisory Committee which then reviews the test results and makes recommendations to DOE for acceptance.

The supplied air suit is tested under the following schedules:

1. *Air Flow*

The suit is tested at air flow rates of 170 and 196 l/m (6 and 7 cfm).

2. *Protection Factors*

The study is conducted in a large test chamber with a polydisperse aerosol challenge. The aerosol has a mass median aerodynamic diameter of about 0.6 μm at a concentration of 25 ± 5 mg/m^3. Three test subjects with varied size and weight perform each of the following exercises for at least 2 min:

a. Standing still, arms hanging downward along the sides of body, normal breathing.
b. Bending forward and touching toes repeatedly.
c. Running in place.
d. Raising arms above head and looking upward repeatedly.
e. Bending knees and squatting repeatedly.
f. Crawling on hands and knees.
g. Standing with arms folded in front of chest and twisting torso from side to side repeatedly.
h. Standing still, arms hanging downward along sides of body, normal breathing.

The average fit factor for each of the above exercises is reported. The range is between 1000 and above 20,000.

3. *Aerosol Penetration (Air Off)*

After the above exercises have been performed, the test operator stops the flow of the breathing air. The test subject continues to stand still with her/his arms hanging downward along her/his sides and breathing normally. The test operator observes the increase in penetration of aerosol into the covering of the device in the breathing zone of the subject, and records the time required for the aerosol penetration to reach values of 0.05, 0.1, and 1%. The operator then restores the breathing air flow to the covering of the device and the test subject remains in the same stance as before. The aerosol penetration is measured until it returns to approximately the value occurring before the flow of breathing air was terminated. This time is defined as "clear out time" (Air On). The air on time varies from 2 to 3 min.

4. Escape Test

Immediately upon the shutoff of the air flow, the test subject begins removing the suit. Removal time is recorded, which is usually less than 6 s.

5. Supplied-Air Hose Test

The air hose and couplings are subjected to the following tests:

a. Strength of connection of the supplied air hose to the covering.
b. Crush resistance of the supplied-air hose.
c. Nonkinkability of supplied-air hose.
d. Strength of the supplied-air hose and couplings.
e. Rapid pull test.

6. Noise Test

The level of noise generated by the flow of air through the device is measured. A test subject wears the entire suit, and the level of noise inside the head portion of the covering at the ear of the test subject is measured at the two air flow rates. The highest acceptable noise level is 80 dBA.

REFERENCES

1. Respiratory Protective Devices; Test for Permissibility; Fees. Code of Federal Regulations, Title 30 Part 11, Mineral Resources, 1993.
2. NIOSH, Notice of Acceptance of Applications for Approvals of Air-purifying Respirators with End-of-Service-Life Indicators (ESLI). Criteria for Certification of End-of-Service-Life Indicators, *Federal Register*, 49, No. 140, 29270, July 19, 1984.
3. Revoir, W. H., Test Report, ANSI Z-88 Ad Hoc Subcommittee for Respirator Test and Approval, February 1981.
4. Revoir, W. H., Test Report, ANSI Z-88 Ad Hoc Subcommittee for Respirator Test and Approval, July 1981.
5. Wilmes, D. P., Final Report to NIOSH. ANSI Z88 Ad Hoc Subcommittee for Respirator Test and Approval, January 8, 1982.
6. Notice of Proposed Rulemaking for Certification of Respiratory Protective Devices. 42 CFR 84, National Institute for Occupational Safety and Health, *Federal Register*, 52, 32401, August 27, 1987.

7. Revision of Tests and Requirements for Certification of Respiratory Protective Devices. Second Notice of Proposed Rulemaking, National Institute for Occupational Safety and Health, September 16, 1988.
8. CDC, Draft Guideline for Preventing the Transmission of Tuberculosis in Health-Care Facilities. Centers for Disease Control and Prevention, *Federal Register*, 58 (195), 52810, October 12, 1993.
9. NIOSH, Respiratory Protective Devices; Proposed Rule. National Institute for Occupational Safety and Health, *Federal Register*, 59, 26850, May 24, 1994.
10. ISEA, Comments Submitted at NIOSH Informal Hearing on the Revision of 42 CFR 84. Industrial Safety Equipment Association (ISEA), Washington, D.C., May 20, 1994.
11. Brown, R. C., Wake, D., Blackford, D. B., and Bostock, G. J., Effect of industrial aerosols on the performance of electrically charged filter material, *Ann. Occup. Hyg.*, 32, 271, 1988.
12. Smith, D. L., Johnston, O. E., and Lockwood, W. T., The efficiency of respirator filters in a coke oven atmosphere, *Am. Ind. Hyg. Assoc. J.*, 40, 1030, 1979.
13. Kennedy, E. R. and Reed, L. D., Evaluation of penetration of pesticide respirator cartridge, NIOSH Report, October 1, 1981.
14. NFPA, Open-circuit self-contained breathing apparatus for firefighting, NFPA 1981, National Fire Protection Association (NFPA), Quincy, MA, 1992.
15. Johnson, J. S., da Roza, R., and McCormack, C. E., Evaluation of a commercial SCBA's compliance to the NFPA 1981 standard for firefighters and measurement of simulated workplace protection factors at high work rates, Lawrence Livermore National Laboratory, Presented at the American Industrial Hygiene Conference, Boston, MA, 1992.
16. Bradley, O., Acceptance testing procedures for airline supplied-air suits, Los Alamos National Laboratory, LA-10156-MS, 1984.

14 FILTER PERFORMANCE I

I. FILTER PENETRATION TESTS CONDUCTED BY THE LASL

Early in 1970, Los Alamos Scientific Laboratory (LASL) conducted a study to determine the efficiency of various types of Bureau of Mines (BM) approved particulate filters.[1] The purpose of this study was to develop quality control procedures for NIOSH to verify performance of approved particulate filters.

Both dioctyl phthalate (DOP) [di-2-ethylhexyl phthalate (DEHP)] and NaCl aerosols were selected for evaluation. Thermally generated monodisperse DOP and cold generated polydisperse DOP were used to test the fume filters. The size for the monodisperse DOP was 0.3 μm which was the aerosol used by the BM to test high-efficiency particulate air (HEPA) filters. The polydisperse DOP had a mass median aerodynamic diameter (MMAD) between 1.1 and 1.2 μm and a standard deviation (σ_g) of 1.76. Because the charges of the electrostatic type dust and mist filters (e.g., resin felt) would be degraded by the oil mist challenge, polydisperse NaCl aerosol was used to test dust and mist filters. The NaCl aerosol was generated by a Wright nebulizer. By varying the air pressure, different size aerosols were generated. The test NaCl aerosol had MMADs of 0.4, 0.8, and 1.2 μm with σ_g between 1.6 and 2. Dust, fume and mist (fume) and dust/mist filters made by AO, MSA, and Willson were selected for testing. The flow rate varied from 16 to 42.5 l/m.

The results of dust and fume filter penetration test for various types of aerosols are shown in Tables 1 and 2. The results indicate that filter penetration is a function of particle size and flow rate. The results of dust and mist filters test with the NaCl aerosol challenge are shown in Table 3. In order to find the correlation between the initial filter penetration with the NaCl aerosol and the final penetration measured on the BM silica dust test, the BM silica dust test was performed on the same filter. It was very difficult to correlate these two tests. Furthermore, there was a wide variation in filter penetration from the same lot. Quantitative fit testing (QNFT) using the 0.8 μm NaCl aerosol was conducted on various dust and mist filters. The fit testing and filter penetration results are listed in Table 4. It appears that the filter penetration is a function of flow rate. For quantitative fit testing, the total penetration is the sum of filter and facepiece leakage. However, there is no correlation between the results of the filter penetration test and the QNFT.

Because the BM silica dust test is a combination efficiency and loading test, additional tests were conducted to determine the effect of NaCl loading on the filter penetration of approved fume, dust and mist filters. As a filter becomes more heavily loaded with an aerosol, the particles begin to become a filter aid and the filter

Table 1 Comparison of Fume Filter Penetration by Several Aerosols

	Percent penetration versus flow rate (l/m)									
	DOP (0.3 μm)		DOP (1.1 μm)			NaCl (0.33 μm)			NaCl (0.39 μm)	
LASL Stock filter	16	32	16	32	42.5	16	32	42.5	16	32
WELSH 7500-7										
#15	18.5	19.0	4.4	5.8	7.0	9.5	13.5	15.0	10.5	15.3
#16	14.0	15.5	4.6	4.8	5.7	8.0	12.0	13.5	7.5	13.7
#17	19.0	22.0	5.2	7.0	8.1	11.3	16.0	17.5	10.0	13.3
MSA Type "S"										
#13	1.4	2.0	0.2	0.4	0.5	0.8	1.4	1.8	0.7	1.2
#14	1.5	2.1	0.2	0.4	0.5	0.9	1.7	2.0	0.5	1.1
#15	2.0	2.7	0.3	0.5	0.6	1.1	1.9	2.3	0.8	1.4
WILLSON R-361										
#10	12.0	15.0	4.7	6.2	7.4	8.0	11.0	12.7	7.0	8.9
#11	10.5	11.0	3.7	5.4	6.6	7.1	9.5	10.5	6.3	8.0
#12	10.0	10.0	3.9	5.3	5.8	6.5	8.0	9.0	10.3	12.8
AO R-56 (new type)										
#1	4.9	5.8	0.6	0.9	1.1	2.5	3.8	4.3	5.2	7.3
#2	9.1	9.8	1.3	2.0	2.3	4.6	6.6	7.6	3.3	4.3
#3	5.5	6.3	0.6	1.1	1.3	2.7	4.2	4.9	4.5	6.6

Note: The 0.3 μm DOP is thermally generated, the 1.1 μm AMMD DOP is air generated and polydispersed with $\sigma_g = 1.76$. The 0.33 μm MMD NaCl aerosol was from a BS 4400 design nebulizer with $\sigma_g = 1.93$. The 0.39 μm MMD NaCl aerosol was generated with a Wright nebulizer at 24 psi with $\sigma_g = 1.73$.

becomes more efficient and the resistance to air flow increases. The results of the NaCl aerosol loading tests on the fume, dust and mist filters are shown in Figures 1 and 2. The results indicate that filter penetration decreases with dust loading. However, the magnitude of change is much greater in the dust and mist filters than the fume filters. For the electrostatic type dust and mist filters, the aerosol penetration increases to a maximum at which point the efficiency starts to increase due to the

Table 2 Results of NaCl Aerosol Tests on Bureau of Mines Approved Dust and Fume Filters

	NaCl penetration (%)			
	16 l/m		32 l/m	
Manufacturer and filter type	Range	Average	Range	Average
Dust filters				
AO R30	2.3–2.9	2.6	4.8–5.7	5.3
AO R30 (Special)	1.5–2.5	1.9	3.3–4.9	4.0
MSA Dustofoe 66	0.3–1.8	1.2	0.8–3.8	2.7
MSA Type F	4.4–8.9	6.5	8.5–15.1	11.5
Welsh 7500-6	5.1–6.9	6.1	9.1–12.4	10.7
Willson R-10	0.3–0.6	0.4	0.8–1.4	0.9
Willson R-560	0.1–0.2	0.1	0.2–0.6	0.3
Fume filters				
AO R56	3.3–5.2	4.2	4.3–7.3	5.7
MSA Type S	0.5–0.9	0.8	1.1–1.7	1.4
Welsh 7500-7	7.5–13.7	10.8	13.3–18.0	15.3
Willson R-361 (LASL)	6.3–10.3	7.9	8.0–12.8	10.1
Willson R-361	9.2–14.3	11.8	12.0–17.2	14.8
Willson R-11	0.3–3.0	1.6	1.0–6.4	3.2

Note: Dust filters tested with 0.82 μm MMD NaCl, $\sigma_g = 1.58$. Fume filters tested with 0.39 μm MMD NaCl, $\sigma_g = 1.73$. These data represent six filters of each type.

From Reference 1.

Table 3 Comparison of Sodium Chloride Aerosol Test on One Type of Dust Filter Versus Silicon Oxide Dust Test on Two Types of Dust Respirators (Same Type Filter)

		Summary of tests on groups of filters and respirators					
		% Penetration		16 l/m		32 l/m	
Type test, filter, or respirator	Filter no.	16 l/m	32 l/m	Range	Average	Range	Average
NaCl							
AO R-30 Filter	1A	1.9	4.0	1.5–2.5	1.9	3.3–4.9	4.0
	1B	1.7	3.7				
	2A	1.8	4.1				
	2B	1.8	3.9				
	3A	1.9	4.1				
	3B	1.8	3.8				
	4A	1.7	3.7				
	4B	2.5	4.9				
	5A	2.3	4.9				
	5B	1.5	3.3				
SiO_2							
AO R-3030			0.71			0.56–0.75	0.67[a]
respirator with			0.83				
single R-30			0.65				
filter			0.37				
AO R-6030			0.87				
respirator with			0.56				
two R-30 filters			0.29	0.29–0.53	0.46[a]		
			0.53				
			0.41				
			0.29				
			0.76				
			0.47				

[a] 90-min test. Single filter respirator equivalent to 32 l/m; dual filter respirator equivalent to 16 l/m, because both respirators were tested at a flow of 32 l/m.

From Reference 1.

NaCl loading, which acts as a filters aid. The penetration for the dust and mist filters eventually reduced to 1% but at different loading level. The AO filters were tested on the same lot; however, it showed a wide variation in filter loading. A test was

Table 4 Comparison of NaCl Aerosol Filter Penetration Test Results with Quantitative NaCl Tests on Dust Respirators (Resin-Impregnated Wool Felt Filters)

		Average NaCl penetration (%)		
		Filter only[c]		
Respirator	Filter	16 l/m	32 l/m	Man tests[d]
AO 3030[a]	R-30	2.6	5.3	5.3–7.0
AO 6030[b]	R-30	2.6	5.3	2.0–2.5
MSA Dustfoe 66[a]	66	1.2	2.7	1.9–3.2
MSA Type F[b]	Type F	6.5	11.5	5.2–7.0
Welsh 7506[b]	7500-6	6.1	10.7	4.0–6.2
Willson R 560[a]	R-560	0.1	0.3	0.3–0.4
Willson 1210[b]	R-10	0.4	0.9	0.3–0.4

[a] Single cartridge respirator.

[b] Dual cartridge respirator.

[c] Average NaCl penetrations on six filters.

[d] Range of integrated average penetrations on 2 to 6 test subjects.

From Reference 1.

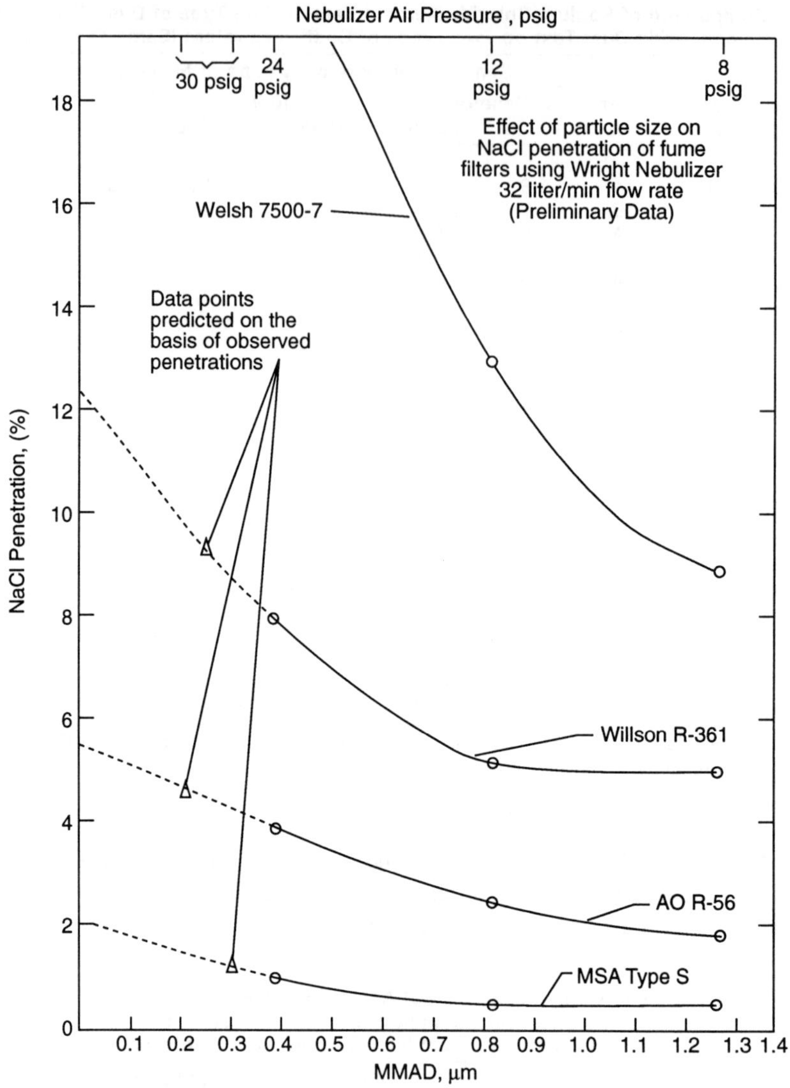

Figure 1 Effect of particle size on NaCl penetration of fume filters using Wright nebulizer (32 l/m flow data). (From Reference 1.)

performed on nondegradable dust and mist filters made by Pulmosan which indicated that this filter had an initial penetration of 22%, but the penetration decreased rapidly until the penetration was only 1% at a dust loading of 58 mg of NaCl.

II. EFFECT OF SODIUM CHLORIDE AEROSOL ON FILTER PERFORMANCE

The first project conducted by the ANSI Z88 ad hoc Subcommittee for Respirator Test and Approval was to determine whether a submicrometer size sodium chloride (NaCl) aerosol was suitable to evaluate the performance of particulate filters.[2] A

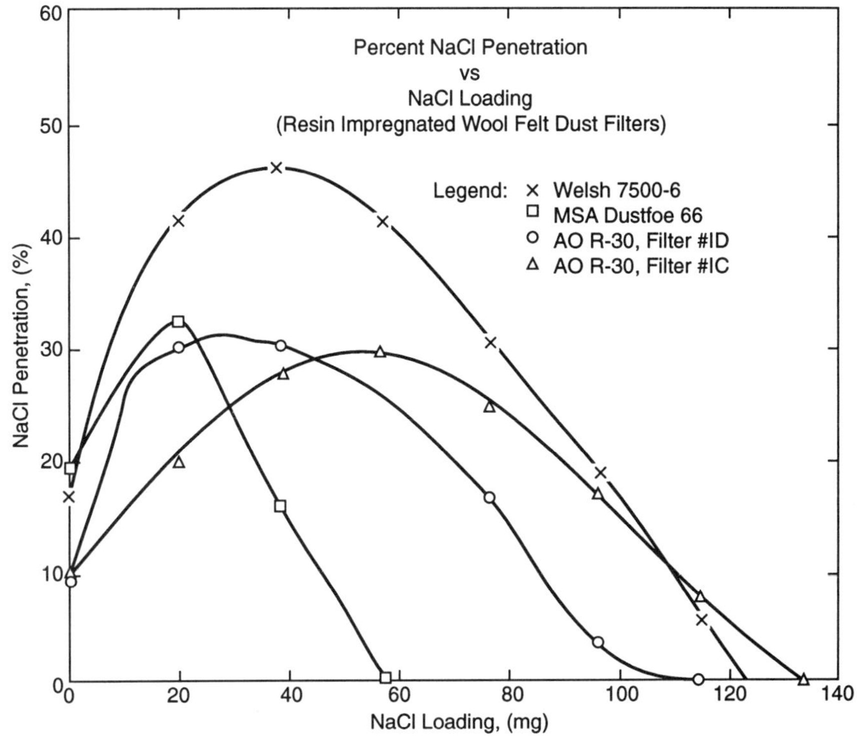

Figure 2 Percent NaCl penetration versus NaCl loading (resin-impregnated wool felt dust filters). (From Reference 1.)

monodisperse sodium chloride aerosol having a particle size of about 0.12 μm was selected for testing. Apparatus for generating this aerosol and for measuring the effectiveness of particulate-filtering devices for removing the particles of this aerosol from air have been developed by the Japanese Institute of Industrial Health. This type of instrument was available to the Subcommittee for performing filter evaluations.

In order to evaluate the performance of the monodisperse sodium chloride filter testing system, tests were carried out on four different types of respirator filter materials. For comparison purposes, tests were also conducted to determine the performance of the these fibrous materials using the testing aerosols prescribed in the 30 CFR 11 for approving particulate filters. These testing conditions are described below.

A. Filters

Four different types of fibrous respirator filter materials were used. Ten specimens of each type of filter material were selected for testing. Each filter had an effective surface area of 100 cm^2.

B. Testing Aerosols and Testing Concentrations

1. A polydisperse silica dust for testing dust and mist filters at a testing concentration of 55 mg/m^3.

Table 5 Comparison Test of Filter Penetration between Sodium Chloride Aerosol and Other Testing Aerosol Listed in 30 CFR 11

Aerosol type	Filter no.	Efficiency (%) Min.	Max.	Avg.	Std. dev.	Press drop (mm H_2O) Min.	Max.	Avg.	Std. dev.
DEHP mist	1	93.60	96.50	92.40	0.75	9.01	9.28	9.10	0.16
	2	53.60	59.20	55.80	2.06	3.42	3.62	3.53	0.05
	3	70.50	73.70	70.46	1.73	4.40	4.60	4.52	0.06
	4	99.993	99.999	99.983	0.006	32.80	31.60	0.73	
Pb fume	1	99.45	99.96	99.63	0.15	8.96	124.50		
	2	94.34	99.43	99.10	0.20	3.49	87.10		
	3	98.28	99.56	99.33	0.53	4.31	68.40		
	4	99.78	99.99	99.87	0.08	30.20	213.50		
SiO_2 dust	1	99.82	99.94	99.88	0.05	8.98	35.30		
	2	98.08	99.29	99.28	0.25	3.47	22.90		
	3	99.39	99.65	99.51	0.08	4.40	17.40		
	4	99.88	99.99	99.94	0.04	30.10	60.10		
NaCl	1	93.42	94.40	93.90	0.33	8.70	9.60	9.20	0.31
		>99.99	>99.99	>99.99		88.00	109.00	96.79	69.30
	2	57.37	60.78	58.95	0.91	3.32	3.63	3.53	0.08
		>99.99	>99.99	>99.99		45.40	62.30	50.40	3.67
	3	69.65	74.41	72.00	71.97	4.32	4.61	4.43	0.09
		>99.99	>99.99	>99.99		44.80	52.00	47.80	2.91
	4	99.94	>99.99	99.96	0.02	30.60	32.40	31.50	0.61
		>99.99	>99.99	>99.99		>135	>135		

Note: Challenge aerosol concentration: DEHP mist — 100 mg/m³; Pb fume — 16 mg/m³; SiO_2 dust — 55 mg/m³; NaCl aerosol — 30 mg/m³. For NaCl test, the value on second row is penetration at the end of the test period.

From Reference 2.

2. A polydisperse lead oxide fume for testing fume filters at a testing concentration of 15 mg/m³.
3. A monodisperse di-2-ethylhexyl phthalate (DEHP) aerosol for testing high-efficiency particulate air (HEPA) filters at a concentration of 100 mg/m³.
4. A monodisperse sodium chloride aerosol at a testing concentration of 30 mg/m³.

C. Flow Rate (Nonpulsating)

1. Thirty two l/m for silica dust, lead fume, and sodium chloride.
2. Forty two l/m for DEHP aerosol.

D. Sampling Time

1. Five to ten seconds for DEHP.
2. Ninety minutes for other test aerosols.

E. Penetration Measurement

1. Instantaneous for DEHP.
2. Continuous for sodium chloride.
3. Integrated for silica dust and lead fume.

The test results of means, ranges, and standard deviations of filter penetration and resistance measurements are presented in Table 5. The summary of the filter test results as a function of time are shown in Table 6.

The monodisperse sodium chloride aerosol testing system has certain advantages over the polydisperse silica dust and lead fume test aerosols listed in 30 CFR 11. The

Table 6 Summary of Filter Test Results

	Time, min								
	0	1	15	30	45	60	75	90	Integrated
Filter #1									
NaCl		93.95	98.78	99.08	>99.99	>99.99	>99.99	>99.99	
DEHP	92.35								
SiO_2									99.88
PbO_2									99.62
Filter #2									
NaCl		58.95	78.94	94.35	98.82	99.97	99.99	>99.99	
DEHP	55.7								
SiO_2									99.28
PbO_2									99
Filter #3									
NaCl		71.97	90.4	98.63	99.74	99.94	99.99	>99.99	
DEHP	70.46								
SiO_2									99.51
PbO_2									94.33
Filter #4									
NaCl		>99.99	>99.99	>99.99	>99.99	>99.99	>99.99	>99.99	
DEHP	99.993								
SiO_2									99.94
PbO_2									99.87

From Reference 2.

NaCl aerosol test system permits the measurement of variations in the aerosol filter efficiency throughout the test period whereas the silica dust and the lead fume test system allows only an integrated measurement of penetration over the test period. For example, the mean aerosol filter efficiency for fibrous filtering material No. 2 varies from 58.95% at the start of a 90-min test to 99.99% at the end of the test period, when the NaCl aerosol is used. For the same filter material, the integrated efficiency is 99.28% for the silica dust at the end of a 90-min test period, and the integrated filer efficiency for the lead fume is 99.10% at the end of a 90 min test period.

The results of the silica dust and the lead fume tests indicate that the No. 2 filter material is highly efficient for removing solid aerosol particles from air; but the results of the NaCl filter penetration test indicate that the No. 2 filter material is not very efficient in removing solid aerosol particles from air at the beginning of the test, and only becomes quite efficient in removing aerosols from air after it has retained a certain amount of solid aerosol particles and has become partially plugged by this retained particulate matter. If this respirator filter material is used in a workplace for protection against highly toxic aerosols with a very low permissible exposure limit (PEL) such as lead or cadmium, the respirator wearer will be constantly exposed to the toxic material because the low PEL for such highly toxic dust prevents the increase of the filter efficiency by plugging of the filter material.

The test results also show that the values of "initial" aerosol particle penetrations of the fibrous materials are almost identical for the monodisperse solid sodium chloride aerosol and the monodisperse liquid DEHP aerosol. Since the NaCl test aerosol is essentially monodisperse and in the respirable size range which is most difficult for fibrous particulate-filtering materials to remove from air, the NaCl aerosol test system would permit the variation of aerosol particle filtering efficiency of a particulate-filtering respirator to be measured from the beginning to the end of a test period.

The ANSI Z88 ad hoc Subcommittee for Respirator Test and Approval agreed that the monodisperse solid sodium chloride aerosol test system met the objectives for a solid aerosol test system for use in evaluating the performance of particulate filtering nonpowered air-purifying respirators designed to protect workers against inhalation of airborne solid particulate matter such as dust particles and fumes.

The subcommittee also conducted investigations to determine the performance of various types of approved nonpowered air-purifying respirator filters.[3] The purpose of this investigation would be to obtain test data to compare the performance of the respirator filters when tested using the previously mentioned monodisperse solid sodium chloride aerosol with the performance of these respirators when tested using various polydisperse solid aerosols listed in 30 CFR 11. Respirators approved by MSHA/NIOSH for protection of persons against inhalation of dusts and mists having permissible exposure limits not less than 0.05 mg/m^3 and 2 million particles per cubic foot (mppcf) would be tested using both a 90-min monodisperse sodium chloride aerosol test and a 90-min polydisperse silica dust aerosol test.

Respirators approved for protection against inhalation of dusts, fumes, and mists having a permissible exposure limit less than 0.05 mg/m^3 and 2 mppcf (generally called "high-efficiency" particulate-filtering respirators) would be tested using both a 90-min monodisperse sodium chloride aerosol test and a 90-min polydisperse silica dust aerosol test. Respirators approved for protection against inhalation of dusts, fumes, and mists having a permissible exposure limit not less than 0.05 mg/m^3 and 2 mppcf would be tested using both a 90-min monodisperse sodium chloride aerosol test and a 312-min polydisperse lead oxide fume aerosol test. Respirators approved for protection against inhalation of organic vapors and sprays of paints, enamels, and lacquers would be tested using a 90-min monodisperse sodium chloride aerosol test and a 156-min polydisperse cellulose nitrate lacquer spray aerosol test. Respirators approved for protection against inhalation of pesticides would be tested using a 90-min monodisperse sodium chloride aerosol test, a 90-min polydisperse silica dust aerosol test, and a 90-min polydisperse lead oxide fume aerosol test. It was decided that three different makes and models of respirators of each mentioned type would be tested. Also, it was decided that three specimens of each selected make and model respirator would be tested.

All of the respirators used in the test program are nonpowered air-purifying respirators. These respirators had been approved by MSHA and NIOSH under the provisions of 30 CFR 11. Brief descriptions of these respirators are given in Table 7.

Each specimen respirator was sealed to a form which was mounted inside a test chamber containing the aerosol test atmosphere. During an aerosol test, the test atmosphere was passed into a specimen respirator at a nonpulsating volumetric flow of 32 l/m. Just prior to performing the aerosol test, and just after completion of the aerosol test, the resistance offered by the specimen respirator to a nonpulsating volumetric air flow of 85 l/m was measured. In the case of the monodisperse sodium chloride aerosol test, however, the resistance offered by the specimen respirator to the nonpulsating volumetric flow of the test atmosphere into the respirator of 32 l/m was measured at 15-min time intervals throughout the 90-min aerosol test period. Later, these resistance values were converted into resistance values for a nonpulsating volumetric air flow of 85 l/m. The 32 l/m flow rate for passing the test atmosphere into the specimen respirator during an aerosol test, and the 85 l/m flow rate for measuring the resistance offered to air flow by the specimen respirator, were used

Table 7 Descriptions of Respirators Selected for Testing

Type of respirator	Identification	Description
Dust/Mist	A-l	Quarter-mask facepiece, single replaceable particulate-filtering element having small area
Dust/Mist	A-2	Half-mask disposable item, single nonreplaceable particulate-filtering element having large area
Dust/Mist	A-3	Quarter-mask facepiece, single replaceable particulate-filtering element having medium area
High-efficiency dust/mist/fume	B-l	Half-mask facepiece, pair of replaceable particulate-filtering cartridges
High-efficiency dust/mist/fume	B-2	Half-mask facepiece, pair of replaceable particulate-filtering cartridges
High-efficiency dust/mist/fume	B-3	Half-mask facepiece, pair of replaceable particulate-filtering cartridges
Pesticide	C-l	Half-mask facepiece, pair of replaceable combination particulate filtering and vapor-removing cartridges
Pesticide	C-2	Half-mask facepiece, single replaceable particulate-filtering element attached by clip to a single vapor-removing cartridge
Pesticide	C-3	Half-mask facepiece, pair of replaceable combination high-efficiency particulate-filtering and vapor-removing cartridges
Non-high-efficiency dust/mist/fume	D-l	Half-mask facepiece, pair of replaceable particulate-filtering cartridges
Non-high-efficiency dust/mist/fume	D-2	Half-mask disposable item, single nonreplaceable particulate filtering element
Non-high-efficiency dust/mist/fume	D-3	Half-mask facepiece, pair of replaceable particulate-filtering cartridges
Paint spray	E-l	Half-mask facepiece, single replaceable particulate filtering element attached by clip to single vapor removing cartridge
Paint spray	E-2	Half-mask facepiece, pair of replaceable particulate filtering elements attached by clips to pair of replaceable vapor-removing cartridges
Paint spray	E-3	Half-mask facepiece, pair of replaceable combination particulate filtering and vapor removing cartridges
Paint spray	E-4	Half-mask facepiece, pair of replaceable combination high-efficiency particulate filtering and vapor-removing cartridges

From Reference 3.

because these are the low rates now listed in 30 CFR 11 for evaluating the performance of nonpowered particulate-filtering respirators.

Three specimens of each make and model respirator used in the test program were subjected to the monodisperse sodium chloride aerosol test. The monodisperse sodium chloride aerosol tests were performed with the use of aerosol generating and aerosol measuring apparatus made by the Sibata Chemical Appliance Manufacturing Company, Limited, of Tokyo, Japan. The test atmosphere consisted of a suspension of monodisperse sodium chloride aerosol particles in air. The sodium chloride particles in the test atmosphere have a mass median aerodynamic diameter (MMAD) of 0.12 μm and a standard deviation (SD) of 1.4. The concentration of sodium chloride aerosol particles in the test atmosphere was 30 mg particulate matter per cubic meter of air. The apparatus records instantaneous values of the efficiency of the specimen respirator in removing the aerosol particles from the test atmosphere passing into the specimen respirator. These instantaneous aerosol particle filtering efficiencies were recorded at 5 min intervals during the test period. An arbitrary test period of 90 min was selected for use in the respirator test program.

Three specimens of each make and model of the dust and mist respirators, three specimens of each make and model of the high-efficiency dust/mist/fume respirators, and three specimens of each make and model of the pesticide respirators were subjected to the polydisperse silica dust aerosol test. The procedures for performing this test are described in detail in Section 11.140-4 of Subpart K of 30 CFR 11. The test atmosphere consists of a suspension of polydisperse silica dust particles in air. The silica dust particles in the test atmosphere have a geometric mean size of 0.4 to 0.6 μm (by optical microscopy) with a standard geometric deviation not exceeding 2. The concentration of silica dust particles in the test atmosphere is 50 to 60 mg particulate matter per cubic meter of air. The test period is 90 min. During a test, the silica dust particulate matter which penetrates the specimen respirator is collected and measured by gravimetric means in units of milligrams. According to Section 11.140-4 of Subpart K of 30 CFR 11, a nonpowered air-purifying respirator approved by MSHA and NIOSH should not permit more then 1.5 mg of the silica dust particulate matter to penetrate.

According to Section 11.140-9 of Subpart K of 30 CFR 11, the initial inhalation resistance to air flow and the final inhalation resistance to air flow offered by a nonpowered dust/mist respirator approved by NIOSH and MSHA, and offered by a nonpowered high-efficiency dust/mist/fume respirator approved by NIOSH and MSHA, should not exceed 30.0 mm water column and 50.0 mm water column, respectively. According to Section 11.183-1 of Subpart M of 30 CFR 11, the initial inhalation resistance to air flow and the final inhalation resistance to air flow offered by a nonpowered pesticide respirator approved by NIOSH and MSHA should not exceed 50.0 mm water column and 70.0 mm water column, respectively.

Three specimens of each make and model of the pesticide respirators were subjected to a polydisperse lead oxide fume aerosol test having a test period of 90 min. The procedures for carrying out this test are described in detail in Section 11.183-5 of Subpart M of 30 CFR 11. Three specimens of each make and model of the nonhigh-efficiency dust/mist/fume respirators were subjected to a polydisperse lead oxide fume aerosol test having a test period of 312 min.

The procedures for performing this test are described in detail in Section 11.140-6 of Subpart K of 30 CFR 11. The test atmosphere consists of a suspension of polydisperse lead oxide fume particles in air. The lead oxide fume aerosol is generated by impinging an oxygen/natural gas flame on the surface of molten lead. The concentration of lead oxide fume particles in the test atmosphere is 15 to 20 mg elemental lead (Pb) particulate matter per cubic meter of air. During a test, the lead oxide fume particulate matter that penetrates the specimen respirator is collected and then is measured by chemical analysis in units of milligrams of elemental lead (Pb). According to Section 11.183-5 of Subpart M of 30 CFR 11, a nonpowered pesticide respirator approved by NIOSH and MSHA, which has been subjected to the 90-min polydisperse lead oxide fume aerosol test, should not permit more than 0.43 mg lead oxide fume particulate matter chemically analyzed as elemental lead (Pb) to penetrate.

According to Section 11.140-6 of Subpart K of 30 CFR 11, a nonpowered nonhigh-efficiency dust/mist/fume respirator approved by NIOSH and MSHA, which has been subjected to the 312-min polydisperse lead oxide fume aerosol test, should not permit more than 1.5 mg lead oxide fume particulate matter chemically analyzed as elemental lead (Pb) to penetrate. According to Section 11.183-1 of Subpart M of 30 CFR 11, the initial inhalation resistance to air flow and the final inhalation resistance to air flow offered by a nonpowered pesticide respirator approved by NIOSH and MSHA should not exceed 50.0 mm water column and 70.0 mm water column, respectively. According to Section 11.140-9 of Subpart K of 30 CFR 11, the initial inhalation resistance to air flow and the final inhalation resistance to air flow offered by a nonpowered nonhigh-efficiency dust/mist/fume respirator approved by NIOSH and MSHA should not exceed 30.0 mm water column and 50.0 mm water column, respectively.

Three specimens of each make and model of the paint spray respirators were subjected to the polydisperse cellulose nitrate lacquer spray aerosol test. The procedures for performing this test are described in detail in Section 11.162-5 of Subpart L of 30 CFR 11. The test atmosphere consists of a suspension of polydisperse cellulose nitrate lacquer spray particles in air. This test atmosphere is created by atomizing a mixture of cellulose nitrate lacquer and lacquer thinner. The concentration of the cellulose nitrate lacquer spray particles in the test atmosphere was 95 to 125 mg particulate matter per cubic meter of air. The test period is 156 min. During a test, the cellulose nitrate lacquer spray particulate matter which penetrates the specimen respirator is collected and then measured by gravimetric means in milligrams.

According to Section 11.162-5 of Subpart L of 30 CFR 11, a nonpowered paint spray respirator approved by NIOSH and MSHA should not permit more than 5.0 mg of the cellulose nitrate lacquer spray particles to penetrate. According to Section 11.162-1 of Subpart L of 30 CFR 11, the initial inhalation resistance to air flow and the final inhalation resistance to air flow offered by a nonpowered paint spray respirator approved by NIOSH and MSHA should not exceed 50.0 mm water column and 70.0 mm water column, respectively.

The results of the aerosol tests and the resistance to air flow measurements for the various types of nonpowered air-purifying respirators are presented in Tables 8 through Table 12. All tests were performed with a monodisperse NaCl aerosol (0.12 μm mmAD) at a flow rate of 32 l/m.

Table 8 contains the test data for the nonpowered dust/mist particulate filtering respirators.

Table 9 presents the test data for the nonpowered high-efficiency dust/mist/fume particulate-filtering respirators.

Tables 10(a) and (b) present the test data for the nonpowered pesticide particulate-filtering and vapor-removing respirators.

Table 11 provides the test data for the nonpowered dust/mist/fume particulate filtering respirators.

Table 12 provides the test data for the nonpowered paint spray particulate filtering and vapor-removing respirators.

In addition, the results of the monodisperse solid sodium chloride aerosol tests for the various types of nonpowered air-purifying respirators are presented in Figures 3 through 11. The filter resistance was measured at an air flow rate of 85 l/m.

Figures 3 through 7 illustrate the variations in efficiency for the various types of nonpowered air-purifying respirators for removing the respirable size monodisperse

Table 8 Summary and Comparison of Test Results for Dust/Mist Respirators Using Monodisperse Sodium Chloride Aerosol and Polydisperse Silica Dust Aerosol

Aerosol	Resp. ident.	Resp. type	Ri initial res. (mm H_2O)	R_f final res. (mm H_2O)	ΔR R_f–Ri (mm H_2O)	Mi Particle load into resp. (mg)	Ei Initial resp. effic. (%)	E_1 Lowest resp. effic. (%)	E_f Final resp. effic. (%)	Em Mean resp. effic. (%)	M_p Particle load penetrating resp. (mg)	M_r Particle load penetrating resp. (mg)	$ΔR/M_r$
Monodisperse sodium chloride	A-1	Dust/mist	13.0	27.1	14.1	86.4	95.4	65.9	69.3	73.8	22.6	63.8	0.22
			13.6	31.6	18.0	86.4	96.8	70.5	73.7	77.0	19.9	66.5	0.27
			12.5	26.0	13.5	86.4	94.6	58.0	63.7	67.2	28.3	58.1	0.23
										72.7			0.24
										mean			mean
	A-2	Dust/mist	6.4	8.0	1.6	86.4	98.9	87.8	87.8	94.1	5.1	81.3	0.020
			6.1	7.5	1.4	86.4	98.8	87.1	87.1	93.7	5.4	81.0	0.017
			6.7	8.3	1.6	86.4	99.0	89.1	89.1	94.6	4.7	81.7	0.20
										94.1			0.019
										mean			mean
	A-3	Dust/mist	19.4	48.4	29.0	86.4	84.7	84.7	99.3	94.9	4.4	82.0	0.35
			19.7	47.9	28.2	86.4	84.1	84.1	95.6	92.0	6.9	79.5	0.35
			21.8	53.5	31.7	86.4	86.7	86.7	99.9	97.1	2.5	83.9	0.38
										94.7			0.36
										mean			mean
Polydisperse silica dust	A-1	Dust/mist	16	25	9	161.9				99.7	0.43	161.5	0.057
			16	26	10	146.6				99.5	0.90	145.7	0.069
			16	25	9	146.6				99.5	0.69	145.9	0.062
										99.5			0.063
										mean			mean
	A-2	Dust/mist	9	11	2	161.0				99.7	0.50	160.5	0.012
			8	10	2	156.1				99.2	1.20	154.9	0.013
			10	12	2	146.6				99.7	0.52	146.1	0.014
										99.5			0.013
										mean			mean
	A-3	Dust/mist	21	30	9	149.9				99.3	1.00	148.9	0.060
			20	30	10	162.4				99.6	0.59	161.8	0.062
			24	40	16	152.4				99.7	0.48	151.9	0.105
										99.5			0.076
										mean			mean

From Reference 3.

Table 9 Summary and Comparison of Tests Results for High-Efficiency Dust/Mist/Fume Respirators Tested Using Monidisperse Sodium Chloride Aerosol and Polydisperse Silica Dust Aerosol

Aerosol	Resp. ident.	Resp. type	Ri initial res. (mm H_2O)	R_f final res. (mm H_2O)	ΔR R_f–Ri (mm H_2O)	Mi Particle load into resp. (mg)	Ei Initial resp. effic. (%)	E_1 Lowest resp. effic. (%)	E_f Final resp. effic. (%)	Em Mean resp. effic. (%)	M_p Particle load penetrating resp. (mg)	M_r Particle load penetrating resp. (mg)	$ΔR/M_r$
Monodisperse sodium chloride	B-1	High-effic. dust/mist/fume	25.3	29.0	3.7	86.4	99.97	99.97	100.0	100.0	0.0	86.4	0.043
			26.6	30.6	4.0	86.4	100.0	100.0	100.0	100.0	0.0	86.4	0.046
			25.0	29.3	4.3	86.4	100.0	100.0	100.0	100.0	0.0	86.4	0.050
										100.0 (mean)			0.046 (mean)
	B-2	High-effic. dust/mist/fume	21.3	25.0	3.7	86.4	99.99	99.99	100.0	100.0	0.0	86.4	0.043
			23.4	26.6	3.2	86.4	100.0	100.0	100.0	100.0	0.0	86.4	0.037
			23.4	26.6	3.2	86.4	99.98	99.98	100.0	100.0	0.0	86.4	0.037
										100.0 (mean)			0.039 (mean)
	B-3	High-effic. dust/mist/fume	17.6	19.2	1.6	86.4	99.99	99.99	100.0	100.0	0.0	86.4	0.019
			17.3	18.9	1.6	86.4	100.0	100.0	100.0	100.0	0.0	86.4	0.019
			17.6	19.4	1.8	86.4	99.99	99.99	100.0	100.0	0.0	86.4	0.021
										100.0 (mean)			0.020 (mean)
Polydisperse silica dust	B-1	High-effic. dust/mist/fume	23	30	7	161.0				99.6	0.72	160.3	0.044
			22	25	3	152.1				99.7	0.60	151.3	0.020
			22	26	4	156.1				99.5	0.81	155.3	0.026
										99.6 (mean)			0.030 (mean)
	B-2	High-effic. dust/mist/fume	24	26	2	161.9				99.3	1.17	160.7	0.012
			24	26	2	161.0				99.5	0.89	160.1	0.012
			24	26	2	162.4				99.3	1.13	161.3	0.012
										99.4 (mean)			0.012 (mean)
	B-3	High-effic. dust/mist/fume	16	18	2	149.8				99.5	0.70	149.1	0.013
			16	18	2	162.4				99.9	0.10	162.3	0.012
			16	18	2	152.4				99.2	1.22	151.2	0.012
										99.5 (mean)			0.013 (mean)

From Reference 3.

Table 10(a) Summary and Comparison of Tests Results for Pesticide Particulate-Filtering/Vapor-Removing Respirators Using Monodisperse Sodium Chloride Aerosol and Polydisperse Silica Dust Aerosol

Aerosol	Resp. ident.	Resp. type	Ri initial res. (mm H_2O)	R_f final res. (mm H_2O)	ΔR R_f–Ri (mm H_2O)	Mi Particle load into resp. (mg)	Ei Initial resp. effic. (%)	E_1 Lowest resp. effic. (%)	E_f Final resp. effic. (%)	Em Mean resp. effic. (%)	M_p Particle load penetrating resp. (mg)	M_r Particle load penetrating resp. (mg)	$ΔR/M_r$
Monodisperse sodium chloride	C-1	Pesticide	27.4	33.8	6.4	86.4	96.9	96.9	99.7	98.9	1.0	85.4	0.075
			28.2	34.6	6.4	86.4	97.0	97.0	98.8	98.3	1.5	84.9	0.075
			26.6	35.4	8.8	86.4	97.2	97.2	99.9	99.4	0.5	85.9	0.102
										98.9 mean			0.084 mean
	C-2	Pesticide	41.0	87.3	46.3	86.4	99.6	99.2	99.7	99.4	0.5	85.9	0.54
			37.5	70.8	33.3	86.4	99.1	98.9	99.8	99.3	0.6	85.8	0.39
			38.8	76.1	37.3	86.4	98.9	98.7	99.6	99.1	0.8	85.6	0.44
										99.3 mean			0.46 mean
	C-3	Pesticide	34.6	37.5	2.9	86.4	100.0	100.0	100.0	100.0	0.0	86.4	0.034
			32.5	36.7	4.2	86.4	99.99	99.99	100.0	100.0	0.0	86.4	0.049
			34.1	36.7	2.6	86.4	100.0	100.0	100.0	100.0	0.0	86.4	0.030
										100.0 mean			0.038 mean
Polydisperse silica dust	C-1	Pesticide	28	32	4	146.6				99.4	0.9	145.7	0.027
			27	29	2	124.4				99.6	0.5	123.9	0.016
			25	30	5	146.3				99.8	0.3	146.0	0.034
										99.6 mean			0.026 mean
	C-2	Pesticide	44	52	8	156.1				99.6	0.6	155.5	0.051
			38	48	10	161.9				98.6	2.3	159.6	0.063
			42	49	7	152.1				98.9	1.7	150.4	0.047
										99.0 mean			0.054 mean
	C-3	Pesticide	33	35	2	146.6				99.5	0.7	145.9	0.014
			35	37	2	124.4				99.0	1.2	123.2	0.016
			36	38	2	146.3				99.4	0.8	145.5	0.014
										99.3 mean			0.015 mean

Table 10(b) Summary and Comparison of Test Results for Pesticide Particulate-Filtering/Vapor-Removing Respirators Using Monodisperse Sodium Chloride Aerosol and Polydisperse Lead Oxide Fume Aerosol

Aerosol	Resp. ident.	Resp. type	Ri initial res. (mm H_2O)	R_f final res. (mm H_2O)	Mi ΔR R_f–Ri (mm H_2O)	Ei Particle load into resp. (mg)	E_1 Initial resp. effic. (%)	E_f Lowest resp. effic. (%)	Em Final resp. effic. (%)	M_p Mean resp. effic. (%)	M_r Particle load penetrating resp. (mg)	Particle load penetrating resp. (mg)	$\Delta R/M_r$
Monodisperse sodium chloride	C-1	Pesticide	27.4	33.8	6.4	86.4	96.9	96.9	99.7	98.9	1.0	85.4	0.075
			28.2	34.6	6.4	86.4	97.0	97.0	98.8	98.3	1.5	84.9	0.075
			26.6	35.4	8.8	86.4	97.2	97.2	99.9	99.4	0.5	85.9	0.102
										98.9			0.084
										mean			mean
	C-2	Pesticide	41.0	87.3	46.3	86.4	99.6	99.2	99.7	99.4	0.5	85.9	0.54
			37.5	70.8	33.3	86.4	99.1	98.9	99.8	99.3	0.6	85.8	0.39
			38.8	76.1	37.3	86.4	98.9	98.7	99.6	99.1	0.8	85.6	0.44
										99.3			0.46
										mean			mean
	C-3	Pesticide	34.6	37.5	2.9	86.4	100.0	100.0	100.0	100.0	0.0	86.4	0.034
			32.5	36.7	4.2	86.4	99.99	99.99	100.0	100.0	0.0	86.4	0.049
			34.1	36.7	2.6	86.4	100.0	100.0	100.0	100.0	0.0	86.4	0.030
										100.0			0.038
										mean			mean
Polydisperse lead oxide fume	C-1	Pesticide	28.0	38.0	10.0	47.9				99.9	0.07	47.8	0.21
			31.0	44.0	13.0	48.9				99.9	0.04	48.9	0.27
			28.5	40.0	11.5	47.9				99.9	0.16	47.7	0.24
										99.9			0.24
										mean			mean
	C-2	Pesticide	41.5	59.0	17.5	49.8				99.8	0.09	49.7	0.35
			43.0	62.0	19.0	48.9				97.3	1.3	47.6	0.40
			40.0	57.0	17.0	47.9				89.1	0.42	47.5	0.36
										98.7			0.37
										mean			mean
	C-3	Pesticide	39.0	41.0	2.0	49.8				99.94	0.03	49.8	0.040
			36.5	40.5	4.0	48.9				99.96	0.02	48.9	0.082
			34.5	42.0	7.5	47.9				99.96	0.02	47.9	0.157
										99.95			0.093
										mean			mean

From Reference 3.

Table 11 Summary and Comparison of Test Results for Non-High-Efficiency Dust/Mist/Fume Particulate-Filtering Respirators Using Monodisperse Sodium Chloride Aerosol and Polydisperse Lead Oxide Fume Aerosol

Aerosol	Resp. ident.	Resp. type	Ri initial res. (mm H_2O)	R_f final res. (mm H_2O)	ΔR R_f–Ri (mm H_2O)	Mi Particle load into resp. (mg)	Ei Initial resp. effic. (%)	E_1 Lowest resp. effic. (%)	E_f Final resp. effic. (%)	Em Mean resp. effic. (%)	M_p Particle load penetrating resp. (mg)	M_r Particle load penetrating resp. (mg)	$\Delta R/M_r$
Monodisperse sodium chloride	D-1	Non-high effic. dust/mist/fume	8.3	10.4	2.1	86.4	98.1	98.1	99.9	99.4	0.5	85.9	0.024
			9.6	11.4	1.8	86.4	99.0	99.0	99.9	99.7	0.3	86.1	0.021
			6.9	8.8	1.9	86.4	98.6	98.6	99.9	99.5	0.4	86.0	0.022
										99.5 mean			0.022 mean
	D-2	Non-high effic. dust/mist/fume	6.9	9.0	2.1	86.4	98.6	94.2	94.2	96.4	3.1	83.3	0.025
			8.8	11.7	2.9	86.4	98.9	97.6	97.6	98.2	1.6	84.8	0.034
			8.5	10.9	2.4	86.4	98.8	97.2	97.2	98.0	1.7	84.7	0.028
										97.5 mean			0.029 mean
	D-3	Non-high effic. dust/mist/fume	20.2	26.1	5.9	86.4	97.8	97.6	98.2	97.9	1.8	84.6	0.070
			19.7	24.5	4.8	86.4	98.5	95.6	96.1	96.0	3.5	82.9	0.058
			19.4	23.9	4.5	86.4	98.2	94.7	95.6	95.2	4.1	82.3	0.055
										96.4 mean			0.061 mean
Polydisperse lead oxide fume	D-1	Non-high effic. dust/mist/fume	10.0	26.5	16.5	182.3				99.97	0.05	182.2	0.091
			10.0	21.0	11.0	176.4				99.98	0.04	176.4	0.062
			11.0	20.0	9.0	178.9				99.99	0.02	178.9	0.055
										99.98 mean			0.068 mean
	D-2	Non-high effic. dust/mist/fume	9.5	29.0	19.5	182.3				99.8	0.35	182.0	0.107
			8.5	26.0	17.5	176.4				99.3	1.30	175.1	0.100
			10.5	27.0	16.5	178.9				99.4	1.00	177.9	0.093
										99.5 mean			0.100 mean
	D-3	Non-high effic. dust/mist/fume	17.5	34.5	17.0	182.3				99.96	0.07	182.2	0.093
			19.0	39.0	20.0	176.4				99.96	0.07	176.3	0.113
			20.5	40.0	19.5	178.9				99.97	0.05	178.9	0.109
										99.96 mean			0.105 mean

From Reference 3.

Table 12 Summary and Comparison of Test Results for Paint Spray Particulate-Filtering/Vapor-Removing Respirators Using Monodisperse Sodium Chloride Aerosol and Polydisperse Cellulose Nitrate Lacquer Spray Aerosol

Aerosol	Resp. ident.	Resp. type	Ri initial res. (mm H_2O)	R_f final res. (mm H_2O)	ΔR R_f–Ri (mm H_2O)	Mi Particle load into resp. (mg)	Ei Initial resp. effic. (%)	E_1 Lowest resp. effic. (%)	E_f Final resp. effic. (%)	Em Mean resp. effic. (%)	M_p Particle load penetrating resp. (mg)	M_r Particle load penetrating resp. (mg)	$\Delta R/M_r$
Polydisperse sodium chloride	E-1	Paint spray	24.5	28.5	4.0	86.4	94.1	71.5	72.3	77.6	19.4	67.0	0.060
			25.5	30.3	4.8	86.4	94.0	72.3	72.9	77.5	19.5	66.9	0.072
			24.2	28.7	4.5	86.4	94.1	70.5	70.9	77.3	19.6	66.8	0.067
										77.5 mean			0.066 mean
	E-2	Paint spray	34.6	40.7	6.1	86.4	99.7	98.1	98.1	98.8	1.0	85.4	0.071
			31.9	37.5	5.6	86.4	99.7	96.2	96.2	97.8	1.9	84.5	0.066
			31.9	35.1	3.2	86.4	96.2	95.1	95.8	95.5	3.9	82.5	0.039
										97.4 mean			0.059 mean
	E-3	Paint spray	26.9	34.3	7.4	86.4	96.7	96.7	98.7	98.0	1.7	84.7	0.087
			27.7	35.1	7.4	86.4	95.8	95.8	97.6	97.2	2.4	84.0	0.088
			26.1	35.6	9.5	86.4	97.4	97.4	99.0	98.5	1.3	85.1	0.112
										97.9 mean			0.096 mean
	E-4	Paint spray (hi-effic filter type)	33.0	36.2	3.2	86.4	99.9	99.9	100.0	100.0	0.0	86.4	0.037
			35.4	38.3	2.9	86.4	99.9	99.9	100.0	100.0	0.0	86.4	0.036
			33.0	36.2	3.2	86.4	99.9	99.9	100.0	100.0	0.0	86.4	0.037
										100.0 mean			0.037 mean
Polydisperse cellulose nitrate lacquer spray	E-1	Paint spray	29.2	41.5	12.3	546.7				98.98	0.1	546.6	0.023
			25.5	48.0	22.5	549.9				99.91	0.5	549.4	0.041
			26.0	42.0	16.0	550.0				98.98	0.1	549.9	0.029
										99.96 mean			0.031 mean
	E-2	Paint spray	38.0	74.2	36.2	565.9				99.98	0.1	565.8	0.064
			32.2	70.0	37.8	552.6				99.91	0.5	552.1	0.068
			31.5	63.0	31.5	545.2				99.98	0.1	545.1	0.058
										99.96 mean			0.063 mean
	E-3	Paint spray	30.0	45.2	15.2	558.0				99.93	0.4	557.6	0.027
			29.5	48.0	18.5	553.4				99.98	0.1	553.3	0.033
			26.5	38.8	12.3	564.7				99.98	0.1	564.6	0.022
										99.96 mean			0.027 mean
	E-4	Paint spray (hi-effic filter type)	36.0	40.5	4.5	552.6				99.95	0.3	552.3	0.0081
			35.0	42.2	7.2	565.1				99.96	0.2	564.9	0.0127
			34.0	39.0	5.0	559.7				99.96	0.2	559.5	0.0090
										99.96 mean			0.0099 mean

From Reference 3.

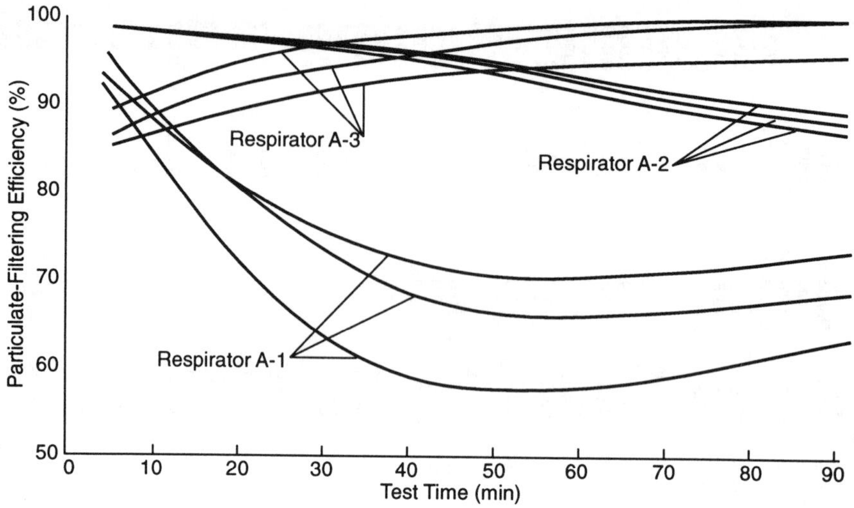

Figure 3 Dust/mist particulate-filtering respirators. (From Reference 3.)

sodium chloride aerosol particles from air. These figures demonstrate that the particulate-filtering efficiency of the respirators varies with test time. There is no figure which illustrates the variation of the particulate-filtering efficiency of the specimens of the nonpowered high-efficiency dust/mist/fume particulate-filtering respirators since all of these specimen respirators were essentially 100% efficient in removing the sodium chloride aerosol particles from air.

Figures 8 through 11 illustrate the variations of the resistance to air flow offered by the various types of nonpowered air-purifying respirators as these respirators became loaded with the respirable size monodisperse solid sodium chloride aerosol particles. These figures illustrate that the resistance to air flow of the respirators varies with test time and with aerosol particle load.

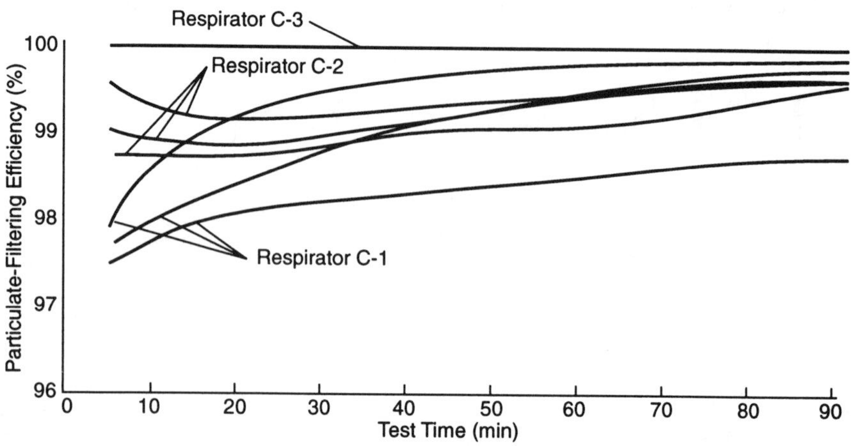

Figure 4 Pesticide particulate-filtering/vapor-removing respirators. (From Reference 3.)

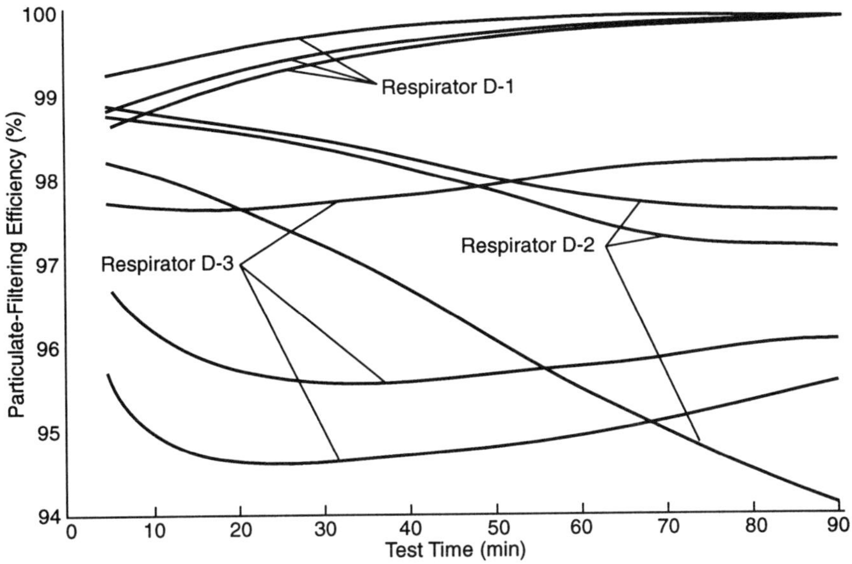

Figure 5 Dust/mist/fume particulate-filtering respirators. (From Reference 3.)

Table 13 contains information which attempts to correlate the respirator particulate-filtering efficiencies and the respirator resistances to air flow, for the tests involving the monodisperse solid sodium chloride aerosol and for the tests involving the various polydisperse solid aerosols currently listed in 30 CFR 11. This table presents values for respirator particulate filtering efficiency ratios and for respirator resistance to air flow ratios. In the case of the respirator particulate-filtering efficiency ratios, if the value is less than 1.0, then the monodisperse sodium chloride aerosol test is more severe than the other polydisperse aerosol tests. In the case of the respirator resistance to air flow ratios, if the value is greater than 1.0, then the

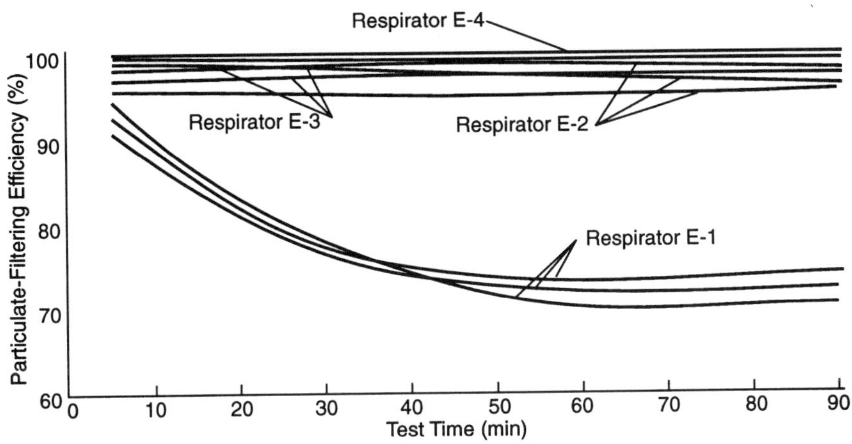

Figure 6 Paint spray particulate-filtering/vapor-removing respirators. (From Reference 3.)

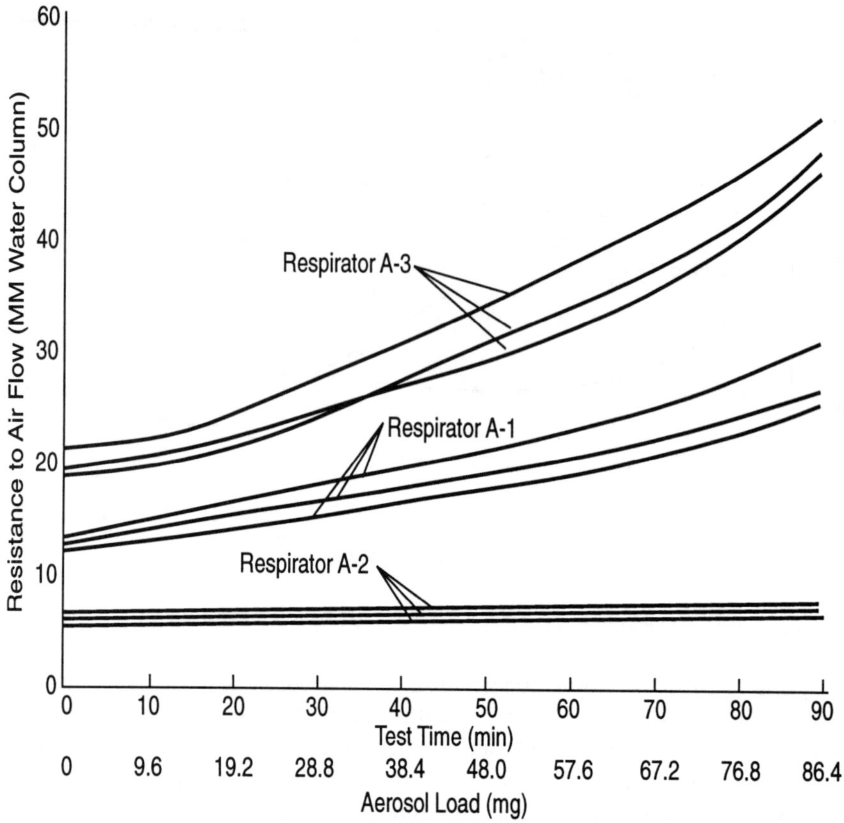

Figure 7 Dust/mist particulate-filtering respirators. (From Reference 3.)

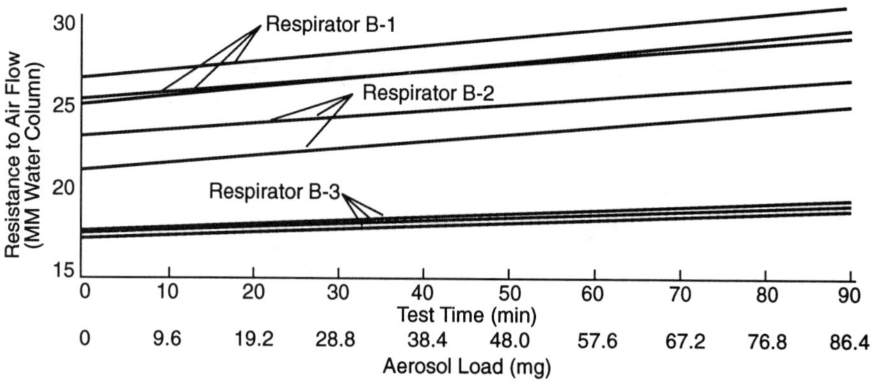

Figure 8 High efficiency dust/mist/fume particulate-filtering respirators. (From Reference 3.)

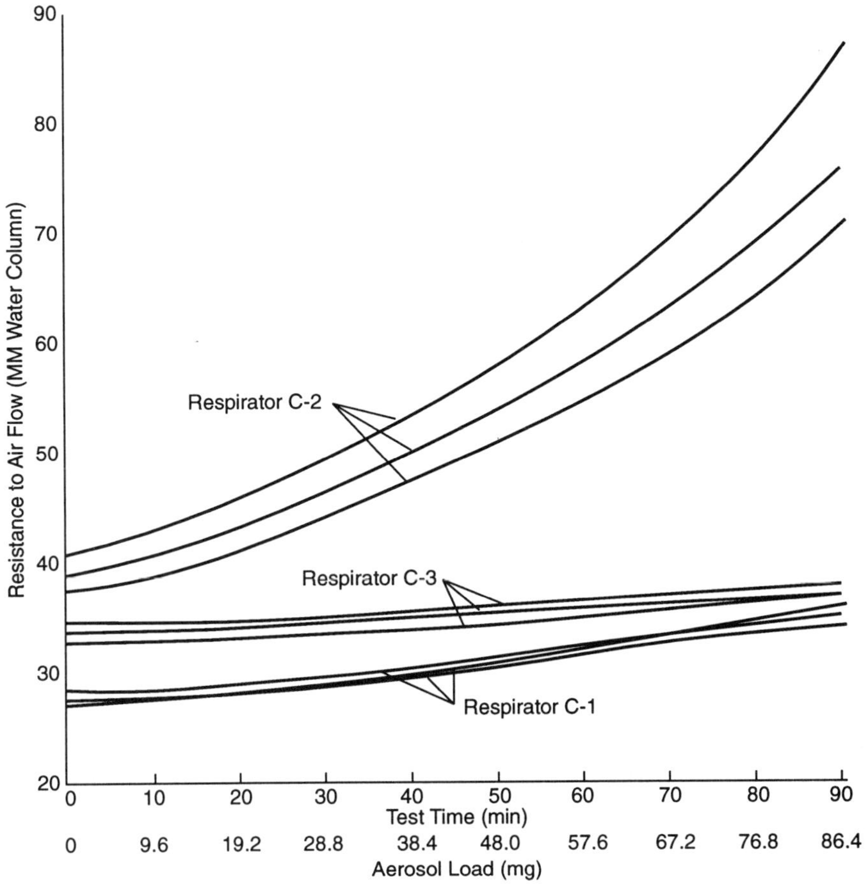

Figure 9 Pesticide particulate-filtering/vapor-removing respirators. (From Reference 3.)

monodisperse sodium chloride aerosol test is more severe than the other polydisperse aerosol tests.

The test results point out the obvious advantage of the monodisperse solid sodium chloride (MNaCl) aerosol test system over the various polydisperse solid aerosol test system currently listed in 30 CFR 11 for evaluating the performance of air-purifying respirators designed for protection against inhalation of solid airborne particulate matter. The MNaCl test system permits the measurement of variation in filter penetration throughout a test period, while the aerosol test systems currently listed in 30 CFR 11 allows only the measurement of the integrated filter penetration. The test data obtained from the various polydisperse aerosol tests carried out in accordance with the provision of 30 CFR 11 show that the overall aerosol particulate-filtering efficiencies for almost all of the specimens of the respirators tested are greater than 99%; but, the test data obtained in the MNaCl aerosol tests show that the instantaneous aerosol particulate-filtering efficiencies of most of the specimens of

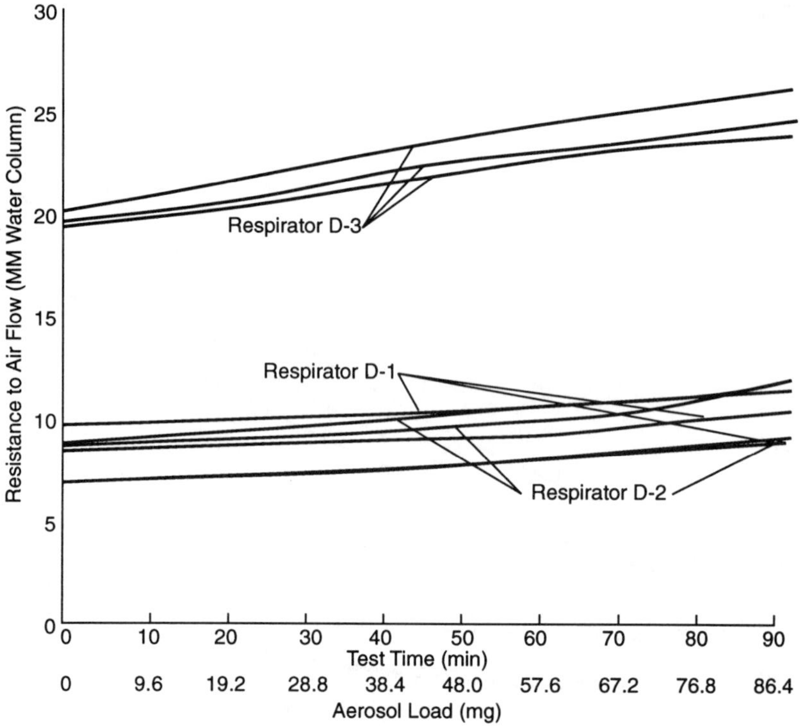

Figure 10 Dust/mist/fume (not high efficiency) particulate-filtering respirators. (From Reference 3.)

the respirators vary during the test period, and some of these variations in instantaneous aerosol particulate-filtering efficiencies are quite great.

These variations in instantaneous aerosol particulate-filtering efficiencies are illustrated in Figures 3 through 6. For example, while the overall aerosol particulate-filtering efficiencies for the A-1 dust/mist particulate-filtering respirator specimens are about 99.5% for the 90-min polydisperse silica dust aerosol tests, the instantaneous aerosol particulate-filter efficiencies of these respirator specimens for the MNaCl aerosol test are initially about 95%. They then gradually decrease to a low range of 58 to 70% followed by a very gradual increase to a range of 64 to 74% by the end of the 90-min test period.

The test data obtained in the monodisperse NaCl aerosol tests indicate that there is a wide variation in the efficiencies of most types of approved nonpowered air-purifying respirators for removing respirable size solid aerosol particles from air. These test data reveal that some respirators become more efficient in removing respirable sized solid aerosol particles from air as they become loaded with retained aerosol particulate matter. On the other hand, some respirators become less efficient in removing respirable sized solid aerosol particles from air as they become loaded with retained aerosol particulate matter. The polydisperse solid aerosol tests performed in accordance with the current provisions of the 30 CFR 11 do not reveal any significant differences in the efficiencies of these approved respirators.

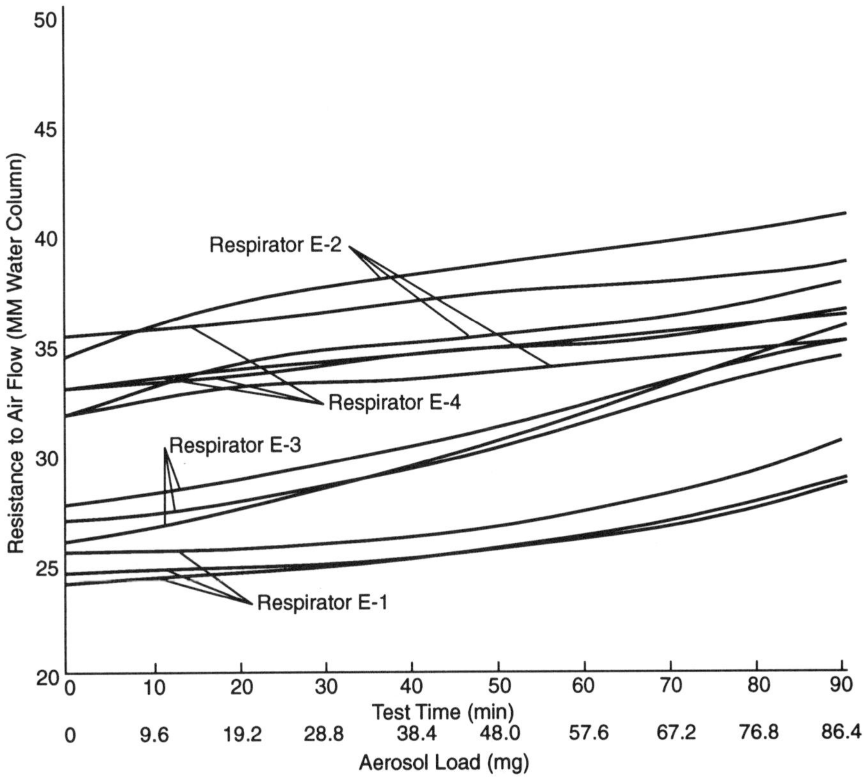

Figure 11 Paint spray particulate-filtering/vapor-removing respirators. (From Reference 3.)

The test data obtained in the MNaCl tests indicate that this aerosol can be used to determine the loading capacity of approved particulate filter respirators. The graphs of Figures 7 through 11 illustrate the variation of the resistance to air flow offered by various nonpowered air-purifying respirators as they become loaded with retained solid NaCl aerosol particles.

The respirator particulate-filtering efficiency ratios listed in Table 13 indicate that the MNaCl aerosol is as severe as the various polydisperse solid aerosols currently listed in 30 CFR 11 for use in evaluating the performance of nonpowered air-purifying respirator filters. However, these ratios may be misleading because they utilize the mean respirator particulate-filtering efficiencies for the MNaCl aerosol tests, which generally are equivalent to the overall respirator particulate-filtering efficiencies for the polydisperse solid aerosol tests. Because the monodisperse solid NaCl test system can measure the instantaneous aerosol particulate-filtering efficiency of respirator which can vary widely during a test period, this aerosol test system is considered to be superior to the aerosol test systems currently listed in 30 CFR 11 for evaluating the performance of nonpowered air-purifying respirator filters.

A review of the respirator resistance to air flow ratios given in Table 13 indicates that the MNaCl aerosol is more severe than the various polydisperse solid aerosol currently listed in 30 CFR 11 for evaluating the aerosol particle loading capacity of

Table 13 Correlations of Respirator Particulate-Filtering Efficiencies and Respirator Resistances for Various Aerosols

Resp. type	Resp. ident.	Efficiency ratio	Resistance ratio
		$NaCl/SiO_2$[a]	**$NaCl/SiO_2$[e]**
Dust/mist	A-1	0.73	3.8
	A-2	0.95	1.5
	A-3	0.95	4.7
Hi-Effic dust/mist/fume	B-1	1.00	1.5
	B-2	1.01	3.3
	B-3	1.01	1.5
Pesticide	C-1	0.99	3.2
	C-2	1.00	8.2
	C-3	1.01	2.5
		NaCl/PbO (90)[b]	**NaCl/PbO (90)[f]**
Pesticide	C-1	0.99	0.35
	C-2	1.01	1.24
	C-3	1.00	0.41
		NaCl/PbO (312)[c]	**NaCl/PbO (312)[g]**
Non-hi-effic dust/mist/fume	D-1	1.00	0.32
	D-2	0.98	0.29
	D-3	0.96	0.58
		NaCl/CNL[d]	**NaCl/CNL[h]**
Paint spray	E-1	0.78	2.1
	E-2	0.97	0.94
	E-3	0.98	3.6
	E-4	1.00	3.7

[a] $NaCl/SiO_2$ = E_m for NaCl aerosol/E_m for SiO_2 dust aerosol.

[b] NaCl/PbO (90) = E_m for NaCl aerosol/E_m for PbO fume aerosol (90-min test).

[c] NaCl/PbO (312) = E_m for NaCl aerosol/E_m for PbO fume aerosol (312-min test).

[d] NaCl/CNL = E_m for NaCl aerosol/E_m for C nitrate lacquer spray aerosol.

[e] $NaCl/SiO_2$ = $\Delta R/M_r$ for NaCl aerosol/$\Delta R/M_r$ for SiO_2 dust aerosol.

[f] NaCl/PbO (90) = $\Delta R/M_r$ for NaCl aerosol/$\Delta R/M_r$ for PbO fume aerosol (90-min test).

[g] NaCl/PbO (312) = $\Delta R/M_r$ for NaCl aerosol/$\Delta R/M_r$ for PbO fume aerosol (312-min test).

[h] NaCl/CNL = $\Delta R/M_r$ for NaCl aerosol/$\Delta R/M_r$ for C nitrate lacquer spray aerosol.

From Reference 3.

nonpowered air-purifying respirator filters, except for the polydisperse solid lead oxide fume aerosol. The resistance to air flow ratios presented in Table 13 indicate that the polydisperse solid lead oxide aerosol is somewhat more severe than the NaCl aerosol in determining the aerosol particle loading capacity of a nonpowered air-purifying respirator filter. Figures 7 through 11 do, however, illustrate that the MNaCl aerosol can be used to evaluate the aerosol particle loading capacity of nonpowered air-purifying respirators for most aerosol applications.

The test results conclude that the monodisperse sodium chloride aerosol test system is considered to meet the requirements for a solid aerosol particle test system for use in evaluating the performance of nonpowered air-purifying respirators designed to protect persons against inhalation of airborne solid particulate matter. This aerosol is essentially monodisperse and has a respirable size which is most difficult for various particulate-filtering materials to remove from air. This test system permits the variation of aerosol particulate-filtering efficiency and the variation of the inhalation resistance of a nonpowered air-purifying respirator to be measured throughout the test period.

III. EFFICIENCY OF APPROVED RESPIRATOR FILTERS AGAINST LEAD AEROSOLS

The OSHA standard for occupational exposure to lead, 29 CFR 1910.1025(f)(2), required the use of high efficiency particulate air (HEPA) filters for nonpowered half-mask negative pressure air-purifying respirators.

On December 28 1978, the Minnesota Mining and Manufacturing (3M) Company, a disposable respirator manufacturer which had no HEPA filter available for sale, petitioned OSHA for reconsidering the prohibition of the use of half-mask negative pressure respirators equipped with dust/mist or fume filters for protection against lead exposure. 3M claimed that there were no indications in the lead standard records that HEPA filters should be required, and also that the dust/mist, or fume filters were inadequate.

An administrative stay on the use of HEPA filters for protection against lead was granted.[4] However, OSHA believed that the dust/mist or fume filters may not provide adequate protection against lead particles which have a very low permissible exposure limit (PEL) of 50 μg/m^3.

Silica dust with a projected diameter of 0.6 μm is used to certify dust/mist filters. The challenge concentration is 55 mg/m^3 and the test duration lasts 90 min. The maximum allowable filter penetration is 1%, which is measured only at the end of the 90 min test. The fume filters are tested at a lower concentration (15 mg/m^3) but at a longer test period (312 min). The overall filter loading and method of penetration measurement is similar to the dust/mist filter test. The high challenge concentration may permit less efficient filters to pass the test. A filter may have a high initial penetration, but, as the test proceeds, the penetration reduces when the filter becomes "plugged" with deposited dust. For an air contaminant with a very low PEL such as lead, the worker may be constantly exposed to lead because the filter may not have an opportunity to become "plugged" and improve its efficiency.

In mid 1979, OSHA requested that NIOSH conduct a study to determine the real time efficiency of approved dust/mist and fume filters.[5] It was agreed that a lead aerosol would be used as the challenge aerosol for filters at concentrations of 1 and 10 mg/m^3. A light scattering photometer was used for penetration determination, because the instrument could measure filter penetration continuously. The lead test aerosol for the dust/mist filter had a mass median aerodynamic diameter (MMAD) of 2.2 μm, which was converted from the projected diameter. The lead test aerosol for the fume filters had a MMAD of 0.6 μm.

A low aerosol concentration (1 mg/m^3) and a high aerosol concentration (10 mg/m^3) were selected for testing. The test times were set at 30 and 90 min. Both types of filters were tested as received and preconditioned at 85% relative humidity for 8 h. The flow rate was set at 32 l/m continuously. Five filters from each manufacturer were tested at both concentrations. However, preconditioned filters were only tested at the high concentration for the dust/mist filters and at low concentration for the fume filters.

The most commonly used dust/mist and fume filters were selected by NIOSH for testing. The test results are summarized in Tables 14 and 15. The filter efficiency is expressed as at least 95% of the tested filters have a probability of 0.95 of exceeding. The filter efficiency was reported at 30 and 90 min test intervals. A majority of fume

Table 14 Real Time Efficiency of Approved Fume Filters

Manufacturer model no.	ARL (min) 30	ARL (min) 90	ARH (min) 30	ARH (min) 90	RHL (min) 30	RHL (min) 90
AO R-51	99.16	99.10	99.11	99.15	99.25	[a]
CESCO RC-19	96.91	97.34	97.95	96.06	97.08	97.35
HS Cover F-108	98.93	98.71	98.81	98.12	98.90	98.98
Glendale	98.94	98.73	99.01	97.46	98.39	98.45
MSA Type S	99.46	99.37	99.20	98.86	99.36	99.53
Norton 7507	99.16	99.19	99.19	99.31	98.77	98.03
3M 9920	98.96	98.46	99.01	98.42	98.74	98.31
Willson R-11	97.76	[a]	99.37	98.99	99.08	99.11
Survivair 1040	98.78	98.70	97.42	96.88	98.50	98.59

Note: ARL, Filters tested as "received" at low concentration; ARH, Filters tested as "received" at high concentration; RHL, Preconditioned filters tested at low concentration.

[a] Data from one of the two 90-min tests were invalid.

From Reference 5.

filters have efficiencies close to 99% under all test conditions. However, the efficiency of dust/mist filters are much lower during the 90 min test. The Glendale, the Pulmosan, and the 3M 8710 had efficiency far less than 99%, which is a requirement for certification. The OSHA assigned protection for a half-mask negative pressure respirator is 10 which is based on a facepiece leakage of 10% and a maximum filter penetration of 1%. A respirators may not provide adequate protection if the filter penetration exceeds 1%.

IV. EFFICIENCY OF APPROVED RESPIRATOR FILTERS AGAINST LATEX AEROSOLS

The collection efficiency of dust/mist filters was evaluated by Brosseau et al.[6] The test aerosol was monodisperse latex spheres with a size range between 0.102

Table 15 Real Time Efficiency of Approved Dust/Mist Filters

Manufacturer model no.	ARL (min) 30	ARL (min) 90	ARH (min) 30	ARH (min) 90	RHH (min) 30	RHH (min) 90
AO R-50	98.74	98.81	98.70	98.41	98.17	98.42
CESCO RC-15	99.87	99.64	99.60	99.58	99.59	99.63
Glendale F-10	96.09	96.73	96.49	[a]	96.01	96.32
MSA Type F	98.46	98.52	98.34	98.46	98.59	98.47
Norton 7506	98.50	98.61	98.83	98.91	98.47	98.13
Pulmosan	99.05	98.75	93.62	94.21	98.14	98.29
Scott 642D	97.56	97.68	98.38	98.48	97.74	97.34
3M 8710	97.05	97.51	96.08	96.30	95.82	95.45
3M 9900	99.71	99.84	99.87	99.87	99.87	99.91
3M 9910	99.29	99.28	98.91	98.94	99.02	99.01
Willson R-10	99.21	98.93	99.42	99.24	99.20	99.05
Survivair 1010	99.44	99.49	99.29	99.25	99.47	99.31

Note: ARL, Filters tested as "received" at low concentration; ARH, Filters tested as "received" at high concentration; RHL, Preconditioned filters tested at high concentration.

[a] Data from one of the two 90-min tests were invalid.

From Reference 5.

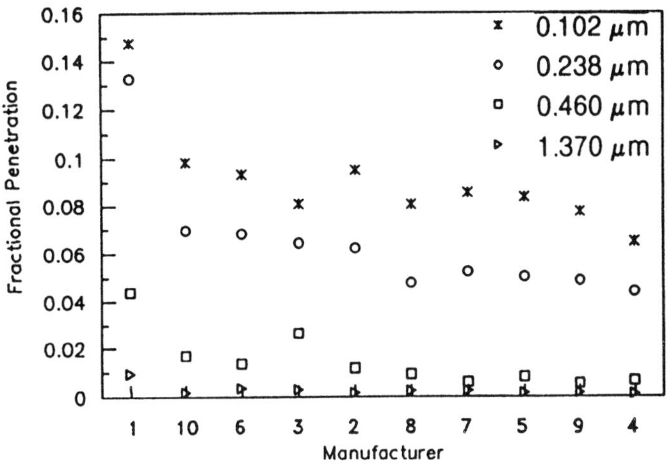

Figure 12 Penetration (four sizes) ranked by manufacturer.

and 2.02 μm. The test protocol described in the ASTM Special Technical Bulletin 975 was selected for conducting this study. One modification was that although the ASTM method physically measures the upstream concentration of the filter, in this study the upstream concentration was measured by diverting the flow around the filter.

Filters made by ten manufacturers were selected for testing. Nine were made of resin-impregnated wool and polypropylene felt and one was made of electret material. Each filter was cut to fit the filter holder used for testing. Three filters were selected randomly from the same lot for testing. The average air flow through the filter was 2700 ml/min. Two series of experiments were performed using latex spheres as the challenge. The first series tested large aerosols 0.460, 0.643, 0.803, 1.37, and 2.02 μm in size. The charges of the aerosol were removed through a neutralizer before testing. Upstream and downstream concentrations of aerosol were measured sequentially for 1 min periods. Three upstream and two downstream samples were taken for each latex sphere size at one minute intervals. Penetration was calculated by dividing the average downstream count by the average upstream count. The particle measurement was performed with a laser forward-scattering photometer. In the second series of experiments, the same 30 filters were tested with three smaller aerosols 0.238, 0.173, and 0.102 μm in size.

The filter penetration results for the four selected aerosol sizes were shown in Figure 12. Penetration ranged from 0.5% at the largest size (2.02 μm), to 1% at the medium particle size of 0.46 μm. Penetration increased to 10% at the smallest size of 0.102 μm. Except for manufacturer 1, most filters showed similar performance. Filters from manufacturers 4 and 9 seemed to perform somewhat better than the rest. The test results indicated that approved dust/mist filters do not have the same collection efficiency over different particle size range and that some filters performed better than others. One factor which the authors did not investigate is the influence of collection efficiency of any charges the aerosol may carry and the potential degradation of the electrostatic filters after storage under adverse environmental conditions such as elevated temperature and humidity. The results of this study confirm that the silica dust test used in the certification of dust/mist filters can not differentiate filters with different collection efficiencies.

V. EFFICIENCY OF COMMON MATERIALS AGAINST EMERGENCY EXPOSURES

Emergency conditions such as fire, explosions, or nuclear reactor malfunctions can expose the public to toxic aerosols. Cooper et al.[7,8] conducted a study to evaluate the effectiveness of readily available materials such as sheets, towels, or disposable masks in providing emergency protection. Materials selected for testing were:

1. A single use respirator (3M-8710).
2. White cotton handkerchief (57/in. by 64/in. thread count).
3. Thin toweling (Broadway Terryweave 90% cotton, 10% polyester).
4. Thick toweling (wash cloth, Terryweave, 88% cotton, 12% polyester).
5. Shirt materials (40% polyester, 60% cotton, 46/in. by 45/in. thread count.
6. Bed sheet (100% cotton, 87/in. by 74/in. thread count).

The respirator is available at hardware or department stores, and other materials are readily available at home or at work. All fabrics were tested dry, and some were also tested wet.

Polydisperse mineral oil droplets were atomized with a nebulizer. The filter material was mounted on a holder inside a test chamber. The oil mist aerosol concentration was measured by an optical particle counter for the upstream and downstream of the filter. Three face velocities of 1.5, 5.0, and 15 cm/s were selected for testing. The dry test was conducted at the room humidity (20 to 35% RH). For the wet test, the fabrics was soaked in distilled water for 10 min and then squeezed dry. Humidity was maintained at 80 to 95% RH during the test. The pressure drop across the fabric at each flow velocity and penetration were recorded. Particle penetration was measured by averaging two upstream and two downstream concentration readings for each particle size range. The reproducibility of the penetration measurements was determined and found to have a standard deviation between 0.01 and 0.06 on four replicates of penetration measurements under an average condition.

Penetration tests on vapor protection were measured on the thick toweling and the bed sheet. Methyl iodide at a concentration of 15 ppm (100 μl/l) was selected as the challenge concentration. The fabric materials were exposed to methyl iodide at 1, 3, 10, and 30 min time intervals. Iodine was generated by passing air through a tube filled with iodine crystals. Tests were also performed on test fabrics wetted with water or a baking soda (sodium bicarbonate) solution.

The penetration results of test materials at a face velocity of 1.5 cm/s and pressure drop of 50 Pa (this corresponds to 10 l/m air flow through an area of 110 cm^2, a 12-cm circle or a rectangular mask of 9×12 cm) for aerosol, methyl iodide, and iodine are summarized in Table 16. It appears that the single use respirator performed best. For the fabric materials, the thicker toweling provided better protection than the thin toweling, shirt, and handkerchief fabric. The toweling did better wet than dry; however, the sheet and the handkerchief did better dry than wet. The sheet

Table 16 Estimated Penetration Through Expedient Respiratory Protection Materials at 50 Pa (0.2 in H_2O) Pressure Drop and 1.5 cm/s Face Velocity

Material	No. layers	Aerosol particle diameter (μm) 0.4	1	5	I_2[b]	CH_3I[b]
Dry						
3M respirator[a] #8710	2	.03	.004	<.01		
sheet	20	.66	.64	.020	1.0	0.6[c]
shirt	15	.54	.59	.070		
thin towel	20	.53	.41	.015		
thick towel	6	.24	.13	<.01		0.6[c]
handkerchief	14	.61	.54	.032		
Wet						
sheet	6	.91	.88	.22	.45	.8[c]
					.15[d]	1.0[d]
shirt	6	1.0	.51	<.02		
thick towel	4	.20	<.01	<.01	.21	1.0
					.10[d]	
handkerchief	2	.98	.95	.37		

[a] Available commercially in single-layer thickness.

[b] Taken from tests at 1.0 cm/s, assuming penetration is the product of single-layer penetrations.

[c] Not shown to be statistically different from 1.00.

[d] Wetted with 5% by weight baking soda solution.

From References 7 and 8.

or towel wetted with baking soda solution or the towel wetted with water could reduce iodine penetration by a factor of 10. It appears that a single use mask of twice the thickness currently in use could produce reduction at a factor of 30 for the test aerosols while the wetted, higher-quality towel could achieve a reduction of at least five for the aerosols and iodine vapor.

The authors concluded that available materials can provide substantial reductions in concentrations of aerosols and certain water soluble gases or vapors at pressure drops acceptable for respiratory protection during emergencies. However, in practice, leakage around the faceseal would significantly reduce the protection provided.

VI. SEARCH FOR SUBSTITUTES TO REPLACE DEHP

After the National Toxicology Program (NTP) announced the preliminary test results of an animal feeding study on di-2-ethylhexyl phthalate (DEHP or DOP) that hepatocellular carcinomas or neoplastic nodules in mice and rats in 1981[9] the NIOSH prohibited the use of DEHP as a fit testing agent in spite of the fact that similar carcinogenic developments were also observed on the control group of test animals. The Occupational Safety and Health Administration had switched to sodium chloride as the challenge agent for performing fit testing for the OSHA compliance officers.

Hinds et al. conducted a study to search for a substitute for DOP. One primary requirement was that the agent must have a low toxicity.[10] The purpose of this study was to compare the aerosol particle size distribution produced by the Laskin aerosol generator using substitute materials with that produced using DEHP. The substitute

Table 17 Some Properties of DOP and Substitute Materials

Material	Specific gravity	Viscosity (cP)	Boiling point (°C)	Refractive index
Corn oil Fisher USP	0.918	[a]	[b]	1.464
Di(2-ethylhexyl)phthalate Hatcol DOP	0.983	82 @ 20°C	350	1.485
Di(2-ethylhexyl)sebacate Uniflex DOS	0.915	17.4 @ 25°C	240	1.448
Mineral oil Arcoprime 200	0.861	52.4 @ 38°C	95% > 360	1.471
Polyethylene glycol Union Carbide PEG 400	1.128	105 @ 25°C	[b]	1.465

[a] A natural product with varying properties.

[b] Boils over a broad temperature range.

From Reference 10.

materials were selected to match the properties of DEHP as closely as possible with special emphasis given to viscosity and vapor pressure. Other properties considered were density, nonhygroscopicity, flash point, refractive index, and commercial availability. A suitable material must have a low vapor pressure at ambient temperature so that the submicrometer particles produced would not evaporate appreciably during the lifetime of the aerosol. Four candidates, corn oil, di-2-ethylhexyl sebacate (DEHS), mineral oil, and polyethylene glycol 400 (PEG 400) were selected for testing. The physical properties of these agents and DEHP are shown in Table 17.

The available Laskin nozzle aerosol generator can be equipped with up to six nozzles with an operating air pressure up to 175 kPa (25 psig). The system used in this investigation utilized a single nozzle for aerosol generation. The air flow through the nozzle was 21 l/m at a pressure of 140 kPa (20 psig). Only a 3 l/m fraction of the generator output was diluted with 240 l/s (500 cfm) of air for particle size determination. The overall dilution ratio was 4800:1. The aerosol particle size was measured by an ASAS-X optical particle counter which measures size range between 0.09 and 3 μm. Factors selected for investigation included the number of jet nozzles, the depth of immersion of the liquid feed collar, line air pressure, and the amount of dilution air needed for size measurement.

The results indicated that no significant variations of particle size were observed between single and four jet nozzles. Four nozzle immersion depths of 0.64, 1.3, 1.9, and 2.5 cm were investigated. The study showed little difference in the count median diameter (CMD) from 0.25 to 0.26 μm. Size distribution were measured at five dilutions over the concentration range of 10^3 to 10^5 particles/ml. The measured CMD gradually increased to above 10^4 particle/ml. To minimize the coincidence error, all measurements were carried out at a concentration of 3,500 ± 500 particles/ml. This concentration corresponded to a theoretical loss of count due to coincidence of about 5%. Aerosol size measurements were determined at four pressures: 70, 105, 140, and 175 kPa (10, 15, 20 and 25 psig). The test results are shown in Figure 13, and Table 18. The mass median diameter (MMD) was calculated from the count data.

In general, the variation in median size and size distribution between the substitute materials and the DEHP was within ± 25%. All showed the same trend of decreasing CMD with increasing air pressure. Corrections were made to compensate

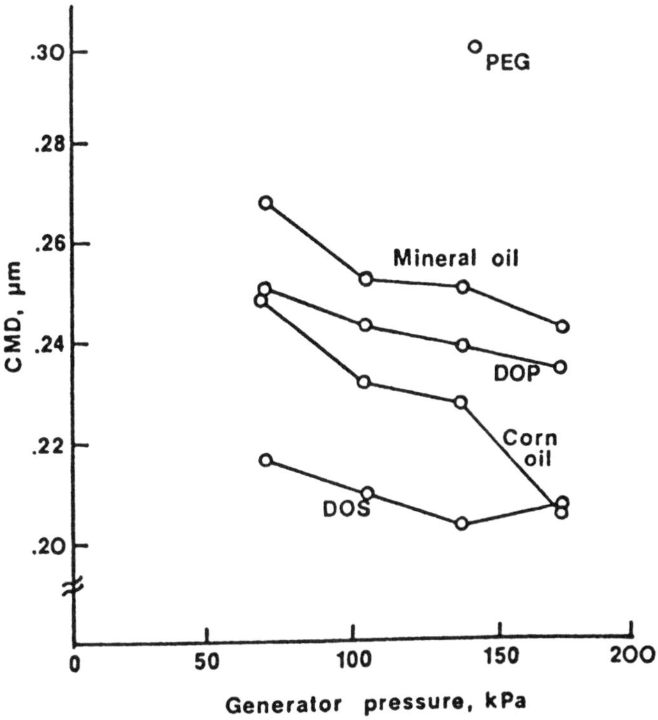

Figure 13 Effect of generator pressure on CMD for five materials. (From Reference 10.)

for reduced coagulation time prior to dilution due to higher air pressure. Among the substitute materials tested, mineral oil and corn oil provided the closest match to the size distribution of the DEHP. Mineral oil also provided the closest match for refractive index. Corn oil became the commonly used substitute for DEHP for commercially available portable fit testing instruments and was also recommended by NIOSH. However, corn oil can become rancid upon prolonged exposure; this requires frequent cleaning of the test chamber to prevent the buildup of unpleasant odors and to prevent bacteria growth which could interfere with aerosol generation.

Table 18 Size Comparison of Generator Output at 140 kPa (20 psi)

Material	CMD (µm)	$D_{84\%}$[a] / $D_{50\%}$	MMD[b] (µm)
DOP	0.25	1.53	0.70
DOS	0.22	1.53	0.57
Mineral oil	0.27	1.52	0.70
Corn oil	0.25	1.53	0.77
PEG	0.30	1.48	0.83

[a] Eighty-four percentile size divided by the 50% percentile size; approximately equal to the geometric standard deviation (GSD).

[b] Calculated from count data.

From Reference 10.

Table 19 Range of Minimum NaCl Initial Instantaneous Filter Efficiency[a] for Commercial Filters in the "Worst Case" Size Region

Flow rate (l/m)	Manufacturer	Dust and mist	Paint, lacquer, and enamel mist	Dust, fume, and mist	High efficiency
16	A	87–88	92–93	98–99	—
	B	88–89	92–93	—	—
	C	84–85	87–88	98–99	>99.999
32	B	82–83	85–86	—	—
42.5	A	79–80	83–84	94–95	99.997–99.998
	C	71–72	78–79	95–96	99.996–99.997
64	A	77–78	79–80	92–93	99.991–99.992
	B	73–74	79–80	—	99.995–99.996
	C	70–71	75–76	93–94	99.995–99.996
85	A	69–70	77–78	91–92	99.977–99.980
	B	73–74	78–79	—	99.993–99.994
	C	69–70	67–68	89–90	99.986–99.987
	D	67–68	86–87	87–88	—

[a] Estimated minimum efficiency range from filter efficiency versus particle-size plots.

From Reference 11.

Many Department of Energy installations do not use the corn oil due to this problem. For example, Los Alamos National Laboratory uses DEHS, and the Lawrence Livermore National Laboratory uses PEG 400 for performing respirator fit testing.

VII. THE WORSE CASE AEROSOL FOR FILTER CERTIFICATION

In order to support the proposed revision of respirator testing and certification regulations, 42 CFR 84, NIOSH conducted a series of tests on filter efficiency evaluation.[11-13] NIOSH was looking for a "worse case challenge aerosol" which would give the maximum penetration or the minimum efficiency. This method should be able to differentiate between *good*, *medium*, and *low* efficiency filters. A filter tester was specifically developed for this purpose and the device may be used for testing and approving particulate filters. The tester is the TSI Filter Efficiency Test System which consists of a continuous condensation nuclei counter capable of measuring concentrations between 10^7 to 10^{-2} particles/ml.

The test aerosols consisted of a solid sodium chloride aerosol with a count mean diameter (CMD) between 0.03 and 0.24 μm, a geometric standard deviation between 1.4 and 1.6, a liquid polydisperse DOP aerosol with a CMD between 0.03 and 0.3 μm, and a geometric standard deviation between 1.6 and 1.8. The particle size for these aerosols was determined by a mobility analyzer. Approved dust/mist (D/M), dust/fume/mist (DF/M), paint, lacquer, and enamel mists (PLEM) and HEPA filters both of mechanical and electrostatic types from four manufacturers were selected for testing.[11] The range of minimum initial instantaneous filter efficiency for the NaCl and DOP aerosols are shown in Tables 19 and 20, respectively. The particle size ranges, which provide the maximum filter penetration for the NaCl and DOP aerosol challenge, are shown in Tables 21 and 22, respectively.

The test results indicated that filter efficiency decrease with an increase of air flow. Except for the HEPA filter, the filter efficiencies for other types of filters are lower than that required for approval which is 99%. There is also a difference in

Table 20 Range of Minimum DOP Initial Instantaneous Filter Efficiency[a] for Commercial Filters in the "Worst Case" Size Region

Flow rate (l/m)	Manufacturer	Dust and mist	Paint, lacquer, and enamel mist	Dust, fume, and mist	High efficiency
16	A	88–89	88–89	98–99	>99.999
	B	87–88	91–92	—	>99.999
	C	84–85	87–88	98–99	99.998–99.999
32	B	84–85	86–87	—	99.998–99.999
42.5	A	79–80	82–83	94–95	99.997–99.998
	B	80–81	85–86	—	—
	C	72–73	78–79	95–96	99.994–99.995
64	A	73–75	77–78	92–93	99.987–99.988
	B	74–75	80–82	—	99.990–99.991
	C	67–68	72–73	91–93	99.977–99.978
85	A	74–75	76–77	87–88	99.978–99.979
	B	70–71	75–76	—	99.983–99.984
	C	69–70	70–71	86–87	99.989–99.990
	D	67–69	84–85	85–86	—

[a] Estimated minimum efficiency range from filter efficiency versus particle-size plots.

From Reference 11.

filter efficiency among different manufacturers. The type of aerosol challenge (solid or liquid) has no significant effect on initial instantaneous filter efficiency. For the most penetrating particle size for different types of filters, the particle size range is larger for the HEPA filters than other types of filters. One possible explanation is that the mechanical type HEPA filter has much larger surface area than other type of filters (400 cm^2 versus 150 cm^2). The larger surface area tends to reduce the face velocity of the filter, resulting in a shift toward a larger particle size to achieve maximum penetration. The particle size which the maximum penetration was observed varies among manufacturers for the same filter type. The authors suggested that the ideal method to determine the minimum filter efficiency

Table 21 Particle Size[a] Range of Minimum NaCl Initial Instantaneous Filter Efficiency for Commercial Respirator Filters

Flow rate (l/m)	Manufacturer	Dust and mist CMD[a] (μm)	Pant, lacquer, and enamel mist CMD[a] (μm)	Dust, fume, and mist CMD[a] (μm)	High efficiency CMD[a] (μm)
16	A	<0.055	0.07–0.11	0.06–0.10	—
	B	<0.055	ND[b]	—	—
	C	<0.055	0.085–0.12	0.10–0.14	0.16–0.20
32	B	0.06–0.10	0.08–0.12	—	—
42.5	A	<0.06	0.06–0.10	0.045–0.085	0.17–0.21
	C	<0.05	0.045–0.075	0.08–0.12	0.15–0.19
64	A	<0.06	0.06–0.095	0.05–0.08	0.16–0.20
	B	<0.06	0.10–0.14	—	0.14–0.18
	C	<0.05	0.05–0.09	0.07–0.11	0.14–0.18
85	A	0.06–0.10	0.065–0.10	0.05–0.08	0.16–0.21
	B	0.06–0.10	ND	—	0.14–0.19
	C	0.06–0.10	0.055–0.085	0.07–0.12	0.12–0.15
	D	0.04–0.07	0.03–0.07	0.05–0.09	—

[a] Estimated particle size of minimum filter efficiency from DMPS experimental or converted from EAA data correlation.

[b] ND, not distinguishable.

From Reference 13.

Table 22 Particle Size[a] Range of Minimum DOP Initial Instantaneous Filter Efficiency for Commercial Respirator Filters

Flow rate (l/m)	Manufacturer	Dust and mist CMD[a] (μm)	Pant, lacquer, and enamel mist CMD[a] (μm)	Dust, fume, and mist CMD[a] (μm)	High efficiency CMD[a] (μm)
16	A	0.05–0.08	0.08–0.12	0.06–0.10	ND[b]
	B	0.06–0.10	0.09–0.13	—	ND
	C	0.04–0.08	0.085–0.125	0.11–0.15	0.125–0.165
32	B	0.055–0.095	0.09–0.13	—	0.09–0.13
42.5	A	<0.04	0.10–0.14	0.04–0.08	0.14–0.18
	B	0.04–0.08	0.085–0.125	—	—
	C	<0.04	0.095–0.135	0.10–0.14	0.13–0.17
64	A	<0.04	0.08–0.12	0.05–0.09	0.11–0.15
	B	0.03–0.06	0.07–0.11	—	0.12–0.16
	C	<0.04	0.05–0.09	0.11–0.15	0.12–0.16
85	A	<0.04	0.06–0.09	0.07–0.10	0.095–0.135
	B	0.04–0.08	0.04–0.08	—	0.135–0.175
	C	0.03–0.07	0.06–0.10	0.08–0.12	0.09–0.13
	D	0.03–0.06	0.055–0.095	0.07–0.10	—

[a] Estimated particle size of minimum filter efficiency from DMPS experimental or converted from EAA data correlation.

[b] ND, not distinguishable.

From Reference 13.

would be one which evaluates filter versus particle size and conducts all future tests at the most penetrating size.

This filter testing method only measures filter efficiency by particle counting; however, the current respirator certification method measures the mass efficiency. The air flow rate (64 l/m) is also significantly higher than the certification flow rate (32 l/m). Filter penetration values obtained from these two methods are not necessarily the same. Because this study only measures the initial filter efficiency, it does not consider the particle loading and charging effect. The mechanical type filter would become more efficient after particle loading. However, the efficiency of electrostatic type filters would become lower after particle loading and neutralization of the charges built up on the filter. One major advantage of this method is that the real time measurement of filter efficiency would differentiate the filter performance in the same class of approved filters.

The second series of tests was to determine the effect of humidity pretreatment on filter efficiency[12] since studies conducted by the Los Alamos Scientific Laboratory, and Ackley[14,15] have demonstrated that when electrostatic filters are stored under high temperature and humidity environment, the filter penetration increase after storage. The same filters used in the previous tests were selected for testing. Filters in the "as received" condition were placed in an environmental chamber at 38°C and 85% relative humidity for up to 42 days. Filters were removed and tested for penetration with the NaCl and the DOP aerosols at 1, 7, 14, 28, and 42 days interval. The results indicated that electrostatic type filters showed approximately 2 to 6% reduction in efficiency depending on the preconditioning time. The efficiency of mechanical type HEPA filters have no detectable reduction in efficiency, and the minimum efficiency remains higher than 99.97%.

The last series of tests in the NIOSH study investigated the effect of particle charging on initial instantaneous penetration of filters.[13] The silica dust and lead fume

Table 23 Initial Instantaneous Filter Efficiency for "Worst-Case" Size Aerosols and Silica Dust and Lead Fume Aerosols

		Silica dust data[a]		Lead fume data[b]		Worst-Case Data	
Filter type	Flow rate (l/m)	# Filters tested	% Efficiency (standard deviation)	# Filters tested	% Efficiency (standard deviation)	% Minimum efficiency	Particle size region (μm)[c]
DM	16	22	98.12 (0.67)	10	91.64 (1.33)	87–89	0.04–0.08
	85	6	96.03 (0.58)	7	80.11 (0.86)	69–75	0.04–0.10
PLEM	16	20	99.36 (0.07)	10	91.82 (0.82)	88–93	0.07–0.12
	85	6	96.11 (0.25)	5	81.63 (1.29)	76–78	0.06–0.10
DFM	16	10	99.42 (0.16)	13	98.23 (0.29)	98–99	0.06–0.10
	85	6	98.49 (0.18)	7	89.13 (0.98)	87–92	0.05–0.10

[a] Silica dust size of 0.48 μm, determined by SEM.

[b] Lead fume size of 0.15 μm, determined by DMPS.

[c] Count mean diameter, determined by DMPS.

From Reference 13.

aerosols used in certifying D/M and DFM filters were selected as the challenge. The silica dust or lead fume aerosols generated in the certification test chambers were withdrawn and diluted to a concentration ≈ 2×10^4 particles/ml, and used for testing. DM, DFM, and paint spray filters from a single of Manufacturer A's production lot were selected for testing. The filters were tested at flow rates of 16 and 85 l/m. The TSI filter efficiency test system was used for initial instantaneous filter efficiency measurement. The reading was taken after 2 min exposure to the challenge aerosol. For comparison purposes, the filters were also challenged with the "worse case" aerosols described above. The results are shown in Table 23. The same test was also conducted when the charges built up on the filter have been neutralized by passing through a Kr-85 radiation source. These results are shown in Table 24.

The test results indicated that the lead fume is more penetrating than the silica dust, and the "worse case aerosol" has significantly reduced the efficiency of these filters. There is also a notable reduction in filter efficiency when air flow increases. The test data in Table 24 show that there is no significant difference in filter penetration between the "original" and "neutralized" silica dust and lead fume. It should be noted that the test results are inconclusive since the low challenge aerosol concentration (20,000 particles/ml) and short test time (2 min) may not be sufficient to observe the effect of charge neutralization on the filters.

Table 24 Initial Instantaneous Filter Efficiency Data for "Charged" and "Neutralized" Silica Dust and Lead Fume Aerosols

		Silica dust data[a]			Lead fume data[b]		
		Charged[a]	Neutralized		Charged[a]	Neutralized	
Filter type	Flow rate (l/m)	% Efficiency (standard deviation)	# Filters tested	%Efficiency (standard deviation)	% Efficiency (standard deviation)	# Filters tested	% Efficiency (standard deviation)
DM	16	98.12 (0.67)	21	98.67 (0.40)[b]	91.64 (1.33)	10	93.12 (0.90)[b]
	85	96.03 (0.58)	7	94.35 (1.40)[b]	80.11 (0.86)	5	80.93 (1.34)
PLEM	16	99.36 (0.07)	24	98.83 (0.21)[b]	91.82 (0.82)	10	92.98 (0.75)[b]
	85	96.11 (0.25)	5	91.04 (0.65)[b]	81.63 (1.29)	5	79.96 (1.22)[b]
DFM	16	99.42 (0.16)	10	99.33 (0.16)	98.23 (0.29)	13	97.78 (0.50)[b]
	85	98.49 (0.18)	5	95.53 (0.15)[b]	89.13 (0.98)	7	88.50 (1.01)

[a] Charged data from Table 23.

[b] Significant difference between charged and neutralized data.

From Reference 13.

REFERENCES

1. Hyatt, E. C., Pritchard, J. A., Richards, C. P., Geoffrion, L. A., and Kressin, E. K., Respirator research and development related to quality control. LASL Project R-037. Quarterly Report — July 1 through September 30, 1971. Los Alamos Scientific Laboratory. LA-4908PR, 1972.
2. Revoir, W. H., Test Report, ANSI Z-88 ad hoc Subcommittee for Respirator Test and Approval, February 1981.
3. Revoir, W. H., Test Report, ANSI Z-88 ad hoc Subcommittee for Respirator Test and Approval, July 1981.
4. OSHA, Partial administrative stay and correction to the standard. Occupational Safety and Health Administration, (44 FR 5446-5448), January 26, 1979.
5. Myers, W. R. and Allender, J., Efficiency of MSHA/NIOSH approved dust, fume and mist class filters and dust and mist class filters against lead aerosol, NIOSH, February 12, 1982.
6. Brosseau, L. M., Evans, J. S., Ellenbecker, M. J., and Feldstein, M. L., Collection efficiency of respirator filter challenged with monodisperse latex aerosols, *Am. Ind. Hyg. Assoc. J.*, 50, 544, 1989.
7. Cooper, D. W., Hinds, W. C., and Price, J. M., Emergency respiratory protection with common materials, *Am. Ind. Hyg. Assoc. J.*, 44, 1, 1983.
8. Cooper, D. W., Hinds, W. C., and Price, J. M., Emergency respiratory protection with common materials. Report (NUREG/CR-2272) to Sandia Laboratories, Albuquerque, NM, June 1981.
9. NTP, Carcinogenesis bioassay of Di(2-Ethyl Hexyl) Phthalate. National Toxicology Program (NTP) DHHS Publication No. (NIH) 82-1773, Carcinogenesis Testing Program, National Institutes of Health, Bethesda, MD, 1982.
10. Hinds, W. C., Macher, J. M., and First, M. W., Size distribution of aerosols produced by the Laskin Aerosol Generator using substitute materials for DOP, *Am. Ind. Hyg. Assoc. J.*, 44, 495, 1983.
11. Stevens, G. A. and Moyer, E. S., Worse case aerosol testing parameters. I. Sodium chloride and dioctyl phthalate aerosol filter efficiency as a function of particle size and flow rate, *Am. Ind. Hyg. Assoc. J.*, 50, 257, 1989.
12. Moyer, E. S. and Stevens, G. A., Worse case aerosol testing parameters. II. Efficiency dependence for commercial respirator filters on humidity pretreatment, *Am. Ind. Hyg. Assoc. J.*, 50, 265, 1989.
13. Moyer, E. S. and Stevens, G. A., Worse case aerosol testing parameters. III. Initial penetration of charged and neutralized lead fume and silica dust aerosols through clean, unloaded respirator filters, *Am. Ind. Hyg. Assoc. J.*, 50, 271, 1989.
14. Ackley, M. W., Degradation of electrostatic filters at elevated temperature and humidity. World Filtration Congress III. Croydon, England, 169, Uplands Press, 1982.
15. Douglas, D. D., Revoir, W. R., Davis, T. O., Pritchard, J. A., Lowry, P. L., Hack, A. L., Richards, C. P., Geoffrion, L. A., Wheat, L. D., Bustos, J. M., and Hesch, P. R., Respirator studies for the National Institute for Occupational Safety and Health, July 1, 1974 through June 30, 1975, Los Alamos Scientific Laboratory, LA-6386-PR, 1976.

15 FILTER PERFORMANCE II

I. EFFECT OF TEMPERATURE AND HUMIDITY ON FILTER EFFICIENCY

The performance of particulate removing filters should not be affected by elevated temperature and humidity. Even if a respirator manufacturer used protective packaging to prevent contact of a particulate filter with an atmosphere of elevated temperature and humidity during shipment and storage, once the respirator's user removed the respirator filter from the protective package, the filter would be exposed to any atmosphere existing in the workplace. The respirator filter may be exposed to the workplace atmosphere for several days, even for several months. If the efficiency of particulate filters deteriorates during storage, the user would be at risk of overexposure to toxic air contaminants.

The effect of temperature and humidity on the efficiency of approved dust/mist filters during storage was investigated by the Los Alamos Scientific Laboratory (LASL).[1] The test condition was set at 32.3°C and 90% relative humidity. Three different types of dust/mist filters from five manufacturers with the following compositions were selected for testing:

A. Filter composed of fine glass fibers and coarse synthetic organic fibers.
B. Filter composed of electrostatic felt (pressed wool fiber felt).
C. Filter composed of electrostatic felt (needled wool fiber and synthetic organic fiber felt).
D. Filter composed of electrostatic felt (pressed wool fiber felt).

A total of 50 filters were exposed to the test condition up to 21 days. The filters were removed periodically from the test chamber and the filter penetration was determined by a polydisperse NaCl aerosol with a mass median aerodynamic diameter of 0.6 μm and a standard deviating of 2.2. The challenge concentration was 15 mg/m^3. The test results are shown in Figure 1 and Table 1.

The test results plotted in Figure 1 indicate that the type A filter permitted a very high initial penetration by the NaCl aerosol, but the penetration values remained fairly constant through the test period. The performance of filters made of electrostatic felt varied widely. The type B filter showed a steep increase in penetration after exposure at high temperature and humidity. There was less of an effect on the penetration on types C and D filters when they were exposed under adverse storage conditions. The test results given in Table 1 indicate that the filter penetration of dust/

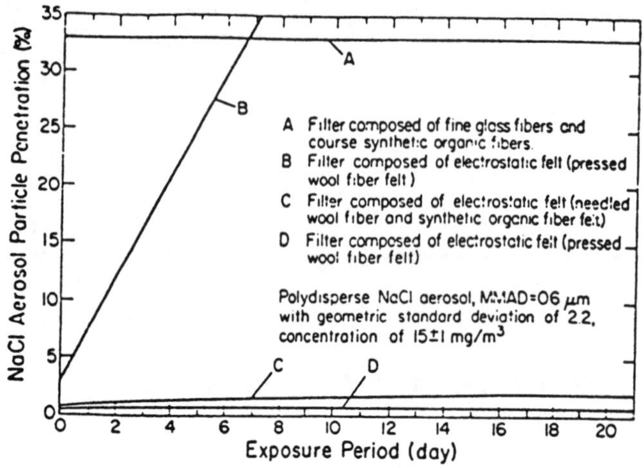

Figure 1 Average effect of exposure of dust-mist respirator filters to air at 90% RH and 32°C (90°F) on NaCl aerosol particle penetration. (From Reference 1.)

mist filters composed of electrostatic felt from manufacturer A increased very rapidly with exposure of the filters to high humidity conditions, and after an exposure of only seven days the penetration values were extremely high. However, the penetration for other filters composed of electrostatic felt, wool fiber felt, and synthetic wool fiber felt, increased by only a moderate rate of 6% or less.

The effect of temperature and humidity on the dust/mist filters was further studied by the LASL.[2] Three types of electrostatic felt filters, 100% wool, 50% wool/

Table 1 **NaCl Aerosol Particle[a] Penetrations Through Dust-Mist Respirator Filters After Exposure to High-Humidity Conditions[b]**

Mfg.	Filter material	Exposure period (days)	Range of NaCl particle pen., %	Mean NaCl particle pen., %	Standard dev. of NaCl particle pen., %	Max. NaCl particle pen. for 99% of filters, %[d]
A	EF-PW	0	2.2–5.3	3.49	0.78	5.31
		7[c]	2.5–80.0	34.4	18.97	78.60
B	EF-NWS	0	1.0–3.0	1.69	0.44	2.72
		21	0.66–4.7	2.04	0.90	4.14
C	EF-NWS	0	0.35–1.5	0.74	0.27	1.37
		21	1.30–3.1	2.00	0.50	3.17
D	EF-NWS	0	0.50–1.9	0.95	0.31	1.67
		21	1.0–72	2.69	1.42	6.00
E	EF-PW	0	0.10–7.7	0.45	1.10	3.01
		21	0.10–5.2	0.67	1.02	3.05

[a] Polydisperse NaCl aerosol, MMAD = 0.6 μm with standard deviation of 2.2, concentration of 15 ± mg/m³.

[b] Air at 32.2°C (90°F) and 90% RH.

[c] Performance of filters deteriorated so rapidly that the high-humidity exposure was ended at 7 days.

[d] Calculated value which = sum of mean initial NaCl particle penetration + 2.33 × standard deviation of initial NaCl particle penetration.

Note: EF, Electrostatic felt; PW, Pressed wool fiber felt; NWS, Needled wool fiber and synthetic organic fiber felt.

From Reference 1.

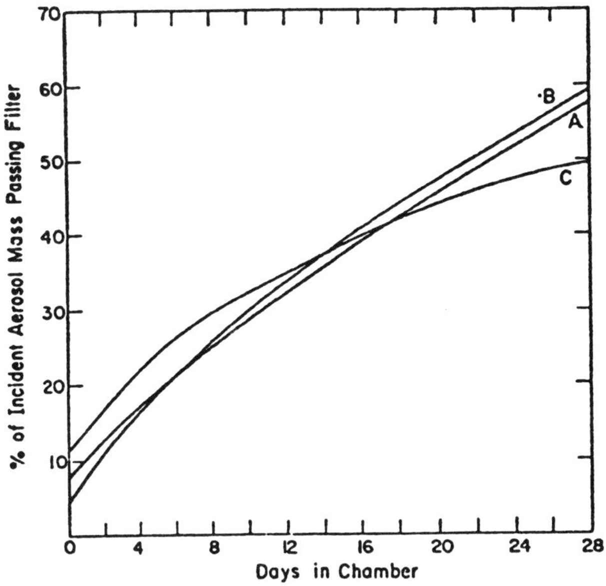

Figure 2 Effects of storage at 32°C, 90% relative humidity on NaCl aerosol penetration for a 32 l/m test flow rate. (From Reference 2.)

50% orlon, and 50% wool/50% acrylic, were selected for study. All filters had an effective area of 35 cm^2. These filters were exposed to air (32°C and 90% relative humidity) up to 28 days. Periodically, the filters were removed for penetration determination at a seven day interval. The test aerosol was NaCl similar to the one used in the previous study, and the air flow was set at 32 and 77 liters per minute (l/m). The test results are presented in Figures 2 and 3.

Aerosol: Polydisperse NaCl, mmad = 0.6 μm, σ_g = 2
Concentration = 15 mg/m^3
Air flew through filter during penetration measurement: 77 l/m
Filters stored in chamber maintained at 90°F, 90% relative humidity

A – Resin-impregnated pressed felt, 100% wool, 35-cm^2-area filter
B – Resin-impregnated needled felt, 50% wool, 50% orlon, 35-cm^2-area filter
C – Resin-impregnated needled felt, 50% wool, 50% acrylic, 35-cm^2-area filter

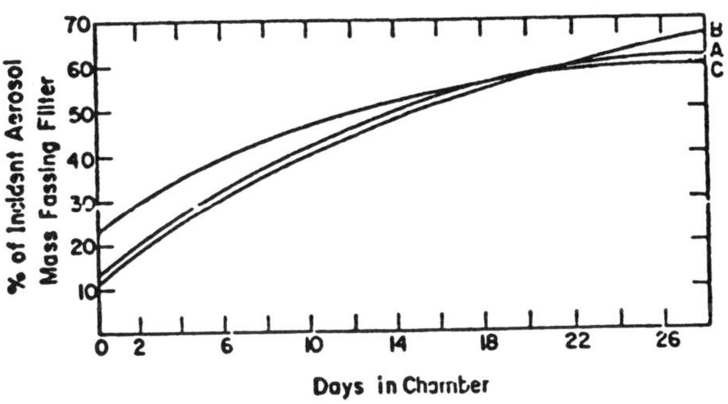

Figure 3 Effects of storage at 32°C, 90% relative humidity on NaCl aerosol penetration of electrostatic felt filters for a 77 l/m flow rate. (From Reference 2.)

Table 2 Effects of Exposure to Elevated Temperatures on Aerosol Penetration

Test flow[a] rate (l/m)	22°C (72°F)	50°C (122°F)	88°C (190°F)	116°C (240°F)	143°C (290°F)
16	5	5	5	27	30
32	10	10	12	44	44
38	10	10	11	42	47
77	16	16	20	55	60
Test flow[b] rate (l/m)	**22°C (72°F)**	**50°C (122°F)**	**88°C (190°F)**	**116°C (240°F)**	**143°C (290°F)**
16	1.0	1.0	9.0	25	30
32	3.5	3.5	27	44	44
38	3.5	3.5	22	41	43
77	10.0	10.0	40	55	55

[a] Filter material: Resin-impregnated pressed felt, 100% wool; 35-cm^2 area; Aerosol: Polydisperse NaCl, MMAD = 0.6 μm, σ_g = 2; Concentration = 15 mg/m^3; % of Incident aerosol mass; Passing filter after 4 h; Exposure to indicated temperature.

[b] Filter material: Resin-impregnated needled felt, 50% wool; 50% Orlon, 35-cm^2 area; Aerosol: Polydisperse NaCl, MMAD = 0.6 μm, σ_g = 2; Concentration = 15 mg/m^3; % of Incident aerosol mass; Passing filter after 4 h; Exposure to indicated temperature.

From Reference 2.

The results indicate that there was not a significant effect of flow rate on filter penetration. At a flow rate of 77 l/m, the filters had an initial penetration between 10 and 25%. There was a large increase in aerosol penetration to 60% upon exposure to high temperature and humidity.

The effect of filter penetration on elevated temperature alone was also studied with the three types of filters. All eight discs of these filters were exposed for four hours at temperatures of 50°, 88°, 116°, and 143°C in a drying oven. After exposure at high temperature, the filter penetration was determined with the same NaCl aerosol used previously. The air flow rates of 16, 32, 38, and 77 l/m were selected for testing. The test results are shown in Table 2.

The results indicate that high temperature caused some loss of electrostatic charge, and a consequent increase in aerosol penetration. The all wool filter showed an increase in penetration after exposed to a temperature of 116°C and higher. The other two types of filters showed increased penetration at a temperature of 88°C and higher. Generally, the penetration increased from 5% to 50% after exposure to high temperature.

The effect of adverse environmental conditions such as temperature and humidity on the performance of filter materials was studied by the Air-Purifying Working Group of the ANSI Z88 ad hoc Subcommittee on Respirator Testing and Approval.[3]

Materials from both a wool/acrylic electrostatic felt filter and a polypropylene electret fiber filter were selected for testing. The elevated temperature and humidity atmosphere was passed through the filter material, having an effective area of 70 cm^2 at a nonpulsating continuous air flow of 32 l/m. The temperature was 35°C and relative humidity was 85%. Two different elevated temperature/humidity exposure time periods of 6 and 24 h were selected. The exposed filter specimens were exposed to two types of aerosols. The first type of aerosol was the silica dust prescribed in 30 CFR 11 for approval dust/mist filters. The second type of aerosol was a NaCl aerosol

Table 3 Silica Dust Aerosol Particle Penetration Tests Before and After Exposure to Elevated Temperatures and Humidity Conditions of a Mechanical/Electrical Type Particulate Filtering Material

Sample	Areal density (gm/M²)	Thickness (inch)	Aerosol conc. (mg/M³)	Aerosol load (mg)	Penetration (mg)	Inhalation resistance Initial (mm H_2O)	Inhalation resistance Final (mm H_2O)
A. As received condition — No conditioning							
1	294	0.164	55.7	160.8	0.1	7.2	9.0
2	297	0.152	55.7	160.8	0.1	8.5	10.8
3	305	0.161	57.7	166.3	0.2	7.0	8.8
4	291	0.152	57.7	166.3	0.1	8.0	9.8
5	299	0.151	57.7	166.3	0.1	8.8	10.8
B. After exposure to elevated temperature and humidity[a]							
1	285	0.152	54.3	154.2	0.2	7.0	8.2
2	300	0.165	54.3	154.2	0.1	7.4	7.5
3	301	0.164	57.5	166.9	0.1	6.8	9.0
4	300	0.157	57.5	166.9	0.1	7.8	9.5
5	293	0.157	57.5	166.9	0.1	6.5	8.8

Note: Material: Polypropylene fiber electret material (electrical charges are permanently embedded into fibers); Test sample area: 70 cm²; Aerosol: Polydisperse silica dust aerosol; Aerosol particle size: 0.62 μm, 1.4 geometric SD; Chamber conditions: 21°C, 20% RH; Test flow: 32 l/m continuous nonpulsating flow; Test time: 90 min.

[a] Exposure conditions: 35°C, 85% RH for 24 h at a continuous flow rate of 32 l/m through test sample prior to aerosol test.

From Reference 3.

having a concentration of 15 mg/m³ and an aerodynamic mass median diameter of 0.6 μm. The test period was 90 min for both types of aerosols. In order to compare the effect of elevated temperature and humidity on the efficiency of these filters some of these filter specimens were not exposed to elevated temperature and humidity. The test results are summarized in Tables 3 through 6, and Figures 4 and 5.

Table 3 presents the results for the 90-min polydisperse silica dust particle aerosol tests carried out on specimens of a polypropylene fiber electret material. Section A of the table presents the results for specimens which were not exposed to the atmosphere having an elevated temperature and an elevated humidity; section B presents the results for specimens which had been exposed to an elevated temperature and humidity of passing air set at 35°C with a relative humidity of 85% for 24 h.

Table 4 presents the results for the 90-min polydisperse silica dust particle aerosol tests carried out on specimens of a wool/acrylic fiber electrostatic felt material. Section A of the table provides the results for specimens which were not exposed to the atmosphere having an elevated temperature and an elevated humidity. Section B of the table presents the results for specimens which had been exposed to an elevated temperature and humidity of passing air set at 35°C with a relative humidity of 85%, for 24 h.

Table 5 presents the results for the 90-min polydisperse sodium chloride aerosol tests carried out on specimens of a polypropylene fiber electret material. Section A of the table shows the results for specimens which were not exposed to the atmosphere having an elevated temperature and elevated humidity; section B presents the results for specimens which had been exposed to air at an elevated temperature and humidity of passing air at 35°C with a relative humidity of 85% for 24 h.

Table 6 presents the results for the 90-min polydisperse sodium chloride particle aerosol tests carried out on specimens of a wool/acrylic fiber electrostatic felt

Table 4 Silica Dust Aerosol Particle Penetration Tests Before and After Exposure to Elevated Temperature and Humidity Conditions of an Electrostatic Felt Type Particulate Filtering Material

Sample	Areal density (oz/yd²)	Thickness (in.)	Aerosol conc. (mg/M³)	Aerosol load (mg)	Penetration (mg)	Inhalation resistance Initial (mm H_2O)	Inhalation resistance Final (mm H_2O)
A. As received conditions — No conditioning							
1	12.0	0.130	55.7	160.8	0.1	7.0	15.5
2	11.8	0.124	55.7	160.8	0.1	6.7	16.8
3	11.9	0.124	55.7	160.8	0.1	7.0	18.0
4	12.1	0.126	57.7	166.3	0.1	6.8	13.0
5	11.6	0.120	57.7	166.3	0.1	7.0	19.5
B. After exposure to elevated temperature and humidity							
1	11.7	0.123	54.3	154.2	0.7	6.3	15.5
2	11.9	0.126	54.3	154.2	1.2	6.4	14.8
3	11.6	0.116	54.3	154.2	0.5	6.0	14.0
4	11.6	0.123	57.5	166.9	0.3	6.5	17.5
5	11.8	0.124	57.5	166.9	0.1	6.8	18.5

Note: Material: Wool/acrylic fiber electrostatic felt material; Test sample area: 70 cm²; Aerosol: Polydisperse silica dust aerosol; Aerosol particle size: 0.62 μm, 1.4 geometric SD; Chamber condition: 21°C, 20% RH; Test flow: 32 l/m continuous nonpulsating flow; Test time: 90 min.

Exposure conditions: 35°C, 85% RH for 24 h at a continuous flow rate of 32 l/m per minute through test sample prior to aerosol test.

From Reference 4.

Table 5 Sodium Chloride Aerosol Particle Penetration Tests Before and After Exposure to Elevated Temperature and Humidity Conditions of a Mechanical/Electrical Type Particulate Filtering Material

Time (min)	Penetration (%) Sample no. 1	2	3	4	5
A. As received condition — No conditioning (aerosol conc.: 16.6 mg/m³)					
0	0.14	0.16	0.05	0.15	0.15
10	0.08	0.20	0.15	0.18	0.22
20	0.18	0.29	0.21	0.21	0.36
30	0.33	0.36	0.30	0.24	0.52
40	0.47	0.44	0.46	0.26	0.72
50	0.54	0.48	0.62	0.34	1.0
60	0.70	0.63	0.85	0.41	1.2
70	0.84	0.71	1.1	0.55	1.6
80	1.2	0.83	1.5	0.67	1.8
90	1.3	0.92	1.7	0.73	2.2
B. After exposure to elevated temperature and humidity					
0	0.14	0.10	0.10	0.13	0.18
10	0.10	0.11	0.12	0.14	0.22
20	0.10	0.15	0.18	0.20	0.34
30	0.25	0.26	0.25	0.27	0.39
40	0.35	0.38	0.30	0.38	0.47
50	0.40	0.50	0.38	0.49	0.58
60	0.45	0.70	0.48	0.63	0.72
70	0.54	0.90	0.56	0.77	0.83
80	0.70	1.0	0.70	0.89	0.90
90	0.75	1.4	0.77	1.10	0.97

Note: Material: Polypropylene fiber electret material; Test sample area: 70 cm²; Test sample areal density: 300 g/m³; Test sample thickness: 0.160 in.; Aerosol: Polydisperse sodium chloride aerosol; Aerosol particle size: 0.60 μm, 2.2 geometric SD; Chamber condition: 26°C, 22% RH; Test flow: 32 l/m continuous nonpulsation flow.

Exposure conditions: 35°C, 85% RH for 24 h at a continuous flow rate of 32 l/m through test sample prior to aerosol test (aerosol conc.: 17.5 mg/m³).

From Reference 3.

Table 6 Sodium Chloride Aerosol Particle Penetration Tests Before and After Exposure to Elevated Temperature and Humidity Conditions of an Electrostatic Felt Particulate Filtering Material

	Penetration (%)				
Time (min) Sample no.	1	2	3	4	5
A. As received condition — No conditioning (aerosol conc.: 16.6 mg/m³)					
0	1.2	0.9	0.5	0.6	
10	1.1	1.3	0.9	0.8	
20	1.6	2.5	1.1	1.3	
30	2.0	3.4	1.3	2.2	
40	2.3	4.7	1.8	3.2	
50	2.9	5.4	2.9	4.0	
60	3.7	7.3	3.1	5.3	
70	4.0	8.1	3.5	6.9	
80	4.8	9.1	4.2	7.6	
90	5.6	9.6	4.5	8.6	
B. After exposure to elevated temperature and humidity					
0	4.0	3.5	5.0	4.5	3.8
10	4.8	3.8	5.5	5.7	4.1
20	4.8	4.2	7.0	7.4	5.3
30	6.0	5.0	10.0	8.5	6.0
40	7.5	5.8	12.0	11.0	6.7
50	10.0	7.0	13.0	13.0	8.1
60	10.5	8.0	14.0	14.0	9.3
70	12.0	9.5	17.0	16.0	10.4
80	12.5	10.0	18.0	17.5	11.0
90	13.0	10.5	18.5	18.5	12.0

Note: Material: Wool/acrylic fiber electrostatic felt material; Test sample area: 70 cm²; Test sample areal density: 12.0 oz/yd²; Test sample thickness: 0.125 in.; Aerosol: Polydisperse sodium chloride aerosol; Aerosol particle size: 0.60 μm, 2.2 geometric SD; Chamber conditions: 26°C, 22% RH; Test flow: 32 l/m continuous nonpulsating flow.

Exposure conditions: 35°C, 85% RH for 24 h at a continuous flow rate of 32 l/m through test sample prior to aerosol test (aerosol conc.: 17.5 mg/m³).

From Reference 3.

material. Section A of the table gives the results for specimens which were not exposed to the atmosphere having an elevated temperature and elevated humidity; section B presents the results for specimens which had been exposed to air at an elevated temperature and an elevated humidity of passing air at 35°C with a relative humidity of 85% for 24 h.

Figure 4 illustrates the change in the penetration of the polydisperse sodium chloride aerosol particles through the specimens of the polypropylene fiber electret material, with an increase in test time for material which had not been exposed to the elevated temperature/humidity atmosphere and for material which had been exposed to the elevated temperature/humidity atmosphere.

Figure 5 illustrates the change in penetration of the polydisperse sodium chloride aerosol particles through the specimens of the wool/acrylic fiber electrostatic felt material, with an increase in test time for material which had not been exposed to the elevated temperature/humidity atmosphere and for material which had been exposed to the elevated temperature/humidity atmosphere.

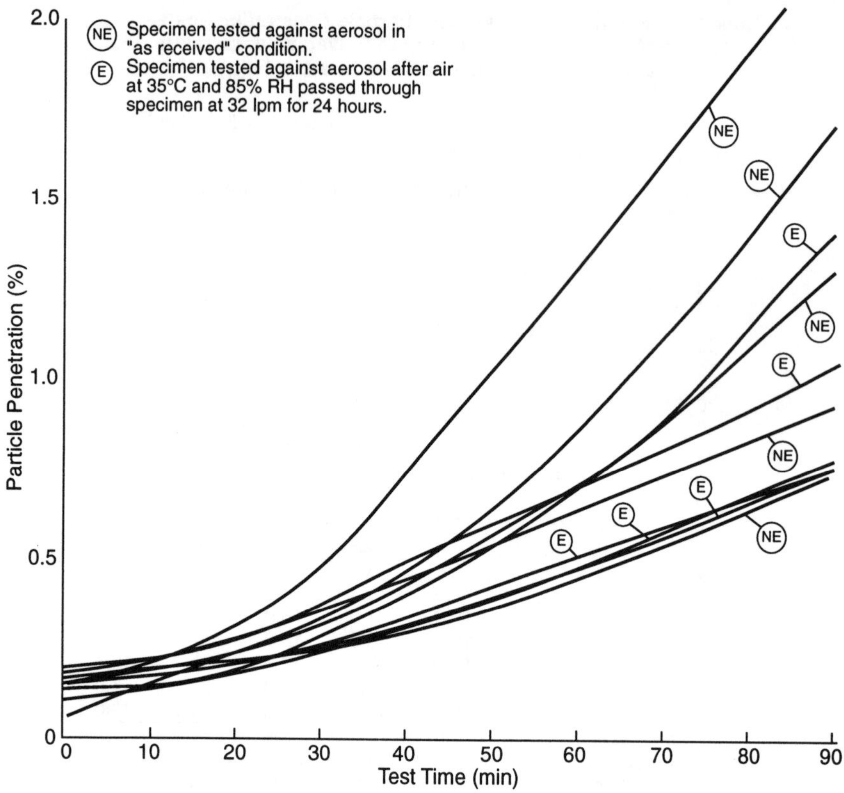

Figure 4 Electret filter degradation due to NaCl aerosol. (From Reference 4.)

The test results indicate that the exposure of a polypropylene fiber electret material to the elevated temperature/humidity atmosphere does not have an adverse effect on the performance of this material in removing the silica dust aerosol particles from air. But, the exposure of a wool/acrylic fiber electrostatic felt material to the elevated temperature/humidity atmosphere results in a slight increase in the penetration of this material by the silica dust aerosol particles. The test results demonstrate that a continual passage of the sodium chloride aerosol particles into the polypropylene fiber electret material results in a continual increase in the penetration, and that this occurs to the same extent for both the material that had not been exposed to the elevated temperature/humidity atmosphere and the material that had been exposed to the elevated temperature/humidity atmosphere.

The test results also indicate that a continual passage of the sodium chloride aerosol particles into the wool/acrylic fiber electrostatic felt material results in a continual increase in the penetration, and that this occurs in the material that had and had not been exposed to the elevated temperature/humidity atmosphere. However, the test results show that the exposure of the wool/acrylic fiber electrostatic felt material to the elevated temperature/humidity atmosphere causes a significant increase in the penetration of this material by the sodium chloride aerosol particles.

This study concluded that the efficiency of the commonly used electrostatic felt filter material would be affected by elevated temperature and humidity, and that the

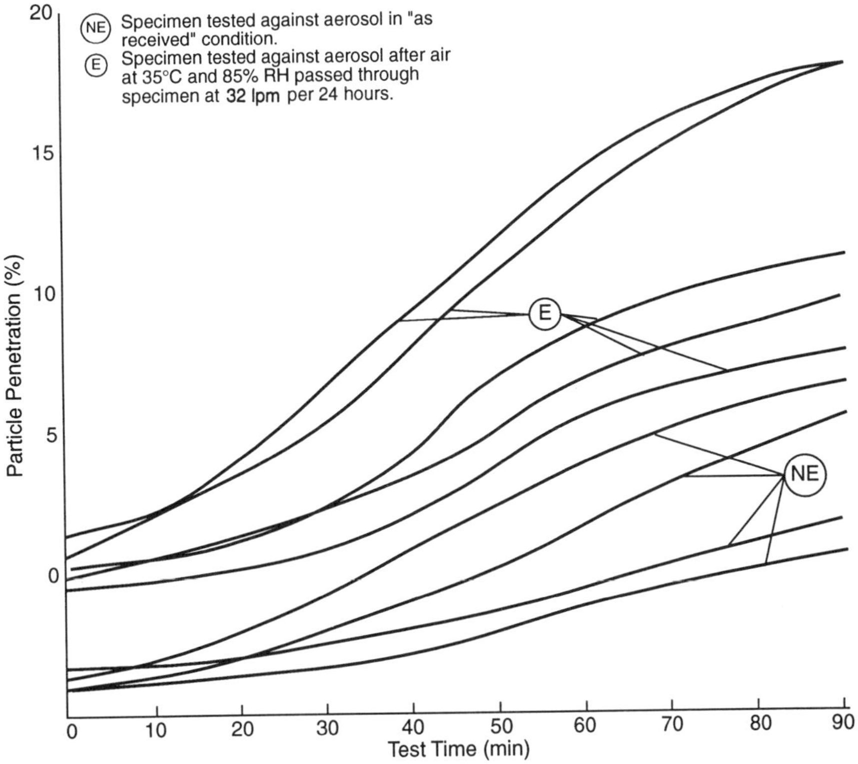

Figure 5 Degradation of electrostatic felts. (From Reference 4.)

certification test for particulate removing filters should include a preconditioning test that would expose the particulate-filtering respirators to an elevated temperature/humidity atmosphere for a given period of time before conducting the filter penetration test.

The effect of storage conditions on electrostatic filters was studied by Ackley.[4] Five different types of electrostatic filters having the following composition were selected for investigation:

A. Electrostatic resin impregnated wool with 45% wool and 55% acrylic.
B. Electrostatic resin impregnated wool with 45% wool and 55% polypropylene.
C. Electrostatic wool felt with 45% wool and 55% polypropylene.
D. Electret with permanently charged polypropylene.
E. Electrostatic spun polymer with polycarbonate.

In the above filters, enhanced aerosol collection efficiency is achieved by localized, nonuniform electrostatic fields. The degree and stability of the electrostatic enhancement of these filters were the subject of this investigation.

Monodisperse di-2-ethylhexyl phthalate (DEHP) aerosol with a particle size of 0.3 μm at a concentration of 100 mg/m^3, and a polydisperse NaCl with a mass median aerodynamic diameter of 1.0 μm with a standard deviation of 1.9 at a concentration of 10 mg/m^3, were selected for evaluation.

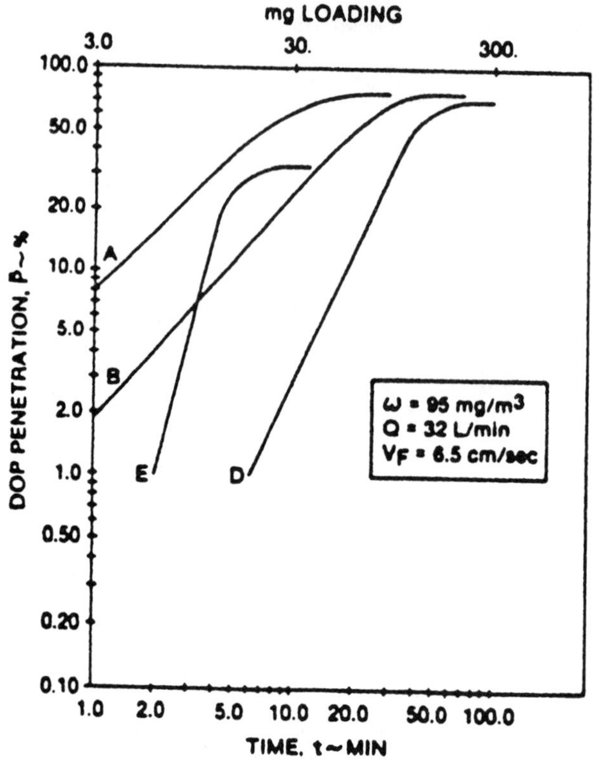

Figure 6 Degradation of electrostatic filters by DOP.

The study consisted of short-term (up to 14 days) and long-term (up to 180 days) environmental testing. Three test conditions were employed in the short-term testing: 21°C and 90% RH (0.014 g H_2O/g of air); 65°C and 0% RH; and 65°C and 70% RH (0.107 g H_2O/g of air). Only one condition, 65°C and 70% RH was used for the long-term filter evaluation.

The short-term test results are illustrated in Figures 6 through 9 (ω, challenge concentration; Q, flow rate; V_F, face velocity). Figure 6 shows that when filter types A, B, D, and E were subjected to the DEHP aerosol challenge, all four filters degraded rapidly until only the mechanical collection ability of the filter remained. All the electrostatic and the electret filters performed similarly with mechanical efficiencies between 20 to 30%. The spun polymer filter (Type E) demonstrated a higher mechanical efficiency at 68%. The degradation appeared to be a function of the challenge aerosol concentration, surface adsorption and wettability.

The DEHP penetration tests shown in Figure 7 indicate that the Type A filter degraded within four hours after exposed at 65°C and 70% RH. The penetration of Type B filter had a lesser effect by exposure to high temperature and humidity. The Type B filter was subject to all three test conditions mentioned above. The test results shown in Figure 8 indicate that the first two test conditions had a similar effect on the filter. However, exposure at 65°C and 70% RH caused a rapid degradation in less than 12 h. Type B filter was challenged by the NaCl aerosol at two face velocities. The results in Figure 9 indicate that at lower face velocity, the penetration varied

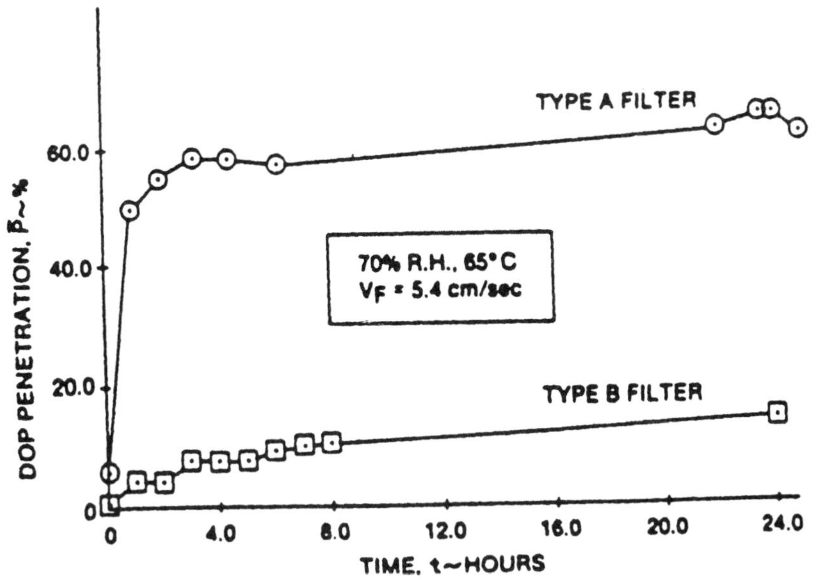

Figure 7 Comparison of degrdation of Type A and Type B filters.

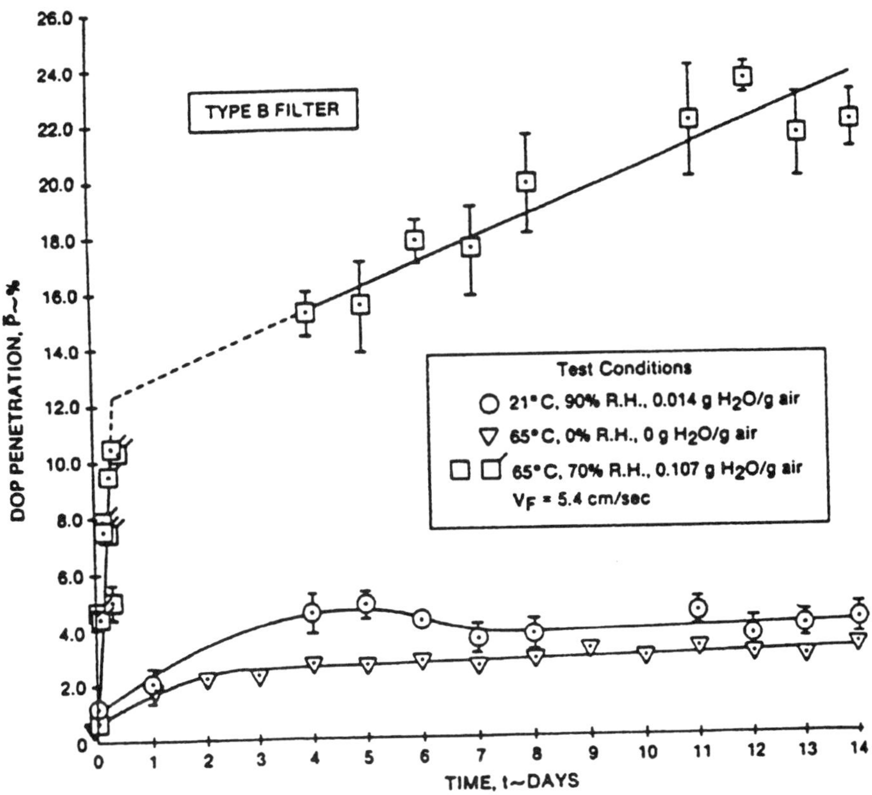

Figure 8 Short-term degradation of Type B filters. (From Reference 4.)

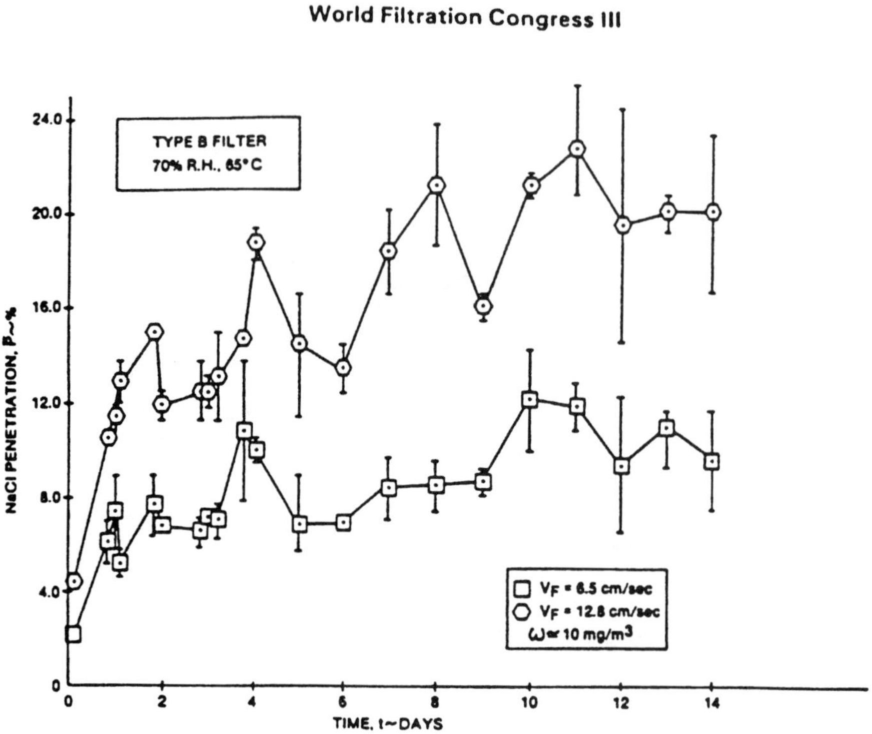

Figure 9 Degradation of Type B filters — NaCl penetration. (From Reference 4.)

between 4 and 8%. However, at higher velocity, the penetration increased to 20%. When comparing the test results with the DEHP challenge, it appeared that filter penetration is a function of particle size and face velocity.

The results of the long-term degradation test measured with the DEHP aerosol are shown in Figure 10. A portion of the electret and electrostatic filters were placed in Saranex® and Type 1 military bags to evaluate protection against degradation. The results indicated that the resin wool (Type B) filter is more susceptible to degradation at this test condition than other type of filters. The DEHP penetration increased to 9% after 154 days. The moisture barrier bags did not have a significant effect on the penetration of the Type B filter. The average DEHP penetration of Type D filter increased from 0.08 to 0.18% during the first 28 days, and the penetration fluctuated to a mean of 0.1% during the remaining 154-day test. Type E filter was tested for 91 days in the test chamber. Most of the average penetration was only slightly higher than the initial penetration of 0.24%. However, the penetration of individual samples varied between 0.09 and 1.5%. It is apparent that the spun polymer filter is less susceptible to adverse environmental conditions, while the resin impregnated filter is more susceptible to adverse environmental conditions.

II. PERFORMANCE OF PARTICULATE FILTERS AGAINST LIQUID AEROSOLS

The ANSI Z88 ad hoc Subcommittee for Respirator Test and Approval conducted an investigation to determine the performance of various types of particulate

World Filtration Congress III

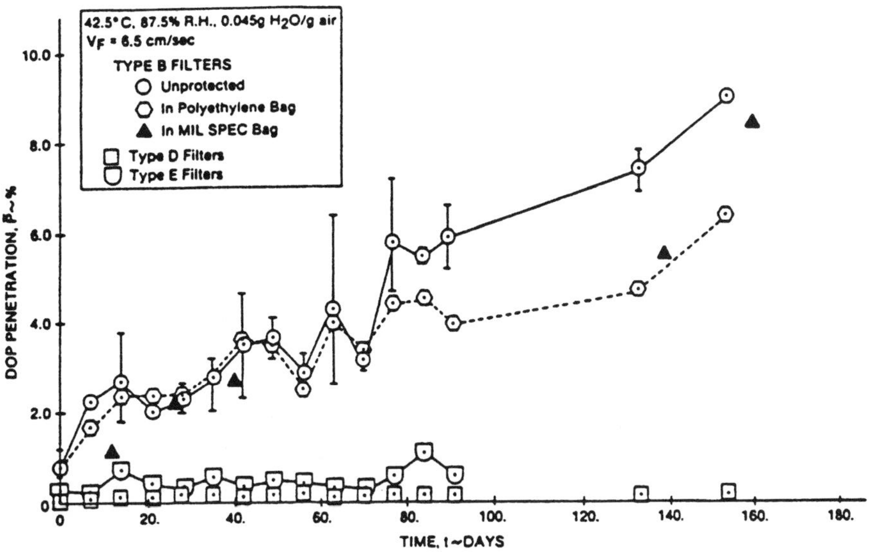

Figure 10 Long-term degradation of electrostatic filters. (From Reference 4.)

filtering devices used in nonpowered air-purifying respirators for removing liquid aerosol particles from air.[5]

The respirator filters used in the test program have been approved by MSHA/NIOSH under the provisions of 30 CFR 11. Three filters, each from the dust/mist, fume, and the high-efficiency particulate air (HEPA), were selected for testing. A description of these devices is given in Table 7.

The test aerosol was a monodisperse liquid aerosol consisting of a suspension of 0.3 μm size di(2-ethylhexyl) phthalate (DEHP) having a concentration of 100 mg/m³. During the test, the test atmosphere was passed through the specimen filter at a continuous flow rate of 32 liters per minute (l/m) if the device was designated to be

Table 7 Particulate Filters Selected for Testing

Type of filter device	Identification	Description
Dust/mist filter	J-1	Single element small electrostatic felt
	J-2	Medium single element paper type material
	J-3	Single element large electret type material
Fume filter	K-1	Dual element paper type material with convoluted construction
	K-2	Dual element multilayer with resin-bonded fiberglass and electrostatic felt material
	K-3	Single element electret material
HEPA filter	L-1	Dual element paper type material with convoluted construction
	L-2	Dual element paper type material with corrugated cylinder construction
	L-3	Dual element paper type material with convoluted construction

From Reference 5.

used singly in a respirator, and at a continuous flow rate of 16 l/m if the device was designated to be used as a pair in a respirator. A light scattering photometer was used to continuously monitor the filter penetration throughout the 60-min test period. Just prior to carrying out an aerosol test, and just after completion of an aerosol test, the filter resistance was measured at a continuous flow rate of 85 l/m if the device was designated to be used singly in a respirator, and at a continuous flow rate of 42.5 l/m if the device was designated to be used as a pair in a respirator.

The results of filter penetration and the resistance to air flow measurements of various types of filters are shown in Table 8. Also, the changes in filter penetration as a function of time are shown in Figures 11 and 12. The data indicate that passing a relatively high concentration of the monodisperse liquid DEHP aerosol particles into the various types of filters does not increase the resistance to air flow provided by these devices. This was expected since liquid DEHP aerosol particles deposited on the fibers of the fibrous materials in the devices form a thin film which surrounds the fibers and does not significantly reduce the porosities of these fibrous materials.

The test results indicate that passage of the liquid DEHP particles, which employ an electrostatic felt type fibrous material and an electret fibrous material, degrades the performance of these fibrous materials for removing liquid DEHP aerosol particles from air. The penetration of the liquid DEHP aerosol particles through the J-1 dust/mist electrostatic felt fibrous material, the J-3 dust/mist electret type fibrous material, the K-2 dust/mist/fume electrostatic felt type fibrous material, and the K-3 dust/mist/fume electret type fibrous material increases significantly as the liquid DEHP aerosol particles are passed into the devices during the test periods.

The data and the graphs demonstrate that the passage of the liquid DEHP aerosol particles into the mechanical type filtering materials such as the J-2 dust/mist filter and the K-1 dust/mist/fume filter, both of which employ paper type fibrous material, does not degrade the performance of this type of particulate filter for removing liquid DEHP aerosol particles from air. A study of the test data indicates that at about one-third of the test period, the penetration of the liquid DEHP aerosol particles through the K-2 electrostatic felt type resin-bonded glass fiber filter starts to become greater than the penetration of DEHP aerosol particles through the J-2 paper type dust/mist particulate-filtering device.

The degradation of the performance of the electrostatic felt type filter material and the electret type filter material for removing liquid DEHP particles from air by the retention of liquid DEHP aerosol particles was expected. The film of deposited liquid DEHP aerosol particles in the electrostatic felt type fibrous material discharges the electrically charged resin granules of this fibrous material and this causes a loss of electromagnetic field of force in the spaces between the fibers of the filter which are used by the fibrous material to remove aerosol particles from air. The film of deposited liquid DEHP in the electret type fibrous material covers and masks the permanently embedded electric charges in the fibers of the filter material. This causes a reduction of electromagnetic field of force in the spaces between the fibers of the filter material which are used by the filter material to remove aerosol particles from air. The paper type filter material utilizes mechanical forces to remove aerosol particles from air. Thus, the film of deposited liquid DEHP in the paper type filter material does not degrade the performance of this type of filter material in removing aerosol particles from air.

Table 8 Penetrations of Particulate-Filtering Devices Used in Nonpowered Air-Purifying Respirators by Monodisperse Liquid 0.3 μm DEHP Aerosol Particles and Resistances to Air Flow Offered by Devices

Particulate-filtering device identification	Type of particulate-filtering device	Penetration of particulate-filtering device by monodisperse liquid DEHP aerosol particles (percent of DEHP particles in test atmosphere)							Resistance to air flow offered by particulate-filtering device	
		0 Min (%)	10 Min (%)	20 Min (%)	30 Min (%)	40 Min (%)	50 Min (%)	60 Min (%)	Initial (mm H_2O)	Final (mm H_2O)
J-1	Dust/mist	13	26	49	51	55	56	56	14	15
		11	28	52	54	56	57	57	15	15
J-2	Dust/mist	19	21	21	21	21	21	20	23	24
		23	25	24	24	24	24	24	23	24
J-3	Dust/mist	11	15	21	31	37	39	41	8	8
		10	3	19	29	34	37	39	9	9
K-1	Dust/mist/fume	3	3	3	2	3	3	3	12	12
		4	3	3	4	3	3	3	11	11
K-2	Dust/mist/fume	4	11	21	26	30	31	32	20	21
		3	12	22	27	32	33	33	20	20
K-3	Dust/mist/fume	3	4	7	11	16	19	21	8	8
		3	5	8	13	17	20	22	7	7
L-1	High-efficiency dust/mist/fume	.002	.003	.003	.003	.004	.004	.004	21	22
		.005	.006	.006	.003	.006	.007	.008	21	21
L-2	High-efficiency dust/mist/fume	.009	.009	.009	.009	.009	.009	.009	18	19
		.001	.001	.001	.001	.001	.001	.001	17	18
L-3	High-efficiency dust/mist/fume	.015	.015	.014	.015	.015	.015	.015	22	24
		.009	.009	.009	.009	.009	.009	.009	23	24

From Reference 5.

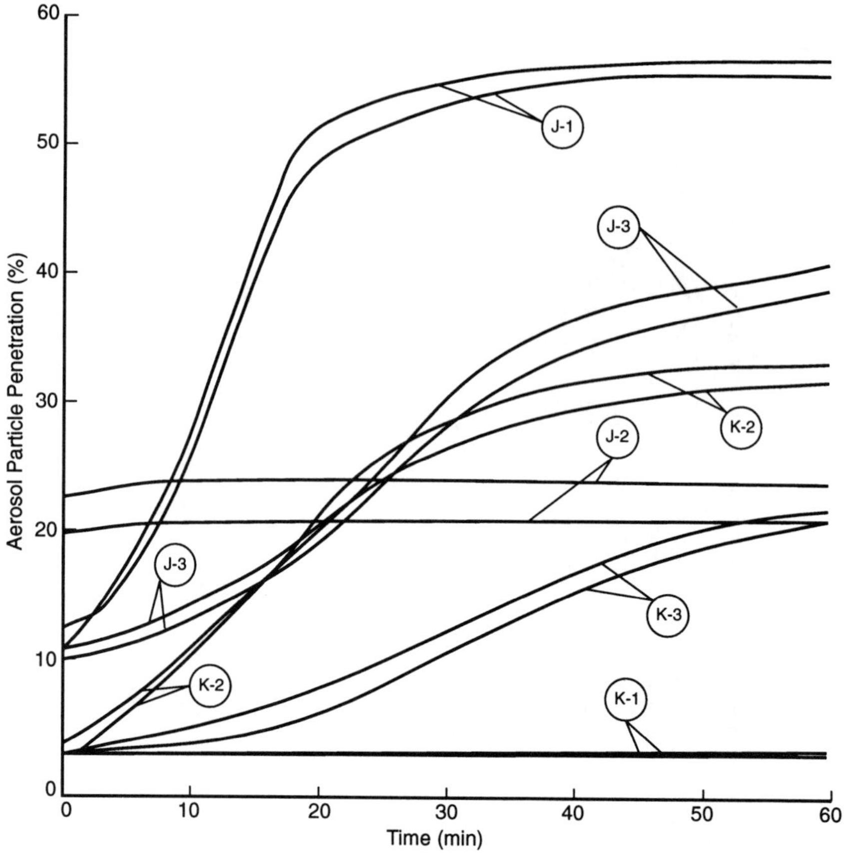

Figure 11 Filter penetration for electrostatic filters against liquid aerosols. (From Reference 5.)

The data indicate that the mechanical type high-efficiency dust/mist/fume particulate filtering devices are extremely efficient for removing monodisperse DEHP aerosol particles of 0.3 μm size from air. This was expected since these devices were developed and designed to permit penetrations of the mentioned aerosol particulate

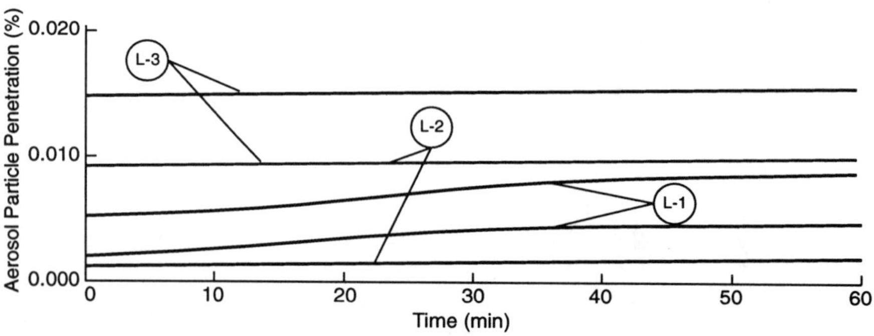

Figure 12 Filter penetration for HEPA filters against liquid aerosols. (From Reference 5.)

Table 9 Combination Filter Cartridges Selected for Testing

Type of device	Identification	Description
Dust/mist filter permanently attached to organic vapor cartridge	X-1	Particulate filtering element having small area attached permanently to the sorbent cartridge; particulate filter element composed of electrostatic felt type fibrous material
Dust/mist filter attached by removal clip to organic vapor cartridge	X-2	Particulate filtering element having small area attached by removal clip to cartridge containing sorbent granules; particulate filtering element composed of electrostatic felt type fibrous material. Device used in a pair in a respirator.
Dust/mist filter attached by removal clip to organic vapor cartridge	Y-1	Particulate filtering element having a convoluted construction with large area attached permanently to cartridge containing sorbent granules; particulate filtering element composed of a paper type fibrous material. Device used in a pair in a respirator.
High-efficiency dust/mist/fume filter permanently attached to organic vapor cartridge	Y-2	Particulate filtering element having a convoluted construction with large area attached permanently to cartridge containing sorbent granules; particulate filtering element composed of a paper type fibrous material.

From Reference 5.

matter not to exceed 0.02% of the concentration of the aerosol particles in the test atmosphere. The test data indicate that for a 60-min test period only one of the three makes and models of the HEPA filter, L-1, was degraded in performance for removing monodisperse submicrometer DEHP liquid aerosol particles from air. However, the degradation of the performance of the L-1 devices was extremely slight.

III. EFFECT OF SOLVENT EXPOSURE ON FILTER PERFORMANCE

The ANSI Z88 ad hoc Subcommittee for Respirator Test and Approval conducted an investigation to determine whether exposing of particulate filtering devices used in air-purifying respirators in air containing an organic vapor would have an effect on the performance of the devices for removing aerosol particles from air.[5]

Four types of particulate filters were selected for testing. All of these devices are components of combination particulate filtering and organic vapor removing devices used in non-powered air-purifying respirators. All these respirators have been approved by MSHA/NIOSH under the provisions of 30 CFR 11. Brief descriptions of these respirators are given in Table 9.

A specimen of each make and model of device was exposed to air containing 1000 ppm carbon tetrachloride at a temperature of 25°C and a relative humidity of 50%, at a continuous flow of 32 l/m if the device is used singly and at a continuous flow rate of 16 l/m if the device is used in a pair. The time of exposure was 60 min. After the organic vapor exposure, the device was tested for aerosol penetration. The test aerosol consisted of monodisperse NaCl particles having a size of 0.12 μm and a concentration of 30 mg/m^3. The aerosol generating and measurement device was the same as used in previous filter penetration tests. The flow rates used in the aerosol penetration test were the same as those used in the organic vapor exposures.

Also, another specimen containing a dust/mist approval which had been exposed to above mentioned organic vapor test was then given a 90-min polydisperse silica dust test in accordance with the requirements prescribed in Subpart K of 30 CFR 11. A specimen device containing a high-efficiency dust/mist/fume filtering unit which had been exposed to the above mentioned organic vapor was given a 60-min monodisperse DEHP aerosol penetration test in accordance with the requirements prescribed in Subpart K of 30 CFR 11. Another specimen device containing a high-efficiency dust/mist/fume filtering unit which had been exposed to the above mentioned organic vapor test was given a 60-min monodisperse NaCl aerosol penetration test in the manner described above. In addition, all of the mentioned aerosol particle penetration tests were performed on specimen devices which had not been given the mentioned organic vapor exposure.

The test results of the aerosol penetration test for the combination particulate filtering and organic vapor removing devices are shown in Table 10. The data indicate that the 60-min exposure of 1000 ppm CCl_4 vapor does not have significant effect on the performance of the particulate-filtering section of the devices for removing aerosol particles.

IV. FACTORS AFFECTING THE PERFORMANCE OF PERMANENT CHARGED FILTERS

Permanently charged electrofibrous (electret) filters are made from fibers of high-resistance polymers with permanent electrostatic charges placed on or near the polymer surface. The permanent charges increase the efficiency of the filter media by providing electrical capture mechanisms in addition to the mechanical capture without increasing air resistance. Usually, the efficiency of permanently charged particulate filters is not significantly affected by high humidity or temperature. However, the electrical enhancement can be reduced or even negated by neutralizing the fiber charge. The mechanisms which could reduce the effectiveness of permanently charged filters were investigated by Biermann et al.[6]

The first series of tests was a loading test. A NaCl aerosol with a mass median aerodynamic diameter (MMAD) of 0.8 μm was used to simulate solid particle loading, and a dioctyl sebacate (DOS) aerosol with a MMAD of 0.9 μm was used to simulate liquid loading. A Filtrete filter with a weight of 200 gm/m^2 was selected for testing. The results of the loading test are shown in Figure 13. The filter efficiency decreases initially and is then followed by a rapid increase. The decrease in efficiency is resulted by the neutralization of the charged particles depositing on the filter. The subsequent increase in efficiency is due to mechanical capture of new particles by previously deposited particles. The point at which efficiency increases is demonstrated by a rapid increase in pressure drop.

For the DOS aerosol loading test, there is a continual decrease in efficiency accompanied by little change in pressure drop. The decrease in filter efficiency is due to the neutralization of filter charges by deposited DOS particles. However, the liquid DOS aerosol does not form a layer of deposited particles on the fiber, which would increase its mechanical collection efficiency.

Table 10 Penetrations of Combination Particulate-Filtering and Organic Vapor-Removing Devices Used in Nonpowered Air-Purifying Respirators by Various Aerosols Previously Exposed to Air Containing an Organic Vapor and Not Previously Exposed to Air Containing an Organic Vapor

Device identification	Type of particulate-filtering section	Organic vapor exposure	Penetration of device by monodisperse sodium chloride aerosol or monodisperse DEHP aerosol (percent of aerosol particles in test atmosphere) 0 Min (%)	10 Min (%)	20 Min (%)	30 Min (%)	40 Min (%)	50 Min (%)	60 Min (%)	Penetration of device by polydisperse silica dust aerosol particles (mg)
X-1	Dust/mist	Yes	10.9	12.3	14.0	15.8	17.9	18.0	16.9	0.45
		No	11.3	13.1	13.7	14.2	15.9	16.6	15.3	0.43
X-2	Dust/mist	Yes	12.8	13.4	14.2	16.1	17.0	17.9	18.6	0.62
		No	13.1	14.8	15.1	16.0	16.8	17.5	18.2	0.61
X-3	Dust/mist	Yes	9.6	12.9	13.1	13.6	13.5	12.9	11.8	0.51
		No	8.9	12.0	12.4	12.6	12.5	11.8	11.0	0.53
Monodisperse liquid DEHP aerosol particle penetration tests										
Y-1	High-efficiency dust/mist/fume	Yes	.005	.005	.006	.005	.005	.005	.005	
		No	.004	.004	.005	.005	.005	.005	.005	
Y-2	High-efficiency dust/mist/fume	Yes	.010	.010	.010	.010	.010	.010	.010	
		No	.089	.009	.009	.009	.009	.009	.009	
Monodisperse solid sodium chloride aerosol particle penetration tests										
Y-1	High-efficiency dust/mist/fume	Yes	.01	.01	.01	.01	.01	.01	.01	
		No	.01	.01	.01	.01	.01	.01	.01	
Y-2	High-efficiency dust/mist/fume	Yes	.01	.01	.01	.01	.01	.01	.01	
		No	.01	.01	.01	.01	.01	.01	.01	

From Reference 5.

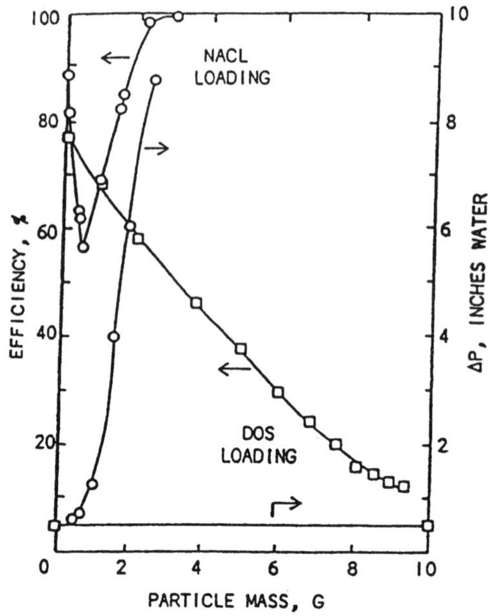

Figure 13 Filtrete 200 g/m² performance with aerosol loading, U = 64 cm/s. (From Reference 6.)

The aerosol penetration as a function of particle diameter for NaCl aerosols is shown in Figure 14. The penetration for the clean filter is compared with three additional penetration curves corresponding to increasing deposits of NaCl particles on the filter media. The penetration for all particle sizes initially increases with loading up to 0.5 g, and then decreases with further aerosol deposition. The point of maximum penetration has shifted to smaller sizes with increasing particle loading because the dominant collection mechanism has shifted from electrical capture to mechanical capture by particle deposits that develop on the fibers.

The measurement of aerosol penetration as a function of particle diameter for DOS aerosols on different loading is shown in Figure 15. There is an overall increase in the penetration of all particle sizes, and an apparent shift to the larger particle diameters having maximum penetration. The diameter of maximum penetration shifts to larger particle diameters because of the loss of electric attractive force in the permanently charged filters. The filter efficiency for particles larger than 0.5 μm diameter is the same for an electrofibrous filter with either charged aerosols or neutral aerosols. In contrast, the efficiency for particles smaller than 0.5 μm diameter is strongly dependent on the particle charge.

The efficiency of permanently charged filters may also be affected by environmental conditions, most notably by exposure to water. Filters were immersed in water solutions for 30 to 40 min, then rinsed three times and allowed to dry. The filter efficiency from exposure to different types of water solutions is presented in Figure 16. The filter efficiency before and after exposure to different solutions was measured by a light scattering photometer when the filter was challenged with the NaCl aerosol. The results indicated that there is some reduction in efficiency when the filter is exposed to water and nitric acid, NaCl solutions, or a solution of water and a

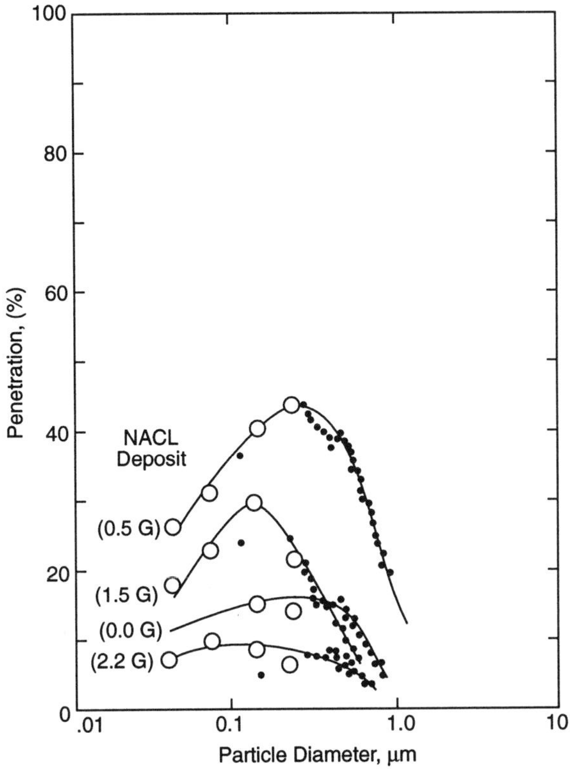

Figure 14 Aerosol penetration at increasing mass loading for NaCl aerosols, Filtrete 200 g/m², U = 64 cm/s. (From Reference 6.)

surfactant. The deterioration was caused by the neutralization of charges readily accessible on the surface of the filter media. A significant reduction of the filter efficiency was evidenced when the filter was treated with a solution containing water, surfactant, and NaCl. This effect was due to the addition of NaCl ions which was able to neutralize the filter charge when the fibers were wetted by a surfactant.

The filter penetration as a function of particle diameter before and after the treatment with water solutions listed above is shown in Figure 17. There is a shift in the diameter of maximum penetration toward larger particle sizes as the permanently charged filter is neutralized.

The effect of organic solvent on filter efficiency was also examined. This may be a consideration when a combination cartridge or canister is used for protection against toxic particulate and gaseous matters. The Filtret media was placed in a sealed container saturated with a mixture of organic solvents for 24 h. The filter was then challenged with the NaCl aerosol. The penetration was measured by a light scattering photometer. The reduction in filter efficiency varied with the challenging solvent. The test results are shown in Figure 18.

There was a minimal reduction in filter efficiency when the filter was treated with saturated hydrocarbons such as hexane. There is a marked reduction when the filter was treated with cyclic hydrocarbons such as benzene. However, there was a dramatic reduction in filter efficiency when the filter was exposed to unsaturated

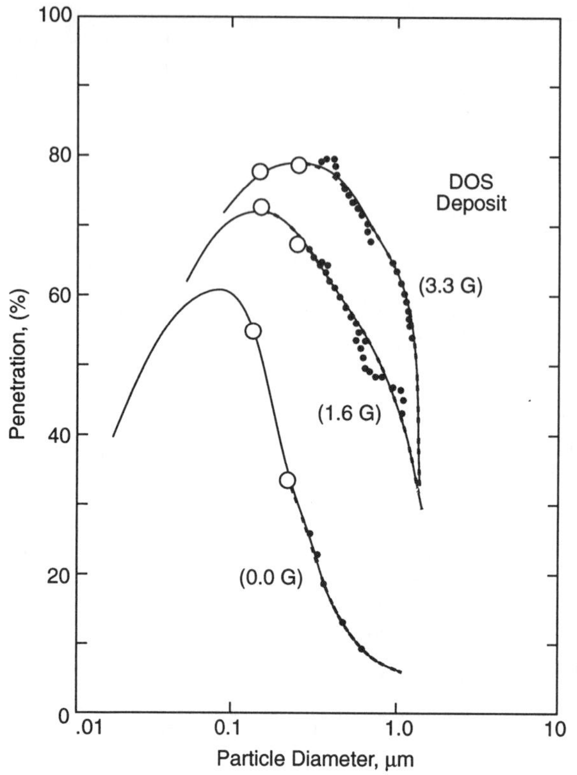

Figure 15 Aerosol penetration at increasing mass loading for DOS aerosols, Filtrete 200 g/m², U = 64 cm/s. (From Reference 6.)

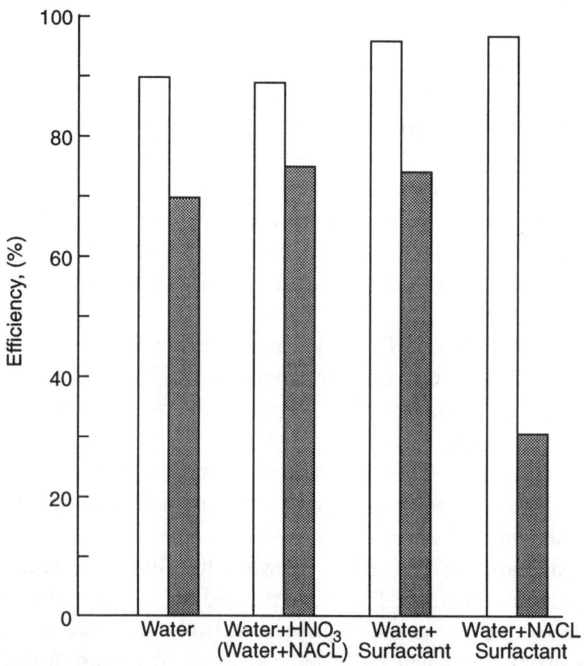

Figure 16 Efficiency of Filtrete 200 g/m² before and after immersing in water solutions, NaCl aerosols, U = 64 cm/s. (From Reference 6.)

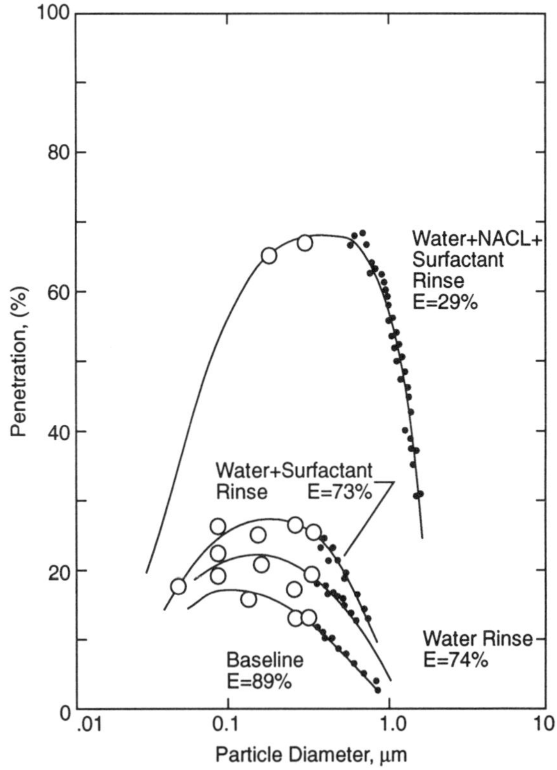

Figure 17 Aerosol penetration of Filtrete 200 g/m² after exposure to various water solutions, U = 64 cm/s. (From Reference 6.)

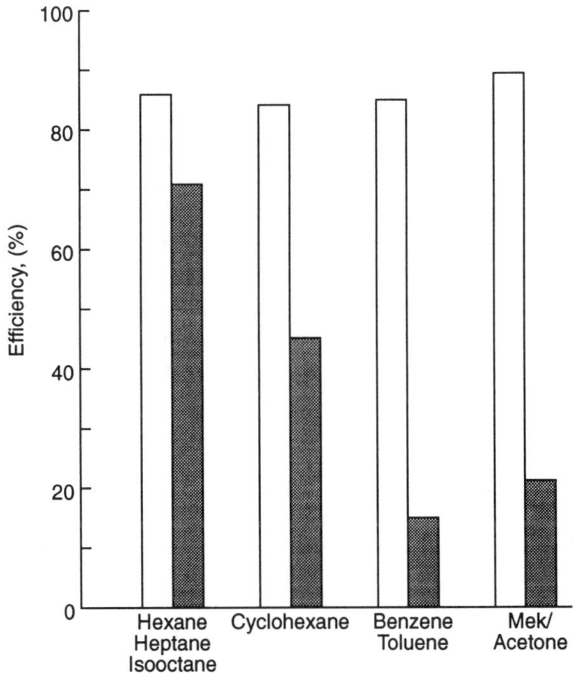

Figure 18 Efficiency of Filtrete 200 g/m² after exposure to organic chemical NaCl, U = 64 cm/s. (From Reference 6.)

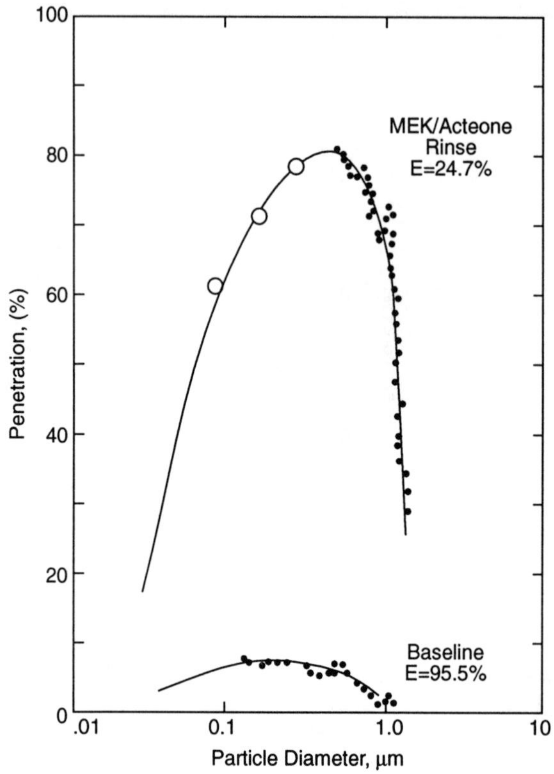

Figure 19 Aerosol penetration of Filtrete 200 g/m² after an MEK/acetone rinse, NaCl, U = 16 cm/s. (From Reference 6.)

hydrocarbons such as ketones. Because all these solvents were not ionic, no neutralization of filter charges occurred. A possible reason for the increased filter penetration was that solvents reacted chemically with the surface charge (presumably present on the surface as free radicals or ions).

The filter penetration measurement before and after solvent rinsing with methyl ethyl ketone is presented in Figure 19. The filter was challenged with the NaCl aerosol and the penetration was measured by a light scattering photometer. Again, the peak penetration, after the filter was rinsed with the ketones, has shifted to larger particle diameters. There was a dramatic increase in penetration after the filter was discharged. The authors concluded that

> Since the filter discharging problem is inherent to permanently charged filters as presently designed and will limit the wide spread use of these filters. Since the filter discharging problem is due to charged aerosols or reactive chemicals, field applications will have to avoid these agents. Thus, permanently charged filters will be ideally suited for filtering neutral or low charged aerosols as may occur in filtering atmospheric aerosols in building ventilation systems. These filters will not perform well in controlling particulate emissions form various industrial processes since the aerosols are generally highly charged.

Table 11 DOP Penetration of Electret HEPA Filters

Approval no.	Filter part no.	DOP leakage (%)	Time (min)
TC-21C-437	9970	0.140	23.5
TC-21C-437	9970	0.148	23.5
TC-21C-437	9970	0.161	23.5
TC-21C-488	2040	1.420	23.5
TC-21C-488	2040	1.520	23.5
TC-21C-488	2040	1.670	23.5
TC-21C-265	7255	0.025	23.5
TC-21C-265	7255	0.018	23.5
TC-21C-265	7255	0.021	23.5
TC-21C-439	7260	0.018	23.5
TC-21C-439	7260	0.011	23.5
TC-21C-439	7260	0.011	23.5

From Reference 7.

Since the late 1980s, two respirators equipped with electret type HEPA filter media were approved by MSHA/NIOSH from a single manufacturer. One is a disposable respirator (TC-21C-437); the other is a thin disk shaped filter without a protective enclosure (TC-21C-488). Because the previous ANSI and the Lawrence Livermore penetration studies on the electret filter media were not conducted on the HEPA filters, NIOSH conducted a DOP aerosol penetration study prescribed in 30 CFR 11 to determine whether the electret type HEPA would degrade upon exposure to DOP mist.[7] For comparison purposes, two "mechanical" type HEPA filters made by the same manufacturer were also evaluated (TC-21C-265 and 439). The test results are shown in Table 11. As expected, the two "electret" type HEPA filters no longer met the acceptable penetration of 0.03% after 23 min of testing. However, there is no change in penetration for the two "mechanical" type HEPA filters. The difference in penetration values between the disk type HEPA filter and the disposable HEPA respirator is due to the latter having more layers of filter media. It then takes longer for the DOP aerosol particles to neutralize the charges built up on the filter media.

A respirator equipped with an electret type filter material may pass the DOP test prescribed in the 30 CFR 11 to be certified as a respirator equipped with a high-efficiency particulate air filter (HEPA) since the DOP test is only performed for 10 s. There may not be any assurance for continued protection provided by the electret type HEPA filters when the respirator is used in the industrial environment. Since HEPA filters are often required by OSHA, NRC, and DOE, for protection against highly toxic or radioactive air contaminants, the degradable HEPA filters would not be suitable for these applications.

V. FILTER AND LEAK CHARACTERISTICS OF FILTERING FACEPIECES

Filtering facepieces (disposable respirators) have been popular for their low cost. Many employers choose these respirators to be in compliance with OSHA regulations. It is commonly known that, in general, disposable respirators do not provide the same level of faceseal and filter efficiency as respirators equipped with elastomeric facepieces. Although there have been many studies conducted to evaluate the

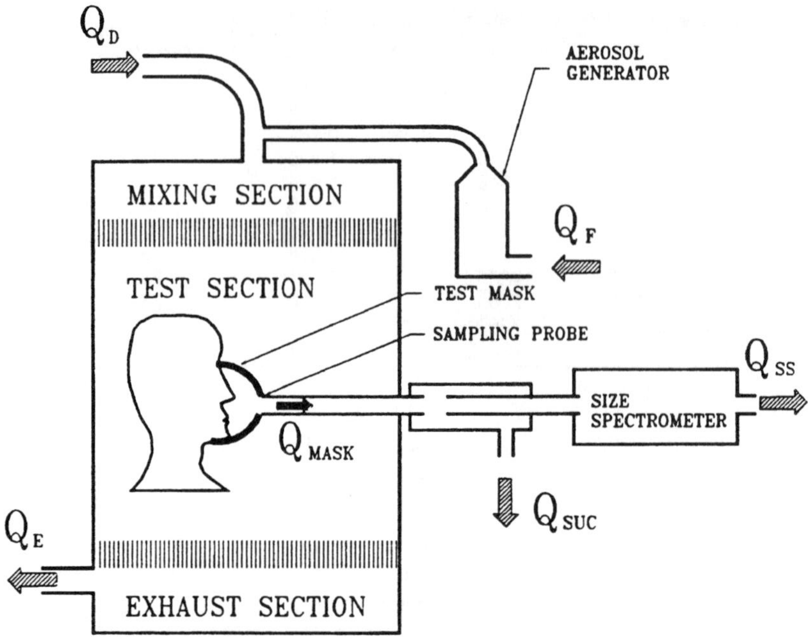

Figure 20 Schematic diagram of experimental setup. Q_D, dilution air flow rate; Q_E, exhaust air flow rate; Q_F, filtered, regulated air flow rate; Q_{SUC}, suction air flow rate; Q_{SS}, sampling flow rate of aerosol size spectrometer; Q_{MASK}, air flow rate through mask. (From Reference 8.)

filter performance of respirators equipped with elastomeric facepieces, there are very little studies to evaluate the efficiency and leak characteristics of filtering facepieces. Chen et al. conducted a series of studies to evaluate the performance of disposable respirators.[8-10]

A test system was developed to evaluate the leak characteristics of filtering facepieces (disposable respirators). The aerosol was generated by a Wright nozzle with several impactors to control the particle sizes. The output aerosol was diluted with air and introduced to a 2 m^3 test chamber which contained a test mannequin. The test setup is shown in Figure 20. The air flow rate through the filtering facepiece varied between 0 and 90 l/m; a sampling rate of 5 l/m gave an overall mask flow rates between 5 and 95 l/m. The tests were performed with corn oil which had a count medium diameter of 2.3 μm and a concentration of 640 particles/ml. The test aerosol had a bimodal distribution with two peaks at approximately 0.75 and 2.8 μm. The purpose of the bimodal distribution was to distinguish the removal mechanism of submicrometer aerosols from that of supermicrometer aerosols (see Figure 21). The particle penetration was measured with a size spectrometer.

The first study was to evaluate filter and leak characteristics of 3M 8715 disposable dust/mist respirators.[8] During the test, the respirator was sealed to a mannequin. The test period was 1 h. To reduce the effect of lung deposition, the aerosol penetration characteristics were measured during inhalation only.

The size distributions for the in-mask aerosol are shown in Figure 22: (A) shows the distribution for the mask sealed to the mannequin, and (B) shows the distribution

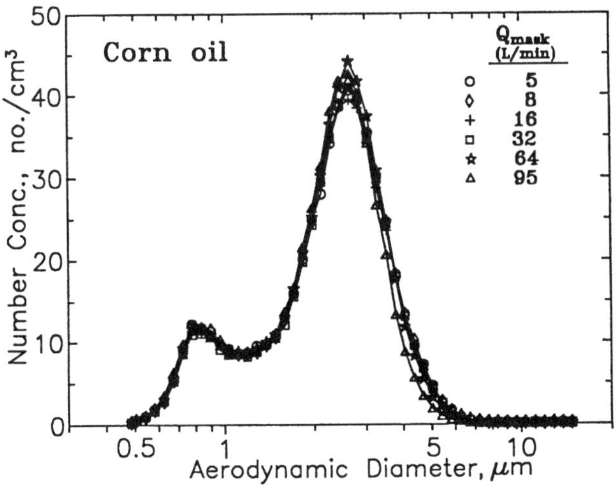

Figure 21 Sampling train performance for Q_{MASK} = 5 to 95 l/m. (From Reference 8.)

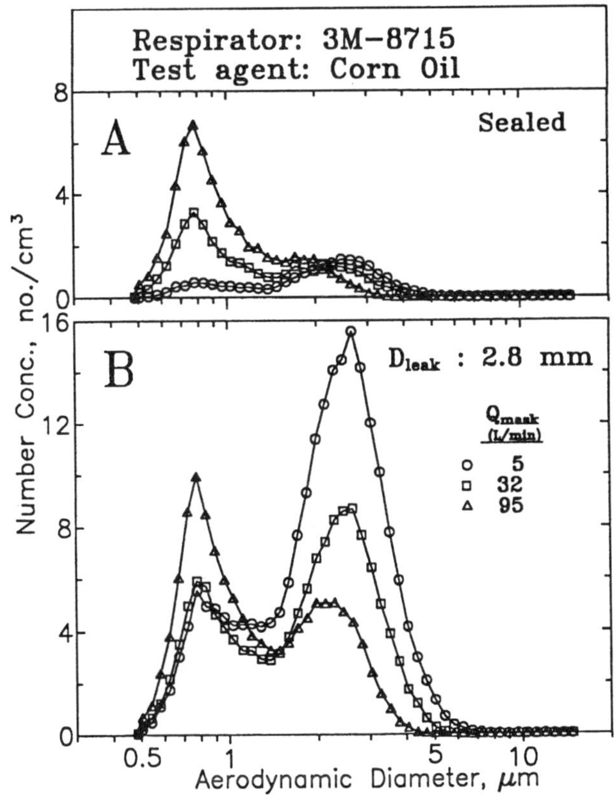

Figure 22 Aerosol size distributions measured inside the mask for the external aerosol of Figure 4. (A) Mask sealed to face; (B) face seal leak hole = 2.8 mm. (From Reference 8.)

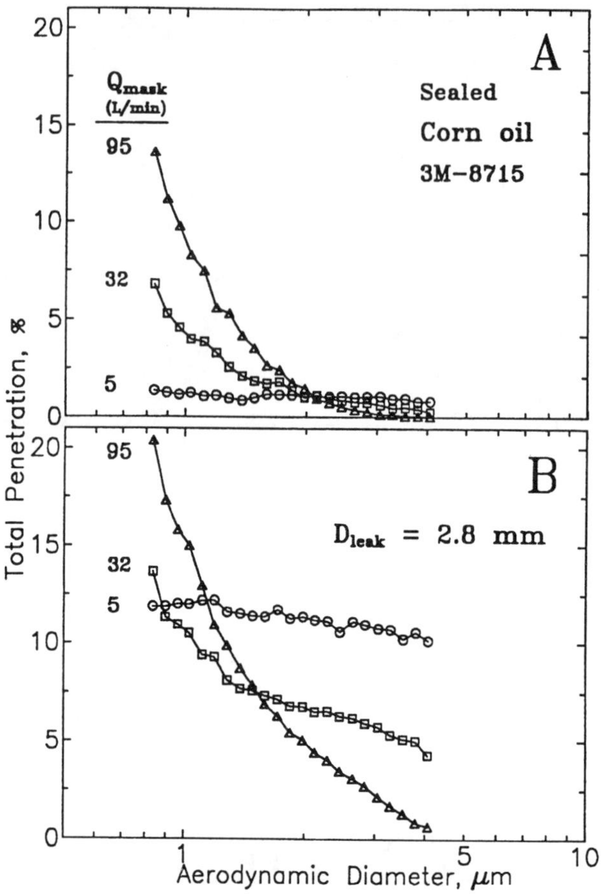

Figure 23 Respirator performances at different flow rates. (A) Mask sealed to face; (B) face seal leak hole = 2.8 mm. (From Reference 8.)

with a leak of 2.8 mm. The aerosol penetration that occurred at different flow rates is shown in Figure 23. Figure 22A shows that the size of the upper size mode reduced more than the lower size mode. Aerosol penetration of the smaller particles in the lower mode increased with air flow. This indicates that electrostatic attraction was the main filtration mechanism. As air flow increased, the time for electrostatic removal was reduced. Hence, the penetration was increased. Aerosol diffusion was similarly time dependent but had a fairly insignificant effect for aerosols larger than 0.5 μm.

The aerosol distribution in the upper size mode varied in the opposite direction. As the air flow increased, the concentration of penetrated aerosols decreased and the mean size of that size range shifted to a smaller size. This indicated that impaction was the primary removal mechanism for the size range. Interception may be important for aerosol particles close to the most penetrating size which was below the size of the measuring instrument. The effect of sedimentation was negligible. When a leak hole of 2.8 mm was added, additional particles entered the respirator cavity and

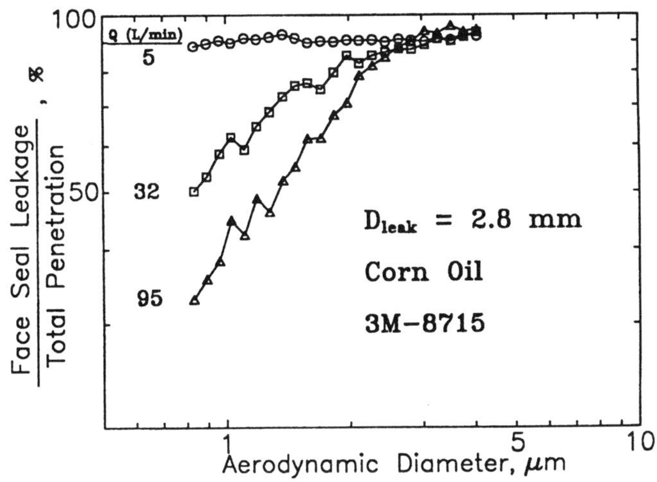

Figure 24 Dependence of percent aerosol leakage on particle size. (From Reference 8.)

increased the aerosol concentration in both size ranges. The most increase occurred in aerosol concentration at the lowest flow rate.

Figure 23A shows that at a low flow rate of 5 l/m aerosol penetration through the filter was 1.5% — regardless of particle size (0.8 to 4 μm). When a leak was added, penetration increased equally for all sizes. At a very low air flow, the aerosols entered the respirator cavity without much removal. At higher air flow, the air velocity in the leak channel and the filter was increased, and removal by impaction occurred in both paths for the larger particle size. Figure 23B shows that for a given leak, the aerosol concentration inside the respirator facepiece was highly flow dependent.

Figure 24 shows the percentage of aerosols entering the respirator cavity through the face seal leak. At a very low air flow (5 l/m), about 95% of aerosols entered the respirator cavity through the leak site, irrespective of particle size. At higher air flow (95 l/m), about 95% of the 4-μm particles entered through the leak site but only about 30% of the 0.8-μm particles. Thus most submicrometer particles entered through the filter material even in the presence of a sizable leak hole.

Faceseal leakage can affect the protection provided by respirators. Leaks can be generally classified as circular or rectangular (slit) in shape. The penetration of aerosols through the leak site depends on the pressure difference. The second series of filter efficiency study on filtering facepieces conducted by Chen and Willeke[9] was to determine the effect of leak shape and filter resistance to faceseal leakage on disposable respirators. The test system was the same as described above[8] except that a Kr-85 radioactive source was added to neutralize the charges of the generated corn oil aerosol and the aerosol was measured by two instruments. The larger aerosols (0.8 to 5 μm) were measured by an aerodynamic particle sizer, and the smaller aerosols (0.1 to 3 μm) were measured by a laser aerosol spectrometer. To minimize the coincidence error in counting, the Challenge aerosol had a concentration of 650 particles/ml with a count mean of 2.3 μm.

One brand each of disposable respirators equipped with a dust/mist filter and a high-efficiency particulate air (HEPA) filter were selected for evaluation. This combination represents low and high pressure drops across the filter media for a given flow rate. The leaks were studied by gluing circular and rectangular tubes into the respirator facepiece about 6.5 cm from the centered sampling probe. Each respirator was sealed to the mannequin by petroleum jelly. A circular hole having a diameter of 3.18 mm with a cross-sectional area of 7.92 mm^2 was chosen as the reference hole size for leak testing. Leakage through leak channels of the same cross-sectional area, but of a smaller effective diameter was tested. The leak combination consisted of a single hole with a diameter of 3.18 mm, four holes each with a diameter of 1.59 mm, and 16 holes each with a diameter of 0.79 mm. The long slit had a size of 0.5 mm × 15.84 mm. The leak channels were 15 mm long with the effective test lengths of 1, 15, and 30 mm. Five different flow rates: 5, 10, 30, 60, and 100 l/m were selected for leak testing.

Figure 25 shows the result of testing with large leak (3.18 mm) and small leak (1.59 mm) holes at different flow rates. The submicrometer aerosol penetration for the DM respirator was highest for the highest flow rates. It is an indication that most of the filtration removal in the DM filter was by electrostatic force. Mechanical removal mechanisms such as impaction and interception dominated in the supermicrometer size range. This mechanism for filtration and leak flow showed a crossover of the curves near 1 μm. For the HEPA respirator, the penetration was low and the penetration curves were relatively flat over a wide range of flow rates. At low leak flow (5 to 10 l/m), low particle size dependence existed over the measured particle size range. At these low flow rates, the pressure differential (Δp) was low and a relatively high portion of the total flow entered the leak. The curves were fairly flat because particles that passed through the relatively large leak channels were nonturbulent. At higher flow rates, reflecting average or higher breathing rates, strong particle size dependence existed in the supermicrometer size range. This was due to impaction losses in the inlet portion of the leak and increased losses in the leak itself where the flow was in the transition or turbulent regime.

For the large leak hole in the DM respirator, aerosol penetration curves for the high flow rate were the same as those for the small leak hole. Only for the two low flow rates were the percentage of aerosol penetration significantly different for different leak sizes. At low flow rates, small particles passing through the filter media were effectively removed by the electrostatic force and the percentage of flow through the leak was increased, as also seen for the HEPA filter. At low leak flow rates (5 to 10 l/m), the penetration for the DM filter was lower than for the HEPA filter.

The effect of leak hole size on penetration was observed through the testing of the HEPA filter respirator because there is no leak through the filter media. Four different hole sizes with effective leak diameters of 0.79, 1.59, 2.38, and 3.18 were investigated. The test results (Figure 26) indicate that penetration was very low at the small hole size. Penetration increased as hole size increased. The penetration of supermicrometer aerosols through leak holes was proportional to the leak size and also to the flow rate. The penetration curves converged at 1 μm. The aerosol penetration of supermicrometer size aerosols fell off rapidly as the flow increased. This decrease was attributed to impaction.

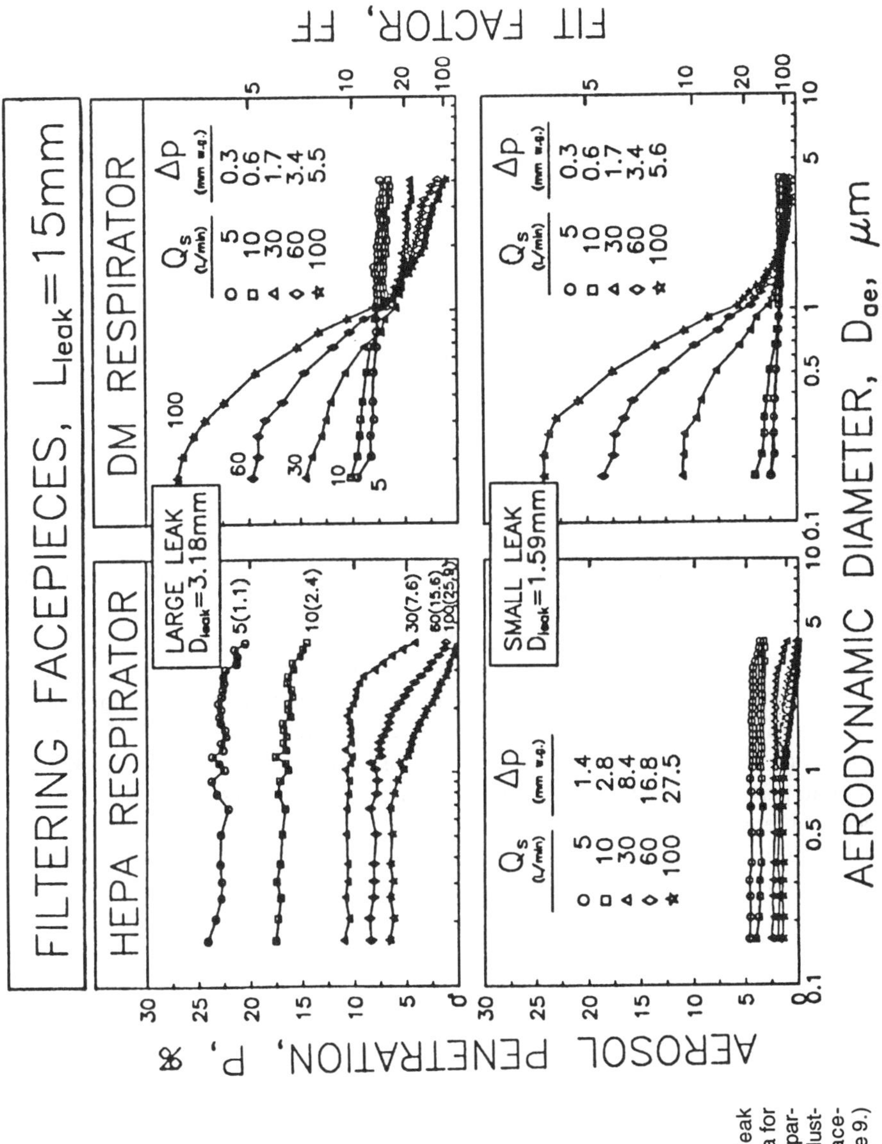

Figure 25 Combined filter and leak aerosol penetration data for typical high efficiency particulate air (HEPA) and dust-mist (DM) filtering facepieces. (From Reference 9.)

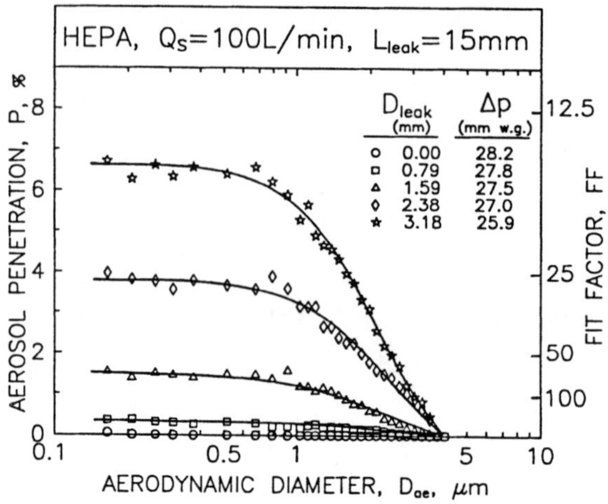

Figure 26 Aerosol penetration dependence on leak size. (From Reference 9.)

The effect of leak length was also investigated on the HEPA respirator at three different lengths: 1, 15, and 30 mm. An increase in leak length resulted in a higher pressure drop and a lower percentage of aerosol penetration. The difference in penetration between long and short leaks was more pronounced at a low flow rate than at a higher flow rate. Aerosol penetration at high flow rate drops off more rapidly with an increase in supermicrometer particle size than at the low flow rate. Penetration decreased with an increase of leak hole length. The leak curves at different particle sizes at the high flow rate was essentially the same irrespective of leak hold length (see Figure 27).

In tests where the respirator leakage had the same effective diameters, pressure drop was smallest for a single hole than 4 holes and the slit. The highest pressure drop occurred at the 16 small holes since the multiple holes and slit increased flow

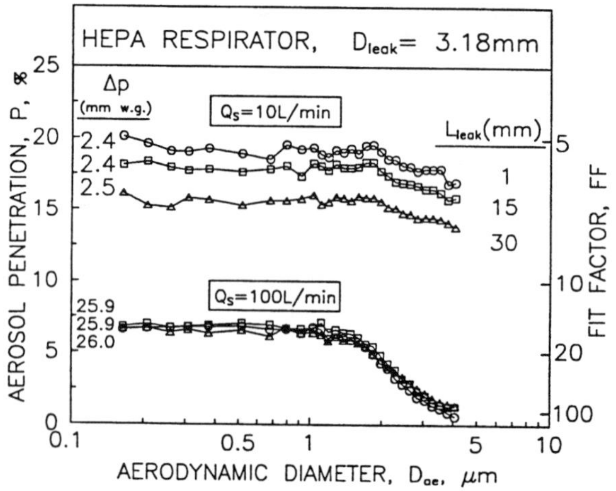

Figure 27 Effect of leak length on aerosol penetration. (From Reference 9.)

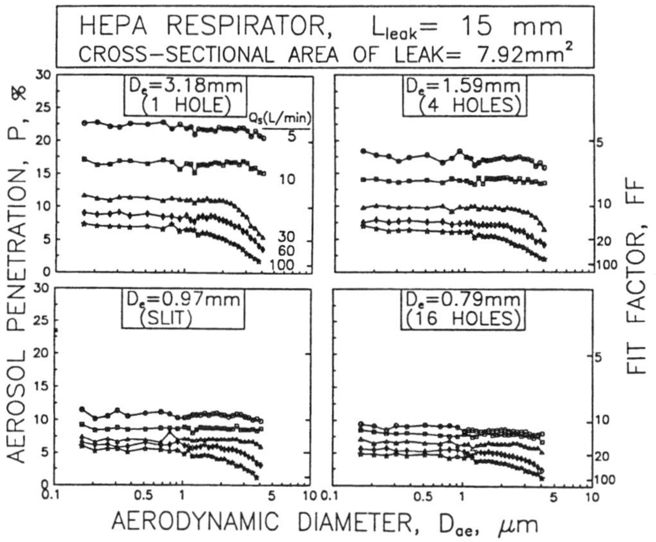

Figure 28 Effect of leak shape on aerosol penetration. (From Reference 9.)

resistance and reduced the potential of leak. Figure 28 shows aerosol penetration through respirators with leaks of four different effective diameters. Aerosol penetration increased with the effective diameter of a leak of a given cross-sectional area because air resistance and aerosol deposition decreased with the increase of effective diameter. A decrease in effective diameter would cause a decrease in the percentage of aerosol penetration and the range of penetration values. The effect of effective leak diameter on aerosol penetration on one fixed flow rate (100 l/m) is shown in Figure 29. Aerosol penetration was moderately dependent upon particle size in the submicrometer region; however, it was strongly dependent on particles larger than

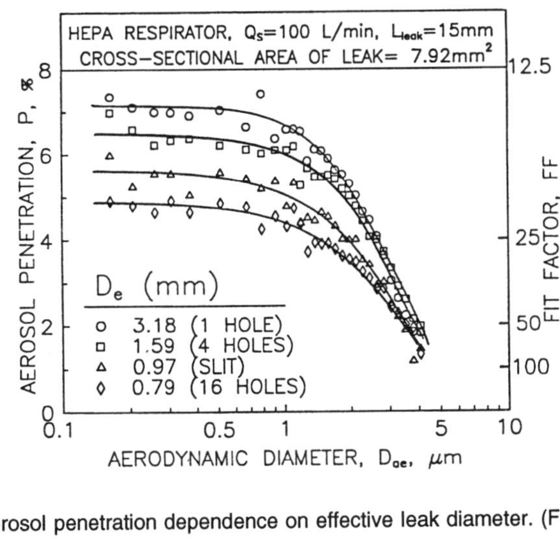

Figure 29 Aerosol penetration dependence on effective leak diameter. (From Reference 9.)

Table 12 Respirators of Four Categories by Four Manufacturers

Respirator category	Respirator code	Respirator type[a]	Manufacturer code	Δp[b] (mm H_2O)	P (%)[c]	q_F[d] (1/cm H_2O)
Nuisance dust	ND-A	FF	A	1.68	83.93	0.0104
	ND-B	FF	B	1.37	85.27	0.0117
Dust, mist	DM-A	FF	A	4.83	21.48	0.0318
	DM-B	FF	A	6.10	26.03	0.0221
	DM-C	FF	C	4.00	46.24	0.0193
	DM-D	FF	B	6.73	62.22	0.0070
	DM-E	CR	A	13.78	10.93	0.0161
	DM-F	CR	D	10.38	12.17	0.0203
Dust, mist, fume	DMF-A	FF	A	12.78	5.01	0.0234
	DMF-B	FF	C	16.50	23.11	0.0089
	DMF-C	CR	D	10.77	8.69	0.0227
High efficiency particulate air filter	HEPA-A	CR	A	25.05	0.0191	0.0342
	HEPA-B	CR	A	28.42	0.0032	0.0360
	HEPA-C	FF	A	26.97	0.0089	0.0346
	HEPA-D	CR	D	23.70	0.0033	0.0436

[a] FF: filtering facepiece; CR: cartridge.

[b] Δp: pressure drop at mask flow of 100 l/m.

[c] P (%): penetration percentage of 0.3-μm aerosol.

[d] q_F: filter quality factor (0.3-μm aerosol, 100 l/m).

From Reference 10.

about 1 μm. The curves for the different effective diameters converged at about 5 μm for the indicated flow rate. The aerosol penetration of submicrometer size aerosols through the circular leak holes and the long slits of equal cross-sectional was about 7% and 5%, respectively; a difference of about 40%.

The authors concluded that if there is a substantial lack of face seal fit and the wearer performs a light workload, the disposable DM respirator may perform better than the HEPA respirator since the high pressure drop across the HEPA filer media pulls a greater amount of aerosols through a leak of given size than the much lower pressure drop across the DM respirator. If the leak size is small, the HEPA respirator always performs better than the DM respirator, as intended. This observation may not be applicable to elastomeric facepieces since the "mechanical" type HEPA filters for elastomeric facepieces have a much larger surface area than the corresponding filtering facepiece. The filtering facepiece equipped with a dust/mist or HEPA filter has the similar surface area. The difference in pressure drop between HEPA and DM filter media for the elastomeric facepiece respirators has a factor of two; however, the difference in pressure drop between a disposable dust/mist filter respirator and a disposable HEPA filer respirator can have a factor of 5.

The third series of the filter efficiency study conducted by Chen et al.[10] was to determine the efficiency of various disposable respirators and filter cartridges of elastomeric facepieces against different sizes of test aerosols. Fifteen filters from four manufacturers were selected for testing: two disposable nuisance dust respirators, four disposable DM respirators and two DM filters, two disposable fume respirators and a fume cartridge, one disposable HEPA respirator, and three HEPA filter cartridges (see Table 12).

The same test system described above was used in this study. The respirator or facepiece was sealed to a mannequin with petroleum jelly. The filter penetration was

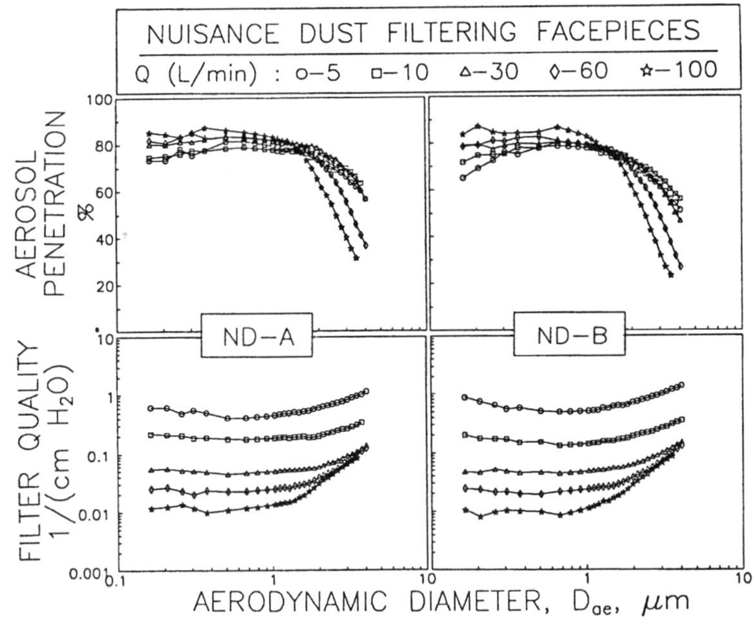

Figure 30 Filter penetration and filter quality of nuisance dust (ND) respirators. (From Reference 10.)

measured by at least three samples of each product. The test aerosol was corn oil with a concentration of 600 particles/ml and a size range between 0.2 and 15 μm. The count mean diameter of the challenge aerosol was 2.3 μm with a geometric standard deviation of 1.7. A Kr-85 radioactive source was utilized to neutralize the test aerosol. The large size aerosols (0.8 to 15 μm) were measured by an aerodynamic particle sizer, while small size aerosols (0.2 to 3 μm) were measured by a laser aerosol spectrometer. The air flow through the filter was continuous at 5, 10, 30, 60, and 100 l/m. The sampling time was set for 3 min. The HEPA filters were tested for 3 to 5 h at 100 l/m only in order to collect a sufficient number of particles for measurement.

The authors proposed a term of filter quality factor, q_F, which correlates filter penetration (P), and flow resistance (Δp). q_F is expressed as:

$$q_F = \ln(1/P)/\Delta p \tag{1}$$

A good respirator should have low filter penetration and low pressure drop. A higher filter quality factor means better performance. However, this factor should only be used to compare filters within the same class, e.g., DM. Table 12 shows the filter quality data for test respirators at a flow rate of 100 l/m.

The test results for the nuisance disposable respirators, DM filters, fume filters, and HEPA filters are shown in Figures 30, 31, 32, 33, and 34, respectively. The two nuisance dust respirators (Figure 30) had similar penetration characteristics. They allowed an 80% penetration of submicrometer aerosols. The penetration dropped off

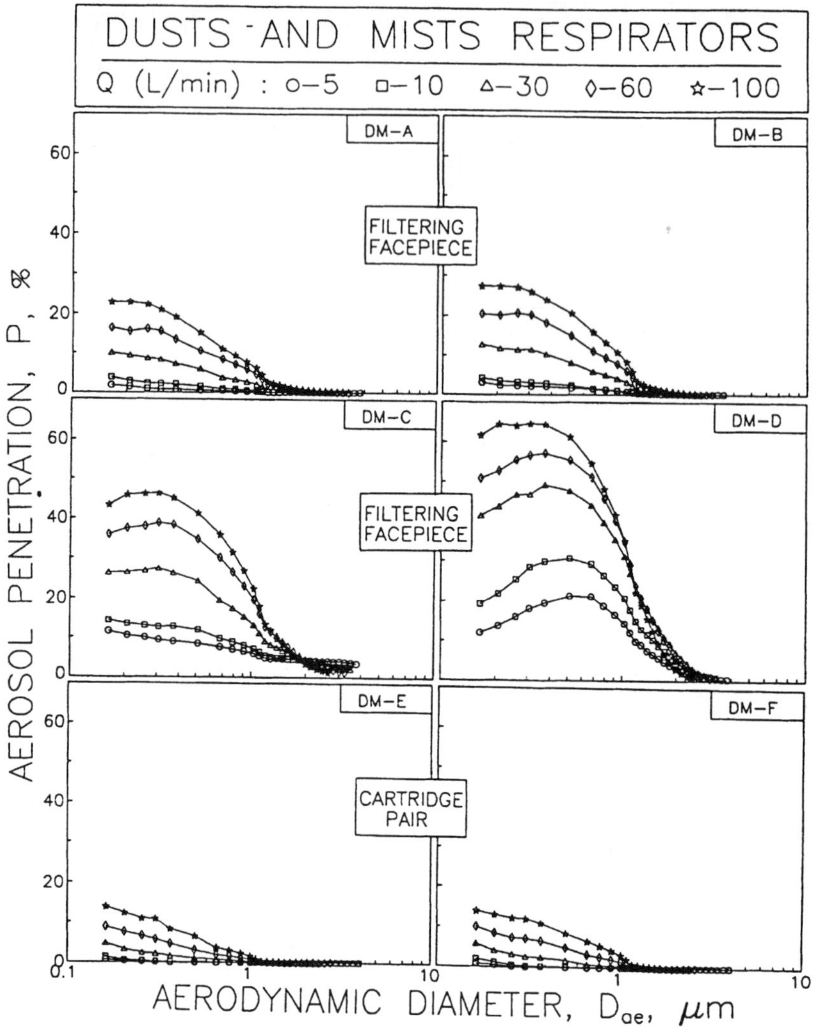

Figure 31 Filter penetration of six dust-mist (DM) respirators. (From Reference 10.)

around 2 μm. The penetration decreased with air flow which indicated that the large aerosols were removed by impaction. The filter quality increased with particle size and decreased with air flow.

For the DM filters (Figure 31), the penetration of the filters was much lower than from the disposable respirators. However, it had the widest variation of penetration among the different filters. The penetration had a stronger flow dependency for submicrometer aerosols than the nuisance dust respirators. This indicates that the aerosols were removed mainly by electrostatic force. The penetration of 0.3 μm aerosol particles ranged from 11% for DM-E to 64% for DM-D when tested at an air flow of 100 l/m. The aerosol penetration converged at 1 μm and diminished at larger particle sizes. The filter quality curves (Figure 32) showed that the disposable DM respirator had a higher filter quality than the two cartridge type filters. However, the

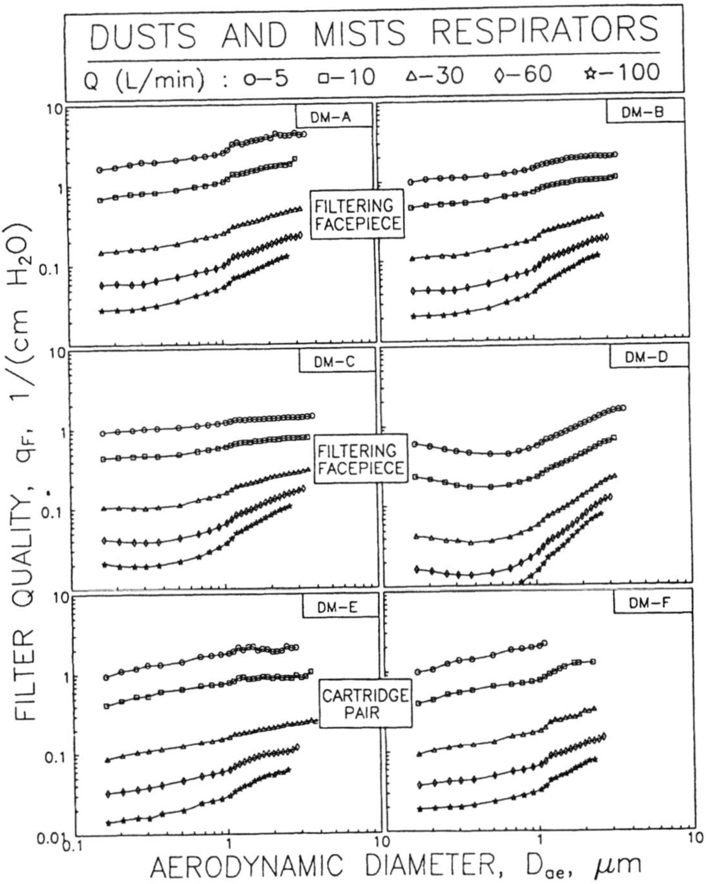

Figure 32 Filter quality q_F of six dust-mist (DM) respirators. (From Reference 10.)

faceseal leakage for the disposable respirators was much worse than the elastomeric facepieces.

The test results for the fume filters (Figure 33) indicated a lower penetration than the DM filters except for the disposable fume respirator B which had exceptional higher penetration similar to DM filters. Filter cartridge C had the best filter quality since it is made of pleated material and has a larger surface area than the disposable fume respirator A.

All tested HEPA filters (Figure 34) met the MSHA/NIOSH certification requirement for penetration (0.03% for a 0.3 μm challenge). Filter cartridge D with the lowest flow resistance showed the highest filter quality factor.

All of the data results from testing with new, unused filters and disposable respirators. For "mechanical" type filters, in which only the mechanical forces of diffusion, interception, and sedimentation remove particles, aerosol loading increases the filter efficiency with time. If the filter material is "electrostatic or electret," the coating of the fibers by particles reduces the electrostatic effects. The removal efficiency decreases with time. When enough loading has occurred, the blocking of

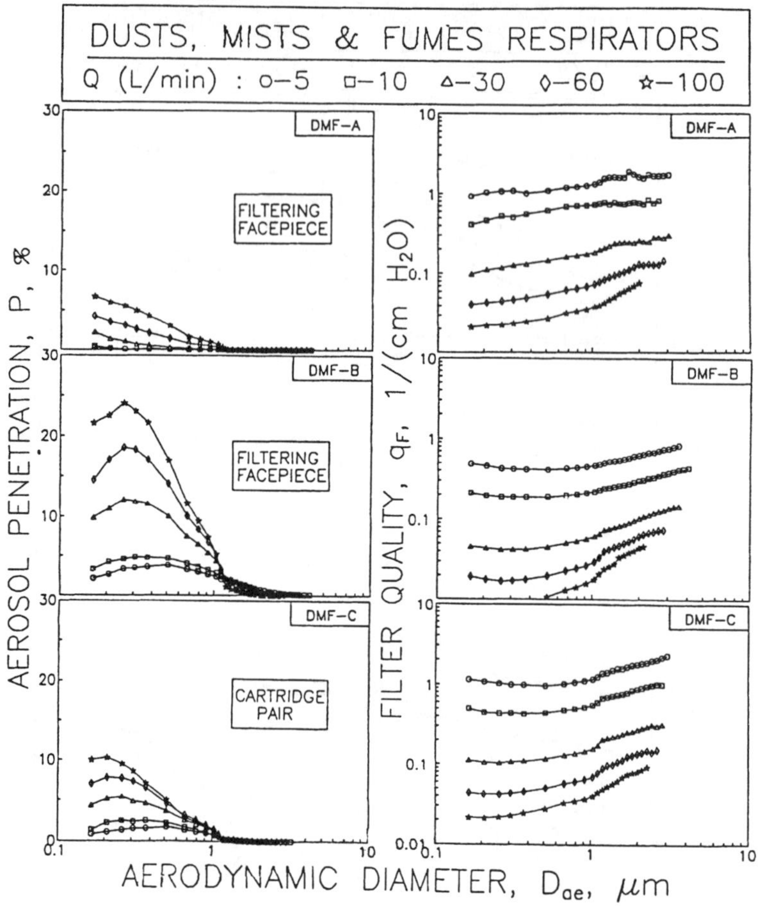

Figure 33 Filter penetration and filter quality q_F of three dust, mist, and fume (DMF) respirators. (From Reference 10.)

the intestinal spaces again increases the collection efficiency. However, the concentration of corn oil and the test time was not sufficient to coat the fibers of the filter and reduce the aerosol collection efficiency.

Two of the DM disposable respirators and one disposable fume filters failed to meet the certification penetration requirement of 1%. However, all of these passed the current certification test. One approved disposable fume respirator had unusually higher penetration which was higher than some DM filters tested. The variation in penetration for the DM filters was as high as 5.5. The authors concluded that by comparing the performance of DM filters, filter performance at 0.2 to 0.6 μm range does not predict filter performance at the supermicrometer size range. Therefore, testing at one submicrometer and one supermicrometer particle size appears to be the minimum necessary for measuring and differentiating the filtration performance of respirators.

Surgical masks have been used to protect the wearer against the inhalation of bacteria or viruses. Questions were raised recently concerning whether these masks

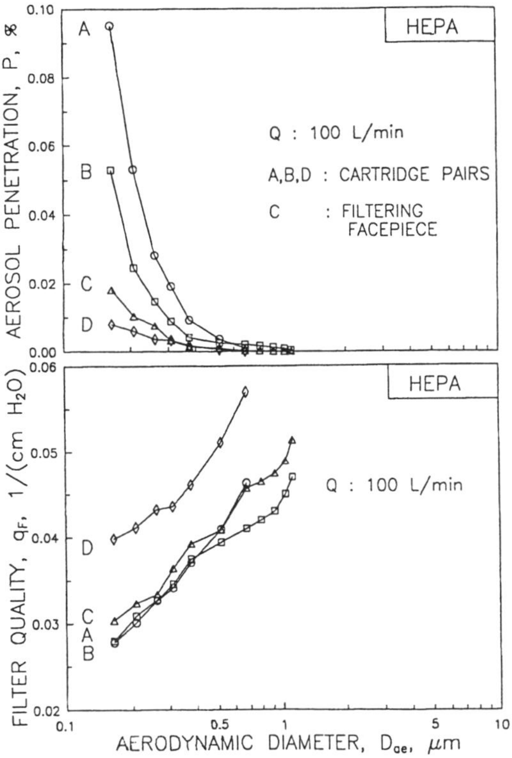

Figure 34 Aerosol penetration and filter quality q_F of four high efficiency particulate air filter (HEPA) respirators. (From Reference 10.)

would provide adequate protection against the small aerosols generated during laser surgery, new diseases such as acquired immunodeficiency syndrome (AIDS) or drug-resistant tuberculosis (TB). Chen and Willeke[11] conducted a study to determine the efficiency of surgical masks against a variety of sizes of aerosols. The test setup was the same one used by the authors to evaluate the efficiency of disposable respirators.[8]

Two surgical masks were selected for this study. Mask A had a thin molded cone that lacked a filter layer. Mask B had a flat shape which contained a layer of filter material. The test results are shown in Figure 35. Mask A, without a filter layer, permitted more aerosol penetration than mask B. The penetration of submicrometer aerosols for mask A was 8D%. The penetration was not dependent upon air flow in the submicrometer range. The penetration dropped for supermicrometer aerosols and was more dependent on air flow. For the 4 μm aerosol particles, the penetration percentage ranged from about 10% at an air flow of 100 l/m to about 55% at 5 l/m.

The aerosol penetration through mask B was less than mask A. The penetration of submicrometer aerosols was strongly air flow dependent. The penetration range for 0.3 μm particles (the most penetrating size) was 25% at an air flow of 5 l/m to about 65% at a flow rate of 100 l/m. The penetration was not flow dependent for supermicrometer aerosols. At low flow rates, there was more time for aerosol

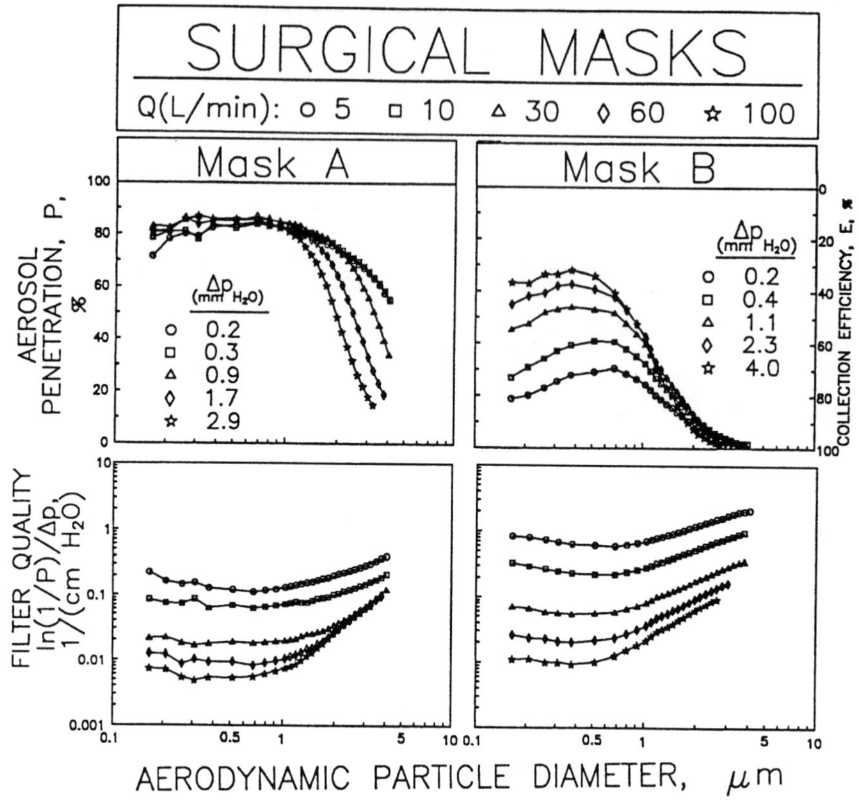

Figure 35 Aerosol penetration and filter quality factor of two surgical masks. (From Reference 11.)

removal by electrostatic force and therefore, less penetration for mask B. For aerosols larger than 1 μm, removal by impaction and interception dominated; therefore, the removal of large particles was greater at higher flow rates as indicated by mask A. The particle removal efficiency for mask B reached 95% for particles larger than 3 μm at all flow rates.

It should be noted that the collection efficiency of electrostatic filter media is dependent on the electric charges built up on the fiber. Once the charges have been neutralized, the filter efficiency would diminish rapidly until sufficient particles collect on the filter to increase its efficiency by mechanical force. The challenge concentration of the corn oil aerosol used in the above studies was not sufficient to neutralize the changes built up on the filter media. In actual use, the efficiency of the electrostatic filters may be reduced since most industrial aerosols are highly charged. Another factor that must be considered is that faceseal leakage was not evaluated in these studies. A respirator which showed low filter penetration may not provide adequate protection during use if the faceseal leakage is significant.

Another study on filter penetration and leak characteristics in filtering facepieces used in the health-care industry was conducted by Weber et al.[12] Eight surgical masks with different filter materials and mask shapes (molded cone and flat) and a DFM

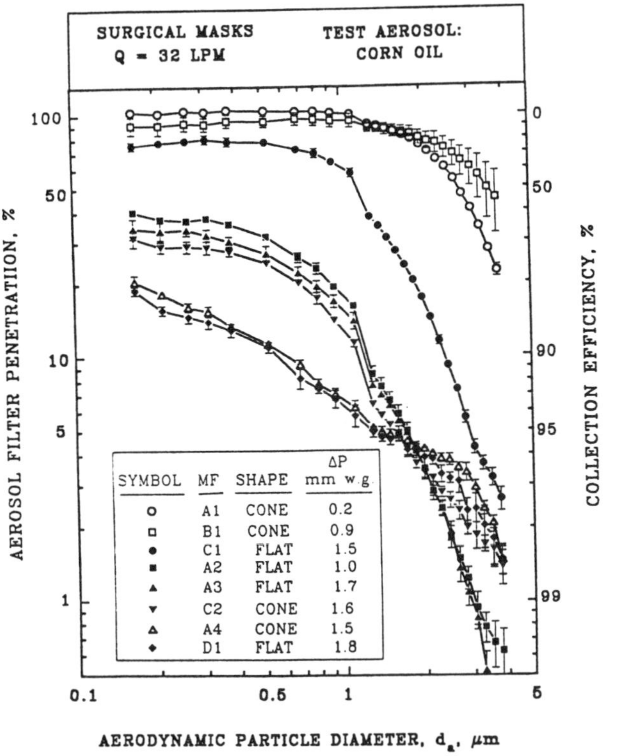

Figure 36 Aerosol penetration through the filter media of eight surgical masks. *Mf*, Manufacturer and model; *ΔP*, pressure drop. (From Reference 12.)

filtering facepiece were selected for evaluation. These masks were coded as A1 to A4, B1, C1 to C3, and D1. All tests were performed on mannequins to study filter penetration, faceseal leakage, and as the effect of leak size. Corn oil aerosol was selected as the challenge. The test aerosol has a particle size range between 0.1 and 4 μm, with a concentration of 800 particles/ml and a geometric standard deviation of 2.0. The same aerosol generation and measurement system used in previous studies was used for filter penetration and faceseal leakage measurements.[8,9] Circular holes were selected for faceseal measurements.

Test results of filter penetration as a function of particle size are shown in Figure 36. The worst performer, A1, had an aerosol filter penetration of approximately 100% for the particle size of 0.2 to 1.0 μm. As the particle size increases, the aerosol penetration decreases. For the same mask, the penetration dropped by half when the particle size increases to 4 μm. Mask A2 made by the same manufacturer shows a filter penetration of 40% at 0.2 μm and about 0.5% at 4 μm. Masks A4 (cone shaped) and D1 (flat shaped) are better performing masks with similar filter efficiency. The penetration is 20% and 1.5% for particle size 2.2 and 4 μm, respectively. Five of eight surgical masks had filtration efficiency of 95% or greater at 2.3 μm particles.

The pressure drop across the masks ranged from 0.2 mm water gauge (wg) for the worst performing mask to 1.8 mm wg for the best performing mask. In general, the pressure drop increased as the penetration decreased. The faceseal leakage was measured by inserting artificial leaks. These leaks do not significantly affect the aerosol penetration for better performing masks such as A4 and D1. The percent penetration for mask D1 without artificial leaks was 20% at 0.15 μm; the penetration was still 20% for 1 and 2 mm diameter leaks but the leak increased to 25% with a 4 mm leak. The pressure drop decreased by 10% when the largest leak (4 mm) was tested. The performance of the cone shaped mask A4 is similar to mask D1. However, the flat shaped mask is likely to cause more faceseal leakage than the cone shaped masks. The aerosol penetration as a function of particle size and leak diameter for masks A4, D1, and an approved DFM filter are shown in Figure 37. The DFM filter has higher pressure drop than two surgical masks (4.3 mm versus 1.8 and 1.5 mm, respectively.

The authors concluded that tested surgical masks provide insufficient protection against submicrometer aerosols. Perimeter leakage preferentially increased the penetration of micrometer-sized versus submicrometer sized particles. Perimeter penetration may be an important factor in the future if more efficient filtration media are used in surgical mask designs.

The electrostatic attraction force can be very significant in improving filter efficiency without significantly increasing pressure drop. However, the filter penetration may increase rapidly when the charges built up on the filter have been neutralized. Chen et al. conducted a study to determine the effect of electrical charges on the efficiency of filtering facepieces.[13] Eight filtering facepieces (FF) consist of 1 nuisance dust, 4 dust/mist (D/M), 2 dust, fume, and mist (DFM) and 1 HEPA were selected for testing. Three filtering facepieces made of the same filtration material were selected to evaluate the cumulative effect of multiple filtration layers: a MSHA/NIOSH approved D/M type with a single filter layer, a CEN P-2 type meeting the European Committee for Standardization (CEN) with 4 filtration layers, and a MSHA/NIOSH approved HEPA type with six filtration layers. The filtering facepieces were sealed to a mannequin to measure air resistance and the penetration of aerosol particles.

In order to neutralize the charges built up on the filters, the respirators were immersed in either isopropanol or static guard solution for 1 h and air dried for 1 h. Based on the effective area of 132 cm^2, the air flow was adjusted to a face velocity of 10 cm/s, corresponding to an air flow of 100 l/m. The filter was loaded with a polydisperse corn oil aerosol having a count mean diameter of 0.6 μm and a geometric standard deviation of 1.8. Each filter was loaded with up to 40 mg/cm^2 of corn oil aerosol. The same filter penetration measuring system described elsewhere was used in this study.[8,9] The aerosol was neutralized to Boltzmann charge equilibrium using a Krypton radioactive source.

The aerosol penetration of tested filtering facepieces without treatment are shown in Figure 38. Filter penetration increases with aerosol loading at a challenge concentration of 5 mg/m^3 and a count mean diameter of 0.3 μm. The penetration curve for the HEPA filter is not shown since it does not change with aerosol loading. Two D/M filtering facepieces which show the best and worst performance as listed

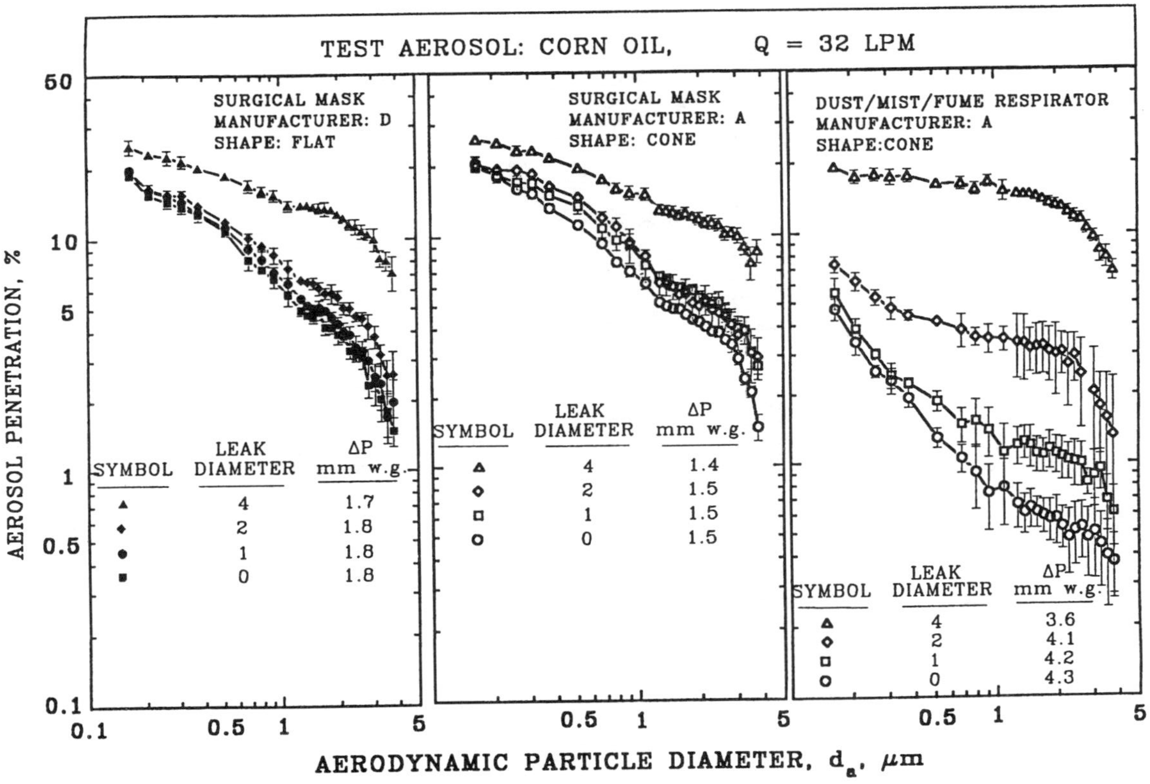

Figure 37 Aerosol penetration and dependence on leak diameter. (From Reference 12.)

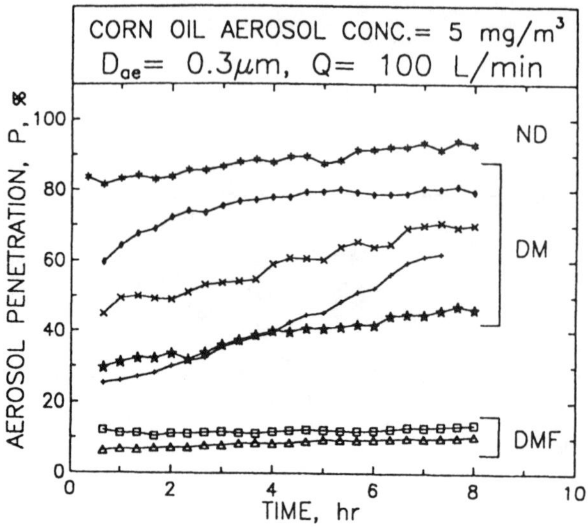

Figure 38 Aerosol penetrations of 7 filtering facepieces. ND, Nuisance dust respirator; DM, dust-mist respirator; DMF, dust-mist-fume respirator. (From Reference 13.)

in Figure 31 have been selected for the charge neutralization study (FF-A and FF-D). The test results are shown in Figure 39. It has been demonstrated that isopropanol and static guard are capable for removing almost all charges built up on the filters. For filters without any chemical treatment, the aerosol penetration first increases due to particle loading because of the neutralization of filter charges due to fiber coating. As the test proceeds, filter penetration decreases because of the increase in filter packing density (ratio of fiber to filter volume). Finally, the filter becomes clogged and penetration is reduced to zero. For FF-A, the maximum penetration is 92% for 0.16 µm aerosols at a filter loading about 38 mg/cm^2 and 37% for 1.19 µm aerosols at a loading of 27 mg/cm^2. The initial increase and subsequent decrease of aerosol penetration also occurs with 2.45 µm aerosols. For this aerosol, the penetration decreases significantly after the maximum penetration point.

For a chemically treated filter, the initial aerosol penetration is very close to the maximum aerosol penetration of an untreated filter after loading. This indicates that most of the charges have been removed by chemical treatment. Static guard appears to be more effective in removing filter charges than isopropanol. No increase in filter penetration was observed for filters treated with static guard. The aerosol penetration at the beginning of a loading cycle is shown as a function of particle size in Figure 40. After the chemical treatment that removed the electrical charges built up on the filter, FF-A and FF-B (FF-D) have almost the same penetration pattern. It demonstrates that FF-A and FF-B have the same mechanical properties: similar fiber diameter and packing density. At the beginning of use, FF-A has much lower filter penetration than FF-B. For submicrometer-sized aerosols, aerosol particle removal by electrical force is 3.5 times greater than by mechanical forces for FF-A. For FF-B, both forces have about the same removal efficiency. For supermicrometer-sized aerosols, mechanical force removal is higher than electrical force removal. At lower face velocity, electrical force removal is expected to be relatively higher for 1 µm

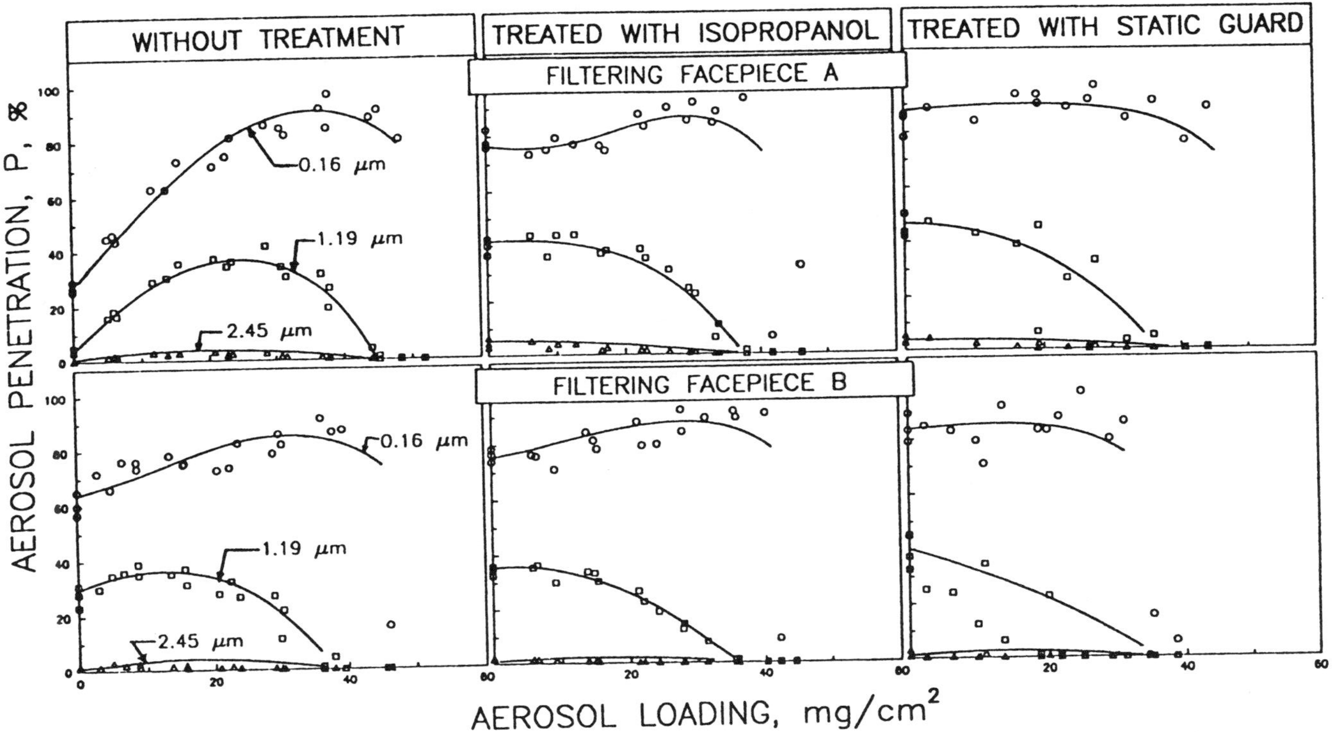

Figure 39 Change in aerosol penetration due to reduction in fiber charge upon loading by aerosols. Face velocity = 10 cm/s. (From Reference 13.)

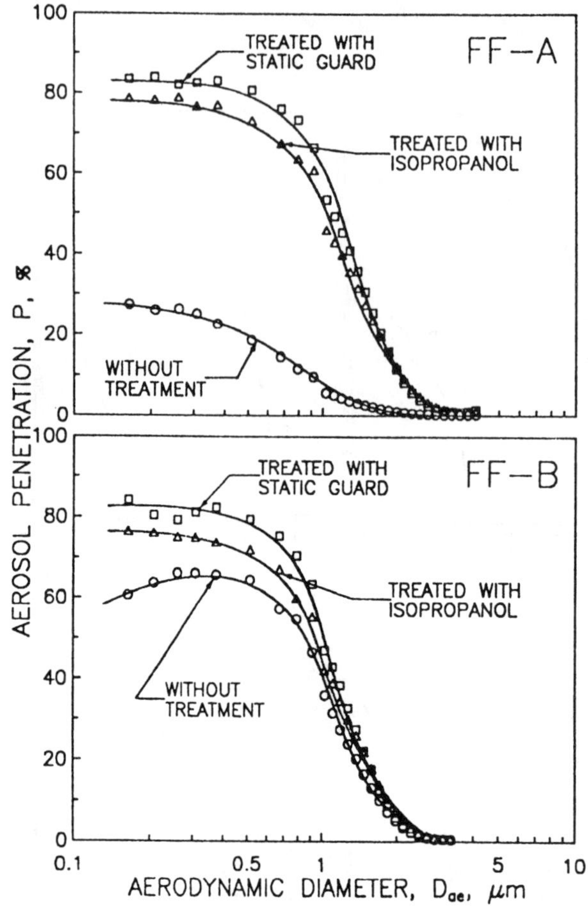

Figure 40 Particle size dependence of penetrated aerosols for dust-mist filtering facepieces at the beginning of a loading cycle with corn oil aerosols. Face velocity = 10 cm/s. (From Reference 13.)

particles because of the particles's increased time of passage through the filter which gives the electrical force more time to act on the particles (see Figure 41).

The effect that the number of filtration layers has on pressure drop and aerosol penetration was also examined. Figure 42 shows that the pressure drop across the filter increased linearly with air flow and the flow is laminar. For example, the pressure drop across the HEPA filter is 1.35 mm H_2O at 5 l/m, 8.1 mm H_2O at 30 l/m, and 27 mm H_2O at 100 l/m. Figure 43 shows that filter penetration exponentially decreases with the number of filtration layers.

VI. FILTER PENETRATION AGAINST BIOAEROSOLS

Under a contract with the 3M Company, a major producer of surgical masks and filtering facepieces, the University of Minnesota conducted a study to evaluate the

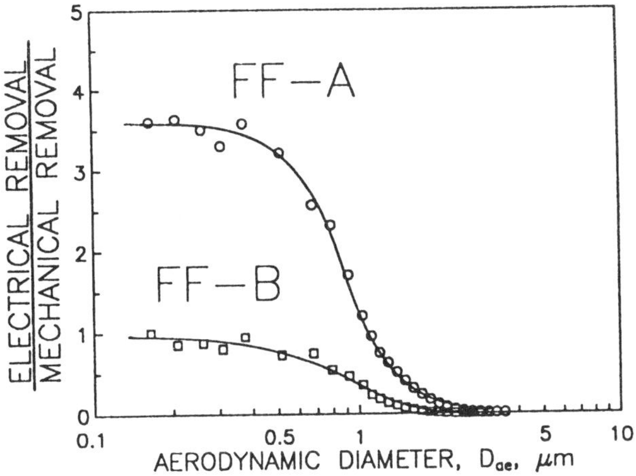

Figure 41 Aerosol particle removal by electrical versus mechanical forces. Face velocity = 10 cm/s. (From Reference 13.)

efficiency of surgical masks and filtering facepieces.[14] The following 3M respirators were selected for testing:

1. Supermicrometer surgical mask, 1812
2. D/M respirator, 1814
3. D/M respirator, 8715
4. DFM respirator, 9920
5. HEPA respirator, 9970

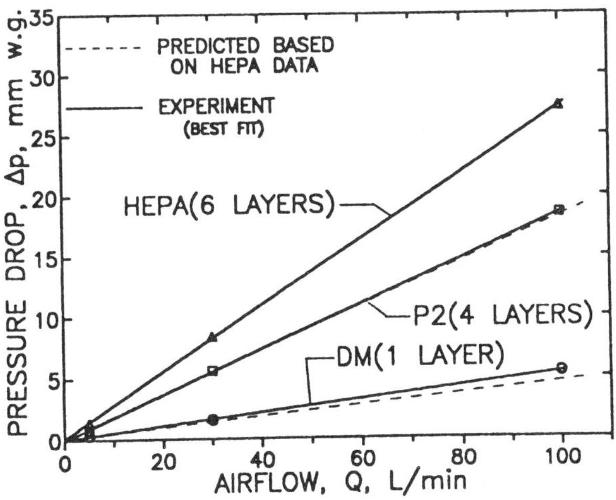

Figure 42 Effect of the number of filtration layers on pressure drop across filtering facepieces. DM, Dust-mist respirator (1 filtration layer); P2, respirator regulated by European Standard (4 filtration layers); HEPA, high efficiency respirator (6 filtration layers). (From Reference 13.)

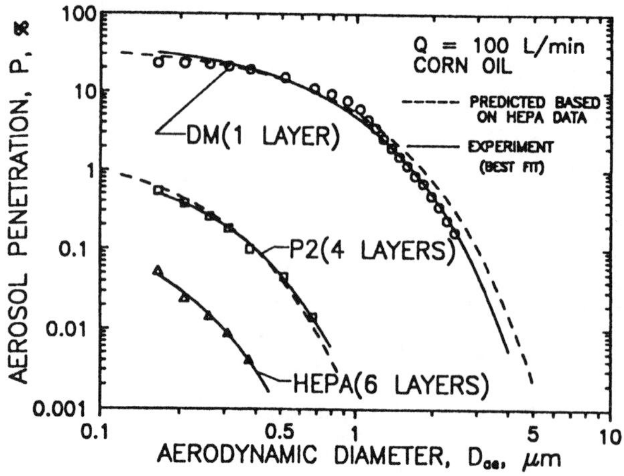

Figure 43 Effect of the number of filtration layers on aerosol penetration. (From Reference 13.)

Except for the surgical mask, all other respirators are approved by MSHA/NIOSH. Both polystyrene latex spheres and a bacteria, *Mycobacteria chelonae abscessus*, which is similar in size to the *M. tuberculosis* were selected for testing. The latex spheres have a count median diameter of 0.804 µm to simulate the size of M. chelonae (0.5 by 2 µm rods). Both viable and nonviable aerosols were generated by a collision nebulizer. The concentration of the viable particles was 10^4 to 10^6 viable particles per m^3. The test aerosols were neutralized by a Kr 85 radioactive source before filter testing. The filter penetration was measured by the Andersen impactor and an Aerodynamic Particle Sizer. The Andersen impactor sampling time was 10 to 30 sec for the upstream samples and 7 to 30 min for downstream samples. The flow rate was 46 l/m. Five tests were conducted for each respirator. The respirator test sequence was selected at random. The test results are shown in Table 13.

The test results indicated that there is no significant difference in filter penetration between the viable and nonviable particles. Lower filter efficiency was observed when the masks were challenged with the latex spheres. As expected, the surgical mask demonstrated the lowest efficiency of approximately 96% and the HEPA filters

Table 13 Mean Efficiency (%) ± Standard Deviation of Selected Masks and Respirators Against *M. Chelonae* and Neutralized Latex Spheres[a]

Type of mask/respirator	*M. chelonae* (Andersen)	*M. chelonae* (APS)	Latex spheres (APS)
Submicron surgical mask (1812)	97.5 ± 0.47	96.8 ± 0.64	95.9 ± 0.29
Dust/mist respirator (1814)	98.6 ± 0.22	97.9 ± 0.41	97.2 ± 0.37
Dust/mist respirator (8715)	97.2 ± 1.64	96.9 ± 1.17	96.3 ± 0.72
Dust/mist/fume respirator (9920)	99.96 ± 0.02	99.88 ± 0.066	99.8 ± 0.092
HEPA respirator (9970)	>99.99 ± <0.01	>99.99 ± <0.01	>99.99 ± <0.01

[a] Based on five tests of each mask/respirator, except that there were 6 tests for the 9920 respirator.

Note: APS, Aerodynamic Particle Sizer.

From Reference 14.

achieved the highest efficiency of >99.99%. The authors concluded that this study only evaluates filter efficiency, not faceseal leakage. Since only one manufacturer's product was evaluated in this study, the test results are not representative of other devices on the market. Since filter penetration is a function of air flow rate and challenge concentration, lower filter efficiency would be expected if tests were conducted at higher challenge concentrations and air flow rates.

Johnson et al. conducted a study to determine the performance of respirators using a viable aerosol challenge.[15] *Bacillus subtilis subsp. niger)* (BG), a simulant to the *M. tuberculosis*, was selected for testing. BG spores are rod shape of 0.7 to 0.8 μm in width and 1.5 to 1.8 μm in length. *M. tuberculosis* rods have a width of 0.3 to 0.6 μm and a length of 1.0 to 4.0 μm. Tests were conducted under simulated breathing conditions. The breathing machine was operated at 15 cycles per minute with a minute volume of 22.5 l/m. A mannequin with a pharyngeal sampling port was connected to the breathing machine. The aerosol challenge was generated from an atomizer. After dilution and mixing, the final challenge aerosol had a concentration of 10^6 particles per liter of air. The following respirators were selected for testing:

1. Approved PAPR equipped HEPA filters, faceshield, and a Tyvek hood (HEPA)
2. Nonapproved Flat fold surgical mask (SM)
3. Approved cup-shaped D/M filtering facepiece (DM)
4. Approved filtering facepiece equipped with a HEPA filter element (HE)

Penetration due to faceseal leakage and filter penetration was separated by repeating the tests after sealing the mannequin and respirator faceseal with tape (1.5 cm of tape overlapped the respirator and 1.5 cm of tape overlapped the mannequin). The test aerosol that penetrated through the respirator was collected in a glass impinger. Sampling time for the chamber was 1 min at a flow rate of 6 l/m. The in-mask sampling time was 3 min at an air flow rate of 12 l/m. The particle size and concentration measurements were performed on an aerodynamic particle sizer. Bioassays were performed by standard plating and counting techniques.

Penetration for the respirator (RPE) and sealed RPE (S-RPE) were calculated as the following:

$$\text{Penetration (\%)} = \frac{\text{CFU penetrating RPE}}{\text{Challenge CFU}} \times 100 \tag{2}$$

The median aerodynamic diameter for the challenge had a range from 0.85 to 0.87 μm. The actual number of aerosolized bacteria used to challenge the RPE in counts per liter, the number of bacteria which penetrated through the unsealed and sealed RPE, and the bacteria penetration through unsealed and sealed respirators are shown in Table 14.

The test results indicated that sealing the facepiece to the mannequin would reduce the bacteria penetration through the respirator. The disposable respirators are less effective in preventing penetration of bacteria challenge than the PAPR. The authors state that the high level of bacteria challenge used in this study (10^5 to 10^6

Table 14 Penetration of RPE by Bacteria under Sealed and Unsealed Conditions

Respirator	Bacteria count/liter Unsealed Challenge	Bacteria count/liter Unsealed In-mask	Percent penetration	Bacteria count/liter Sealed Challenge	Bacteria count/liter Sealed In-mask	Percent penetration
PAPR	1.18×10^6	6.13×10^2	0.05	NT	NT	
HE	1.23×10^5	1.23×10^6	8	1.23×10^5	5.3×10^3	4
DM	1.23×10^6	2.67×10^5	22	1.88×10^6	4.9×10^4	19
SM	1.47×10^6	4.95×10^5	33	2.0×10^6	1.1×10^5	28

Note: NT, not tested.

From Reference 15.

per liter) may be encountered in procedures such as autopsy, abscess irrigation, bronchial lavage, or orthopedic procedures. The use of PAPR equipped with HEPA filters is recommended when there is a high probability of encountering high levels of infectious aerosols.

REFERENCES

1. Held, B., Revoir, W. H., Davis, T. O., Pritchard, J. A., Lowry, P. L., Hack, A., Richards, C. P., Geoffrion, L. A., Wheat, L. D., and Hyatt, E. C., Respirator Studies for the National Institute for Occupational Safety and Health, July 1, 1973 through June 30, 1974, Los Alamos Scientific Laboratory, LA-5805-PR, 1974.
2. Douglas, D., Revoir, W. H., Davis, T. O., Pritchard, J. A., Lowry, P. L., Hack, A., Richards, C. P., Geoffrion, L. A., Wheat, L. D., Bustos, J. M., and Hesch, P. R., Respirator Studies for the National Institute for Occupational Safety and Health, July 1, 1974 through June 30, 1975, Los Alamos Scientific Laboratory, LA-6386-PR, 1976.
3. Revoir, W. H., Air-Purifying Group, ANSI Z-88 ad hoc Subcommittee for Respirator Test and Approval, February 2, 1981.
4. Ackley, M. W., Degradation of Electrostatic Filters at Elevated Temperature and Humidity. World Filtration Congress III. Croydon, England, 169–176, Uplands Press, 1982.
5. Revoir, W. H., Air-Purifying Group, ANSI Z-88 ad Hoc Subcommittee for Respirator Test and Approval Report, September 29, 1981.
6. Biermann, A. H., Lum, B. Y., and Bergman, W., Evaluation of Permanently Charged Electrofibrous Filters, Present at the 17th Department of Energy Nuclear Air Clearing Conference, Denver, CO, 1982.
7. Coffey, C., Personal Communication, NIOSH, October 23, 1992.
8. Chen, C. C., Ruuskanen, J., Pilacinski, W., and Willeke, K., Filter and leak penetration characteristics of a dust and mist filtering facepiece, *Am. Ind. Hyg. Assoc. J.*, 51, 632, 1990.
9. Chen, C. C. and Willeke, K., Characteristics of face seal leakage in filtering facepieces, *Am. Ind. Hyg. Assoc. J.*, 53, 532, 1992.
10. Chen, C. C., Lehtimäki, M., and Willeke, K., Aerosol penetration through filtering facepieces and respirator cartridges, *Am. Ind. Hyg. Assoc. J.*, 53, 556, 1992.
11. Chen, C. C. and Willeke, K., Aerosol penetration through surgical masks, *Am. J. Infect. Control*, 20, 177, 1992.
12. Weber, A., Willeke, K., Merchionl, R., Myoja, T., McKay, R., Donnelly, J., and Liebhaber, F., Aerosol penetration and leakage characteristics of masks used in the health care industry, *Am. J. Infect. Control*, 21, 167, 1993.

13. Chen, C. C., Lehtimake, M., and Willeke, K., Loading and filtration characteristics of filtering facepieces, *Am. Ind. Hyg. Assoc. J.*, 54, 51, 1993.
14. Chen, S. K., Vesley, D., Brosseau, L. M., and Vincent, J. H., Evaluation of single-use masks and respirators for protection of health care workers against mycobacterial aerosols, *Am. J. Infec. Control*, 22, 65, 1994.
15. Johnson, B., Martin, D. D., and Resnick, I. G., Efficacy of selected respiratory protective equipment challenged with *Bacillus subtilis* subsp. *niger*, *Appl. Environ. Microbiol.*, 60, 2184, .

16 SORBENT PERFORMANCE I

I. INTRODUCTION

Chemical cartridges or canisters are used for protection against exposure to gaseous air contaminants such as acid gases, alkaline gases, organic vapors, or vapors of inorganic compounds. The most common application for chemical cartridges and canisters is for protection against a variety of organic vapors. When the gas or vapor contaminant passes through a fresh adsorbent bed of activated carbon, most of the contaminant is removed. As time progresses, more and more contaminant flows through the adsorbent bed without being adsorbed. Eventually, an equilibrium is reached when the challenge concentration equals the effluent concentration. The ratio between the effluent and challenge concentration is defined as breakthrough. It is often expressed in percent. For example, if the challenge concentration is 500 ppm, a 10% breakthrough occurs when the effluent concentration reaches 50 ppm. Service life of a cartridge or canister is defined as the time required to reach a specific effluent concentration.

There are many factors which would change the service life of the sorbent. Some of the important factors are:

1. Relative humidity (RH)
2. Challenge concentration
3. Flow rate
4. Temperature
5. Physical and chemical properties of the challenge
6. Physical characteristics of the sorbent

Generally, the service life of the sorbent decreases with an increase of relative humidity (RH). Studies have indicated that service life is about the same between 0 and 50% RH. Significant reduction in service life of the sorbent is observed at higher relative humidities, because the amount of sorbate which the sorbent can hold depends on the amount of water vapor the sorbent adsorbs. The service life is inversely proportional to the challenge concentration. An increase in temperature and flow rate would reduce the service life of the sorbent. The low boiling point compound challenges would yield a shorter service life than the high boiling point compounds. The size and depth of the sorbent bed, the granular size, weight and pore distribution of the carbon, and the polarity of compounds being adsorbed would also affect the breakthrough time of the sorbent.

A linear relationship exists between the concentration and the service life. A family of straight lines can be obtained when the log of the service life at selected breakthrough ranges is plotted against the log of the concentration. Similarly, a plot of log concentration versus log 10% breakthrough time for various flow rates yields a family of straight lines.

II. ADSORPTION CAPACITY

Adsorption capacity of the carbon may be experimentally determined by volumetrically or gravimetrically measuring its uptake of solvent vapors at complete saturation. However, this method cannot predict the adsorption capacity of untested solvents. The only method that has been successful for this purpose is using various forms of the Polanyi potential theory by plotting the log of the amount of sorbate adsorbed against the adsorption potential.

Dubinin[1] modified the Polanyi potential theory and developed an adsorption isotherm for predicting adsorption capacity in microporous solids. The equation follows:

$$W_s = \rho W_o \exp\left[-\frac{BT^2}{\beta^2}\left[\log(p_s/p)^2\right]\right] \tag{1}$$

where W_s = adsorption capacity per unit weight of carbon (gm/gm).
ρ = solvent density (gm/cm^3).
W_o = Total volume of adsorption space (cm^3).
B = Microporosity constant for the carbon.
T = Temperature (°K).
β = Affinity coefficient of solvent vapor for the activated carbon.
p_s = Saturated vapor pressure of the solvent at temperature T (torr).
p = Equilibrium partial pressure of the solvent (torr).

The constants B and W_o are related to the pore structure of the adsorbents. The affinity coefficient β characterizes the adsorption of a given vapor with respect to another vapor selected as a standard. The usual practice is to assign $\beta = 1$ for a reference compound, and using several values of p_s/p, to solve for B and W_o. The β value can also be determined by the following approximation:

$$\beta \approx \frac{P}{P_s} \approx \frac{v}{v_s} = \frac{\rho_s M}{M_s \rho} \tag{2}$$

where P, P_s = Parachors for the unknown and standard solvent (calculated from Snugen's equation).
v, v_s = Molar volume for the unknown and standard solvent (cm/mole).
M, M_s = Molecular weight for the unknown and standard solvent (gm/mole).

In order to provide adequate protection to a respirator wearer against overexposure to toxic vapors, the sorbent breakthrough information should be available. Since the service life of the sorbent is affected by many variables, it may not be feasible to perform experiments to determine effects of these variables. Numerous equations have been developed to predict the breakthrough time of the sorbent.

Early in 1920, Bohart and Adams[2] studied the breakthrough of chlorine through charcoal. An equation on breakthrough was derived:

$$\frac{C}{C_o} = \frac{a}{a_o} \exp\left[k(C_o t - a_o x / V)\right] \tag{3}$$

where C = breakthrough concentration
C_o = inlet concentration
a = residual adsorption capacity
a_o = initial adsorption capacity
k = adsorption rate constant
x = bed depth
V = flow rate of contaminant
t = breakthrough time

Later, Makelenburg[3,4] proposed that at the breakthrough time, the penetration of gas through the sorbent bed has been negligible, then the weight of gas supplied equals to the weight of gas adsorbed. The weight of gas supplied is equal to the flow rate L times the concentration C_o and the duration of flow, t_b . The weight of gas adsorbed equals the adsorption capacity N_o times the volume of adsorbent A(z-h). Thus

$$t_b L c_o = N_o A(z - h) \tag{4}$$

and

$$t_b = \frac{N_o A(z - h)}{L c_o} \tag{5}$$

where t_b = Breakthrough time (min).
N_o = Saturation capacity of the adsorbent bed (gm/ml).
A = Cross sectional area of the bed (cm^2).
L = Volumetric flow rate (cm^3/min).
c_o = Input concentration (gm/ml).
z = bed depth (cm).
h = "dead layer" depth (cm).

The "dead layer" depth h is assumed to be equal to the critical bed depth I; the critical bed depth is that value below which breakthrough would be instantaneous. The major difference between the various forms of the Mecklenburg equation is in the expression used to calculate I.

In the middle 1940s, Klotz[5] proposed the following equation which was a variation of the Mecklenburg equation. This equation has been used by Nelson[6-13] in his comprehensive study on cartridge breakthrough. This equation was developed for turbulent flow, but it appears to hold for laminar flow through the carbon bed. It should be used for breakthrough less than 20%.

$$t_b = \frac{w_s \rho_c An}{QC_o}\left[z + \frac{1}{a_c \rho_c}\left(\frac{dG}{\eta}\right)^{0.41}\left(\frac{\eta}{\rho_a D}\right)^{0.67} \ln(C_b / C_o)\right] \quad (6)$$

where $C_o = \dfrac{MC_I}{24.1 \times 10^6}$

$G = \dfrac{1000\,\rho_a Q V_v \rho_c}{60\ An}$

z = V/A
t_b = Breakthrough time (min).
w_s = Equilibrium static adsorption capacity per unit weight of carbon (g/g).
ρ_c = Carbon density (g/cm^3).
A = Cross-sectional area of the adsorbent bed (cm^2).
V = Carbon volume (cm^3).
n = Number of cartridges tested.
C_o = Challenge concentration (gm/l).
z = Bed depth (cm).
a_c = Specific surface area (cm^2/g).
d = Diameter of granule (cm).
G = Mass velocity through cartridge (gm/cm^2-s).
η = Viscosity of the air-gas stream (gm/cm-s).
ρ_a = Density of air-vapor stream (gm/cm^3).
D = Diffusion coefficient (cm^2/s).
C_b = Breakthrough concentration (ppm).
C_I = Assault concentration (ppm).
V_v = Void volume (cm^3/g).
Q = Flow rate (l/m).
M = molecular weight.

In their comprehensive study on sorbent adsorption, Smoot et al.[14] pointed out that the Klotz equation gives fairly accurate breakthrough time for several types of compounds. However, low-molecular-weight or low-boiling compounds such as methanol, ethanol, methyl chloride, dichloromethane, and vinyl chloride give much shorter breakthrough times than predicted by the equation. Using the experimental values of w_s from Nelson's Type I cartridges,[10] they made a comparison between experiments and calculated 10% breakthrough time for various compounds. The results are shown in Table 1.

Danby et al.[15] developed the following equation in 1946:

$$t = \frac{1}{kc_o}\left[\ln[\exp(kN_o\lambda / L) - 1] - \ln(c_o / c - 1)\right] \quad (7)$$

Table 1 Comparison of Experimental 10% Breakthrough Times (Minutes) with Times Calculated from the Klotz Equation

	t (exp)	t (calc)	% Deviation
benzene	88.6	86.6	–2.3
toluene	114	105	–7.9
ethylbenzene	105	109	+3.8
m-xylene	116	101	–12.9
cumene	103	89.4	–13.2
mesitylene	105	93.8	–10.7
p-cymene	92.9	87.8	–5.5
methanol	3.2	12.1	+278
ethanol	45.3	76.9	+69.8
2-propanol	81.8	109	+33.3
allyl alcohol	105	130	+23.8
propanol	111	133	+19.8
sec-butanol	121	122	+0.8
butanol	141	142	+0.7
2-pentanol	111	118	+6.3
3-methyl-l-butanol	121	120	–0.8
4-methyl-2-pentanol	96.1	105	+9.3
pentanol	130	126	–3.1
2-ethyl-l-butanol	101	113	+11.9
methyl chloride	0.7	3.3	+371
vinyl chloride	6.6	12.7	+92.4
ethyl chloride	10.7	15.4	+43.9
2-chloropropane	35.9	43.4	+20.9
allyl chloride	44.6	52.7	+18.2
1-chloropropane	34.8	51.0	+46.6
2-chloro-2-methylpropane	52.3	60.8	+16.3
1-chlorobutane	88.1	85.3	–3.2
2-chloro-2-methylbutane	79.3	82.3	+3.8
1-chloropentane	96.6	94.8	–1.9
chlorocyclopentane	106	108	+1.9
chlorobenzene	131	124	–5.3
1-chlorohexane	95.8	96.9	+1.1
o-chlorotoluene	122	117	–4.1
1-chloroheptane	101	91.1	–9.8
3-chloromethylheptane	80.5	80.4	–0.1
dichloromethane	15.8	24.9	+57.6
t-1,2-dichloroethylene	50.3	56.5	+12.3
1,1-dichloroethane	40.1	68.2	+70.1
c-1,2-dichloroethylene	42.6	62.9	+47.7
1,2-dichloroethane	79.7	95.0	+19.2
1,2-dichloropropane	90.3	99.8	+10.5
1,3-dichloropropene	110	111	+0.9
1,4-dichlorobutane	129	119	–7.8
o-dichlorobenzene	132	123	–6.8
chloroform	52.4	71.5	+36.5
methyl chloroform	58.9	80.8	+37.2
trichloroethylene	83.0	97.9	+18.0
1,1,2-trichloroethane	112	118	+5.3
1,2,3-trichloropropane	132	124	–6.1
carbon tetrachloride	90.0	89.9	–0.1
perchloroethylene	129	124	–3.9
1,1,2,2-tetrachloroethane	131	128	–2.3
pentachloroethane	117	112	–4.3
methyl acetate	46.5	60.1	+29.2
vinyl acetate	81.1	92.4	+13.9
ethyl acetate	84.7	81.6	–3.7
isopropyl acetate	85.6	103	+20.3

Table 1 Comparison of Experimental 10% Breakthrough Times (Minutes) with Times Calculated from the Klotz Equation (*continued*)

	t (exp)	t (calc)	% Deviation
propyl acetate	99.0	101	+2.0
allyl acetate	95.6	104	+8.8
sec-butyl acetate	101	91.9	–9.0
butyl acetate	96.9	102	+5.3
isopentyl acetate	88.3	88.3	0.0
1,3-dimethylbutyl acetate	76.0	79.5	+4.6
pentyl acetate	87.3	93.3	+6.9
hexyl acetate	85.3	89.0	+4.3

Note: Average deviation = +4.0 min.

From References 10 and 14.

where k = Adsorption rate constant (ml/g-min).
N_o = Adsorption capacity (ml/g).
λ = Bed depth (cm).
L = Linear flow rate (cm/min).
c_o = Input concentration (g/ml).
c = Effluent concentration at time t (g/ml).
t = Breakthrough time (min).

Wheeler et al.[16] developed an equation which assumes that the rate-controlling removal process is first order and reversible. Jonas et al.[17-21] proposed a modified Wheeler equation for breakthrough prediction:

$$t_b = \frac{w_e}{C_o Q}\left[W - \frac{\rho_\beta Q}{k_v}\ln(C_o / C_b)\right] \tag{8}$$

where $C_o = MC_l/24.2 \times 10^6$

and $k_v \approx 14.4\ v_l^{1/2} d^{-3/2} = 14.4\ (1000\ Q/nA)^{1/2}$

where t_b = Breakthrough time (min).
C_b = Exit concentration (g/cm^3).
C_o = Inlet concentration (g/cm^3).
Q = Volumetric flow rate (cm^3/min).
W = Weight of adsorbent (g).
ρ_β = Bulk density of the packed bed (g/cm^3).
w_e = Adsorption capacity (g/g).
k_v = Adsorption rate constant (min^{-1}).
v_l = Superficial velocity across the particle (cm/min).

Another formula which is similar to the above was proposed by van Dongen:

$$t_b = E_g E_p H \frac{L - d_p}{6E_g k}\ln(C_o / C)\frac{L}{V} - \frac{d_p}{V6E_g k} \tag{9}$$

where t_b = Breakthrough time (s).
E_g = Porosity parameter of charcoal bed.
E_p = Porosity parameter of charcoal bed.
H = Adsorption capacity parameter.
L = Bed depth (cm).
v = Real velocity of gas in bed (cm/s).
k = Adsorption rate constant (cm_4/g-s).
C_o = Input concentration (g/cm^3).
C = Breakthrough concentration (g/cm^3).
d_p = Diameter of sorbent (cm).

The Klotz-Mecklenburg equation and the modified Wheeler equations are used widely by many investigators to predict service life of sorbents.[10-11,22-29]

The most comprehensive study on respirator cartridge service life was conducted by Nelson et al. at the Lawrence Livermore National Laboratory in the early 1970s.[6-13] The data generated has been used extensively by others in sorbent service life studies. The test set up includes compressed air supply, humidifier, solvent injector, test chamber, detector, and a breath simulator. Laboratory compressed air passes over a water reservoir, a heater in the reservoir, activated by a humidity controller, maintains a preset relative humidity. The organic vapor is generated by vaporizing the test solvent which is continuously fed by a syringe pump. The solvent vapor is then combined with the humidified air stream. The solvent vapor concentration is monitored by a flame ionization detector. The vapor laden air is drawn through the cartridge by a breathing simulator. The downstream concentration of the solvent vapor is monitored by another flame ionization detector for breakthrough measurement. A schematic diagram of the experimental setup is shown in Figure 1.

The system is capable to generate organic vapor at concentrations between 1 ppm to several percent. The flow rate can be adjusted from 20 to 250 l/m. The relative humidity range can vary from 5 to 95%. A concentration as low as 100 parts per billion can be detected from the effluent of the cartridge. The details of the system construction, calibration, and the construction of the breath simulator have been reported by Nelson.[7] The system is now commercially available through Miller and Nelson Research Inc., in Monterey, CA.

The difference between steady state flow and pulsation flow was also investigated by Nelson.[9] The pulsating flow test was performed on the mechanical breathing simulator developed previously.[8] Five work rates of 0, 248, 415, 622, 830 and 1107 kgm/min were selected; these correspond to steady state flow rates of 14.0, 20.6, 29.8, 36.7, 53.3, and 71.4 l/m, respectively. Two commercially available organic vapor cartridges were tested in this investigation. The cartridges were vacuum dried and preconditioned at 22°C, 50% and 80% RH. Toluene and five other organic vapors were selected as the challenge at a concentration of 1000 ppm. The test relative humidity was set at 50%. Breakthrough time was measured at 1, 10, 50, and 99%. Nelson concluded that no significant difference was observed between the steady state flow and pulsating flow patterns, even at the heaviest work rate. An increase in humidity could shorten the service life of the sorbent. The cartridge service life is inversely proportional to flow rate for a given test vapor, concentration, and relative humidity.

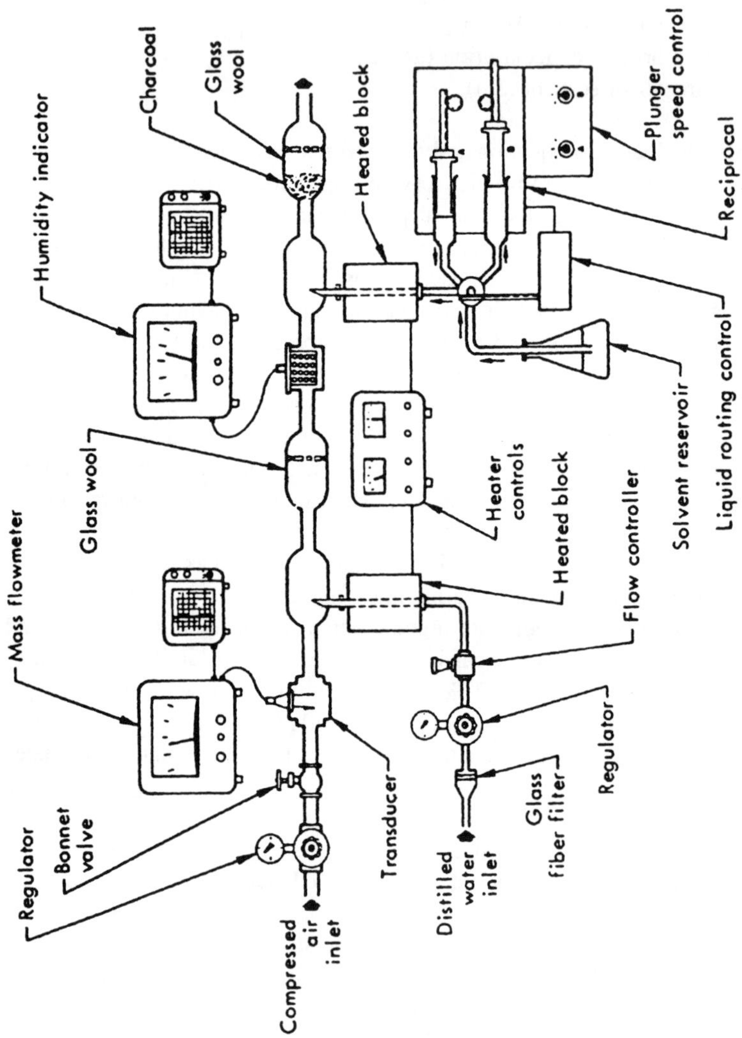

Figure 1 Schematic diagram of the apparatus used to produce test atmospheres of known composition. (From Reference 6.)

III. EFFECT OF CONCENTRATION

The commercially available cartridges are made of coconut or petroleum based activated carbon. Both types were evaluated by Nelson.[10] The cartridges made by MSA (coconut base) and AO (petroleum base) were selected for the majority of his study. He performed the cartridge breakthrough studies on 121 solvent vapors and gases which can be classified as aromatic, alcohols, monochlorides, dichlorides, trichlorides, tetrachlorides, acetates, ketones, alkanes, and amines. The standard test conditions include a solvent concentration of 1000 ppm, 50% RH, 20 to 25°C, and 53.3 l/m continuous flow which is equivalent to a moderately heavy work rate of 830 kgm-m/min. The test results on breakthrough time at 1, 10, and 99% are reported in Table 2.

In order to study the effect of concentration on breakthrough and the effect of humidity on service life of sorbents, Nelson conducted additional tests to determine the effect of these variables.[11] He selected nine solvents and one gas in this study. He used three different types of cartridges (two coconut and one petroleum based carbon). The concentration varies from 50 to 2000 ppm. The results of this study indicates that an asymmetrically S shaped curve is formed by plotting the breakthrough time vs. concentration. A family of curves of methyl acetate shows breakthrough at different concentrations (see Figure 2). Longer breakthrough usually occurred at lower concentrations.

A logarithmic plot of breakthrough times of methyl acetate at different concentrations shows that a straight line forms at a given breakthrough range. If various breakthrough percents of methyl acetate are plotted, a family of straight lines can be obtained as shown in Figure 3. Hence, breakthrough time at lower concentration can be extrapolated. Mathematically, this data can be represented by the empirical Freundlich isotherm and takes the form

$$t_b = aC^b \tag{10}$$

where t_b is the breakthrough time, C is the concentration, a and b are constants for a given set of experimental conditions. Since the family of lines is approximately parallel indicating that the slopes but not the intercepts are essentially constant. It is concluded that once the slope for one breakthrough time–concentration relationship is known, the slope of other percentage can be approximated. The experimental values of the slope (B) and intercept (A) for different solvents obtained from the log plot of concentration and 10% breakthrough time are listed in Table 3. The average slope is – 0.67 with a standard deviation of ± 0.17 for the nine solvents tested. This implies that

$$t_b\, C^{0.67} = \text{constant}$$

or

$$t_b = t_b' \frac{(C')^{0.67}}{C} \tag{11}$$

Therefore, if the breakthrough time at one concentration is known, breakthrough times at other concentrations can be calculated. However, the best results are

Table 2 Breakthrough Time of Selected Hydrocarbons

Solvent	BP (°C)	Vapor pressure at 20°C (torr)	Diffusion coefficient at 25°C (cm²/s)	Volume charcoal (ml)[a]	Weight charcoal (g)[b]	Experimental breakthrough times			Wt. solvent adsorbed per wt. of carbon			Weight water adsorbed at $t_{100\%}$ per weight carbon[e] (g/g)
						$t_{1\%}$ (min)	$t_{10\%}$ (min)	$t_{99\%}$ (min)	$t_{1\%}$[c] (g/g)	$t_{10\%}$[d] (g/g)	$t_{100\%}$ (g/g)	
Aromatics[f]												
Benzene	80.1	72.4	0.0932	146	59.1	73.3	88.6	170	0.214	0.249	0.327	0.035
Toluene	110.6	21.8	0.0849	144	57.5	94.3	114	196	0.334	0.397	0.473	—
Ethyl benzene	136.2	7.08	0.0755	148	55.9	83.7	105	223	0.352	0.432	0.573	0.021
m-Xylene	138.4	6.16	0.0670	148	59.7	98.7	116	193	0.388	0.450	0.536	—
Cumene	152.4	3.34	0.0677	146	60.0	81.2	103	153	0.359	0.447	0.538	0.022
Mesitylene	164.7	1.73	0.0663	147	59.5	85.5	105	189	0.382	0.460	0.565	—
p-Cymene	176.7	1.28	0.0630	148	56.9	75.6	92.9	253	0.394	0.476	0.594	—
Alcohols[f]												
Methanol	64.7	96.8	0.1520	148	55.4	0.2	3.2	47.0	0.0003	0.004	0.018	0.091
Ethanol	78.4	43.6	0.1181	146	55.5	28.0	45.3	207	0.051	0.079	0.168	0.135
Isopropanol	82.3	32.4	0.1013	145	55.7	54.3	81.8	247	0.129	0.188	0.314	0.133
Allyl alcohol	97.0	21.4	0.1021	145	55.4	65.5	105	280	0.152	0.234	0.362	0.117
Propanol	97.1	14.4	0.0993	144	55.7	70.4	111	250	0.168	0.255	0.384	0.065
sec-Butanol	99.5	12.6	0.0891	144	55.9	96.0	121	196	0.281	0.347	0.438	0.050
Butanol	117.7	5.65	0.0861	143	55.5	115	141	235	0.340	0.409	0.512	0.016
2-Pentanol	119.9	4.36	0.0728	147	56.7	86.8	111	277	0.298	0.373	0.517	0.020
3-Methyl-1-butanol	131.2	2.4	0.0709[h]	147	55.0	97.0	121	195	0.344	0.421	0.525	0.047
4-Methyl-2-pentanol	131.8	3.9	0.0653[h]	148	55.7	75.4	96.1	243	0.306	0.382	0.545	0.041
Pentanol	137.9	1.82	0.0716	148	55.0	102	130	208	0.362	0.451	0.552	0.019
2-Ethyl-1-butanol	146.8	1.56	0.0653[h]	150	55.7	76.5	101	257	0.310	0.400	0.582	0.030
Monochlorides[f]												
Methyl chloride	–24.2	3460	0.1140[g]	148	55.7	0.05	0.7	14.6	0.0001	0.001	0.008	0.074
Vinyl chloride	–13.9	2310	0.0995[g]	147	55.7	3.8	6.6	46.3	0.009	0.016	0.038	0.070
Ethyl chloride	12.3	1110	0.0950[g]	148	56.1	5.6	10.7	38.7	0.014	0.026	0.048	0.096
2-Chloropropane	35.2	430	0.0819[g]	148	56.8	26.1	35.9	109	0.080	0.107	0.167	0.018
Allyl chloride	44.5	300	0.0975	141	55.5	30.5	44.6	108	0.093	0.132	0.194	0.040
1-Chloropropane	46.7	277	0.0829[g]	146	55.8	24.5	34.8	141	0.076	0.105	0.196	0.080

2-Chloro-2-methyl-propane	50.8	240	0.0737[g]	147	57.0	37.4	52.3	168	0.134	0.183	0.279	0.010
1-Chlorobutane	77.5	80.0	0.0745[g]	145	55.8	72.3	88.1	145	0.265	0.317	0.391	0.008
2-Chloro-2-methyl-butane	85.7	60.3	0.0675[g]	148	57.0	58.8	79.3	194	0.243	0.320	0.439	—
1-Chloropentane	108.4	23.9	0.0681[g]	147	56.8	74.7	96.6	197	0.310	0.392	0.505	—
Chlorocyclopentane	113	—	0.0718	146	55.6	77.5	106	211	0.322	0.429	0.562	0.050
Chlorobenzene	132	9.11	0.0747	146	55.8	107	131	205	0.478	0.574	0.688	—
1-Chlorohexane	134.5	7.15	0.0631[g]	150	55.8	77.3	95.8	189	0.370	0.449	0.591	0.013
o-Chlorotoluene	159.2	2.66	0.0688	146	55.5	102	122	192	0.514	0.605	0.741	0.005
1-Chloroheptane	159.2	2.2	0.0591[g]	150	55.5	81.5	101	143	0.437	0.531	0.625	0.013
3-(Chloromethyl)-heptane	172	—	0.0557[g]	148	55.7	63.4	80.5	183	0.374	0.465	0.614	0.032
Dichlorides[f]												
Dichloromethane	40.2	346	0.1037	144	55.4	10.1	15.8	63.7	0.034	0.052	0.101	0.132
trans-1,2-Dichloro-ethylene	49	263	0.0828[g]	145	55.7	33.0	50.3	124	0.127	0.187	0.268	0.065
1,1-Dichloroethane	56.5	158	0.0919	143	55.5	23.3	40.1	225	0.092	0.152	0.327	0.034
cis-1,2-Dichloro-ethylene	59.8	142	0.0828[g]	144	55.3	29.8	42.6	165	0.116	0.160	0.298	0.063
1,2-Dichloroethane	83.5	60.2	0.0907	144	55.3	54.0	79.7	186	0.214	0.305	0.456	0.036
1,2-Dichloropropane	96.4	42.3	0.0794	145	55.8	65.0	90.3	200	0.291	0.393	0.554	0.034
cis,trans-1,3-Di-chloropropene	108	—	0.0763[g]	147	56.7	85.5	110	208	0.319	0.402	0.606	0.036
1,4-Dichlorobutane	162	3.10	0.0666[g]	146	55.6	108	129	215	0.505	0.637	0.755	—
o-Dichlorobenzene	180.4	0.86	0.0646[g]	146	57.3	109	132	239	0.618	0.736	0.911	—
Trichlorides[f]												
Chloroform	61.2	159.3	0.0888	144	55.6	33.2	52.4	174	0.158	0.240	0.415	0.050
Methyl chloroform	74	105	0.0794	144	56.2	40.4	58.9	197	0.212	0.299	0.530	0.036
Trichloroethylene	86.5	58.6	0.0875	145	54.8	55.3	83.0	195	0.293	0.428	0.625	0.038
1,1,2-Trichloroethane	113.6	17.5	0.0792	145	56.1	71.8	112	206	0.377	0.568	0.773	0.043
1,2,3-Trichloropropane	156	2.1	0.0672[g]	147	56.2	111	132	223	0.643	0.754	0.915	—
Tetrachlorides[f]												
Carbon tetrachloride	76.8	92.1	0.0828	144	55.4	77.0	90.0	147	0.473	0.545	0.677	—
Perchloroethylene	121.2	14.0	0.0797	146	55.7	107	129	209	0.704	0.835	1.01	—
1,1,2,2,-Tetrachloro-ethane	146	4.73	0.0722	146	55.3	104	131	216	0.809	0.861	1.07	—

Table 2 Breakthrough Time of Selected Hydrocarbons (*continued*)

Solvent	BP (°C)	Vapor pressure at 20°C (torr)	Diffusion coefficient at 25°C (cm^2/s)	Volume charcoal (ml)[a]	Weight charcoal (g)[b]	Experimental breakthrough times $t_{1\%}$ (min)	$t_{10\%}$ (min)	$t_{99\%}$ (min)	Wt. solvent adsorbed per wt. of carbon $t_{1\%}$[c] (g/g)	$t_{10\%}$[d] (g/g)	$t_{100\%}$ (g/g)	Weight water adsorbed at $t_{100\%}$ per weight carbon[e] (g/g)
Pentachlorides[f]												
Pentachloroethane	161	—	0.0673	145	56.1	93.0	117	187	0.742	0.914	1.13	—
Acetates[f]												
Methyl acetate	57.3	170.2	0.0978	146	55.6	32.8	46.5	143.8	0.097	0.133	0.214	0.073
Vinyl acetate	72.5	91.2	0.0793[g]	148	55.6	55.0	81.1	235	0.188	0.269	0.391	0.006
Ethyl acetate	72.2	74.4	0.0861	145	60.6	66.8	84.7	172	0.215	0.267	0.350	0.008
Isopropyl acetate	87.5	47.3	0.0770	146	55.6	64.5	85.6	166	0.262	0.339	0.453	0.036
Isopropenyl acetate	96	30.0	0.0718[g]	145	57.1	80.6	106	193	0.323	0.409	0.507	—
Propyl acetate	101.3	24.9	0.0768	145	55.9	78.8	99.0	164	0.318	0.392	0.507	0.020
Allyl acetate	103.5	—	0.0718[g]	148	56.1	75.8	95.6	246	0.299	0.369	0.517	0.043
sec-Butyl acetate	111.5	—	0.0648	145	55.9	82.8	101	164	0.381	0.456	0.537	—
Butyl acetate	126.1	8.0	0.0672	148	55.0	77.3	96.9	228	0.361	0.443	0.595	0.023
Isopentyl acetate	141.5	3.8	0.0603[g]	148	55.8	70.9	88.3	177	0.366	0.447	0.584	0.020
1,3-Dimethylbutyl acetate	146.2	4	0.0567[g]	149	55.8	60.6	76.0	188	0.346	0.426	0.588	0.025
Pentyl acetate	148.4	2.95	0.0610	148	55.7	72.6	87.3	225	0.375	0.444	0.616	0.014
Hexyl acetate	169	—	0.0567[g]	148	55.5	67.0	85.3	202	0.385	0.480	0.658	—
Ketones[h]												
Acetone	56.2	307	0.1049	164	60.8	37.1	46.0	119	0.078	0.095	0.135	0.109
2-Butanone	79.6	70.6	0.0903	160	66.0	81.9	94.4	239	0.198	0.225	0.295	0.056
2-Pentanone	102.3	30.5	0.0793	159	58.0	104	121	231	0.341	0.392	0.483	0.016
3-Pentanone	102.7	26.8	0.0740[g]	162	58.4	93.5	114	175	0.305	0.365	0.449	0.055
4-Methyl-2-pentanone	115.5	16	0.0677[g]	160	58.1	96.1	111	211	0.367	0.418	0.511	0.007
Mesityl oxide	129.7	7.9	0.0760	169	60.2	122	139	224	0.440	0.495	0.578	—
Cyclopentanone	130.7	—	0.0796[g]	167	64.3	141	161	308	0.408	0.460	0.585	—
2,4-Pentanedione	140.4	6.5	0.0727[g]	176	70.6	130	114	214	0.408	0.447	0.527	0.009
3-Heptanone	147.3	—	0.0628[g]	160	58.1	91.0	105	186	0.396	0.451	0.548	0.014
2-Heptanone	151.2	2.8	0.0628[g]	187	61.9	101	114	237	0.412	0.460	0.560	0.005

Cyclohexanone	155.6	3.4	0.0729[g]	166	64.7	126	144	249	0.423	0.477	0.585	0.003
5-Methyl-3-heptanone	159.5	—	0.0588[g]	165	62.9	86.4	99.3	183	0.389	0.442	0.550	0.024
3-Methyl-cyclohexa-none	168	—	0.0669[g]	162	63.5	101	123	216	0.395	0.472	0.595	—
Diisobutyl ketone	169.4	—	0.0554[g]	162	60.4	70.8	83.3	171	0.369	0.427	0.556	0.005
4-Methylcyclohexa-none	171.3	—	0.0669[g]	172	67.9	111	126	245	0.463	0.520	0.580	0.009
Alkanes[h]												
Pentane	36.1	424	0.0842	166	62.6	60.7	71.3	147	0.155	0.179	0.228	0.020
2,3-Dimethylbutane	58.0	191	0.0689[g]	167	63.5	72.0	81.1	192	0.216	0.241	0.300	0.021
Hexane	68.7	121	0.0732[g]	160	55.0	52.3	64.6	178	0.181	0.220	0.334	0.036
Methylcyclopentane	71.8	110	0.0734[g]	164	66.7	62.2	76.1	174	0.174	0.209	0.278	0.036
Cyclohexane	80.7	77.5	0.0743[g]	161	59.3	68.7	82.3	179	0.216	0.254	0.337	0.024
Cyclohexane	83.3	70.4	0.0765[g]	159	57.3	85.8	100	193	0.272	0.313	0.401	—
2,2,4-Trimethylpentane	96.5	38.6	0.0594[g]	161	57.5	68.3	80.4	166	0.300	0.348	0.440	0.019
Heptane	98.5	35.4	0.0636[g]	160	57.9	78.2	89.8	188	0.299	0.339	0.432	0.019
Methylcyclohexane	100.9	36.2	0.0679[g]	160	57.5	68.5	80.5	191	0.259	0.300	0.440	0.045
1,3,5-Cycloheptatriene	115.5	—	0.0742[g]	168	63.4	121	137	246	0.389	0.435	0.536	0.003
2,2,5-Trimethylhexane	124.5	12.5	0.0559[g]	166	64.2	67.6	80.0	168	0.266	0.310	0.444	0.001
5-Ethylidene-2-nor-bornene	147.5	4.2	0.0644[g]	165	67.4	86.5	101	221	0.341	0.393	0.526	0.007
Cyclooctane	150	—	0.0639[g]	172	70.3	96.8	113	220	0.345	0.397	0.497	0.027
Nonane	150	3.15	0.0559[g]	168	63.5	76.2	89.3	198	0.340	0.393	0.490	—
Decane	174	0.96	0.0530[g]	164	57.7	70.8	81.5	158	0.386	0.439	0.553	0.017
Amines[h]												
Methylamine	–6.7	2160	0.1300[g]	161	58.5	12.4	17.9	91.5	0.015	0.020	0.041	0.150
Dimethylamine	6.7	1285	0.1023[g]	164	70.0	17.1	21.7	94	0.024	0.030	0.050	0.291
Ethylamine	16.6	872	0.1032[g]	161	59.8	40.5	49.7	270	0.068	0.081	0.179	0.268
Isopropylamine	31.5	478	0.0879[g]	160	60.4	65.6	75.8	194	0.142	0.162	0.275	0.136
Propylamine	47.8	247	0.0879[g]	172	65.5	90.0	111	330	0.180	0.217	0.335	0.252
Diethylamine	55.5	188	0.0993	160	62.1	88.0	105	213	0.229	0.269	0.351	0.100
Butylamine	77.5	75	0.0872	162	63.9	110	125	278	0.278	0.312	0.429	0.031
Triethylamine	89.4	46.2	0.0754	160	59.1	81.1	91.0	174	0.307	0.341	0.426	0.032
Dipropylamine	110	18.2	0.0647[g]	170	68.2	93.0	105	226	0.305	0.340	0.439	0.060
Diisopropylamine	110.5	—	0.0647[g]	171	67.0	77.0	87.1	185	0.257	0.288	0.370	0.069
Cyclohexylamine	134	—	0.0664[g]	169	64.3	112	128	279	0.382	0.431	0.564	0.141
Dibutylamine	159	1.61	0.0567[g]	164	64.6	75.5	84.8	186	0.335	0.372	0.480	0.009

Table 2 Breakthrough Time of Selected Hydrocarbons (*continued*)

Solvent	BP (°C)	Vapor pressure at 20°C (torr)	Diffusion coefficient at 25°C (cm²/s)	Volume charcoal (ml)[a]	Weight charcoal (g)[b]	Experimental breakthrough times			Wt. solvent adsorbed per wt. of carbon			Weight water adsorbed at $t_{100\%}$ per weight carbon[e] (g/g)
						$t_{1\%}$ (min)	$t_{10\%}$ (min)	$t_{99\%}$ (min)	$t_{1\%}$[c] (g/g)	$t_{10\%}$[d] (g/g)	$t_{100\%}$ (g/g)	
Miscellaneous												
Methyl iodide[h]	42.4	336	0.0900[g]	160	58.2	11.6	17.7	94.5	0.063	0.092	0.187	0.223
Acrylonitrile[h]	77.3	67.8	0.1059	160	58.0	48.5	61.1	168	0.098	0.121	0.174	0.016
Dibromomethane[h]	98.6	33	0.0826	160	62.8	82	121	279	0.502	0.717	1.06	0.006
Pyridine[h]	115.2	15.4	0.0824[g]	160	64.3	119	134	292	0.324	0.360	0.460	0.076
Epichlorohydrin[f]	116.9	12.5	0.0821[g]	146	56.9	85.5	110	288	0.308	0.387	0.566	0.014
2-Methoxyethanol[f]	124.4	9.0	0.0884	146	55.5	116	145	249	0.352	0.431	0.540	0.096
1,2-Dibromoethane[h]	131.5	8.6	0.0840[g]	161	61.9	141	165	303	0.946	1.09	1.336	—
1-Nitropropane[h]	131.6	7.2	0.0781[g]	160	63.2	143	164	322	0.443	0.504	0.638	—
2-Ethoxyethanol[f]	135.5	—	0.0788	148	55.4	77.0	123	283	0.277	0.426	0.593	0.011
Acetic anhydride	139.6	3.7	0.0755[g]	161	61.0	124	138	200	0.459	0.506	0.592	0.070
2-Methoxyethyl-acetate[f]	144.5	3.6	0.0666[g]	150	55.7	93.3	113.2	303	0.438	0.521	0.707	—
Bromobenzene[h]	156	3.0	0.0661[g]	160	58.1	142	159	297	0.849	0.940	1.126	—
2-Ethoxyethyl acetate[f]	156.3	2.0	0.0619[g]	148	55.8	79.5	96.5	248	0.417	0.497	0.689	0.010

[a] Average volumes for a cartridge pair are 146.3 ± 1.9 ml for the Type 1 and 163.8 ± 4.3 ml for the Type 2.

[b] Average weights for a cartridge pair are 56.1 ± 1.2 g for the Type 1 and 62.2 ± 3.9 g for the Type 2.

[c] Calculated from $W_{1\%} = t_{1\%}\ MQC/24.1 \times 10^6\ W_c$.

[d] Estimated from $W_{10\%} = [t_{1\%} + 0.9\ (t_{10\%} - t_{1\%})]\ [MQC]/24.1\ 10^6\ W_c$.

[e] Calculated from $W_{H_2O} = W_{total} - W_S$.

[f] Used Type 1 cartridges.

[g] Calculated from Gilliland's equation $D_{25} = [0.0043T^{1.5}(1/M_{air} + 1/M_{vap})^{0.5}]/(\nu_{air}^{0.33} + \nu_{vap}^{0.33})^2 P$, where ν_{air} is 29.9 ml/mol and ν_{vap} is calculated from the LeBas approximation (see Reference 11).

[h] Used Type 2 cartridges.

From Reference 10.

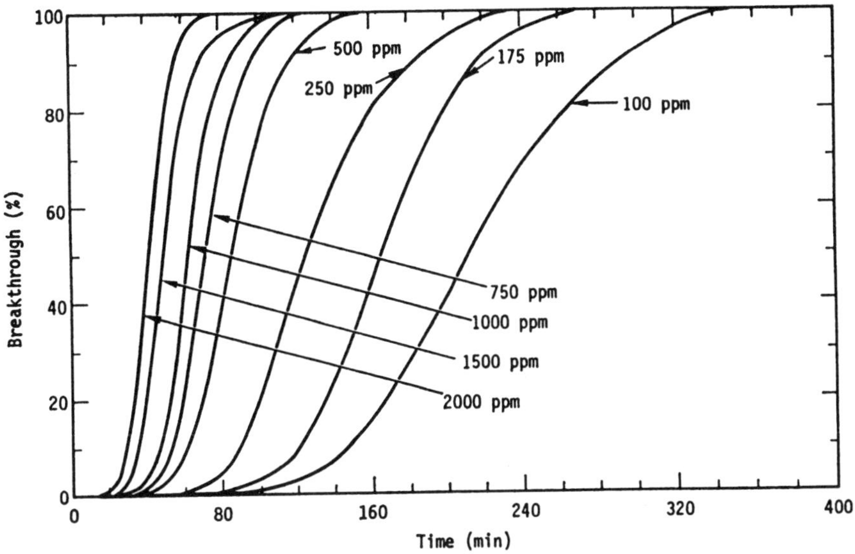

Figure 2 Breakthrough curves for Type 1 cartridges at various concentrations of methyl acetate. (From Reference 11.)

obtained if each individual slope for a given set of conditions is determined experimentally. This value is not appreciably affected by the flow rate; however, it is affected to some degree by relative humidity.

Figure 4 shows that a linear plot of flow rate is obtained when different concentrations of benzene are plotted against breakthrough time on the logarithmic scale. The parallelism of three different flow rates indicates that the ratio of breakthrough

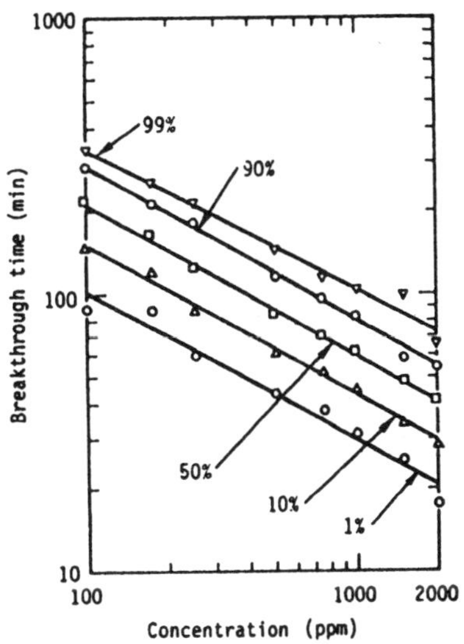

Figure 3 Effect of concentrations on the breakthrough time of methyl acetate at various breakthrough percents. (From Reference 11.)

Table 3 Experimental Values for *a* (Intercept) and *b* (Slope)

Solvent	Cartridge type	A (Intercept)	B (Slope)
Acetone	2	1.04×10^3	–0.452
Acetone	3	4.46×10^3	–0.621
Benzene	1	1.56×10^4	–0.743
Carbon tetrachloride	1	8.04×10^3	–0.671
Dichloromethane	1	3.53×10^2	–0.395
Diethylamine	1	5.36×10^3	–0.729
Hexane	1	3.92×10^4	–0.937
Hexane	3	3.89×10^4	–0.908
Isopropanol	1	5.53×10^3	–0.603
Methyl acetate	1	1.84×10^3	–0.542
Methyl chloroform	1	1.30×10^4	–0.744
	Vapor average		–0.67 ± 0.17
Vinyl chloride	1	2.05×10^2	–0.348
Vinyl chloride	3	2.68×10^2	–0.436
Vinyl chloride	4	2.48×10^2	–0.387
	Gas average		–0.39 ± 0.04

From Reference 11.

times at a given concentrations are inversely proportional to flow rate over the concentration range of 100 to 2000 ppm.

IV. SERVICE LIFE PREDICTION MODELS

Nelson showed that cartridge breakthrough can be predicted mathematically from the following formula:[11]

$$t_b = \frac{10^6 V_m w_s W_c}{MQC} \tag{12}$$

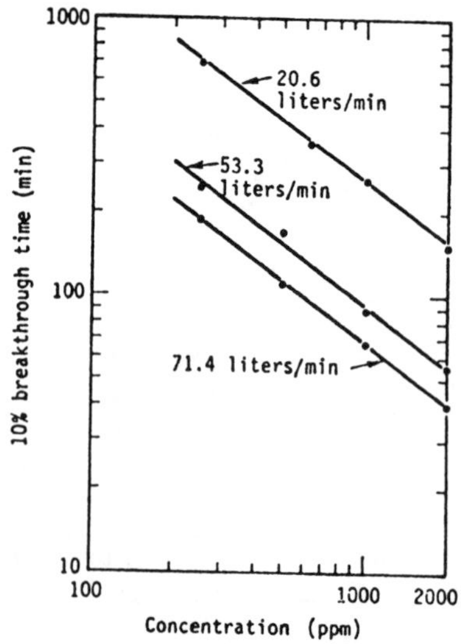

Figure 4 The effect of concentration on the 10% breakthrough time of benzene at various flow rates. (From Reference 11.)

Table 4 Comparison of Calculated and Observed Adsorption Capacity of Type 2 Cartridges for Several Solvents[a]

Solvent	Affinity coefficient, β	Adsorptive capacity, w_s Observed (gm/gm)	Adsorptive capacity, w_s Calculated from Equation 2 (gm/gm)	Deviation from observed (%)
Acetone	0.88	0.135	0.236	+75
2-Butanone	1.10[b]	0.295	0.360	+22
2-Pentanone	1.19[b]	0.483	0.427	−12
2-Heptanone	1.57[b]	0.560	0.493	−12
Pentane	1.08	0.228	0.225	−1.3
Hexane	1.295	0.334	0.319	−4.5
Heptane	1.46	0.432	0.375	−13
Nonane	1.85[c]	0.490	0.434	−11
Decane	2.05[c]	0.553	0.442	−20
Methylamine	0.72[c]	0.041	0.061	+49
Ethylamine	0.91[c]	0.179	0.166	−7.3
Butylamine	1.29[c]	0.429	0.371	−14
Dibutylamine	1.25[c]	0.480	0.466	−2.9

[a] The equilibrium partial pressure, p, at 1000 ppm is 0.76 torr.

[b] Calculated from Equation 3 using molar volumes.

[c] Calculated from parachors using Sugden's equation.

From Reference 10.

where t_b = Initial cartridge breakthrough time (min).
V_m = Molar volume at the system temperature and pressure (24.1 l/mole at 20°C, 1 atm).
w_s = Weight of solvent adsorbed per gram of activated carbon (g/g).
W_c = Weight of activated carbon (g).
M = Molecular weight of the contaminant (g/mole).
Q = Air flow rate (l/m).
C = Upstream gas concentration (ppm).

All the terms can be calculated or measured except w_s. Since the weight of solvent adsorbed is extremely difficult to predict, Equation 12 is useful only if the adsorption weight is determined mathematically. Nelson further stated that estimation of the breakthrough time can be calculated from the adsorption isotherm and the Mecklenburg equation. Nelson showed how w_s can be calculated from the experimental data generated in his cartridge breakthrough studies.[10] Some of the values of β and w_s are shown in Table 4. It should be noted that w_s is the weight adsorbed at total saturation or 100% breakthrough. Initial breakthrough occurs before the equilibrium adsorptive capacity is established. However, the breakthrough time can be estimated from the Mecklenburg equation (Equation 6) once w_s has been determined.

Nelson has developed a simple equation to estimate cartridge service life at a 10% breakthrough:[13]

$$t_{10\%} = \frac{2.4 \times 10^6 w_c w^o_{ad}}{CMQ} \tag{13}$$

and

$$w^o_{ad} = (a + bt) \tag{14}$$

Table 5 Solvent Class Coefficients

Solvent type	Boiling point range (°C)	Coefficients a	Coefficients b
Acetates[a]	50 to 190	–0.050	0.0038
Alcohols[a]	60 to 160	–0.46	0.0071
Alkanes	20 to 200	0.095	0.0022
Alkyl benzenes	80 to 220	0.12	0.0024
Amines[a]	–10 to 220	0.037	0.0033
Ketones	50 to 220	0.034	0.0029
Monochlorides	–30 to 250	0.032	0.0033
Dichlorides	40 to 250	–0.092	0.0048
Trichlorides	60 to 200	–0.080	0.0056
Tetrachlorides	70 to 200	0.19	0.0049

[a] Lower boiling solvents weighted more heavily when determining coefficients.

From Reference 11.

where $t_{10\%}$ = 10% breakthrough time (min).
w^{o}_{ad} = Weight of adsorbed solvent (g).
M = Molecular weight of the solvent vapor.
Q = Flow rate (l/m).
t = Normal boiling point (°C).
a,b = Coefficients determined by solvent type.
C = Concentration (ppm).
w_c = Weight of carbon (m).
M = Molecular weight.

The coefficients of a and b can be determined from the plot of boiling points against the adsorption capacity determined experimentally for a group of solvents such as ketones, or monochlorides. The values of coefficients a (intercept) and b (slope) for several solvent classes are listed in Table 5.

Based on his experimental data, Nelson[13] made calculations of the 10% breakthrough time from Mecklenburg, Wheeler, and Nelson equations on selected solvents and compared these values with the breakthrough time he determined experimentally. The results are presented in Table 6. More volatile solvents show the greatest deviation from the calculated values. This is due primarily to the preferential adsorption of the water vapor present. There is less deviation from the calculated values at lower breakthrough time.

The kinetic adsorption capacity and gas adsorption rate of activated charcoal sorbent bed are described by a linear relationship between gas breakthrough time and sorbent weight. The result is the modified Wheeler equation proposed by Jonas et al.[17-22] which can be rearranged as:

$$t_b = \frac{w_e}{C_o Q}\left[W - \frac{\rho_\beta Q}{k_v}\ln(C_o / C_b)\right] \tag{8}$$

or

$$t_b = \frac{w_e W}{C_o Q} - \frac{w_e \rho_\beta}{k_v C_o}\ln(C_o / C_b) \tag{15}$$

Table 6 Comparison of Experimental and Calculated 10% Breakthrough at 22°C

				10% Breakthrough time			
				Calculated			
Solvent	Concentration (ppm)	Flow rate (l/m)	Test relative humidity (%)	Mecklenburg Eq. (min)	Wheeler Eq. (min)	Eq. 13 (min)	Exp. (min)
Benzene[a]	125	53.3	50	440	418	377	355
	500	53.3	50	169	161	150	134
	2000	53.3	50	59.3	56.4	59.4	41.9
Benzene[c]	1000	53.3	20	114	110	126[d]	101
	1000	53.3	50	114	110	127	101
	1000	53.3	80	114	110	112[d]	87.4
Toluene[a]	1000	20.6	50	328	322	255	288
	1000	36.7	50	180	174	143	164
	1000	53.3	50	121	116	98.7	114
Methanol[a]	1000	53.3	50	8.6	7.9	–0.5	3.2
Isopropanol[a]	500	53.3	50	170	160	77.5	126
	2000	53.3	50	75.1	63.3	30.7	54.7
Butanol[a]	1000	53.3	50	150	143	120	141
Pentanol[a]	1000	53.3	50	137	134	139	130
Vinyl chloride[b]	50	40	50	99.2	96.2	–69.1	77.0
	250	40	50	58.5	56.8	–23.6	52.5
	1000	40	50	32.4	31.4	–9.4	22.7
Ethyl chloride[a]	1000	53.3	50	21.5	18.5	26.5	10.7
1-Chlorobutane[a]	1000	53.3	20	89.9	87.2	75.5[d]	86.3
	1000	53.3	50	89.9	87.2	73.3	87.3
	1000	53.3	90	89.9	87.2	61.0[d]	68.0
Chlorobenzene[a]	1000	53.3	50	132	128	97.9	131
Dichloromethane[a]	500	53.3	50	30.3	28.5	44.5	30.0
	2000	53.3	50	21.1	19.7	17.7	17.3
o-Dichlorobenzene[a]	1000	53.3	50	132	130	124	132
Chloroform[a]	1000	53.3	50	69.9	66.6	51.9	52.4
Methyl chloroform[a]	250	53.3	50	251	242	149	207
	2000	53.3	50	51.3	49.5	37.2	56.1
Trichloroethylene[a]	1000	53.3	50	108	103	72.6	83.0
Carbon tetrachloride[a]	1000	53.3	20	82.4	79.1	89.4[d]	84.9
	1000	53.3	50	82.4	79.1	86.8	68.8
	1000	53.3	90	82.4	79.1	71.3[d]	66.0
Perchloroethylene[a]	1000	53.3	50	128	123	112	129
Methyl acetate[a]	100	53.3	50	373	353	248	146
	1000	53.3	50	73.9	81.8	53.4	45.9
Ethyl acetate[c]	1000	53.3	20	115	111	83.1[d]	88.4
	1000	53.3	65	115	111	79.9[d]	90.4
	1000	53.3	90	115	111	67.2[d]	68.4
Propyl acetate[a]	1000	53.3	50	110	106	72.4	99.0
Butyl acetate[a]	1000	53.3	50	106	104	87.1	96.9
Acetone[a]	100	53.3	50	504	484	499	245
	500	53.3	50	160	154	170	96.7
	1000	53.3	50	94.4	90.8	107	66.3
Acetone[c]	1000	53.3	20	94.4	90.8	110[d]	61.1
	1000	53.3	80	94.4	90.8	94.3[d]	54.5
	1000	53.3	90	94.4	90.8	89.0[d]	53.1
2-Butanone[b]	1000	53.3	50	136	132	103	94.4
DiIsobutyl ketone[b]	1000	53.3	50	97.4	97.2	103	83.3
Pentane[b]	1000	53.3	50	86.2	83.8	67.5	71.3
Hexane[c]	100	53.3	50	646	631	420	565
	500	53.3	50	156	152	144	143
	2000	53.3	50	44.1	43.0	57.0	37.9
Hexane[a]	1000	53.3	0	77.6	75.4	66.7[d]	76.7
	1000	53.3	65	77.6	75.4	66.7[d]	68.1
	1000	53.3	90	77.6	75.4	56.6[d]	64.0
Cyclohexane[b]	1000	53.3	50	124	122	90.4	82.3
Heptane[b]	1000	53.3	50	106	104	86.9	80.5
Methylamine[b]	1000	53.3	50	49.1	46.7	13.4	17.9
Ethylamine[b]	1000	53.3	50	99.9	96.1	57.1	49.7
Diethylamine[a]	250	53.3	50	117	110	179	92.5
	1000	53.3	50	52.7	49.8	71.0	35.6
	2000	53.3	50	33.7	31.9	44.7	20.6
Dipropylamine[b]	1000	53.3	50	141	140	110	105

[a] MSA cartridges, coconut base, 52.2 g/pair.

[b] AO cartridges, coconut base, 62.2 g/pair.

[c] AO cartridges, petroleum base, 70.5 g/pair.

[d] Use humidity correction.

From Reference 13.

where t_b = Breakthrough time (min).
C_o = Inlet concentration (g/ml).
C_b = Exit concentration (g/ml).
k_v = rate coefficient (min^{-1}).
ρ_β = bulk density of packed bed (g/ml).
Q = Volumetric flow rate (ml/min).
W = Weight of adsorbent (g).
w_e = Adsorption capacity (g/g).

The values of C_o, C_b, Q, and W are established by experimental conditions. The value of ρ_β can be determined experimentally as a part of the manufacturer's quality control criteria. C_o and C_b can be preselected. If different weight (W) of sorbent are used to determine the breakthrough time (t_b) for a selected penetration fraction, C_o/C_b, values for w_e and k_v can be calculated from Equation 15 if the flow rate and temperature is fixed. When breakthrough time (t_b) is plotted as a function of sorbent weight (W), a straight line results where the slope and intercept allow calculation of the kinetic adsorption capacity w_e and the adsorption rate constant (k_v). The slope is equal to w_e/C_oQ. The y-axis intercept is equal to

$$-\frac{w_e\rho_\beta}{k_vC_o}\ln\frac{C_o}{C_b} \tag{16}$$

and the x-axis intercept $W_{c'}$ (critical bed weight) is equal to $\rho_\beta\, Q \ln(C_o/C_b)/k_v$. If the slope is known, a value for w_e can be determined. By inserting w_e into the y-axis intercept relationship, the value of k_v can be calculated. Also, k_v can be calculated from the x-axis intercept value.

Moyer made an investigation on the use of the modified Wheeler equation to predict the cartridge breakthrough.[23] He developed a special test cell which could attach a stack of four cartridges. In this arrangement, breakthrough time at four different carbon weight can be obtained from a single test. He used acetone as a testing agent. A plot of breakthrough time at different sorbent weight is shown in Figure 5. By varying concentrations of acetone, the Wheeler constant for the acetone challenge vapor at 1% breakthrough concentration can be calculated. The values are presented in Table 7.

Sansone and Jonas[24,25] showed that once the values of k_v and w_e have been determined, the Dubinin-Radushkevich equation[26,27] can be used to predict adsorption parameters of other untested solvents.

$$\ln W_v = \ln W_o - k(RT \ln P_o/P)^2/\beta^2 \tag{17}$$

and

$$P_e = \frac{n^2-1}{n^2+2}(M/d_L) \tag{18}$$

$$\beta = P_e(\text{vapor})/P_e\ (\text{ref}) \tag{19}$$

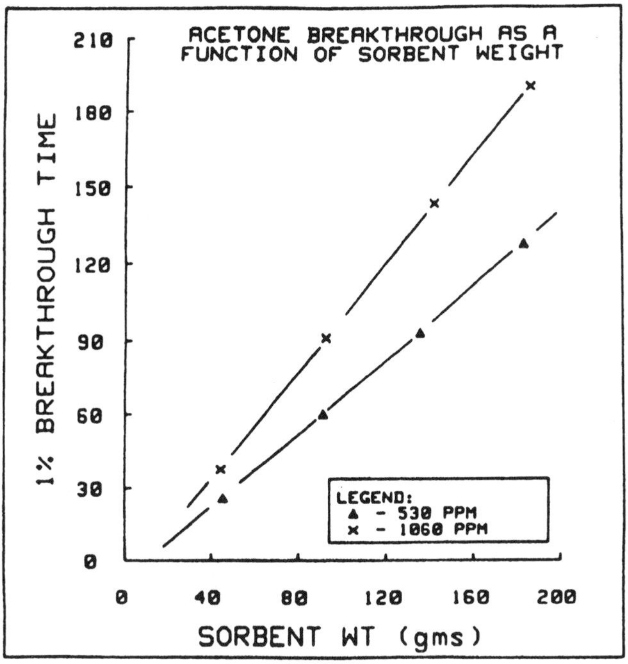

Figure 5 One percent acetone breakthrough time as a function of cartridge sorbent weight. (From Reference 23.)

where W_v = Adsorption space occupied by the condensed adsorbate (cm^3/g).
r = solvent density (g/cm^3).
W_0 = Maximum adsorption space available for condensed adsorbate (cm^3/g).
k = Microporosity constant for the carbon.
T = Temperature (°K).
β = Affinity coefficient of solvent vapor for the activated carbon.
P_0 = Saturated vapor pressure of the solvent at temperature T (torr).
P = Equilibrium partial pressure of the solvent (torr).
R = Gas constant.
M = Molecular weight of solvent.
d_L = Density of solvent liquid.
n = Refractive index of solvent liquid.
P_e = Electronic polarization.

The constants k and W_0 are related to the pore structure of the adsorbents. The affinity coefficient β characterizes the adsorption of a given vapor with respect to another vapor selected as a standard. The adsorption space of a carbon for a vapor is related to the weight of vapor adsorbed per unit weight of carbon by the relationship:

$$W_e = d_L \times W_v \tag{20}$$

If the liquid density (d_L) is known, then adsorption space W_v can be determined from Equation 20.

Table 7 Wheeler Constants for Acetone Challenge Vapor at 1% Breakthrough

Vapor	Lot	Points	Challenge conc. (ppm)	Slope	y Axis intercept	R^2	Average bed bulk density (ρ_β) g/cm³	W_s Kinetic adsorption capacity (g/g)	K_v Rate constant (per min)	W_c Critical bed weight (g)
Acetone	A	8	1060	0.715	−7.665	0.995	0.388	0.116	10670	10.7
Acetone	B	7	1060	0.743	−4.985	0.998	0.404	0.121	17810	6.7
Acetone	A	8	750	0.889	−10.650	0.999	0.393	0.102	9670	12.0
Acetone	B	8	750	0.978	−7.769	0.998	0.401	0.112	14850	8.0
Acetone	A	8	530	1.075	−11.121	0.999	0.393	0.087	11170	10.4
Acetone	B	8	530	1.031	−2.161	0.996	0.395	0.084	55810	2.1

From Reference 23.

Table 8 Comparison of Experimentally Determined with Calculated Adsorption Parameters

Vapor	Adsorption capacity W_t (g/g)			Adsorption rate constant K_t (min)		
	Exptl.	Calc.	% Dev.	Exptl.	Calc.	% Dev.
Benzene	0.404	0.409	−1.2	1029	1031	−0.2
Chloroform	0.728	0.693	−5.1	780	834	−6.5
p-Dioxane	0.476	0.483	−1.4	1083	971	−11.5
Acrylonitrile	0.404	0.375	−7.7	1160	1251	−7.3
1,2-Dichloroethane	0.616	0.583	+5.7	1048	916	−14.4

From Reference 28.

Jonas et al.[28] stated that the product of the adsorption rate constant k_v and the square root of the adsorbate's molecular weight was a constant for a particular carbon. Thus the relationship

$$(k_v\, M^{1/2})_i = (k_v\, M^{1/2})_{ref} \tag{21}$$

Sansone and Jonas[24] showed that once the rate constant (k_v) and adsorption capacity (W_e) obtained from one solvent under a given test condition is determined, the adsorption parameters of other untested solvents can be calculated from the modified Wheeler equation and Equations 17 to 21. They performed tests to determine the breakthrough time and adsorption capacity for carbon tetrachloride, chloroform, p-dioxane, acrylonitrile, and 1,2-dichloroethane. Adsorption parameters such as the adsorption capacity (W_e) and adsorption rate constant (k_v) determined from experimental data deviated from those calculated over a range of −1.4 to 7.7% for the adsorption capacity, and over a range of −7.3 to 14.4% for the adsorption rate constant when the data were used to predict the carbon performance of 31 carcinogenic vapors. The comparison of experimentally determined and calculated adsorption parameters are shown in Table 8. They also used Equation 17 to calculate adsorption parameters for other 26 carcinogens; the results are tabulated in Table 9.

Ackley[29] developed an experimental bed residence time model to characterize the performance of a sorbent bed in removing gaseous contaminants. He claims that the model is applicable to both adsorption and chemisorption processes. The fundamental characteristics produced can be used to predict the performance of respirator cartridges and canisters over a wide range of operating conditions.

The residence time τ is defined as

$$\tau = \frac{d_b}{v_f} = 0.06\frac{V}{Q_{air}} \tag{22}$$

where d_b is the bed depth, v_f is superficial velocity, the sorbent volume is V, and the air flow rate is Q_{air}.

Since the volume of the sorbent is equal to the sorbent weight divided by the bed density, then

$$\tau = \frac{W}{\rho_b Q_{air}} \tag{23}$$

Table 9 Predicted Adsorption Parameters for Carcinogenic Vapors[a]

Vapor	Liquid density[b] (g/cm³)	Refractive index[c]	Adsorption rate constant (min⁻¹)	Adsorption capacity (g/g)
Acetamide	1.159	1.4274	1186	0.494
Acrylonitrile	0.8060	1.3911	1251	0.357
Benzene	0.8761	1.5011	1031	0.409
Carbon tetrachloride	1.5881	1.4607	735	0.741
Chloroform	1.4832	1.4459	834	0.688
bis(Chloromethyl)ether	1.315	1.4346	850	0.608
Chloromethyl methyl ether	1.0605	1.3974	1016	0.480
1,2-Dibromo-3-chloropropane	2.093	1.553	593	0.992
1,1-Dibromoethane	2.0555	1.5128	665	0.962
1,2-Dibromoethane	2.1792	1.5383	665	1.020
1,2-Dichloroethane	1.2492	1.4448	916	0.575
Diepoxy butane (meso)	1.1157	1.4330	982	0.510
1,1-Dimethyl hydrazine	0.791	1.4075	1176	0.359
1,2-Dimethyl hydrazine	0.8274	1.4204	1176	0.375
Dimethyl sulfate	1.332	1.3874	812	0.615
p-Dioxane	1.0333	1.4220	971	0.475
Ethylenimine	0.8321	1.412	1388	0.354
Hydrazine	1.0083	1.4698	1610	0.380
Methyl methane sulfonate	1.2943	1.4140	869	0.595
1-Naphthylamine	1.229	1.6703	762	0.585
2-Naphthylamine	1.0614	1.6493	762	0.506
N-Nitrosodiethylamine	0.9422	1.4386	902	0.442
N-Nitrosodimethylamine	1.0059	1.4368	1059	0.468
N-Nitroso-*N*-methylurethane	1.133	1.4363	793	0.534
N-Nitrosopiperidine	1.0631	1.4933	853	0.501
N-Nitrosodipropylamine	0.9163	1.4437	799	0.434
1,3-Propane sultone	1.393	1.450	825	0.646
β-Propiolacetone	1.1460	1.4118	1074	0.508
Propylenimine	0.802	1.409	1206	0.361
Safrole	1.096	1.5383	716	0.522
Urethane	0.9862	1.4144	966	0.456
Vinyl chloride	0.9114	1.3700	1153	0.404

[a] Predicted values are for the carbon and test conditions specified in the text.

[b] Liquid densities are at 20°C relative to water at 4°C unless otherwise stated.

[c] Refractive indices are for the D line of the sodium spectrum and at 20°C unless otherwise stated.

From Reference 24.

The modified Wheeler equation can be transformed as:

$$t_b = \frac{w_e \rho_b}{C_o}\left[\tau - \frac{1}{k_v}\ln(C_o / C_x)\right] \tag{24}$$

By plotting t_b against τ, the adsorption capacity w_e can be calculated from the slope and the intercept as described elsewhere in this chapter.

Ackley states that breakthrough times must be measured at two different flow rates (for fixed volume beds) for reliable application of the proportionality relationship.

Yoon and Nelson[30,31] proposed a gas adsorption model which is a further modification of the modifier Wheeler equation. The first order rate equation is defined as

$$-\frac{dQ}{dt} = k'QP \tag{25}$$

and

$$k' = \frac{kCF}{w_e} \tag{26}$$

where Q = Probability of adsorption.
P = Probability of breakthrough, P = 1 − Q.
k′ = Proportionality constant (min^{-1}).
w_e = Adsorption capacity of carbon.
C = Contaminant concentration (l/m).

Solving Equation 25, the following relationship is obtained:

$$\ln\frac{Q}{1-Q} = k'(\tau - 1) \tag{27}$$

or

$$\ln\frac{P}{1-P} = -k'(\tau - 1) \tag{28}$$

where τ is the time (min) required to obtain 50% breakthrough. Accordingly, when Q = 1/2 (or P = 1/2), t ≡ τ. Similarly, the probability of breakthrough P is:

$$P = \frac{1}{1 + e^{k'(\tau-1)}} \tag{29}$$

Rearrange Equation 29, the expression for breakthrough time is obtained:

$$t = \tau + \frac{1}{k'}\ln\frac{P}{1-P} \tag{30}$$

since

$$P = \frac{C_b}{C_I}$$

where C_b = Breakthrough concentration.
C_I = Incoming concentration.

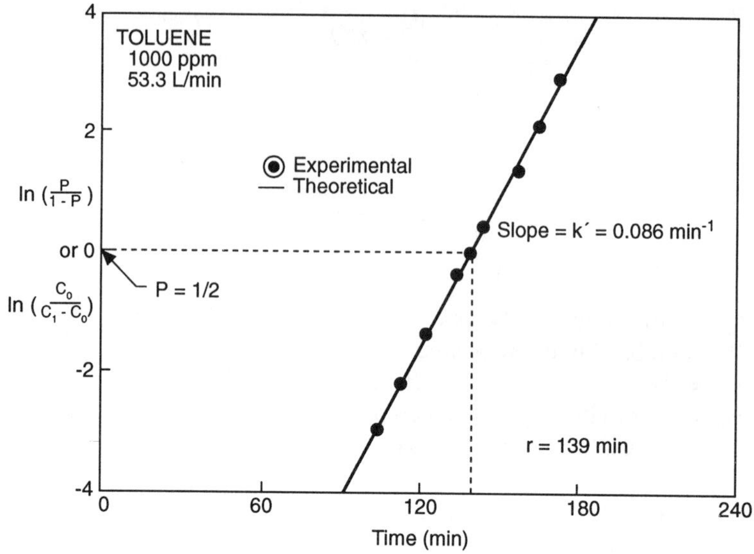

Figure 6 Plot of $\ln\frac{P}{1-P}$ versus time for toluene at 1000 ppm. The solid line is derived from theory. Experimental data are from Reference 10. (From Reference 30.)

The following equation is derived to express breakthrough time in terms of the initial contamination concentration, C_I and the breakthrough concentration, C_b:

$$t = \tau + \frac{1}{k'} \ln \frac{C_b}{C_I - C_b} \tag{31}$$

In order to calculate breakthrough time t using Equation 30 or 31, it is necessary to determine the parameters τ and k'.

According to Equation 28, the values of k' can be obtained from the slope of the plot ln (1/1 – P) or ln $C_b/(C_I - C_b)$ versus the breakthrough time and the value of τ (50% breakthrough time) can be determined as the time t at P = ½ (i.e., $C_b = \frac{1}{2} C_I$). A plot of ln (P/1 – P) versus time using breakthrough data on toluene generated by Nelson[13] is shown in Figure 6. The authors claim that both k' and τ can be obtained from a minimum of two accurate experimental data points. Following the determination of k' and τ, one may generate the complete breakthrough curve for a given set of experimental conditions by applying Equation 31.

Yoon and Nelson used the breakthrough data on toluene, methyl acetate, vinyl chloride, and chlorinated hydrocarbons developed by Nelson[13] to show that there is a good agreement between the theoretical model and the experimental data. Table 10 lists values of τ and k' for methyl acetate and vinyl chloride which are obtained from plots derived from Figure 7. A comparison of theoretical and experimental breakthrough curve for methyl acetate at various concentrations is shown in Figure 7. Figure 8 shows the breakthrough curves of various chlorinated hydrocarbons.

They demonstrate that the model can be used to study effects associated with breakthrough time. Three experimental points are required to determine break-

Table 10 Values of τ, k′, and k for Methyl Acetate and Vinyl Chloride at Various Assault Concentrations

Assault concentration (ppm)	Methyl acetate τ (min/cartridge)	Methyl acetate k′ (min)	Methyl acetate k	Vinyl chloride τ (min/cartridge)	Vinyl chloride k′ (min)	Vinyl chloride k
50	—	—	—	111	0.052	5.77
100	211	0.032	6.75	86.2	0.071	6.12
175	163	0.047	7.66	—	—	—
250	129	0.049	6.32	71.9	0.088	6.33
500	85.0	0.085	7.23	44.9	0.114	5.12
750	70.5	0.102	7.19	—	—	—
1000	62.0	0.120	7.44	34.4	0.146	5.02
1500	49.5	0.132	6.53	26.4	0.207	5.46
2000	38.5	0.149	5.74	19.6	0.286	5.61
Average			6.86 ± 0.64			5.63 ± 0.48

From Reference 31.

through time at different concentrations. Two of the data points must be at the same concentration and the third data point must be obtained at the same breakthrough percentage as one of the two previous points but at a different concentration. For the effect on flow rate, only two experimental data points obtained at the same flow rate and the same breakthrough percentage are needed. However, these data points should be obtained at two different contaminant concentrations in order to determine the effect of concentration on the breakthrough time at various flow rates. In the case of breakthrough percentage effect, three experimental points are required to address the effect of concentration on breakthrough time at given breakthrough percentage. The

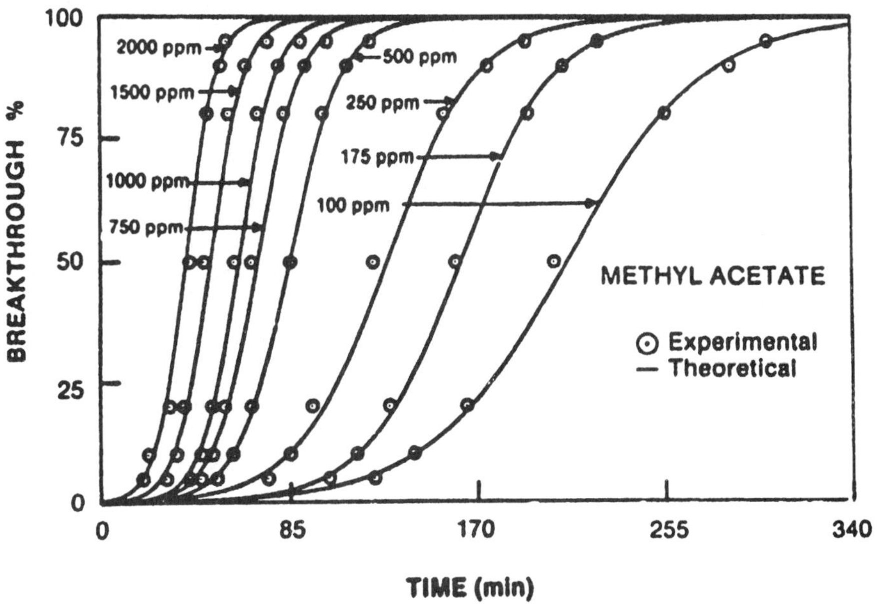

Figure 7 Comparison of theoretical breakthrough curves (solid lines) with experimental data for methyl acetate at various assault concentrations. (From Reference 31.)

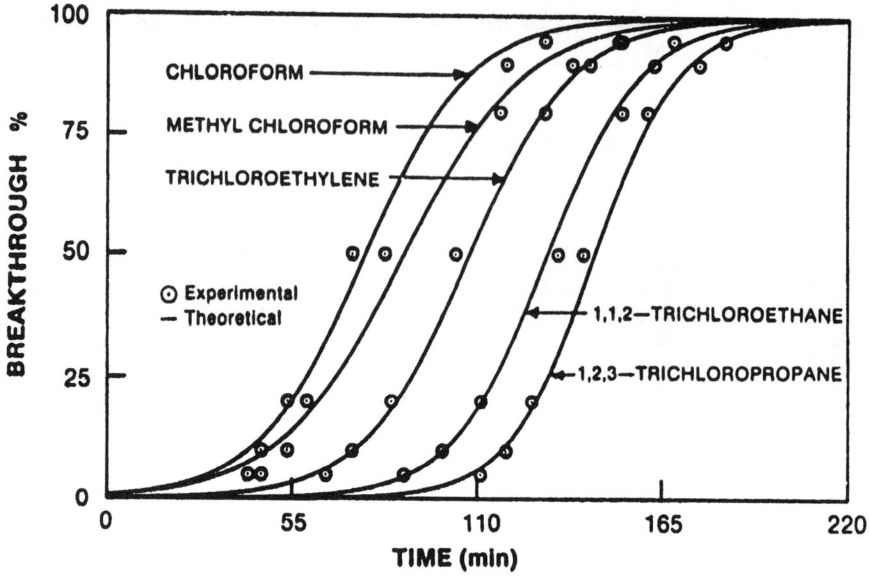

Figure 8 Comparison of theoretical breakthrough curves (solid lines) with experimental data for various trichlorinated hydrocarbon compounds at 1000 ppm, 50% relative humidity, and a flow rate of 53.3 l/m. (From Reference 31.)

three points must satisfy the same conditions as those specified in the concentration effects stated above.

Wood and Moyer[32] made a comparative study on the modified Wheeler equation proposed by Jonas, the Yoon and Nelson model and the residence time model proposed by Ackley. The same modified Wheeler equation can be applied for the following approaches:

1. Varying sorbent bed weight (Jonas).
2. Varying penetration fraction (Yoon and Nelson).
3. Varying bed residence time (Ackley).

Each approach should yield the same values of rate coefficient k_v and adsorption capacity W_e from slopes and intercepts of straight line plots of experimental data. They made an investigation to determine whether three different approaches would yield the same results.

Acetone was selected as the testing solvent at a concentration of 1000 ppm. The flow rates varies from 32 to 115 l/m. By varying the weight of sorbent, the adsorption capacity W_e and the rate coefficient k_v for these three approaches can be calculated from the experimental data. A comparison of the adsorption capacity W_e and the rate coefficient k_v obtained from these approaches are presented on Table 11. There is very little difference in the values of W_e as determined from the three methods. However, there is wide difference in the values of k_v since it is influenced by the flow rate and selected penetration fraction. The results suggested that the curve fitting approach (Yoon and Nelson) may be less valid than the other two at 10% breakthrough. The k_v values calculated by all three methods were approximately propor-

Table 11 Comparison of Modified Wheeler Parameters Obtained by Three Approaches

Average airflow rate (l/m)	Adsorption capacity, W_e (g/g)			Adsorption rate coefficient, k_v (min^{-1})		
	Bed weight variation	Residence time variation	Breakthrough curve fitting	Bed weight variation	Residence time variation	Breakthrough curve fitting
32.2	0.106	0.106	0.092	4730	5050	6580
40.4	0.110	0.108	0.107	4880	5920	6870
63.9	0.116	0.116	0.114	7940	7860	9160
83.0	0.115	0.113	0.109	7440	7860	10410
94.5	0.113	0.113	0.108	8670	8770	13090
112.5	0.109	0.110	0.107	9460	8730	12780
Combined		0.106			11400	
Averaged	0.112	0.111	0.108			

From Reference 32.

tional to the square root of the flow rate. The implication of this dependence on air flow velocity is that breakthrough data at significantly different air flow rates should be examined carefully before combining them to calculate a single W_e and k_v using the Wheeler equation.

The authors concluded that bed weight variation for constant flow rate (or residence time variation for nearly constant flow rate) is the method of choice for data collection and analysis. Data on breakthrough time should be collected at more than one breakthrough fraction, preferably by a factor of 10 and covering the range of interest in practical application. In addition, the effect of flow rate, challenge concentration, humidity, and temperature should also be studied. This would allow application of the results to actual used conditions.

V. EFFECT OF HUMIDITY

There is no other factor than moisture which could impose a profound effect on the service life of the sorbent. Many sorbent studies have focused on this subject.[33-41] The amount of water vapor present in the air is usually expressed as relative humidity. The amount of water vapor present in the air is a function of temperature. Values of water vapor present at different humidities at 20°C are presented in Table 12.

In terms of contaminant concentration, water vapor will generally be present in amount at least comparable to the amount of contaminants or exceed it by several orders of magnitude.

Table 12 Water Content of Air at 20°C

RH, %	mg/m^3	ppm
10	1730	2300
20	3460	4600
30	5190	6920
40	6920	9200
50	8650	11500
60	10400	13840
70	12100	16000
80	13800	18500
90	15600	20800
100	17300	23100

Table 13 Water Vapor Adsorption Under Static and Dynamic Conditions

Sorbent type	Dynamic (g)	Static (g)
I	11.50	0.90
II	13.44	1.95
III	12.44	5.38
IV	14.44	2.93
V	9.50	3.31
VI	5.09	2.06

From Reference 33.

Balieu[33] conducted a study to determine the effect of water vapor on the capacity of the sorbent. At equilibrium conditions, the water uptake at higher humidities could be 50% by weight of the sorbent. Water vapor absorption can occur under static or dynamic conditions. Static adsorption occurs during production or during storage after intermittent use. The rate of uptake is dependent on the rate of diffusion. Dynamic adsorption occurs during use when the air is drawn through the sorbent. The rate of uptake depends on the air flow rate and the size of the sorbent. The water vapor uptake at dynamic condition is much higher than at static condition. A study was conducted to determine the rate of uptake of six sorbent cartridges under static or dynamic conditions.

Under static conditions, the sorbent cartridges were exposed in a climatic chamber for 24 h at 20 ± 2°C and 90% relative humidity. Under dynamic conditions, the sorbent cartridges were exposed for 4 h at 20 ± 2°C, 75 ± 5% relative humidity and at a flow rate of 30 ± 0.5 l/m. The results of water vapor adsorption on the sorbent are listed on Table 13.

The results indicated that even at a lower relative humidity and a shorter equilibration time, substantially large amounts of water vapor was adsorbed under dynamic conditions. Due to the difference in carbon type and capacity, the time to reach the state of equilibrium differed from 15 to 25 h.

The sorbent seldom reaches the state of equilibrium during use, but only takes about 50% of the equilibrium value. The effect of water adsorption on the sorbent service life for several types of respirator cartridges is presented on Table 14. The test conditions were: 5000 ppm, 50 l/m, 75% RH, and 20°C. The sorbent was tested under the following conditions:

1. As received.
2. Humidified under static conditions as described above.
3. Equilibrated under dynamic conditions as described above.

In the humidified state where approximately 25% of the equilibrium capacity has been reached, breakthrough time could be lowered by several orders of magnitude. For cartridges containing carbon only, the lowering of the service life was caused by the competition between adsorption of water molecules and organic vapor. On the other hand, the adsorption capacity can be increased by water vapor when the sorbate is water soluble or when the sorbate is readily hydrolyzed by water. For cartridges containing activated carbon impregnated with inorganic salt of heavy metals such as chromium or copper, the impregnants either react chemically with the contaminants or act as catalyst giving rise to oxidation or decomposition process. The effect of water vapor for this type sorbent is dependent upon the contaminants. For example,

Table 14 Effect of Humidification

Filter type	Test substance	Breakthrough times (min)			Breakthrough concentration (ppm)
		As received	Humidified	Equilibrated	
I (A)	CCL_4	70	59	22	10
I (B)	HCN		40 (35)	23 (5)	10
I (B)	HCN	46 (18)	54 (8)		10
II (A)	CCL_4	62		16	10
II (B)	HCN		45 (42)	33 (28)	10
II (B)	HCN	59 (48)		52 (32)	10
III (A)	CCL_4	47	15	4	10
IV (A)	CCL_4	62	50	9	10
IV (B)	SO_2	19	26	32	5
IV (B)	H_2S	170	129	136	10
IV (E)	SO_2	16	25	11	5
VI (B)	SO_2	21	22	27	5
VI (B)	H_2S	126	16	17	10

From Reference 33.

the service life is usually increased with an acid gas contaminant such as sulfur dioxide or hydrogen chloride.

In general, the adsorption of water vapor by the sorbent is usually low at the beginning. However, due to the high content of water vapor in the air, substantial amounts of water vapor can be taken up and the predominant effect is a lowering of the breakthrough time (and hence the capacity) resulting in many cases in a dangerously low level of protection.

The effect of preconditioning humidity and use of humidity on the cartridge service life has been investigated by Nelson.[12] The cartridges were tested at a concentration of 1000 ppm, at a flow rate of 53.3 l/m, and at 0, 20, 50, 80, and 90% relative humidity. The preconditioning humidity varied between 0% and 80% RH. Seven solvents were chosen in this study. The results indicated that service life does not vary appreciably at preconditioning humidity up to 50% RH. However, when the relative humidity exceeded 50% RH, the initial breakthrough time was shortened considerably. As noted in Figure 9, the characteristic S shaped breakthrough curve

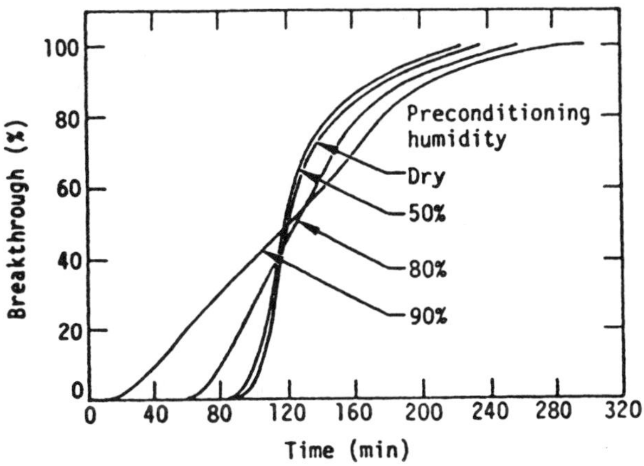

Figure 9 Breakthrough percent as a function of time at various preconditioning humidities. MSA cartridges were tested with 1000 ppm hexane at 53.3 l/m. (From Reference 12.)

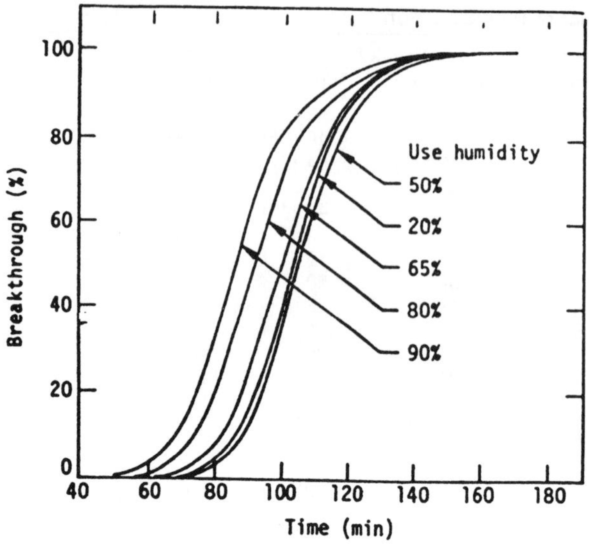

Figure 10 Breakthrough percent as a function of time at various use humidities. MSA cartridges were tested at 1000 ppm with 1-chlorobutane and preconditioned at 50% relative humidity. (From Reference 12.)

for hexane was distorted to almost a straight line at a preconditioning humidity of 90%. This phenomenon may be explained by the competition of water vapor for the available adsorption sites and the inability for the solvent vapor to displace the adsorbed water vapor.

In addition to the water vapor normally taken up by the cartridge during storage, moisture can reenter the cartridge during use. Figure 10 shows the breakthrough curve for 1-chlorobutane at humidities from 20 to 90%. The cartridges were preconditioned at 50% RH. The breakthrough curves for 20 and 50% are practically coincidental. However, higher RH show a substantial reduction in service life of the cartridges. The moisture uptake is almost an exponential function of use humidity rather than a much flatter response shown for preconditioning humidity.

Figure 11 shows a combined effect of preconditioning and use humidity on the breakthrough time for carbon tetrachloride. The cartridges were preconditioned at 50, 65, and 80% RH and tested at 20, 50, 65, 80, or 90% RH. Figure 11(a) shows a slight separation of the breakthrough curves at different use humidities at a preconditioning humidity of 50%. The separation is noticeable and a decrease in breakthrough time is observed at a preconditioning humidity of 65% as shown in Figure 11(b). Figure 11(c) shows the largest separation and breakthrough time decrease.

Correction factors for humidity at 1000 ppm, 53.5 l/m, and 22°C for a pair of cartridges were calculated by Nelson[12] and shown in Table 15. The data are normalized to the 50% preconditioning and use relative humidity. If the breakthrough time is known at the normalized conditions, service life at other humidities may be estimated by using the appropriate correction factor obtained from Table 15. It should be pointed out that all preconditioning was performed under the static condition, i.e., no air was passed through the cartridge(s) during preconditioning. The effect of

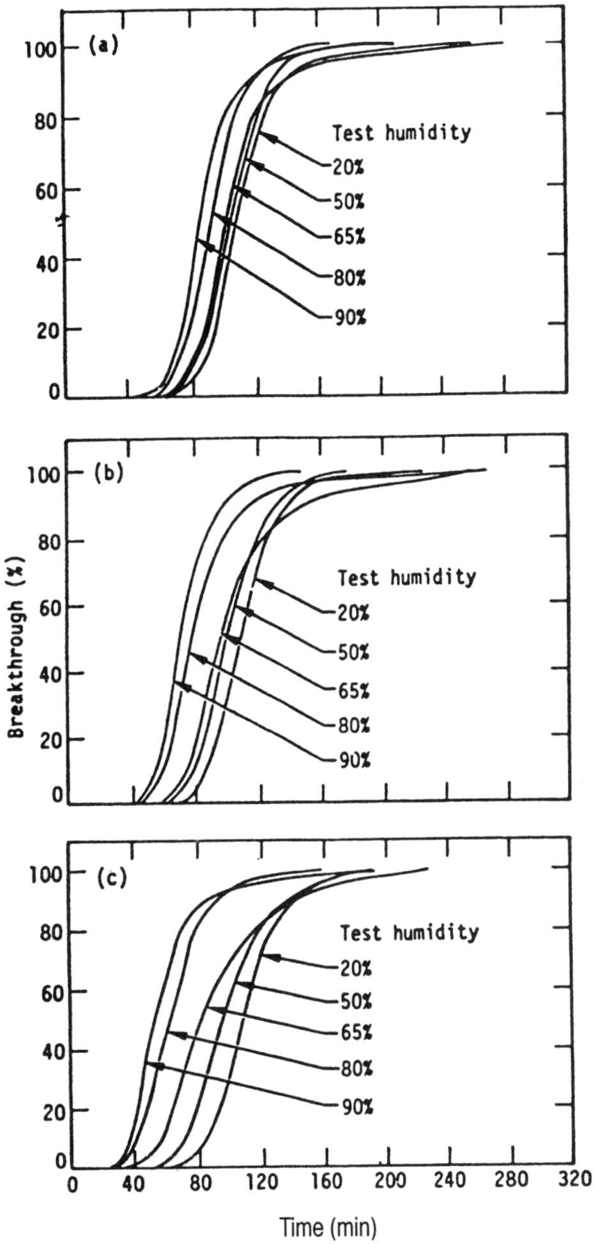

Figure 11 Breakthrough percent as a function of time for various use humidities at (a) 50%, (b) 65%, and (c) 80% preconditioning humidities. MSA cartridges were tested with carbon tetrachloride at 1000 ppm and 53.3 l/m. (From Reference 12.)

preconditioning on sorbent service life would be more pronounced if the preconditioning was performed under dynamic conditions.

Werner[34] conducted a study to determine the effect of humidity on the breakthrough time of trichlorethylene (TCE). He used a custom made column filled with

Table 15 Breakthrough Time Correction Factor

Test relative humdity (%)	Breakthrough time multiplier preconditioning relative humidity (%)					
	0	20	50	65	80	90
0	0.94 (0.08)[a,b]	0.95 (0.04)[a-c]	0.99 (0.07)[a-c]	0.97 (estimated)	0.95 (0.07)[a-c]	0.95 (—)[a]
20	1.02 (0.06)[a,b]	1.02 (0.03)[a,b,d-g]	1.03 (0.04)[a,b,d-g]	1.04 (0.03)[d-g]	1.01 (0.05)[a,b,d-g]	1.00 (0.05)[a,d-g]
50	0.98 (0.07)[a-c]	0.99 (0.03)[a,d,b-g]	1.00 (0.05)[a-g]	0.99 (0.04)[d-g]	0.95 (0.08)[a-g]	0.77 (0.20)[a,c-g]
65	0.97 (0.04)[a,b]	0.98 (0.04)[a,b,d-g]	0.99 (0.05)[a,b,d-g]	0.94 (0.04)[d-g]	0.84 (0.10)[a,b,d-g]	0.66 (0.25)[a,d-g]
80	0.87 (0.06)[a-c,e]	0.91 (0.05)[a,b,d-g]	0.88 (0.04)[a-g]	0.83 (0.09)[d-g]	0.72 (0.16)[a-g]	0.50 (0.27)[a,c-g]
90	0.84 (0.03)[a-c]	0.85 (0.04)[a,b,d-g]	0.83 (0.06)[a,b,d-g]	0.78 (0.09)[d-g]	0.67 (0.13)[a,b,d-g]	0.48 (0.20)[a,d-g]

Note: At various humidities at 1000 ppm, 53.3 l/m, and 22°C for a pair of cartridges containing coconut or petroleum base carbon. The data was normalized to the 50% preconditioning and test relative humidity. The footnotes indicate the test vapor employed and the carbon type. The number in parentheses is the standard deviation.

[a] Isopropanol, coconut base.

[b] Hexane, coconut base.

[c] Benzene, petroleum base.

[d] Acetone, petroleum base.

[e] Carbon tetrachloride, coconut base.

[f] 1-Chlorobutane, coconut base.

[g] Ethyl acetate, petroleum base.

From Reference 12.

activated carbon. The flow rate was set at 7.7 l/m and the temperature was set at 23°C. The challenge concentrations for trichloroethylene were set at 50, 100, 165, and 220 ppm. The relative humidity varied from 5 to 85%. The breakthrough time at 10 and 50% penetration are shown in Table 16.

The data sets were fitted with the Dubinin-Polyanyi (D-P) isotherm. It was found that at each humidity level the effect of water vapor was consistent with the theoretical basis of the D-P equation. The deleterious effect of humidity is less at high TCE concentration than at lower concentrations. If the exploration of the data beyond the experimental range, one may predict that at some high adsorbate concentration the effect of humidity would be negligible. The author concluded that the amount of TCE adsorbed on carbon decreases with an increase of humidity. At low concentrations, the effect of humidity is greater than at high TCE concentrations. He also indicated that two assumptions concerning the effect of humidity on vapor phase activated carbon adsorption should be applied with caution. First, the assumption that relative humidity below 50% has negligible effects on adsorption may be invalid. All levels of relative humidity have an adverse effect at the low TCE concentration tested. Second, applying a "multiplier factor" determined at one adsorbate concentration to account for a given humidity at other adsorbate concentration is tenuous. The effect of humidity on the carbon's adsorption capacity is strongly influenced by the challenge TCE concentration.

Hall et al.[35] made an investigation on the effect of moisture on sorbent service life. They selected the worse case scenario when an adsorbate with poor water

Table 16 TCE Adsorbed for Individual Trials

Relative humidity (%)	TCE Influent mg/mm	Concentration ppm	TCE Adsorbed[a] (g TCE/g carbon)
5	303	52	0.286
25	295	50	0.257
50	293	50	0.180
65	295	50	0.114
85	293	50	0.027
5	602	103	0.334
25	605	103	0.320
50	597	102	0.284
65	599	102	0.160
85	593	101	0.054
5	987	168	0.399
25	995	170	0.403
50	978	167	0.342
65	996	170	0.218
85	986	168	0.098
5	1331	227	0.434
25	1306	223	0.431
50	1356	231	0.370
65	1322	226	0.262
85	1304	222	0.121

[a] Calculations were based on time to 50% TCE breakthrough.

From Reference 34.

solubility was adsorbed on the sorbent bed which had been saturated with water vapors. Carbon tetrachloride and water were simultaneously introduced into the adsorbent bed. The sorbent bed had been equilibrated with water vapor until there was no weight gain. The challenge concentration of CCl_4 was 90 μg/ml and the flow rate was set at 285 ml/min. A series of at least three repetitive experimental trials were conducted at 0, 30, 52, 60, 86, and 90% relative humidity.

The values of the kinetic adsorption capacity, W_e, and the adsorption rate constant, k_v, as described in the modified Wheeler equation are presented in Table 17. When the water vapor concentration exceeded 50% RH, the rate constant decreased in a linear fashion. The kinetic adsorption capacity, W_e, for the carbon tetrachloride-dry carbon system is 0.366 g/g. As the system water vapor concentration is increased, the adsorption capacity showed a decrease which is a linear function of relative humidity. The regression line predicts a minimum W_e of 0.16 g/g, a 45%

Table 17 Adsorption Parameters of Respirator Activated Carbon for Carbon Tetrachloride at Various Humidities

Relative humidity (23°C)	Adsorption rate constant K_v (s^{-1})	Kinetic adsorption capacity, W_e (g/g)
0	65	0.361
30	68	0.286
52	63	0.271
60	48	0.248
86	25	0.203
90	18	0.194

From Reference 35.

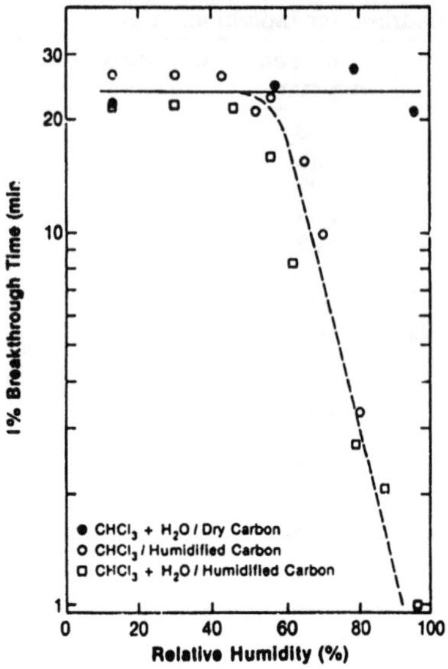

Figure 12 One percent breakthrough time as a function of relative humidity. (From Reference 36.)

reduction from the dry carbon value. The prediction of service life from the modified Wheeler equation based on the dry carbon values of k_v and W_e would overestimate severely the protective capacity of the cartridge.

Jonas et al.[36] conducted a study to determine the effect of moisture on the adsorption of chloroform by activated carbon. Adsorbent beds with various weights were exposed to the chloroform vapor at a flow rate of 285 ml/min and a temperature of 23°C. The challenge concentration was set at 108 μg/ml which represented a relative pressure of 0.0936 at 23°C for $CHCl_3$. The breakthrough study was performed under three conditions: (1) chloroform was introduced into a humidified sorbent bed; (2) chloroform and water vapor were concurrently introduced into a dry adsorbent bed, and (3) both chloroform and water vapor were introduced into a humidified sorbent bed. The test relative humidity varied from 13 to 97%. The 1% breakthrough data at various test conditions is shown in Figure 12.

It appears that the breakthrough time of chloroform was not affected by moisture when the carbon bed was initially dry. There was also no significant change on the breakthrough time for chloroform if the carbon bed had previously been equilibrated with air having a relative humidity of less than 40%. When the relative humidity of the humidified carbon exceeded 40%, there was a marked decrease in breakthrough time no matter whether the inlet chloroform is dry or humidified with water vapor. The effect of increasing the volumetric or superficial flow rates of the challenge vapor could affect both the shape of the breakthrough curve and the magnitude of individual breakthrough time.

Using published data on the humidity effect on sorbent breakthrough time, several investigators developed models that would predict the effects of water on the service life of the sorbent.[37-41]

Wood[37,38] developed a mathematical model to describe the effects of relative humidity on adsorption capacities of charcoal beds for a fixed challenge concentration of a water-immiscible vapor. This model is based on five assumptions: (1) that only adsorbed (or condensed) water and adsorbate affect the capacity for the water-immiscible adsorbate; (2) that the rate of adsorption is not affected significantly by the amount of water present; (3) that water and adsorbate equilibrium exist between gas and solid phases; (4) that there is a fixed concentration of homogeneously distributed micropores which can contain either molecules of water or adsorbate; and (5) that the challenge concentration of adsorbate is kept constant was only valid for the analysis of relative humidity effects only.

Wood demonstrated that the model applies to both dry and preconditioned sorbent beds and provides good correlations between the effect of relative humidity and challenge vapor concentration on adsorption capacities of charcoal beds. Using the published data generated by Nelson,[12] Werner,[34] and unpublished data developed by himself, Wood found that the model works well with this data. The model also works with both 10% (Nelson), and 50% (Werner) breakthrough capacities.

Using the Polanyi Potential Theory and the Dubinin-Radushkevich Model, Underhill[39] developed an equation which would permit a rapid calculation of the effect of water vapor on the adsorption of water-immiscible organic compounds. His equation is based on two assumptions: (1) that the adsorption potential in the presence of completely saturated air is some multiple of the adsorption potential from dry air; and (2) that the adsorption potential decreases linearly with a decrease in the free energy of the water vapor. The data developed by Werner[34] on trichlorethylene have been used to fit the equation. The mean error between the calculated value and Werner's experimental values was 11%, which is in the same range as the experimental error in Werner's measurements.

Based on a model developed to predict service life of sorbents, Yoon and Nelson[40,41] modified their formula to predict the effect of humidity on cartridge breakthrough by adding a theoretical parameter, "a". Using the breakthrough data generated by Nelson[13] on benzene and methyl chloroform, Yoon and Nelson demonstrated that there is generally good agreement between calculated and experimental data for most test conditions. The breakthrough curves at various concentration and humidities for benzene and methyl chloroform are shown in Figures 13 to 16.

At higher breakthrough percentages, the deviation of experimental data from symmetric breakthrough curves increases, particularly at higher test humidities and lower concentrations. The greatest deviation between experimental and calculated values for benzene is observed at 250 ppm and at a test relative humidity of 80% (Figure 14). At higher challenge concentrations (e.g., 1000 ppm), the test humidity has little effect on methyl chloroform breakthrough (see Figure 16). As shown in Figure 15, there is no substantial differences in the breakthrough curves of methyl chloroform at 250 ppm over the humidity range between 0 and 50%. If the humidity is increased to 80%, however, the breakthrough time decreased markedly. A similar phenomenon is also observed for benzene. The decreased breakthrough time observed at high humidity may be attributed to the increased ability of water molecules to compete for adsorption sites.

If water vapor is present, competition between water molecules and contaminant molecules for adsorption sites is a factor as testing proceeds. When the level of relative humidity is 20% or greater, the number of water molecules is much greater

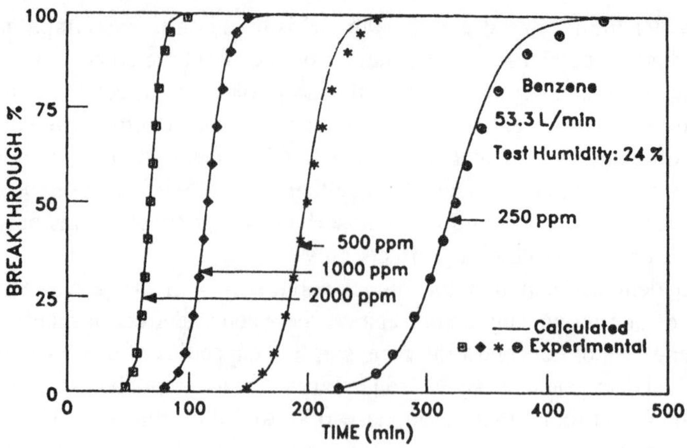

Figure 13 Comparison of calculated breakthrough curves (solid lines) with experimental data for benzene at various assault concentrations and a test humidity of 24%. The preconditioning humidity is 50%. (From Reference 41.)

than that of the chemical contaminant at contaminant concentrations pertinent to this study. For example, at 20% relative humidity and a contaminant concentration of 250 ppm, the ratio of water molecules to contaminant molecules is approximately 22; for 80% relative humidity and a contaminant concentration of 250 ppm, the ratio increases to 88. Therefore, if the contaminant concentration is high, water has very little effect on the adsorption of contaminant molecules even through the water/contaminant concentration molecule ratio is 10 to 20, a situation realized for 1000 to 2000 ppm at 80% relative humidity.

For example, the breakthrough curves for methyl chloroform at 1000 ppm and 80% relative humidity is very similar to the corresponding curves observed at 0, 20, and 50% (refer to Figure 16). As the contaminant concentration is decreased and the test humidity maintained at 80%, the water/contaminant ratio increases and water

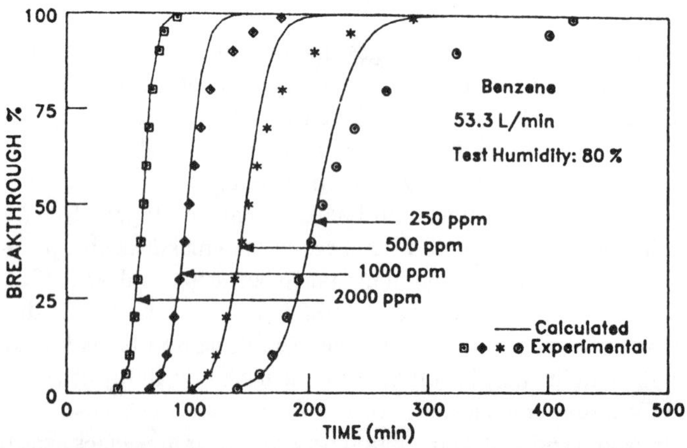

Figure 14 Comparison of calculated breakthrough curves (solid lines) with experimental data for benzene at various assault concentrations and a test humidity of 80%. The preconditioning humidity is 50%. (From Reference 40.)

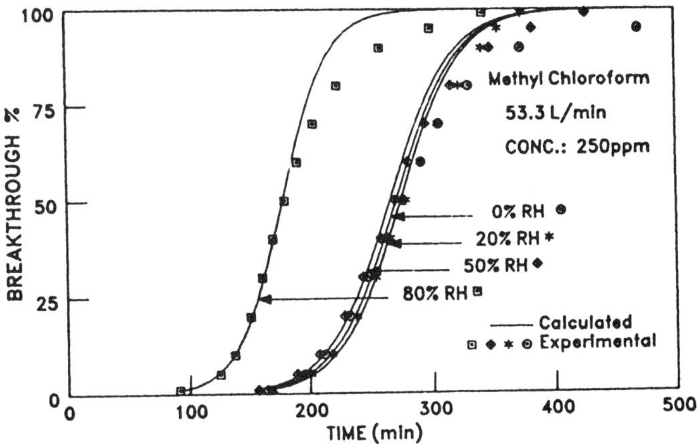

Figure 15 Comparison of calculated breakthrough curves (solid lines) with experimental data for methyl chloroform at various test humidities and at an assault concentration of 250 ppm. The preconditioning humidity is 50%. (From Reference 41.)

competes more effectively for active sites. This accounts for the relatively large deviation between experimental data and theory observed at 250 ppm and/or at 80% test humidity. For breakthrough percentages exceeding 50%, deviation between experiment and theory increases with decreasing contaminant concentration and increasing test humidity.

The above reviews indicate that the presence of high humidity could significantly reduce the service life of the sorbent when it is used for protection against organic vapors, especially when the toxic organic vapor has a very low permissible exposure limit such as benzene (1 ppm). Since most organic vapors have poor odor warning properties, the respirator wearer could be overexposed to toxic vapors when the respirators are used under high humidity condition without knowing that water vapor has significantly reduced the effectiveness of the respirator.

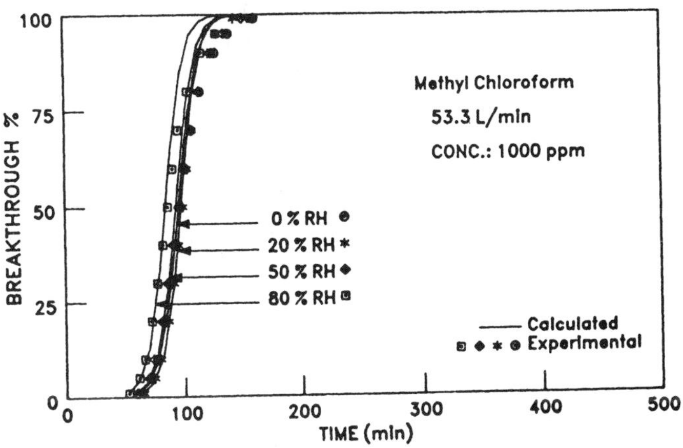

Figure 16 Comparison of calculated breakthrough curves (solid lines) with experimental data for methyl chloroform at various test humidities and at an assault concentration of 1000 ppm. The preconditioning humidity is 50%. (From Reference 41.)

In the U.S., the chemical industry and the petroleum industry, the largest users of chemical cartridges or canisters, are concentrated in the Gulf coast area where relative humidity is as high as 90% in the summer. The relative humidity is also very high in the summer in most parts of the U.S. To ensure that the cartridges or canisters would provide adequate protection when they are used under high temperature and humidity, the respirator certification regulation should require that cartridges and canisters be equilibrated under a dynamic condition and tested at a relative humidity at least 70% in both situations.

VI. THE EFFECT OF DESORPTION

For economic reasons, cartridges and canisters are often used repeatedly until sorbent breakthrough occurs. Since low boiling compounds are generally retained poorly by sorbents, the user may be exposed to the toxic chemical when it is desorbed upon storage. Balieu[42] conducted a study on the effect of desorption of cartridges against methanol. Four different types of cartridges were challenged with 5500 ppm of methanol at 22°C, 75% relative humidity, and a flow rate of 30 l/m to approximately 50% capacity of the sorbent. After the cartridges were stored for 1 h, clean air (22°C, 75% relative humidity, and an air flow of 30 l/m) was passed through them. The effluent concentration was continuously monitored during the desorption test. It was found that breakthrough occurred as short as 8 min. If the same quantity of methanol was adsorbed before the desorption test, the variation of challenge concentration had little effect on the effluent concentration of methanol. The effect of concentration was found to be small. The effluent concentration increased only with an increase in the challenge concentration.

During his cartridge service life study on 1,3-butadiene (C_4H_6), Ackley[43] conducted a series of tests to determine the effect of desorption. He used carbon filled reactor tubes for this series of tests. Three reactor tubes were challenged with 1000 ppm C_4H_6, at a flow rate of 64 l/m until a 10% breakthrough with a residence time $\tau = 0.21$ s was reached. The fourth tube was tested similarly, but it was allowed to achieve only about 50% of the saturation capacity. Immediately upon completing each test, clean air was introduced to the reactor tube ($\tau = 0.21$ s) at a predetermined temperature and relative humidity for 180 min. The effluent from the tube was monitored continuously. Desorption of C_4H_6 occurred readily as shown in Figure 17. Data on C_4H_6 desorption tests are shown in Table 18.

Both the desorption rate and the peak concentration for C_4H_6 increased with increasing temperature and humidity of the clean air purge. For those reactor tubes completely saturated with C_4H_6, desorption was immediate. Desorption was delayed approximately 20 min for the partially saturated bed. The sum of this delay time and the adsorption test time was approximately equal to the breakthrough time of a completely saturated bed of identical composition.

In addition to the service life study for ethylene dibromide (EDB), Nelson[44] also conducted a EDB desorption study on organic vapor cartridges (Norton) and canisters (MSA). Unequilibrated cartridges canisters are challenged with 50 and 200 ppm of EDB respectively at 85% relative humidity, 25°C, and flow rate of 64 l/m. The exposure time was 4 h. Two types of desorption measurements were then undertaken.

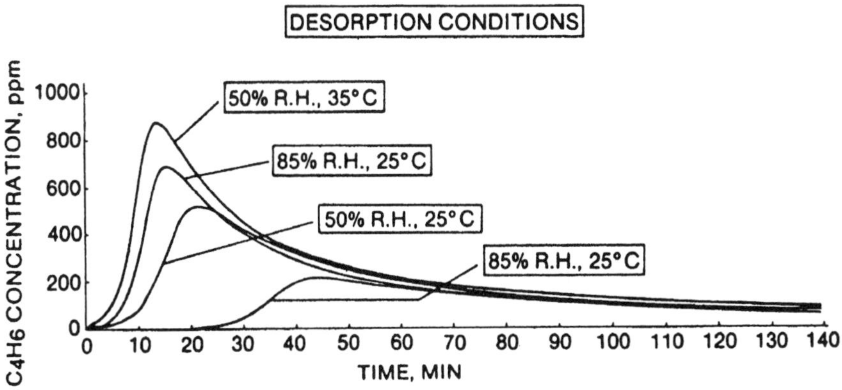

Figure 17 Desorption of 1,3-Butadiene from activated carbon. (From Reference 43.)

The first type was a dynamic desorption study, which involved storing the devices in a sealed bag at 25 ± 5°C. After at least a 72 h storage time, the devices were purged with clean air at 32 l/m, 70% relative humidity, and 25 ± 0.3°C for 2 h. The second type, a static desorption study, involved placing the exposed devices in sealed bags. The cartridges were stored on edge in small bags which were filled with 1.5 l of clean air. The canisters were stored in upright position in larger bags, which were filled with 20 l of clean air. The filled air was at 70% RH and 25°C. Approximately 50 ml of the air was drawn at the end of each day to determine the EDB concentration. An additional pair of bags were filled with clean air as blanks.

The results of dynamic and static desorption tests are shown in Tables 19 and 20. On the dynamic desorption test, the cartridges show an immediate initial response which slowly drops to zero. This is caused by the desorption of EDB from the inlet side of the cartridges and the subsequent readsorption on the outlet side during the 72 h storage period. The canister showed no EDB response after 2 h. The desorption of the canister was not evident because the devices contain inhalation valves and were stored in an upright position. The results of the static desorption test showed that the cartridges initially obtained a relatively high concentration (1.5 ppm) which decayed with time due to adsorption on the cartridge outlet side. The canister, however, measured 150 ppb EDB upon standing. This was probably due to leakage around or through the inhalation valve rather than any migration through the bed.

During the development of the ethylene oxide (EtO) canister, the Mine Safety Appliance Company conducted desorption tests to determine whether EtO would

Table 18 Desorption of C_4H_6 from Carbon

Adsorption	Desorption conditions		
Service life (min)	Temperature (°C)	Relative humidity (%)	% Desorbed in 180 min
34.8	25	50	84
36.5	35	50	102
35.0	25	85	105
18.0	25	85	83

From Reference 43.

Table 19 Results of the Ethylene Dibromide Dynamic Desorption Tests

Run no.	Device	Carbon volume (ml)	Breakthrough conc. (ppb) Initial	1 h	2 h
912	Cartridges	98	400	120	80
913	Cartridges	97	450	130	100
914	Cartridges	98	350	60	40
915	Canister	830	>50	>50	>50
916	Canister	840	>50	>50	>50
917	Canister	810	>50	>50	>50

From Reference 44.

desorb upon storage.[45] In the desorption test, the canister was equilibrated at 85% RH. Other test conditions were 25°C, 64 l/m, and 85% RH. The canister was tested at 1000 ppm EtO for 30 min and stood for 4 h; and tested at 1000 ppm EtO for 30 min and stood overnight or over the weekend. This cycle repeats until 1 ppm breakthrough of EtO was observed. For comparison purpose, the supersize MSA organic vapor canister was also tested under the same test conditions. The results are shown on Table 21.

It appears that no breakthrough of EtO occurred until the 22nd cycle and the actual service life when the canister was exposed to EtO is not much different from the canister tested normally. This is an indication that desorption of EtO was not occurred during storage.

From the data presented above, desorption occurs mostly for low boiling point and low molecular compounds. The sorbent bed depth would minimize the effect of desorption. Since air-purifying respirators are permissible for protecting workers from exposure to carcinogens such as acrylonitrile, benzene, or ethylene oxide, it is prudent that desorption information be made available before permitting these cartridges or canisters for multiday use.

VII. FIELD METHOD FOR PREDICTING SORBENT SERVICE LIFE

All published work on sorbent service life studies is based on laboratory work. Cohen[46] proposed a field method for service life determination. He designed different sizes of respirator carbon tube (RCT) packed with the same activated carbon as the

Table 20 Results of the Ethylene Dibromide Static Desorption Tests

Run no.	Device	Carbon volume (ml)	Bag concentration after x days (ppm) 1	2	3	4	5	6	7	10
918	Cartridges	98	1.5	1.3	1.3	1.1	1.0	0.9	0.7	0.6
919	Cartridges	98	1.5	1.4	0.8	0.8	0.8	0.7	0.7	0.5
920	Cartridges	97	1.4	0.9	0.9	0.7	0.7	0.6	0.5	0.3
			Bag concentration after x days (ppb)							
921	Canisters	830	<50	<50	<50	<50	50	100	100	150
922	Canisters	830	<50	<50	<50	<50	<50	<50	<50	<50
923	Canisters	820	50	<50	50	50	<50	50	100	100

From Reference 44.

Table 21 Desorption Study on Ethylene Oxide Canister

Canister type	Service time (ST), min	No. of cycles	Elapsed time, day	Window change min/cycles
EtO	660	22	15	300 (10)
EtO	660	22	15	240 (8)
EtO	660	22	15	390 (13)
EtO	720	24	16	240 (8)
OV	31	2	(43 ppm leak occurred during test)	
OV	30	2	(75 ppm leak occurred during test)	
OV	31	2	(59 ppm leak occurred during test)	

From Reference 45.

respirator cartridge used in the workplace. The physical properties of the activated carbon used for air sampling (SKC) and for the respirator cartridge (MSA) and various sizes of RCTs used in his study are listed in Table 22.

Experimental work was performed to determine the variability in packing of RCTs and to test the theory of bed residence time proposed by Ackley.[29] Carbon tetrachloride was selected as the challenge with a concentration of 1000 ppm. A breakthrough concentration of 100 ppm (10%) was selected. The test results are shown in Table 23. The author used the term relative standard deviation (RSD), a measure of variability of each set of data, for all variables measured in his experiments. The RSD for the breakthrough time corrected to 1000 ppm CCl_4 varied between 0.4 and 4.9% for the RCTs. The RSD for the cartridges varied between 0.8 and 2.3%. By plotting the breakthrough time against the bed residence time as shown in Figure 18, it shows that the bed residence time model holds true for several different configurations of carbon beds.

The author concluded that RCTs can be used to determine respirator cartridge breakthrough, and can be used to determine the concentration of air contaminants, if a validated air sampling method is available which uses charcoal tubes. This approach may allow the worker exposure monitoring and the cartridge service life determination performed during a single trip by an industrial hygienist.

Table 22 Physical Characteristics of Carbon and Carbon Beds Tested

	Carbon characteristics	
	SKC	MSA
Particle size (cm)	0.04–0.08	0.08–0.12
Surface area (m^2/g)	1120	1000
pH in aqueous solution	9.7	9.0

Physical characteristics of RCTs, cartridge, and charcoal tube

Type	Diameter (cm)	Length (cm)	Volume (cm^3)	Carbon weight (g)	Density (g/cm^3)
Small RCT	1.1	3.3	2.9	1.24	0.43
Small RCT	1.1	8.6	8.2	3.7	0.45
Large RCT	2.5	2.4	11.8	5.2	0.50
Large RCT	2.5	3.3	14.7	7.3	0.44
Cartridge	7.2	2.15	101.0	43.5	0.43
Charcoal tube	0.8	4.0	1.8	1.0	0.55

From Reference 46.

Table 23 Results of Carbon Tetrachloride Testing

Carbon weight (g)	Co[a] (ppm)	t[b] (min × 10^{-3})	Vel. (cm/s)	Tb (min) Uncorr.[c]	Tb (min) Corr.[d]	Wo[e] (g/g)	Tb 1/10[f] (%)	Tb 5/10[g] (%)	N
1.24									
Mean	857	0.57	83.5	22.5	20.3	0.47	66	89	3
RSD (%)	0.3	0.0	0.0	1.3	1.5	2.0	4.0	0.5	
1.24									
Mean	887	1.12	42.9	44.2	41.0	0.49	74	92	3
RSD (%)	4.4	2.1	1.8	3.6	2.4	5.0	6.1	1.0	
1.24									
Mean	1065	2.45	19.6	87.4	91.2	0.53	79.5	93	3
RSD (%)	2.6	0.4	0.5	3.0	2.4	0.3	—	0.9	
3.7									
Mean	804	0.57	251.0	23.6	20.3	0.46	85	97	3
RSD (%)	6.7	0.0	0.0	3.6	1.2	3.1	0.6	0.5	
3.7									
Mean	935	1.09	131.6	42.9	40.9	0.51	87	96	3
RSD (%)	6.0	1.1	1.1	4.1	1.4	1.62	1.1	1.0	
3.7									
Mean	916	2.53	57.1	100.8	93.6	0.50	91	98	3
RSD (%)	10.7	4.5	4.4	7.3	1.5	0.9	1.6	0.5	
5.2									
Mean	1113	0.62	62.8	15.8	17.0	0.40	59	88	3
RSD (%)	2.7	0.0	0.0	1.3	2.5	3.1	4.2	1.4	
5.2									
Mean	854	1.10	35.8	42.9	38.6	0.47	72	90	3
RSD (%)	1.8	0.9	1.8	3.6	2.4	2.7	8.6	1.9	
5.2									
Mean	938	2.65	14.8	97.5	92.0	0.49	77	93	5
RSD (%)	5.1	1.1	9.9	5.7	4.0	0.48	5.9	1.7	
7.3									
Mean	983	0.67	82.8	23.4	23.2	0.48	70	92	3
RSD (%)	0.4	1.9	1.0	1.9	1.6	1.0	2.9	0.5	
7.3									
Mean	833	1.09	50.3	45.1	39.9	0.48	75	93	3
RSD (%)	0.9	0.0	0.0	3.0	3.3	3.4	3.5	0.5	
7.3									
Mean	967	2.49	22.2	95.7	93.5	0.52	82	95	3
RSD (%)	2.7	0.8	0.6	3.2	1.5	0.0	2.5	0.5	
43.5	**(NIOSH cartridge test)**								
Mean	1114	0.70	50.4	22.8	24.4	0.49	80	94	3
RSD (%)	3.9	1.4	1.7	3.6	1.4	1.7	1.1	0.0	
43.5	**(NIOSH cartridge test)**								
Mean	1022	0.98	35.6	33.7	34.2	0.49	82	94	3
RSD (%)	2.7	1.2	1.5	1.3	0.7	0.0	0.6	0.0	
43.5	**(cartridge test)**								
Mean	903	1.94	18.1	72.9	68.1	0.48	88	95	4
RSD (%)	0.8	2.0	1.9	1.5	1.1	1.7	0.5	0.5	
43.5	**(NIOSH cartridge test)**								
Mean	1077	1.98	17.6	62.9	65.9	0.48	86	95	3
RSD (%)	6.9	2.0	0.0	4.0	1.7	3.5	0.5	0.5	
43.5	**(cartridge test)**								
Mean	964	2.88	12.2	103.6	100.4	0.48	90	97	4
RSD (%)	10.2	1.7	1.5	7.8	1.9	2.3	—	—	
1.0	**(charcoal tube)**								
Mean	921	0.45	151.5	19.7	18.6	0.45	84	95	4
RSD (%)	4.4	0.0	0.0	6.0	3.9	3.3	3.0	1.7	
1.0	**(charcoal tube)**								
Mean	976	0.92	73.9	35.9	35.9	0.43	84	94	3
RSD (%)	1.3	0.0	0.0	4.5	4.7	3.3	1.7	1.3	
1.0	**(charcoal tube)**								
Mean	1077	1.55	44.2	56.4	63.8	0.48	86	95	6
RSD (%)	7.7	0.3	0.3	4.7	5.5	7.4	3.4	1.4	

[a] Co = challenge concentration of carbon tetrachloride.

[b] t = bed residence time.

[c] Tb Uncorr. = time to 10% breakthrough.

[d] Tb Corr. = time to 10% breakthrough corrected to a challenge concentration of 1000 ppm of carbon tetrachloride.

[e] Wo = Capacity of carbon measured as grams of carbon tetrachloride adsorbed (at 10% breakthrough) per gram of carbon.

[f] Tb 1/10 = Ratio of time to 1% breakthrough:time to 10% breakthrough.

[g] Tb 5/10 = Ratio of time to 5% breakthrough:time to 10% breakthrough.

From Reference 46.

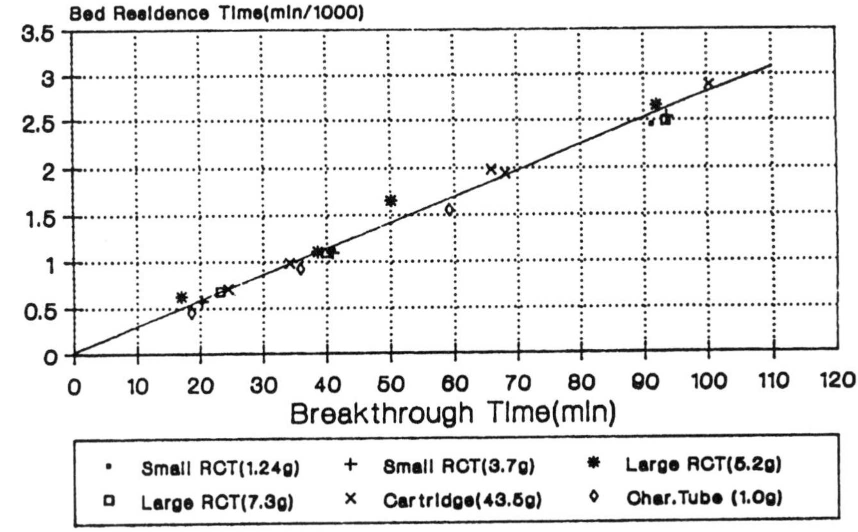

Figure 18 Results of carbon tetrachloride testing: plot of bed residence time versus breakthrough time. (From Reference 46.)

REFERENCES

1. Dubinin, M. M., The potential theory of adsorption of gases and vapors for adsorbents with energetically nonuniform surfaces, *Chem. Rev.*, 60, 235, 1960.
2. Bohart, G. S. and Adams, E. Q., Some aspects of the behavior of charcoal with respect to chlorine, *J. Am. Chem. Soc.*, 42, 523, 1920.
3. Mecklenburg, W., Layer filtration: A contribution to the theory of the gas mask, *Z. für Electrochemie*, 31, 488, 1925.
4. Mecklenburg, W., The layer filtration theory of gas masks, II, *Kollid Zeitschrift*, 52, 88, 1930.
5. Klotz, I. M., The adsorption wave, *Chem. Rev.*, 39, 241, 1946.
6. Ruch, W. E., Nelson, G. O., Lindeken, C. L., Johnsen, R. G., and Hodkins, D. J., Respirator cartridge efficiency studies I. Experimental design, *Am. Ind. Hyg. Assoc. J.*, 33, 105, 1972.
7. Nelson, G. O. and Hodkins, D. J., Respirator cartridge efficiency studies II. Preparation of test atmospheres, *Am. Ind. Hyg. Assoc. J.*, 33, 110, 1972.
8. Nelson, G. O., Johnsen, R. E., Lindeken, C. L., and Taylor, R. D., Respirator cartridge efficiency studies III. A mechanical machine to simulate human respiration, *Am. Ind. Hyg. Assoc. J.*, 33, 745, 1972.
9. Nelson, G. O. and Harder, C. A., Respirator cartridge efficiency studies. IV. Effects of steady and pulsation flow, *Am. Ind. Hyg. Assoc. J.*, 33, 797, 1972.
10. Nelson, G. O. and Harder, C. A., Respirator cartridge efficiency studies. V. Effect of solvent vapor, *Am. Ind. Hyg. Assoc. J.*, 35, 391, 1974.
11. Nelson, G. O. and Harder, C. A., Respirator cartridge efficient studies: VI. Effect of concentration, *Am. Ind. Hyg. Assoc. J.*, 37, 205, 1976.
12. Nelson, G. O. and Harder, C. A., Respirator cartridge efficient studies. VII. Effect of relative humidity and temperature, *Am. Ind. Hyg. Assoc. J.*, 37, 280, 1976.
13. Nelson, G. O. and Correia, A. N., Respirator cartridge efficiency studies. VIII. Summary and conclusions, Am. Ind. Hyg. Assoc. J., 37, 514, 1976.

14. Smoot, D. M. and Smith, D. L., Development of Improved respirator cartridge and canister test methods, NIOSH, July, DHEW (NIOSH) Publication No.77-209, 1977.
15. Danby, C. J., Davoud, J. G., Everett, D. H., Hinshelwood, C. N., and Lodge, R. M., The kinetics of adsorption of gases by granular reagents, *J. Chem. Soc. (London)*, 918, 1946.
16. Wheeler, A. and Robell, A. J., Performance of fixed-bed catalytic reactors with poison in the feed, *J. Catalysts*, 13, 299, 1969.
17. Jonas, L. A., Boardway, J. C., and Meseke, E. L., Prediction of adsorption behavior of activated carbons, *J. Collo. Interface Sci.*, 50, 538, 1975.
18. Jonas, L. A. and Rehrmann, J. A., Predictive equation in gas adsorption kinetics, *Carbon*, 11, 59, 1973.
19. Jonas, L. A. and Rehrmann, J. A., Kinetics of adsorption of organophosphorus vapors from air mixtures by activated carbon, *Carbon*, 10, 657, 1972.
20. Jonas, L. A. and Rehrmann, J. A., The rate of gas adsorption by activated carbon, *Carbon*, 12, 95, 1974.
21. Jonas, L. A. and Svirbely, W. J., The kinetics of adsorption of carbon tetrachloride and chloroform from air mixtures by activated carbon, *J. Catalysis*, 24, 446, 1972.
22. Rehrmann, J. A. and Jonas, L. A., Dependence of gas adsorption rates of carbon granule size and linear flow velocity, *Carbon*, 16, 47, 1978.
23. Moyer, E. S., Organic vapor (OV) respirator cartridge testing — potential Jonas Model applicability, *Am. Ind. Hyg. Assoc. J.*, 48, 791, 1987.
24. Sansone, E. B. and Jonas, L. A., Prediction of activated carbon performance for carcinogen vapors, *Am. Ind. Hyg. Assoc. J.*, 42, 688, 1981.
25. Sansone, E. B., Tewarl, Y. B., and Jonas, L. A., Prediction of vapors from air by adsorption on activated carbon, *Environ. Sci. Technol.*, 12, 1511, 1979.
26. Dubinin, M. M., Physical adsorption of gases and vapors in micropores, *Prog. Surf. Membr. Sci.*, 9, 1, 1975.
27. Reucroft, P. J., Simpson, W. J., and Jonas, L. A., Sorption properties of activated carbon, *J. Phy. Chem.*, 75, 3526, 1971.
28. Jonas, L. A., Tewari, Y. B., and Sansone, E. B., Prediction of adsorption rate constants of activated carbon for various vapors, *Carbon*, 17, 345, 1979.
29. Ackley, M. W., Residence time model for respirator sorbent beds, *Am. Ind. Hyg. Assoc. J.*, 46, 679, 1985.
30. Yoon, H. Y. and Nelson, J. H., Application of gas adsorption kinetics. I. A theoretical modal for respirator cartridge service life, *Am. Ind. Hyg. Assoc. J.*, 45, 509, 1984.
31. Yoon, H. Y. and Nelson, J. H., Application of gas adsorption kinetics. II. A theoretical modal for respirator cartridge service life and its practical application, *Am. Ind. Hyg. Assoc. J.*, 45, 517, 1984.
32. Wood, G. O. and Moyer, E. S., A review of the wheeler equation and comparison of its applications to organic vapor respirator cartridge breakthrough data, *Am. Ind. Hyg. Assoc. J.*, 50, 400, 1989.
33. Balieu, E., Effects of water vapor on the performance of respirator gas and vapor filters, *J. Int. Soc. Respir. Protn*, 1, 95, 1983.
34. Werner, M. D., The effects of relative humidity on the vapor phase adsorption of trichloroethylene by activated carbon, *Am. Ind. Hyg. Assoc. J.*, 46, 585, 1985.
35. Hall, T., Breysse, P., Corn, M., and Jonas, L. A., Effects of adsorbed water vapor on the adsorption rate constant and the kinetic adsorption capacity of the wheeler kinetic model, *Am. Ind. Hyg. Assoc. J.*, 49, 461, 1988.
36. Jonas, L. A., Sansone, E. B., and Farris, T. S., The effect of moisture on the adsorption of chloroform by activated carbon, *Am. Ind. Hyg. Assoc. J.*, 46, 20, 1985.
37. Wood, G. O., A model for adsorption capacities of charcoal beds. I. Relative humidity effects, *Am. Ind. Hyg. Assoc. J.*, 48, 622, 1987.

38. Wood, G. O., A model for adsorption capacities of charcoal beds. II. Challenge concentration effects, *Am. Ind. Hyg. Assoc. J.*, 48, 703, 1987.
39. Underhill, D. W., Calculation of the performance of activated carbon at high relative humidities, *Am. Ind. Hyg. Assoc. J.*, 48, 909, 1987.
40. Yoon, Y. H. and Nelson, J. H., A theoretical study of the effect of humidity on respirator cartridge service life, *Am. Ind. Hyg. Assoc. J.*, 49, 325, 1988.
41. Yoon, Y. H. and Nelson, J. H., Effect of humidity on respirator cartridge service life, *Am. Ind. Hyg. Assoc. J.*, 51, 202, 1990.
42. Balieu, E., Respirator filters in protection against low boiling compounds, *J. Int. Soc. Respir. Protn*, 1, 125, 1983.
43. Ackley, M. W., Chemical cartridge respirator performance: 1,3-butadiene, *Am. Ind. Hyg. Assoc. J.*, 48, 447, 1987.
44. Nelson, G. O., Cartridge and canister ethylene dibromide desorption performance tests, OSHA Ethylene Dibromide Docket, H-111, 1983.
45. Birkner, L. R., Hanlon, R. G., Nelson, G. O., Miller, W. B., Ethylene oxide: Air-purifying respirator sorbent identification, unpublished, 1981.
46. Cohen, H. J., Briggs, D. E., and Garrison, R. P., Development of a field method for evaluating the service lives of organic vapor cartridges — Part I: Results of laboratory testing using carbon tetrachloride, *Am. Ind. Hyg. Assoc. J.*, 50, 486, 1989.

17 SORBENT PERFORMANCE II

I. CARTRIDGE AND CANISTER SERVICE LIFE DETERMINATION

In addition to the cartridge breakthrough data developed by Nelson et al. many investigators also conducted cartridge breakthrough studies. Before the creation of NIOSH, respirators were tested and approved by the Bureau of Mines (BM) under the test schedules prescribed in 30 CFR 11. The test method for chemical cartridges was listed in Schedule 23B, which prescribed carbon tetrachloride as the test agent for organic vapor. Carbon tetrachloride has a threshold limit value (TLV) of 10 ppm and an odor threshold of 50 ppm, making its use hazardous (CCl_4 was not considered as a suspected carcinogen then). The Bureau searched for a safe replacement for CCl_4.[1] Halocarbons were selected because these solvents were relatively innocuous with physical properties similar to CCl_4. Three halocarbons, fluorotrichloromethane (Freon 11), dichloromonofluoromethane (Freon 21), and 1,1,2-trichloro-1,2,2-trifluoromethane (Freon 113), were selected for testing. The test conditions were: 1000 ppm challenge, 32 and 64 liters per minute (l/m) flow rate, room temperature, and 50% relative humidity. The breakthrough time was measured at 0.5% of the challenge.

The test conditions and results are listed in Table 1. The BM Testing Schedule 23B requires that at a challenge concentration of 1000 ppm and a flow rate of 64 l/m, the organic vapor cartridges must have a minimum service life of 50 min for a breakthrough of 5 ppm (0.5%); at the same breakthrough, the organic vapor cartridge must have a minimum service life of 100 min at flow rate of 32 l/m. It indicates that at a breakthrough concentration of 5 ppm (0.5%), the service life for Freon 11 and 21 was 21.8 and 8.8 min, respectively, at a flow rate of 64 l/m. The service life for Freon 113, however, was 40 min at 64 l/m and 83.5 min at 32 l/m. It was concluded that Freon 113 may be used as a testing agent for organic vapor cartridges since its service life is close to the required breakthrough time. For unexplained reasons, it was not considered in the revision of 30 CFR 11 in 1972.

Later, Freedman et al.[2] at the Bureau of Mines conducted studies on the service life of organic vapor cartridges against a variety of volatile organic compounds with boiling points up to 126°C (butyl acetate). The Bureau had expressed concerns that several compounds with toxicities approaching that of CCl_4 could be hazardous if cartridge service lives for these compounds were considerably lower than for CCl_4. The test conditions were the same as those prescribed in the previous BM study. The test conditions and results are listed in Table 2. The results indicate that many lower

Table 1 Comparison of Lives (Min), Chemical Cartridges, Displacer Method, 0.1% Concentration, CCl_4 and Freon 113

64 l/m airflow		32 l/m airflow	
CCl_4	Freon 113	CCl_4	Freon 113
52.9	39.0	134.2	85.1
53.3	43.4	136.0	89.6
54.8	40.4	131.0	88.6
55.5	36.7	—	87.2
56.0	38.7	—	90.1
61.3	41.9	—	80.1
64.6	40.9	—	77.4
60.0	41.5	—	72.1
58.0	39.4	—	86.2
—	37.6	—	77.9
—	40.6	—	85.8
57.4 ± 3.93[a]	40.0 ± 1.96[a,b]	133.7 ± 2.53[a]	83.7 ± 5.87[a,c]

[a] Average and standard deviation.

[b] Decrease in life 17.4 min (30.3%). Average Freon 113/CCl_4 ratio 0.70 ± 0.05.

[c] Decrease in life 50.0 min (37.4%). Average Freon 113/CCl_4 ratio 0.65 ± 0.02.

From Reference 1.

boiling point compounds have service lives far less than the required 50 min. For example, methyl bromide, with a TLV of 20 ppm, had a 0.5% breakthrough time of 1.2 min. The level rose rapidly to 100% in 5 min. According to the BM test schedule, the organic vapor cartridges should provide protection for at least 60 min against 1000 ppm of organic vapor. For methyl bromide, a material having poor odor warning properties and a low TLV, a person wearing these cartridges would be overexposed after 5 min. It may be one of the reasons why MSHA/NIOSH does not approve organic vapor cartridges for protection against chemicals with poor odor warning properties.

As a part of a NIOSH contract for the development of improved respirator cartridge and canister testing methods, Smoot[3] conducted cartridge service life studies for acetone, 2-butanone, ethanol, 2-propanol, ethyl acetate, 3-chloropropene, chloroform, trichloroethylene, carbon tetrachloride, hexane, benzene and 1,2-dibromethane. He used custom-made reusable cartridges filled with 12 × 20 mesh petroleum-base activated carbon in his study. The test condition was set at 25°C, 32 l/m, and 1000 ppm for the 12 solvents. The humidity was set at 0, 50, and 80%. The test results are listed in Table 2.

Henry and Wilhelme[4] conducted a study to determine the service life of organic vapor canisters for protection against acrylonitrile (AN). They used commercially available chin style, industrial size, and super size (originally made for vinyl chloride protection) organic vapor canisters for a 10% breakthrough evaluation at ambient temperature and 50% relative humidity. The challenge concentration varied from 10 to 1000 ppm with relative humidities of 7, 50, and 90%. The flow rate was set at 60 l/m. The results are shown in Table 2. The authors concluded that the service life is proportional to the amount of carbon present in the canister. Under controlled conditions, the experimentally determined breakthrough times for AN compare closely to those estimated from the Nelson equation.

Table 2 Cartridge or Canister Breakthrough Data, 3/96

Author	Type	Brand	Solvent	Concn ppm	Flow Rate lpm	Temp C	RH %	Carb Wt. gm	Breakthrough Time, t1%	t10%	t50%	t99%	Note	Ref
Freedman	Cart	Comm OV	Acetic acid	1000	32	23	50		105				BT@0.5%	2
Freedman	Cart	Comm OV	Acetaldehyde	1000	32	23	50		1.0				BT@0.5%	2
Moyer	Cart	OV	Acetone	530	64	22	50	45	38					11
Moyer	Cart	OV	Acetone	530	64	22	50	48	46					11
Moyer	Cart	OV	Acetone	530	64	22	50	48	45					11
Moyer	Cart	OV	Acetone	530	64	22	50	92	92					11
Moyer	Cart	OV	Acetone	530	64	22	50	93	90					11
Moyer	Cart	OV	Acetone	530	64	22	50	96	104					11
Moyer	Cart	OV	Acetone	530	64	22	50	140	138					11
Moyer	Cart	OV	Acetone	530	64	22	50	143	142					11
Moyer	Cart	OV	Acetone	530	64	22	50	145	151					11
Moyer	Cart	OV	Acetone	530	64	22	50	183	184					11
Moyer	Cart	OV	Acetone	530	64	22	50	187	190					11
Moyer	Cart	OV	Acetone	530	64	22	50	193	198					11
Moyer	Cart	OV	Acetone	750	64	22	50	45	31					11
Moyer	Cart	OV	Acetone	750	64	22	50	47	30					11
Moyer	Cart	OV	Acetone	750	64	22	50	47	38					11
Moyer	Cart	OV	Acetone	750	64	22	50	48	40					11
Moyer	Cart	OV	Acetone	750	64	22	50	92	70					11
Moyer	Cart	OV	Acetone	750	64	22	50	94	74					11
Moyer	Cart	OV	Acetone	750	64	22	50	96	85					11
Moyer	Cart	OV	Acetone	750	64	22	50	96	86					11
Moyer	Cart	OV	Acetone	750	64	22	50	139	111					11
Moyer	Cart	OV	Acetone	750	64	22	50	143	118					11
Moyer	Cart	OV	Acetone	750	64	22	50	144	131					11
Moyer	Cart	OV	Acetone	750	64	22	50	145	137					11
Moyer	Cart	OV	Acetone	750	64	22	50	185	152					11
Moyer	Cart	OV	Acetone	750	64	22	50	189	160					11
Moyer	Cart	OV	Acetone	750	64	22	50	190	174					11
Moyer	Cart	OV	Acetone	750	64	22	50	192	184					11
Moyer	Cart	OV	Acetone	750	64	22	50	194	174					11
Moyer	Cart	OV	Acetone	1060	64	22	50	45	28					11
Moyer	Cart	OV	Acetone	1060	64	22	50	46	26					11
Moyer	Cart	OV	Acetone	1060	64	22	50	47	24					11
Moyer	Cart	OV	Acetone	1060	64	22	50	49	31					11
Moyer	Cart	OV	Acetone	1060	64	22	50	91	60					11
Moyer	Cart	OV	Acetone	1060	64	22	50	92	57					11
Moyer	Cart	OV	Acetone	1060	64	22	50	94	66					11
Moyer	Cart	OV	Acetone	1060	64	22	50	97	66					11
Moyer	Cart	OV	Acetone	1060	64	22	50	136	93					11
Moyer	Cart	OV	Acetone	1060	64	22	50	137	89					11
Moyer	Cart	OV	Acetone	1060	64	22	50	142	104					11
Moyer	Cart	OV	Acetone	1060	64	22	50	146	103					11
Moyer	Cart	OV	Acetone	1060	64	22	50	184	127					11
Moyer	Cart	OV	Acetone	1060	64	22	50	185	120					11
Smoot	Cart	Custom	Acetone	1000	32	25	0	34	43	48	56	82		3
Smoot	Cart	Custom	Acetone	1000	32	25	80	34	38	44	53	109		3
Smoot	Cart	Custom	Acetone	1000	32	25	50	34	42	50	61	105		3
Freedman	Cart	Comm OV	Acetonitrile	1000	32	23	50		16				BT@0.5%	2
Stampfer	Can	MSA-OV	Acrolein	4.9	64	25	<15		1320					7
Stampfer	Can	MSA-OV	Acrolein	5.0	64	26	80		1080					7
Beaumont	Cart	Willson-OV	Acrylonitrile	100	54	22	50		300				BT@1ppm, EQ55%RH	16
Beaumont	Cart	MSA-OV	Acrylonitrile	22	54	22	50		750				BT@1ppm, EQ55%RH	16
Beaumont	Cart	MSA-OV	Acrylonitrile	100	54	22	50		270				BT@1ppm, EQ55%RH	16
Beaumont	Cart	MSA-OV	Acrylonitrile	1000	54	22	50		75				BT@1ppm, EQ55%RH	16
Beaumont	Cart	Willson-OV	Acrylonitrile	55	54	22	50		510				BT@1ppm, EQ55%RH	16
Henry	Can	MSA-OV	Acrylonitrile	11	60	23	7	420		5640				4
Henry	Can	MSA-OV	Acrylonitrile	140	60	23	50	420		1800				4
Henry	Can	MSA-OV	Acrylonitrile	359	60	23	50	420		720				4

Henry	Can	MSA-OV	Acrylonitrile	11	60	23	50	420		5220			4
Henry	Can	MSA-VCM	Acrylonitrile	21	60	23	50	760		9480			4
Henry	Can	MSA-OV	Acrylonitrile	29	60	23	50	420		2700			4
Henry	Can	MSA-OV	Acrylonitrile	10	60	23	90	420		960			4
Henry	Can	MSA-OV	Acrylonitrile	8	60	23	50	220		384			4
Henry	Can	MSA-OV	Acrylonitrile	11	60	23	50	420		5220			4
Henry	Can	MSA-OV	Acrylonitrile	1230	60	23	50	420		420			4
Kennedy	Cart	MSA-OV	Acrylonitrile	141	32	25	0		198	219	230	1%,5%&10%BT	27
Kennedy	Cart	MSA-OV	Acrylonitrile	141	32	25	80		128	143	149	1%,5%&10%BT	27
Kennedy	Cart	Norton-Pest	Acrylonitrile	20	32	25	50		547	579		BT@1&2ppm,deepshell	
Kennedy	Cart	MSA-OV	Acrylonitrile	141	32	25	80		98	111	118	1%,5%&10%BT	27
Kennedy	Cart	MSA-OV	Acrylonitrile	141	32	25	80		88	99	407	1%,5%&10%BT	27
Kennedy	Cart	MSA-OV	Acrylonitrile	141	32	25	80		105	119	125	1%,5%&10%BT	27
Kennedy	Cart	Norton-Pest	Acrylonitrile	141	32	25	80		74	86	92	1%,5%&10%BT	27
Nelson	Cart	NortonOV	Acrylonitrile	200	32	25	50	48	314	360	381	t1%,5% &10%	25
Nelson	Cart	NortonOV	Acrylonitrile	200	32	25	50	48	311	358	382	t1%,5% &10%	25
Nelson	Cart	NortonOV	Acrylonitrile	200	32	25	50	48	750	828	864	t1%,5% &10%	25
Nelson	Cart	NortonOV	Acrylonitrile	200	32	25	50	88	726	798	834	t1%,5% &10%	25
Nelson	Cart	NortonOV	Acrylonitrile	200	32	25	80	88	539	606	642	t1%,5% &10%	25
Nelson	Cart	NortonOV	Acrylonitrile	200	32	25	80	48	234	271	290	t1%,5% &10%	25
Nelson	Cart	NortonOV	Acrylonitrile	200	32	25	80	48	207	255	279	t1%,5% &10%	25
Nelson	Cart	NortonOV	Acrylonitrile	200	32	25	80	88	519	594	630	t1%,5% &10%	25
Nelson	Cart	NortonOV	Acrylonitrile	1000	32	25	50	48	147	165	176	t1%,5% &10%	25
Nelson	Cart	NortonOV	Acrylonitrile	1000	32	25	50	88	276	319	336	t1%,5% &10%	25
Nelson	Cart	NortonOV	Acrylonitrile	1000	32	25	50	88	315	340	355	t1%,5% &10%	25
Nelson	Cart	NortonOV	Acrylonitrile	1000	32	25	50	48	147	166	176	t1%,5% &10%	25
Nelson	Cart	NortonOV	Acrylonitrile	1000	32	25	80	88	282	307	320	t1%,5% &10%	25
Nelson	Cart	NortonOV	Acrylonitrile	1000	32	25	80	48	124	139	145	t1%,5% &10%	25
Nelson	Cart	NortonOV	Acrylonitrile	1000	32	25	80	88	263	295	308	t1%,5% &10%	25
Nelson	Cart	NortonOV	Acrylonitrile	1000	32	25	80	48	123	141	150	t1%,5% &10%	25
Revoir	Cart	NortonOV	Acrylonitrile	20.0	32	25	50		347	379		BT@1ppm&2ppm,shallow	28
Revoir	Cart	NortonOV	Acrylonitrile	20.9	32	25	50		550	583		BT@1ppm&2ppm,deep	28
Revoir	Cart	NortonOV	Acrylonitrile	20.9	32	25	50		210	227		BT@1ppm&2ppm,deep	28
Revoir	Cart	NortonOV	Acrylonitrile	21.1	32	25	50		545	581		BT@1ppm&2ppm,deep	28
Revoir	Cart	NortonOV	Acrylonitrile	21.1	32	25	50		553	581		BT@1ppm&2ppm,deep	28
Revoir	Cart	NortonOV	Acrylonitrile	21.5	32	25	50		552	586		BT@1ppm&2ppm,deep	28
Revoir	Cart	NortonOV	Acrylonitrile	21.7	32	25	50		551	581		BT@1ppm&2ppm,deep	28
Revoir	Cart	NortonOV	Acrylonitrile	22.1	32	25	50		529	560		BT@1ppm&2ppm,deep	28
Revoir	Cart	NortonOV	Acrylonitrile	19.8	32	25	80		199	213		BT@1ppm&2ppm,deep	28
Revoir	Cart	NortonOV	Acrylonitrile	19.8	32	25	80		208	215		BT@1ppm&2ppm,deep	28
Revoir	Cart	NortonOV	Acrylonitrile	20.2	32	25	80		204	215		BT@1ppm&2ppm,deep	28
Revoir	Cart	NortonOV	Acrylonitrile	20.3	32	25	80		203	219		BT@1ppm&2ppm,deep	28
Revoir	Cart	NortonOV	Acrylonitrile	20.5	32	25	80		200	213		BT@1ppm&2ppm,deep	28
Revoir	Cart	NortonOV	Acrylonitrile	94.6	32	25	50		422	436		BT@1ppm&2ppm,deep	28
Revoir	Cart	NortonOV	Acrylonitrile	95.9	32	25	50		478	491		BT@1ppm&2ppm,deep	28
Revoir	Cart	NortonOV	Acrylonitrile	96.0	32	25	50		473	487		BT@1ppm&2ppm,deep	28
Revoir	Cart	NortonOV	Acrylonitrile	96.9	32	25	50		474	485		BT@1ppm&2ppm,deep	28
Revoir	Cart	NortonOV	Acrylonitrile	97.3	32	25	50		242	251		BT@1ppm&2ppm,deep	28
Revoir	Cart	NortonOV	Acrylonitrile	97.3	32	25	50		247	253		BT@1ppm&2ppm,deep	28
Revoir	Cart	NortonOV	Acrylonitrile	97.3	32	25	50		487	499		BT@1ppm&2ppm,deep	28
Revoir	Cart	NortonOV	Acrylonitrile	97.9	32	25	50		251	257		BT@1ppm&2ppm,deep	28
Revoir	Cart	NortonOV	Acrylonitrile	98.6	32	25	50		461	475		BT@1ppm&2ppm,deep	28
Revoir	Cart	NortonOV	Acrylonitrile	99.3	32	25	50		250	256		BT@1ppm&2ppm,deep	28
Revoir	Cart	NortonOV	Acrylonitrile	99.3	32	25	50		248	254		BT@1ppm&2ppm,deep	28
Revoir	Cart	NortonOV	Acrylonitrile	99.7	32	25	50		242	249		BT@1ppm&2ppm,deep	28
Silverstein	ChCan	MSA-OV	Acrylonitrile	100	32	25	50		60				26
Silverstein	Cart	MSA-OV	Acrylonitrile	100	32	25	50		520	573		BT@1&2ppm	26
Silverstein	Cart	MSA-OV	Acrylonitrile	20	32	25	50		618	649		BT@1&2ppm	26
Stampfer	Can	MSA-OV	Acrylonitrile	101	64	23	<15		1078				7
Stampfer	Can	MSA-OV	Acrylonitrile	98	64	26	80		488				7
Beaumont	Cart	AO-OV	Benzene	100	54	22	50		540			BT@1ppm,EQ55%RH	16
Beaumont	Cart	Willosn-OV	Benzene	100	54	22	50		660			BT@1ppm,EQ55%RH	16
Beaumont	Cart	MSA-AG	Benzene	100	54	22	50		330			BT@1ppm,EQ55%RH	16
Beaumont	Cart	MSA-OV	Benzene	110	54	22	50		540			BT@1ppm,EQ55%RH	16
Beaumont	Cart	AO-OV	Benzene	60	54	22	50		720			BT@1ppm,EQ55%RH	16
Beaumont	Cart	AO-OV	Benzene	10	54	22	50		3210			BT@1ppm,EQ55%RH	16

Bollinger	Cart	Binks-OV	Benzene	17	64	25	85	43	225					12
Bollinger	Cart	Binks-OV	Benzene	17	64	25	85	43	222					12
Bollinger	Cart	Binks-OV	Benzene	17	64	25	85	43	207					12
Freedman	Cart	Comm OV	Benzene	1000	32	23	50		63				BT @0.5%	2
Nelson	Cart	NortonOV	Benzene	100	32	25	50	47	1158	1230	1272		t1%,5% &10%	25
Nelson	Cart	NortonOV	Benzene	100	32	25	50	87	2136	2340	2424		t1%,5% &10%	25
Nelson	Cart	NortonOV	Benzene	100	32	25	50	47	1098	1194	1254		t1%,5% &10%	25
Nelson	Cart	NortonOV	Benzene	100	32	25	80	87	1548	1758	1848		t1%,5% &10%	25
Nelson	Cart	NortonOV	Benzene	100	32	25	80	47	708	816	870		t1%,5% &10%	25
Nelson	Cart	NortonOV	Benzene	100	32	25	80	47	732	846	906		t1%,5% &10%	25
Nelson	Cart	NortonOV	Benzene	500	32	25	50	47	301	323	336		t1%,5% &10%	25
Nelson	Cart	NortonOV	Benzene	500	32	25	50	87	636	678	690		t1%,5% &10%	25
Nelson	Cart	NortonOV	Benzene	500	32	25	50	47	343	366	373		t1%,5% &10%	25
Nelson	Cart	NortonOV	Benzene	500	32	25	50	87	648	690	702		t1%,5% &10%	25
Nelson	Cart	NortonOV	Benzene	500	32	25	80	87	545	579	600		t1%,5% &10%	25
Nelson	Cart	NortonOV	Benzene	500	32	25	80	47	242	269	287		t1%,5% &10%	25
Nelson	Cart	NortonOV	Benzene	500	32	25	80	87	531	567	590		t1%,5% &10%	25
Nelson	Cart	NortonOV	Benzene	500	32	25	80	47	274	293	304		t1%,5% &10%	25
Smoot	Cart	Custom	Benzene	1000	32	25	0	34	78	89	101	127		3
Smoot	Cart	Custom	Benzene	1000	32	25	50	34	87	97	111	160		3
Smoot	Cart	Custom	Benzene	1000	32	25	80	34	79	86	101	183		3
Stampfer	Can	MSA-OV	Benzene	498	64	25	<15		547					7
Stampfer	Can	MSA-OV	Benzene	495	64	26	80		387					7
Coffey	Cart	MSA-OV	Butadiene,1,3 -	10	64	20	85		206	211	223	246	BT@2,3,5&10ppm	22
Coffey	Cart	MSA-OV	Butadiene,1,3 -	10	64	20	85		240	249	260	282	BT@2,3,5&10ppm	22
Coffey	Cart	MSA-OV	Butadiene,1,3 -	10	64	20	85		206	211	217	227	BT@2,3,5&10ppm	22
Coffey	Cart	MSA-OV	Butadiene,1,3 -	20	64	20	50		353	398	479	563	BT@2,3,5&10ppm	22
Coffey	Cart	MSA-OV	Butadiene,1,3 -	20	64	20	50		409	432	460	521	BT@2,3,5&10ppm	22
Coffey	Cart	MSA-OV	Butadiene,1,3 -	20	64	20	50		334	355	395	575	BT@2,3,5&10ppm	22
Coffey	Cart	MSA-OV	Butadiene,1,3 -	20	64	20	85		180	183	185	196	BT@2,3,5&10ppm	22
Coffey	Cart	MSA-OV	Butadiene,1,3 -	20	64	20	85		162	172	179	242	BT@2,3,5&10ppm	22
Coffey	Cart	MSA-OV	Butadiene,1,3 -	20	64	20	85		251	252	259	267	BT@2,3,5&10ppm	22
Coffey	Cart	MSA-OV	Butadiene,1,3 -	20	64	20	85		212	215	222	244	BT@2,3,5&10ppm	22
Coffey	Cart	MSA-OV	Butadiene,1,3 -	50	64	20	50		339	348	360	417	BT@2,3,5&10ppm	22
Coffey	Cart	MSA-OV	Butadiene,1,3 -	50	64	20	50		310	315	322	339	BT@2,3,5&10ppm	22
Coffey	Cart	MSA-OV	Butadiene,1,3 -	50	64	20	50		394	400	411	424	BT@2,3,5&10ppm	22
Coffey	Cart	MSA-OV	Butadiene,1,3 -	50	64	20	85		152	157	163	173	BT@2,3,5&10ppm	22
Coffey	Cart	MSA-OV	Butadiene,1,3 -	50	64	20	85		195	199	216	220	BT@2,3,5&10ppm	22
Coffey	Cart	MSA-OV	Butadiene,1,3 -	50	64	20	85		130	134	140	154	BT@2,3,5&10ppm	22
Ackley	Cart	Scott-OV	Butadiene,1,3-	100	64	25	50		132.8					8
Ackley	Cart	Scott-OV	Butadiene,1,3-	100	32	25	50		240.7					8
Ackley	Cart	Scott-OV	Butadiene,1,3-	100	32	25	50		260.0					8
Ackley	Cart	Scott-OV	Butadiene,1,3-	100	64	25	50		142.0					8
Ackley	Cart	Scott-OV	Butadiene,1,3-	100	32	25	50		245.1					8
Ackley	Cart	Scott-OV	Butadiene,1,3-	1000	64	25	0		41.8				EQ 0%RH	8
Ackley	Cart	Scott-OV	Butadiene,1,3-	1000	64	25	50		38.3					8
Ackley	Cart	Scott-OV	Butadiene,1,3-	1000	32	25	50		75.2					8
Ackley	Cart	Scott-OV	Butadiene,1,3-	1000	64	34	50		29.6					8
Ackley	Cart	Scott-OV/AG	Butadiene,1,3-	1000	32	25	50		67.3					8
Ackley	Cart	Scott-OV	Butadiene,1,3-	1000	32	25	50		46.6				EQ 85%RH	8
Ackley	Cart	Scott-AG	Butadiene,1,3-	1000	32	25	50		67.3					8
Ackley	Cart	Scott-AG	Butadiene,1,3-	1000	64	25	50		31.1					8
Ackley	Cart	Scott-OV	Butadiene,1,3-	1000	64	34	50		30.4					8
Ackley	Cart	Scott-AG	Butadiene,1,3-	1000	32	25	50		63.4					8
Ackley	Cart	Scott-OV/AG	Butadiene,1,3-	1000	32	25	50		66.9					8
Ackley	Cart	Scott-OV/AG	Butadiene,1,3-	1000	32	25	50		61.7					8
Ackley	Cart	Scott-OV	Butadiene,1,3-	1000	64	25	50		37.3					8
Ackley	Cart	Scott-OV/AG	Butadiene,1,3-	1000	64	25	50		33.0					8
Ackley	Cart	Scott-OV	Butadiene,1,3-	1000	32	25	50		48.1				EQ 85%RH	8
Ackley	Cart	Scott-OV	Butadiene,1,3-	1000	32	25	50		73.9					8
Ackley	Cart	Scott-OV	Butadiene,1,3-	1000	64	25	50		37.3					8
Ackley	Cart	Scott-OV	Butadiene,1,3-	1000	32	25	50		68.7					8
Ackley	Cart	Scott-AG	Butadiene,1,3-	1000	64	25	50		32.0					8
Ackley	Cart	Scott-OV/AG	Butadiene,1,3-	1000	64	25	50		31.3					8
Ackley	Cart	Scott-OV	Butadiene,1,3-	1000	64	25	85		6.3				EQ-85%RH	8
Ackley	Cart	Scott-OV	Butadiene,1,3-	1000	64	25	85		5.8				EQ-85%RH	8

Bollinger	Cart	Binks-OV	Butadiene,1,3-	75	64	25	85	43	92				12
Bollinger	Cart	Binks-OV	Butadiene,1,3-	75	64	25	85	43	72				12
Bollinger	Cart	Binks-OV	Butadiene,1,3-	75	64	25	85	43	55				12
Freedman	Cart	Comm OV	Butane	1000	32	23	50		11			BT@0.5%	2
Freedman	Cart	Comm OV	Butanol	1000	32	23	50		92			BT@0.5%	2
Smoot	Cart	Custom	Butanone,2-	1000	32	25	0	34	87	96	109	150		3
Smoot	Cart	Custom	Butanone,2-	1000	32	25	50	34	83	93	106	148		3
Smoot	Cart	Custom	Butanone,2-	1000	32	25	80	34	80	87	98	126		3
Freedman	Cart	Comm OV	Butyl Acetate	1000	32	23	50		60				BT@0.5%	2
Freedman	Cart	Comm OV	Butyl Formate	1000	32	23	50		76				BT@0.5%	2
Smoot	Cart	Custom	Carbon Tet	1000	32	25	0	34	86	93	109	146		3
Smoot	Cart	Custom	Carbon Tet	1000	32	25	50	34	78	86	100	136		3
Smoot	Cart	Custom	Carbon Tet	1000	32	25	80	34	69	77	90	154		3
Simon	Cart	ScottAGESP	Chlorine Dioxide	500	64	25	50	112	8	20	3	21	see text for BT	9
Simon	Cart	Scott-AG	Chlorine Dioxide	500	64	25	50	90	22	42	46	60	see text for BT	9
Simon	Cart	MSA-OVAG	Chlorine Dioxide	500	64	25	50	101	452	55	61	69	see text for BT	9
Simon	Cart	Scott-642OV	Chlorine Dioxide	500	64	25	50	89	56	66	58	67	see text for BT	9
Simon	Can	Scott-631	Chlorine Dioxide	500	64	25	50	217	152	166	119	163	see text for BT	9
Simon	Can	Scott-631	Chlorine Dioxide	500	64	25	50	270	210	250	165	192	see text for BT	9
Simon	Cart	3M-AG	Chlorine Dioxide	500	64	25	50	101	65	84	74	844	see text for BT	9
Simon	Cart	3M-OV	Chlorine Dioxide	500	64	25	50	81	37	45	39	43	see text for BT	9
Simon	Cart	AO-AG	Chlorine Dioxide	500	64	25	50	na	49	67	68	82	see text for BT	9
Simon	Cart	Scott-OVAG	Chlorine Dioxide	500	64	25	50	115	68	82	59	73	see text for BT	9
Simon	Cart	Scott-AG	Chlorine Dioxide	500	64	25	50	131	79	98	63	94	see text for BT	9
Simon	Can	Scott-184	Chlorine Dioxide	500	64	25	50	538	200	260	194	221	see text for BT	9
Simon	Cart	MSA-AG	Chlorine Dioxide	500	64	25	50	119	59	73	53	61	see text for BT	9
Simon	Cart	3M-OVAG	Chlorine Dioxide	500	64	25	50	95	36	63	36	50	see text for BT	9
Freedman	Cart	Comm OV	Chloroform	1000	32	23	50		24				BT@0.5%	2
Smoot	Cart	Custom	Chloroform	1000	32	25	80	34	48	56	67	98		3
Smoot	Cart	Custom	Chloroform	1000	32	25	0	34	56	65	76	108		3
Smoot	Cart	Custom	Chloroform	1000	32	25	50	34	56	65	76	108		3
Stampfer	Can	MSA-OV	Chloroform	525	64	24	<15		437					7
Stampfer	Can	MSA-OV	Chloroform	494	64	27	80		274					7
Smoot	Cart	Custom	Chloroprene,3-	1000	32	25	0	34	47	55	66	125		3
Smoot	Cart	Custom	Chloroprene,3-	1000	32	25	50	34	51	56	65	114		3
Smoot	Cart	Custom	Chloroprene,3-	1000	32	25	80	34	40	44	52	73		3
Stampfer	Can	MSA-OV	CMME	1.0	64	26	80			34			BT@30%	7
Stampfer	Can	MSA-OV	CMME	1.0	64	24	<15		119					7
Pers Comm	Cart	F-H	Cyanogen Chloride	4	16	25	80			44			mg/L, BT@0.04UG/L	
Pers Comm	Cart	HF-C	Cyanogen Chloride	4	32	25	80		111				mg/L, BT@0.04UG/L	
Pers Comm	Cart	R-I	Cyanogen Chloride	4	32	25	80		69				mg/L, BT@0.04UG/L	
Pers Comm	Cart	I-F	Cyanogen Chloride	4	16	25	80		21				mg/L, BT@0.04UG/L	
Pers Comm	Cart	HF-C	Cyanogen Chloride	4	32	25	80		126				mg/L, BT@0.04UG/L	
Stampfer	Can	MSA-OV	Dibromomethane,1,2-	6.5	64	26	80		1080					7
Stampfer	Can	MSA-OV	Dibromomethane,1,2-	6.5	64	22	<15		1140					7
Smoot	Cart	Custom	Dibromomethane,1,2-	1000	32	25	50	34	129	139	158	212		3
Smoot	Cart	Custom	Dibromomethane,1,2-	1000	32	25	80	34	106	120	142	220		3
Smoot	Cart	Custom	Dibromomethane,1,2-	1000	32	25	0	34	128	138	152	198		3
Pers Comm	Cart	F-H	DMMP	4	16	25	20		269				mg/L, BT@0.04UG/L	
Pers Comm	Cart	I-H	DMMP	3	32	25	20		278				mg/L, BT@0.04UG/L	
Pers Comm	Cart	Riot Cntl	DMMP	3	32	25	20		54				mg/L, BT@0.04UG/L	
Pers Comm	Cart	R-I	DMMP	3	32	25	20		733				mg/L, BT@0.04UG/L	
Pers Comm	Cart	A-H	DMMP	3	32	25	20		287				mg/L, BT@0.04UG/L	
Pers Comm	Cart	HF-C	DMMP	3	32	25	20		372				mg/L, BT@0.04UG/L	
Stampfer	Can	MSA-OV	Epichlorohydrin	250	64	24	80		765					7
Stampfer	Can	MSA-OV	Epichlorohydrin	251	64	23	<15		1080					7
Freedman	Cart	Comm OV	Ethanol	1000	32	23	50		11				BT@0.5%	2
Smoot	Cart	Custom	Ethanol	1000	32	25	0	34	39	49	74	155		3
Smoot	Cart	Custom	Ethanol	1000	32	25	50	34	38	60	94	199		3
Smoot	Cart	Custom	Ethanol	1000	32	25	80	34	38	55	87	180		3
Bollinger	Cart	Binks-OV	EEA, 2-	80	64	25	85	43	156					12
Bollinger	Cart	Binks-OV	EEA, 2-	78	64	25	85	43	207					12
Bollinger	Cart	Binks-OV	EEA, 2-	89	64	25	85	43	154					12
Bollinger	Cart	Binks-OV	Ethoxyethanol,2-	80	64	25	85	43	98					12
Bollinger	Cart	Binks-OV	Ethoxyethanol,2-	77	64	25	85	43	95					12
Bollinger	Cart	Binks-OV	Ethoxyethanol,2-	76	64	25	85	43	99					12

Nelson	Cart	NorthOV	Ethoxyethanol,2-	10	25	25	80		1110					32
Nelson	Cart	NorthOV	Ethoxyethanol,2-	10	25	25	80		1080					32
Nelson	Cart	NorthOV	Ethoxyethanol,2-	100	25	25	80		1164					32
Nelson	Cart	NorthOV	Ethoxyethanol,2-	100	25	25	80		1224					32
Nelson	Cart	NorthOV	Ethoxyethanol,2-	500	25	25	20		330					32
Nelson	Cart	NorthOV	Ethoxyethanol,2-	500	25	25	80		294					32
Nelson	Cart	NorthOV	Ethoxyethanol,2-	500	25	25	80		282					32
Nelson	Cart	NorthOV	EEA, 2-	10	25	25	80		>1104					32
Nelson	Cart	NorthOV	EEA, 2-	10	25	25	80		>1104					32
Nelson	Cart	NorthOV	EEA, 2-	100	25	25	80		1104					32
Nelson	Cart	NorthOV	EEA, 2-	100	25	25	80		18.4					32
Nelson	Cart	NorthOV	EEA, 2-	500	25	25	20		276					32
Nelson	Cart	NorthOV	EEA, 2-	500	25	25	80		228					32
Nelson	Cart	NorthOV	EEA, 2-	500	25	25	80		210					32
Freedman	Cart	Comm OV	Ethyl Acetate	1000	32	23	50		60				BT@0.5%	2
Moyer	Cart	OV	Ethyl Acetate	750	32	25	50	36	74	77	79		BT@5,10&15ppm	50
Moyer	Cart	OV	Ethyl Acetate	750	32	25	50	72	176	179	181		BT@5,10&15ppm	50
Moyer	Cart	OV	Ethyl Acetate	750	32	25	50	108	280	282	285		BT@5,10&15ppm	50
Moyer	Cart	OV	Ethyl Acetate	750	32	25	50	144	390	394	396		BT@5,10&15ppm	50
Moyer	Cart	OV	Ethyl Acetate	750	32	25	80	36	69	72	74		BT@5,10&15ppm	50
Moyer	Cart	OV	Ethyl Acetate	750	32	25	80	73	160	163	168		BT@5,10&15ppm	50
Moyer	Cart	OV	Ethyl Acetate	750	32	25	80	109	254	257	259		BT@5,10&15ppm	50
Moyer	Cart	OV	Ethyl Acetate	750	32	25	80	145	345	348	350		BT@5,10&15ppm	50
Moyer	Cart	OV	Ethyl Acetate	1000	32	25	50	37	59	62	63		BT@5,10&15ppm	50
Moyer	Cart	OV	Ethyl Acetate	1000	32	25	50	72	139	142	143		BT@5,10&15ppm	50
Moyer	Cart	OV	Ethyl Acetate	1000	32	25	50	110	227	229	231		BT@5,10&15ppm	50
Moyer	Cart	OV	Ethyl Acetate	1000	32	25	50	145	345	348	318		BT@5,10&15ppm	50
Moyer	Cart	OV	Ethyl Acetate	1000	32	25	80	36	55	62	63		BT@5,10&15ppm	50
Moyer	Cart	OV	Ethyl Acetate	1000	32	25	80	73	133	135	136		BT@5,10&15ppm	50
Moyer	Cart	OV	Ethyl Acetate	1000	32	25	80	109	207	210	211		BT@5,10&15ppm	50
Moyer	Cart	OV	Ethyl Acetate	1000	32	25	80	146	285	287	289		BT@5,10&15ppm	50
Moyer	Cart	OV	Ethyl Acetate	1500	32	25	50	37	46	48	49		BT@5,10&15ppm	50
Moyer	Cart	OV	Ethyl Acetate	1500	32	25	50	74	105	111	112		BT@5,10&15ppm	50
Moyer	Cart	OV	Ethyl Acetate	1500	32	25	50	111	172	175	177		BT@5,10&15ppm	50
Moyer	Cart	OV	Ethyl Acetate	1500	32	25	50	147	241	244	245		BT@5,10&15ppm	50
Moyer	Cart	OV	Ethyl Acetate	1500	32	25	80	37	39	41	42		BT@5,10&15ppm	50
Moyer	Cart	OV	Ethyl Acetate	1500	32	25	80	73	94	96	97		BT@5,10&15ppm	50
Moyer	Cart	OV	Ethyl Acetate	1500	32	25	80	110	152	154	155		BT@5,10&15ppm	50
Moyer	Cart	OV	Ethyl Acetate	1500	32	25	80	146	212	214	215		BT@5,10&15ppm	50
Moyer	Cart	OV	Ethyl Acetate	2000	32	25	50	37	35	36	37		BT@5,10&15 ppm	50
Moyer	Cart	OV	Ethyl Acetate	2000	32	25	50	74	86	87	88		BT@5,10&15 ppm	50
Moyer	Cart	OV	Ethyl Acetate	2000	32	25	50	111	139	140	141		BT@5,10&15 ppm	50
Moyer	Cart	OV	Ethyl Acetate	2000	32	25	50	158	197	198	199		BT@5,10&15 ppm	50
Moyer	Cart	OV	Ethyl Acetate	2000	32	25	80	36	34	35	36		BT@5,10&15 ppm	50
Moyer	Cart	OV	Ethyl Acetate	2000	32	25	80	73	81	83	83		BT@5,10&15 ppm	50
Moyer	Cart	OV	Ethyl Acetate	2000	32	25	80	109	128	129	130		BT@5,10&15 ppm	50
Moyer	Cart	OV	Ethyl Acetate	2000	32	25	80	146	181	183	183		BT@5,10&15 ppm	50
Moyer	ChCan	OV	Ethyl Acetate	1000	32	50	50	202	127	132	135		BT@5,10&15 ppm	50
Moyer	ChCan	OV	Ethyl Acetate	1000	32	50	50	199	125	130	133		BT@5,10&15 ppm	50
Moyer	ChCan	OV	Ethyl Acetate	1000	32	50	80	199	99	107	111		BT@5,10&15 ppm	50
Moyer	ChCan	OV	Ethyl Acetate	1000	32	50	80	192	78	82	84		BT@5,10&15 ppm	50
Moyer	ChCan	OV	Ethyl Acetate	1000	32	50	80	190	98	109	114		BT@5,10&15 ppm	50
Moyer	ChCan	OV	Ethyl Acetate	2000	32	50	50	186	56	58	60		BT@5,10&15 ppm	50
Moyer	ChCan	OV	Ethyl Acetate	2000	32	50	80	184	44	46	48		BT@5,10&15 ppm	50
Moyer	ChCan	OV	Ethyl Acetate	5000	32	50	50	182	25	25	26		BT@5,10&15 ppm	50
Moyer	ChCan	OV	Ethyl Acetate	5000	32	50	80	183	17	17	18		BT@5,10&15 ppm	50
Moyer	ChCan	OV	Ethyl Acetate	7500	32	50	50	201	20	21	22		BT@5,10&15 ppm	50
Moyer	ChCan	OV	Ethyl Acetate	7500	32	50	80	203	11	17	18		BT@5,10&15 ppm	50
Smoot	Cart	Custom	Ethyl Acetate	1000	32	25	0	34	61	84	99	146		3
Smoot	Cart	Custom	Ethyl Acetate	1000	32	25	50	34	63	87	102	137		3
Smoot	Cart	Custom	Ethyl Acetate	1000	32	25	80	34	65	79	93	119		3
Freedman	Cart	Comm OV	Ethyl Bromide	1000	32	23	50		17				BT@0.5%	2
Freedman	Cart	Comm OV	Ethyl Formate	1000	32	23	50		28			58	BT@0.5%	2
Freedman	Cart	Comm OV	Ethyl iodide	1000	32	23	50		84				BT@0.5%	2
Nelson	Cart	Norton-OV	Ethylene Dibromide	5	64	25	85		>480				BT@0.1%, EQ25%,85%	13
Nelson	Cart	MSA-OV	Ethylene Dibromide	5	64	25	85		>480				BT@0.1%	13

Nelson	Cart	Norton-OV	Ethylene Dibromide	5	64	25	85		>480				BT@0.1%, EQ25%,85%	13
Nelson	Cart	Norton-OV	Ethylene Dibromide	5	64	25	85		>480				BT@0.1%	13
Nelson	Cart	MSA-OV	Ethylene Dibromide	5	64	25	85		>480				BT@0.1%, EQ25%,85%	13
Nelson	Cart	Norton-OV	Ethylene Dibromide	5	64	25	85		>480				BT@0.1%, EQ25%,85%	13
Nelson	Cart	Norton-OV	Ethylene Dibromide	5	64	25	85		>480				BT@0.1%, EQ25%,85%	13
Nelson	Cart	MSA-OV	Ethylene Dibromide	5	64	25	85		>480				BT@0.1%, EQ25%,85%	13
Nelson	Cart	MSA-OV	Ethylene Dibromide	5	64	25	85		>480				BT@0.1%	13
Nelson	Cart	MSA-OV	Ethylene Dibromide	5	64	25	85		>480				BT@0.1%	13
Nelson	Cart	Norton-OV	Ethylene Dibromide	5	64	25	85		>480				BT@0.1%, EQ25%,85%	13
Nelson	Cart	MSA-OV	Ethylene Dibromide	5	64	25	85		>480				BT@0.1%	13
Nelson	Cart	MSA-OV	Ethylene Dibromide	5	64	25	85		>480				BT@0.1%, EQ25%,85%	13
Nelson	Cart	MSA-OV	Ethylene Dibromide	5	64	25	85		>480				BT@0.1%	13
Nelson	Cart	MSA-OV	Ethylene Dibromide	5	64	25	85		>480				BT@0.1%, EQ25%,85%	13
Nelson	Cart	MSA-OV	Ethylene Dibromide	5	64	25	85		>480				BT@0.1%, EQ25%,85%	13
Nelson	Cart	Norton-OV	Ethylene Dibromide	5	64	25	85		>480				50ppb or 0.1%	13
Nelson	Cart	Norton-OV	Ethylene Dibromide	5	64	25	85		>480				50ppb or 0.1%	13
Nelson	Cart	Norton-OV	Ethylene Dibromide	5	64	25	85		>480				50ppb or 0.1%	13
Nelson	Cart	Norton-OV	Ethylene Dibromide	5	64	25	85		>480				50ppb or 0.1%	13
Nelson	Cart	Norton-OV	Ethylene Dibromide	35	64	25	85		492	564	600		BT@50,100,150 ppb	13
Nelson	Cart	Norton-OV	Ethylene Dibromide	50	64	25	85		439	480	496		BT@50,100,150 ppb	13
Nelson	Cart	Norton-OV	Ethylene Dibromide	100	64	25	85		285	300	339		BT@50,100,150 ppb	13
Nelson	Cart	Norton-OV	Ethylene Dibromide	200	64	25	85		143	156	163		BT@50,100,150 ppb	13
Nelson	Cart	Norton-OV	Ethylene Dibromide	200	64	25	85		151	163	170		BT@50,100,150 ppb	13
Nelson	Can	MSA-OV	Ethylene Dibromide	50	64	25	85		>480				50ppb or 0.1%	13
Nelson	Can	MSA-OV	Ethylene Dibromide	50	64	25	85		>480				50ppb or 0.1%	13
Nelson	Can	MSA-OV	Ethylene Dibromide	50	64	25	85		>480				50ppb or 0.1%	13
Nelson	Can	MSA-OV	Ethylene Dibromide	50	64	25	85		>480				50ppb or 0.1%	13
Nelson	Can	MSA-OV	Ethylene Dibromide	50	64	25	85		>480				50ppb or 0.1%	13
Nelson	Can	MSA-OV	Ethylene Dibromide	100	64	25	85		516	558	594		BT@50,100,150 ppb	13
Nelson	Can	MSA-OV	Ethylene Dibromide	200	64	25	85		396	420	438		BT@50,100,150 ppb	13
Nelson	Can	MSA-OV	Ethylene Dibromide	2000	64	25	85		51	90	116		BT@50,100,150 ppb	13
Nelson	Can	MSA-OV	Ethylene Dibromide	2000	64	25	85		48	67	84		BT@50,100,150 ppb	13
Freedman	Cart	Comm OV	Ethylene oxide	1000	32	23	50		0.45				BT @0.5%	2
MSA	Can	MSA-EtO SS	Etylene Oxide	100	64	25	85		2011				BT @1ppm, 85% EQ	24
MSA	Can	MSA-EtO SS	Etylene Oxide	100	64	25	85		2121				BT @1ppm, 85% EQ	24
MSA	Can	MSA-EtO SS	Etylene Oxide	100	64	25	85		2460				BT @1ppm, 85% EQ	24
MSA	Can	MSA-EtO SS	Etylene Oxide	100	64	25	85		2140				BT @1ppm, 85% EQ	24
MSA	Can	MSA-EtO SS	Etylene Oxide	1000	64	25	25		1775				BT @1ppm, 25% EQ	24
MSA	Can	MSA-EtO SS	Etylene Oxide	1000	64	25	25		1725				BT @1ppm, 25% EQ	24
MSA	Can	MSA-EtO SS	Etylene Oxide	1000	64	25	25		1926				BT @1ppm, 25% EQ	24
MSA	Can	MSA-EtO SS	Etylene Oxide	1000	64	25	25		1926				BT @1ppm, 25% EQ	24
MSA	Can	MSA-EtO SS	Etylene Oxide	1000	64	25	85		811				BT @1ppm, 85% EQ	24
MSA	Can	MSA-EtO SS	Etylene Oxide	1000	64	25	85		653				BT @1ppm, 85% EQ	24
MSA	Can	MSA-EtO SS	Etylene Oxide	1000	64	25	85		767				BT @1ppm, 85% EQ	24
MSA	Can	MSA-EtO SS	Etylene Oxide	1000	64	25	85		769				BT @1ppm, 85% EQ	24
MSA	Can	MSA-EtO SS	Etylene Oxide	2000	64	25	25		190				No BT after 190,25%EQ	24
MSA	Can	MSA-EtO SS	Etylene Oxide	2000	64	25	25		315				No BT after315,25%EQ	24
MSA	Can	MSA-EtO SS	Etylene Oxide	5000	64	25	50		262				No BT after 262'	24
MSA	Can	MSA-EtO SS	Etylene Oxide	5000	64	25	50		330				No BT after 330'	24
MSA	Can	MSA-EtO SS	Etylene Oxide	5000	64	25	50		241				1ppm BT @241'	24
MSA	Can	MSA-EtO SS	Etylene Oxide	5000	64	25	85		300				No BT after 300',85%EQ	24
MSA	Can	MSA-EtO SS	Etylene Oxide	5000	64	25	85		186				1ppm BT@186',85%EQ	24
Bollinger	Cart	MSA-AG	Formaldehyde	15	64	25	85	43	393					12
Bollinger	Cart	MSA-AG	Formaldehyde	15	64	25	85	43	234					12
Bollinger	Cart	MSA-AG	Formaldehyde	15	64	25	85	43	282					12
Henry	Cart	MSA-OV	Formaldehyde	0.5	60	23	50	39		114	192		BT@100%	5
Henry	Cart	Norton-AG	Formaldehyde	1.4	60	23	50	60		3375	>4800		BT@100%	5
Henry	Cart	MSA-OV	Formaldehyde	1.4	60	23	50			118	170		BT@100%	5
Henry	Cart	Norton-exp	Formaldehyde	1.4	60	23	50	55		2820	>4800		BT@100%	5
Henry	Cart	MSA-OV Can	Formaldehyde	1.5	60	23	50	245		2405	5019		BT@100%	5
Henry	Cart	Norton-OVAG	Formaldehyde	1.8	60	23	50	54		278	1237		BT@100%	5
Henry	Cart	Willson-OV	Formaldehyde	2.4	60	23	50	56		30	36		BT@100%	5
Henry	Cart	MSA-OV	Formaldehyde	2.6	60	23	7			83	113		BT@100%	5
Henry	Cart	MSA-OV	Formaldehyde	2.6	60	23	50			67	159	390	BT@100%	5
Henry	Cart	MSA-OV	Formaldehyde	2.6	60	23	50	49		342	1260		BT@100%	5
Henry	Cart	MSA-OVAG	Formaldehyde	2.6	60	23	85			125	600		BT@100%	5

Freedman	Cart	Comm OV	Heptane	1000	32	23	50		61				BT@0.5%	2
Moyer	Cart	OV	Heptane	800	32	50	50	45	91	94	96		BT@5,10&15 ppm	50
Moyer	Cart	OV	Heptane	800	32	50	50	91	212	215	217		BT@5,10&15 ppm	50
Moyer	Cart	OV	Heptane	800	32	50	50	137	346	349	352		BT@5,10&15 ppm	50
Moyer	Cart	OV	Heptane	800	32	50	80	47	75	78	80		BT@5,10&15 ppm	50
Moyer	Cart	OV	Heptane	800	32	50	80	94	175	178	180		BT@5,10&15 ppm	50
Moyer	Cart	OV	Heptane	800	32	50	80	141	287	291	294		BT@5,10&15 ppm	50
Moyer	Cart	OV	Heptane	1000	32	50	50	47	70	73	74		BT@5,10&15 ppm	50
Moyer	Cart	OV	Heptane	1000	32	50	50	94	173	176	178		BT@5,10&15 ppm	50
Moyer	Cart	OV	Heptane	1000	32	50	50	141	277	281	282		BT@5,10&15 ppm	50
Moyer	Cart	OV	Heptane	1000	32	50	50	187	383	396	388		BT@5,10&15 ppm	50
Moyer	Cart	OV	Heptane	1000	32	50	80	47	67	69	7[illegible]		BT@5,10&15 ppm	50
Moyer	Cart	OV	Heptane	1000	32	50	80	94	151	154	15[illegible]		BT@5,10&15 ppm	50
Moyer	Cart	OV	Heptane	1000	32	50	80	141	250	253	25[illegible]		BT@5,10&15 ppm	50
Moyer	Cart	OV	Heptane	1000	32	50	80	188	344	374	349		BT@5,10&15 ppm	50
Moyer	Cart	OV	Heptane	1250	32	50	50	47	60	62	64		BT@5,10&15 ppm	50
Moyer	Cart	OV	Heptane	1250	32	50	50	94	144	147	148		BT@5,10&15 ppm	50
Moyer	Cart	OV	Heptane	1250	32	50	50	141	235	238	239		BT@5,10&15 ppm	50
Moyer	Cart	OV	Heptane	1250	32	50	50	188	329	331	332		BT@5,10&15 ppm	50
Moyer	Cart	OV	Heptane	1250	32	50	80	47	46	49	50		BT@5,10&15 ppm	50
Moyer	Cart	OV	Heptane	1250	32	50	80	94	124	129	131		BT@5,10&15 ppm	50
Moyer	Cart	OV	Heptane	1250	32	50	80	141	210	212	213		BT@5,10&15 ppm	50
Moyer	Cart	OV	Heptane	1250	32	50	80	188	293	296	297		BT@5,10&15 ppm	50
Moyer	Cart	OV	Heptane	1500	32	50	50	47	51	53	54		BT@5,10&15 ppm	50
Moyer	Cart	OV	Heptane	1500	32	50	50	93	122	124	126		BT@5,10&15 ppm	50
Moyer	Cart	OV	Heptane	1500	32	50	50	140	197	199	200		BT@5,10&15 ppm	50
Moyer	Cart	OV	Heptane	1500	32	50	50	187	274	277	278		BT@5,10&15 ppm	50
Moyer	Cart	OV	Heptane	1500	32	50	80	47	38	41	24		BT@5,10&15 ppm	50
Moyer	Cart	OV	Heptane	1500	32	50	80	94	85	88	89		BT@5,10&15 ppm	50
Moyer	Cart	OV	Heptane	1500	32	50	80	140	131	133	135		BT@5,10&15 ppm	50
Moyer	ChCan	OV	Heptane	2500	32	50	50	201	59	62	63		BT@5,10&15 ppm	50
Moyer	ChCan	OV	Heptane	2500	32	50	80	202	57	60	62		BT@5,10&15 ppm	50
Moyer	ChCan	OV	Heptane	5000	32	50	50	199	26	27	28		BT@5,10&15 ppm	50
Moyer	ChCan	OV	Heptane	5000	32	50	80	207	26	28	29		BT@5,10&15 ppm	50
Moyer	ChCan	OV	Heptane	7500	32	50	50	202	19	20	21		BT@5,10&15 ppm	50
Moyer	ChCan	OV	Heptane	7500	32	50	80	203	16	17	18		BT@5,10&15 ppm	50
Freedman	Cart	Comm OV	Hexane-N	1000	32	23	50		55				BT@0.5%	2
Smoot	Cart	Custom	Hexane-N	1000	32	25	0	34	69	76	87	123		3
Smoot	Cart	Custom	Hexane-N	1000	32	25	50	34	72	79	88	113		3
Smoot	Cart	Custom	Hexane-N	1000	32	25	80	34	60	66	77	134		3
Moyer	Cart	OV	Hexane-N	500	32	25	50	47	109	112	115		BT@5,10&15ppm	50
Moyer	Cart	OV	Hexane-N	500	32	25	50	95	259	263	266		BT@5,10&15ppm	50
Moyer	Cart	OV	Hexane-N	500	32	25	50	142	414	419	421		BT@5,10&15ppm	50
Moyer	Cart	OV	Hexane-N	500	32	25	80	48	84	87	89		BT@5,10&15ppm	50
Moyer	Cart	OV	Hexane-N	500	32	25	80	96	191	194	196		BT@5,10&15ppm	50
Moyer	Cart	OV	Hexane-N	500	32	25	80	144	307	310	313		BT@5,10&15ppm	50
Moyer	Cart	OV	Hexane-N	750	32	50	50	47	76	79	80		BT@5,10&15ppm	50
Moyer	Cart	OV	Hexane-N	750	32	50	50	95	182	186	187		BT@5,10&15ppm	50
Moyer	Cart	OV	Hexane-N	750	32	50	50	141	294	298	300		BT@5,10&15ppm	50
Moyer	Cart	OV	Hexane-N	750	32	50	50	188	407	410	412		BT@5,10&15ppm	50
Moyer	Cart	OV	Hexane-N	750	32	50	80	45	64	66	67		BT@5,10&15ppm	50
Moyer	Cart	OV	Hexane-N	750	32	50	80	91	147	149	151		BT@5,10&15ppm	50
Moyer	Cart	OV	Hexane-N	750	32	50	80	137	231	233	234		BT@5,10&15ppm	50
Moyer	Cart	OV	Hexane-N	750	32	50	80	184	316	318	320		BT@5,10&15ppm	50
Moyer	Cart	OV	Hexane-N	1000	32	50	50	46	62	64	66		BT@5,10&15ppm	50
Moyer	Cart	OV	Hexane-N	1000	32	50	50	93	139	141	143		BT@5,10&15ppm	50
Moyer	Cart	OV	Hexane-N	1000	32	50	50	139	224	227	228		BT@5,10&15ppm	50
Moyer	Cart	OV	Hexane-N	1000	32	50	50	186	308	311	313		BT@5,10&15ppm	50
Moyer	Cart	OV	Hexane-N	1000	32	50	80	46	50	51	53		BT@5,10&15ppm	50
Moyer	Cart	OV	Hexane-N	1000	32	50	80	94	117	120	121		BT@5,10&15ppm	50
Moyer	Cart	OV	Hexane-N	1000	32	50	80	139	188	191	192		BT@5,10&15ppm	50
Moyer	Cart	OV	Hexane-N	1000	32	50	80	187	262	264	269		BT@5,10&15ppm	50
Moyer	Cart	OV	Hexane-N	1500	32	50	50	48	44	45	46		BT@5,10&15ppm	50
Moyer	Cart	OV	Hexane-N	1500	32	50	50	95	106	108	109		BT@5,10&15ppm	50
Moyer	Cart	OV	Hexane-N	1500	32	50	50	140	173	175	176		BT@5,10&15ppm	50
Moyer	Cart	OV	Hexane-N	1500	32	50	50	186	242	243	244		BT@5,10&15ppm	50

Moyer	Cart	OV	Hexane-N	1500	32	50	80	46	40	41	42	BT@5,10&15ppm	50
Moyer	Cart	OV	Hexane-N	1500	32	50	80	92	96	97	98	BT@5,10&15ppm	50
Moyer	Cart	OV	Hexane-N	1500	32	50	80	139	156	157	158	BT@5,10&15ppm	50
Moyer	Cart	OV	Hexane-N	1500	32	50	80	187	216	218	219	BT@5,10&15ppm	50
Moyer	ChCan	OV	Hexane-N	1000	32	50	50	209	106	110	112	BT@5,10&15ppm	50
Moyer	ChCan	OV	Hexane-N	1000	32	50	80	201	97	102	104	BT@5,10&15ppm	50
Moyer	ChCan	OV	Hexane-N	2000	32	50	50	210	55	57	59	BT@5,10&15ppm	50
Moyer	ChCan	OV	Hexane-N	2000	32	50	80	211	54	56	57	BT@5,10&15ppm	50
Moyer	ChCan	OV	Hexane-N	5000	32	50	50	202	29	30	31	BT@5,10&15ppm	50
Moyer	ChCan	OV	Hexane-N	5000	32	50	80	203	33	34	35	BT@5,10&15ppm	50
Moyer	ChCan	OV	Hexane-N	7500	32	50	50	200	21	22	23	BT@5,10&15ppm	50
Moyer	ChCan	OV	Hexane-N	7500	32	50	80	201	21	23	24	BT@5,10&15ppm	50
MSA	Cart	MSA-OV/AG	Hydrogen Sulfide	1000	64	25	50		57			BT@0.1%	23
MSA	Cart	MSA-AG	Hydrogen Sulfide	1000	64	25	50		96			BT@0.1%	23
MSA	Cart	MSA-AG	Hydrogen Sulfide	1000	64	25	50		>240			BT@0.1%,EQ85%RH	23
MSA	Cart	MSA-OV/AG	Hydrogen Sulfide	1000	64	25	50		160			BT@0.1%,EQ85%RH	23
MSA	Cart	MSA-AG	Hydrogen Sulfide	1000	64	25	50		>240			BT@0.1%,EQ25%RH	23
MSA	Cart	MSA-AG	Hydrogen Sulfide	1000	64	25	50		102			BT@0.1%	23
MSA	Cart	MSA-AG	Hydrogen Sulfide	1000	64	25	50		>180			BT@0.1%, EQ85%RH	23
MSA	Cart	MSA-AG	Hydrogen Sulfide	1000	64	25	50		>210			BT@0.1%, EQ85%RH	23
MSA	Cart	MSA-AG	Hydrogen Sulfide	1000	64	25	50		93			BT@0.1%	23
MSA	Cart	MSA-OV/AG	Hydrogen Sulfide	1000	64	25	50		179			BT@0.1%,EQ25%RH	23
MSA	Cart	MSA-OV/AG	Hydrogen Sulfide	1000	64	25	50		173			BT@0.1%,EQ25%RH	23
MSA	Cart	MSA-OV/AG	Hydrogen Sulfide	1000	64	25	50		164			BT@0.1%,EQ85%RH	23
MSA	Cart	MSA-OV/AG	Hydrogen Sulfide	1000	64	25	50		63			BT@0.1%	23
MSA	Cart	MSA-OV/AG	Hydrogen Sulfide	1000	64	25	50		87			BT@0.1%	23
MSA	Can	MSASS-HS	Hydrogen Sulfide	300	64	25	50		>12000			BT@0.1%	23
MSA	Can	MSASS-HS	Hydrogen Sulfide	300	64	25	50		>12000			BT@0.1%	23
MSA	Can	MSASS-HS	Hydrogen Sulfide	1000	64	25	50		>2100			BT@0.1%	23
MSA	Can	MSASS-HS	Hydrogen Sulfide	1000	64	25	50		>2160			BT@0.1%,EQ85%RH	23
MSA	Can	MSASS-HS	Hydrogen Sulfide	1000	64	25	50		>2340			BT@0.1%	23
MSA	Can	MSASS-HS	Hydrogen Sulfide	1000	64	25	50		>2100			BT@0.1%,EQ25%RH	23
MSA	Can	MSASS-HS	Hydrogen Sulfide	1000	64	25	50		>2160			BT@0.1%	23
MSA	Can	MSASS-HS	Hydrogen Sulfide	1000	64	25	50		>2340			BT@0.1%	23
MSA	Can	MSASS-HS	Hydrogen Sulfide	10000	64	25	50		294			BT@0.1%	23
MSA	Can	MSASS-HS	Hydrogen Sulfide	10000	64	25	50		246			BT@0.01%,CEQ85%RH	23
MSA	Can	MSASS-HS	Hydrogen Sulfide	10000	64	25	50		240			BT@0.01%,CEQ85%RH	23
MSA	Can	MSASS-HS	Hydrogen Sulfide	10000	64	25	50		300			BT@0.01%	23
MSA	Can	MSASS-HS	Hydrogen Sulfide	10000	64	25	50		264			BT@0.01%	23
MSA	Can	MSASS-HS	Hydrogen Sulfide	10000	64	25	50		246			BT@0.01%,EQ85%RH	23
MSA	Can	MSASS-HS	Hydrogen Sulfide	10000	64	25	50		234			BT@0.01%,EQ25%RH	23
MSA	M/P	MSA	Hydrogen Sulfide	1000	64	25	50		180			BT@0.01%,EQ25%RH	23
MSA	M/P	MSA	Hydrogen Sulfide	1000	64	25	50		46			BT@0.1%	23
MSA	M/P	MSA	Hydrogen Sulfide	1000	64	25	50		208			BT@0.1%,EQ25%RH	23
MSA	M/P	MSA	Hydrogen Sulfide	1000	64	25	50		>145			BT@0.1%,EQ85%RH	23
MSA	M/P	MSA	Hydrogen Sulfide	1000	64	25	50		>205			BT@0.1%,EQ85%RH	23
MSA	M/P	MSA	Hydrogen Sulfide	1000	64	25	50		47			BT@0.1%	23
MSA	M/P	MSA	Hydrogen Sulfide	1000	64	25	50		47			BT@0.1%	23
Nelson	Can	Scott-OV	Iodine	10	64	25	50		>120				21
Nelson	Can	Scott-OV	Iodine	11	32	25	25		>120			BT@0.01ppm,EQ25%RH	21
Nelson	Can	Scott-OV	Iodine	11	32	25	85		>120			BT@0.01ppm,EQ25%RH	21
Nelson	Can	Scott-OV	Iodine	11	32	25	25		>120			BT@0.01ppm,EQ25%RH	21
Nelson	Can	Scott-OV	Iodine	11	64	25	50		>120			BT@0.01ppm	21
Nelson	Can	Scott-OV	Iodine	11	32	25	85		>120			BT@0.01ppm,EQ 25%RH	21
Nelson	Can	Scott-OV	Iodine	12	64	25	50		>120			BT@0.01ppm	21
MSA	Cart	MSA-Hg	Mercury	21.5	64	20	50		701			BT@0.05mg/m3,EQ25%RH	17
MSA	Cart	MSA-Hg	Mercury	21.5	64	20	50		847			BT@0.05mg/m3,EQ25%RH	17
MSA	Cart	MSA-Hg	Mercury	21.5	64	20	50		615			BT@0.05mg/m3,EQ25%RH	17
MSA	Cart	MSA-Hg	Mercury	21.5	64	20	50		431			BT@0.05mg/m3,EQ25%RH	17
MSA	Cart	MSA-Hg	Mercury	21.5	64	20	50		723			BT@0.05mg/m3,EQ25%RH	17
MSA	Cart	MSA-Hg	Mercury	21.5	64	20	50		670			BT@0.05%	17
MSA	Cart	MSA-Hg	Mercury	21.5	64	20	50		631			BT@0.05 mg/m3	17
MSA	Cart	MSA-Hg	Mercury	21.5	64	20	50		547			BT@0.05 mg/m3	17
Freedman	Cart	Comm OV	Methanol	1000	32	23	50		2.2			BT@0.5%	2
Hitchcock	S/G	8-20 mesh	Methanol	2000	64	23	41	80		15	19		6
Hitchcock	S/G	8-20 mesh	Methanol	2000	64	23	41	93		18	23		6

Hitchcock	S/G	12-28 mesh	Methanol	2000	64	23	75			21				6
Hitchcock	S/G	8-20 mesh	Methanol	2000	64	23	41	86		16	20			6
Hitchcock	S/G	12-28 mesh	Methanol	2000	64	23	90			16				6
Hitchcock	S/G	8-20 mesh	Methanol	2000	64	23	41	72		11	14			6
Hitchcock	S/G	12-28 mesh	Methanol	2000	64	23	91			16				6
Hitchcock	DR		Methanol	2000	64	23	45	105		1	1			6
Hitchcock	SL		Methanol	2000	64	23	45	105		0	1			6
Hitchcock	S/G	12-28 mesh	Methanol	2000	64	23	65			23				6
Hitchcock	S/G	8-20 mesh	Methanol	2000	64	23	45	105		10			EQ85%RH	6
Hitchcock	S/G	12-28 mesh	Methanol	2000	64	23	44			28				6
Hitchcock	S/G	12-42 mesh	Methanol	2000	64	23	45	105		28				6
Hitchcock	S/G	12-28 mesh	Methanol	2000	64	23	30			26				6
Hitchcock	S/G	12-28 mesh	Methanol	2000	64	23	45	105		26	32			6
Hitchcock	S/G	8-20 mesh	Methanol	2000	64	23	41	108		19	25			6
Hitchcock	S/G	8-20 mesh	Methanol	2000	64	23	45	105		24				6
Hitchcock	S/G	12-28 mesh	Methanol	2000	64	23	90			18				6
Hitchcock	A/C		Methanol	2000	64	23	45	105		4	10			6
Hitchcock	S/G	12-28 mesh	Methanol	2000	64	23	45			29				6
Hitchcock	M/S		Methanol	2000	64	23	45	105		0	1			6
Hitchcock	S/G	8-20 mesh	Methanol	2000	64	23	41	115		18	24			6
Hitchcock	M/S		Methanol	2000	64	23	45	105		0	2			6
Hitchcock	AS		Methanol	2000	64	23	45	105		0	0			6
Hitchcock	S/G	8-20 mesh	Methanol	2000	64	23	41	102		19	25			6
Hitchcock	S/G	12-28 mesh	Methanol	2000	64	23	35			28				6
Hitchcock	A/A		Methanol	2000	64	23	45	105		4	7			6
Bollinger	Cart	Binks-OV	MEA,2-	66	64	25	85	43	110					12
Bollinger	Cart	Binks-OV	MEA,2-	66	64	25	85	43	110					12
Bollinger	Cart	Binks-OV	MEA,2-	69	64	25	85	43	108					12
Nelson	Cart	NorthOV	Methoxyethanol,2-	10	25	25	80		1080					32
Nelson	Cart	NorthOV	Methoxyethanol,2-	10	25	25	80		1080					32
Nelson	Cart	NorthOV	Methoxyethanol,2-	100	25	25	80		1044					32
Nelson	Cart	NorthOV	Methoxyethanol,2-	100	25	25	80		1020					32
Nelson	Cart	NorthOV	Methoxyethanol,2-	500	25	25	20		366					32
Nelson	Cart	NorthOV	Methoxyethanol,2-	500	25	25	80		300					32
Nelson	Cart	NorthOV	Methoxyethanol,2-	500	25	25	80		294					32
Nelson	Cart	NorthOV	MEA, -2	10	25	25	80		>1080					32
Nelson	Cart	NorthOV	MEA, -2	10	25	25	80		>1080					32
Nelson	Cart	NorthOV	MEA, -2	100	25	25	80		1122					32
Nelson	Cart	NorthOV	MEA, -2	100	25	25	80		1086					32
Nelson	Cart	NorthOV	MEA, -2	500	25	25	20		288					32
Nelson	Cart	NorthOV	MEA, -2	500	25	25	80		270					32
Nelson	Cart	NorthOV	MEA, -2	500	25	25	80		252					32
Bollinger	Cart	Binks-OV	Methoxyethanol,2-	66	64	25	85	43	99					12
Bollinger	Cart	Binks-OV	Methoxyethanol,2-	69	64	25	85	43	71					12
Bollinger	Cart	Binks-OV	Methoxyethanol,2-	68	64	25	85	43	97					12
Freedman	Cart	Comm OV	Methyl Acetate	1000	32	23	50		22				BT@0.5%	2
Freedman	Cart	Comm OV	Methyl Bromide	1000	32	23	50		1.2			5		2
Freedman	Cart	Comm OV	Methyl Formate	1000	32	23	50		2.6				BT@0.5%	2
Freedman	Cart	Comm OV	Methyl iodide	1000	32	23	50		22				BT@0.5%	2
Moyer	Cart	OV-MANU A	Methyl Isocyanate	300	64	25	20	66	36.0	42.2	44.0	49.8	BT@1,5,10&32.8ppm	30
Moyer	Cart	OV-MANU B	Methyl Isocyanate	536	64	25	20	100	14.5	28.8	37.6	46.3	BT@1,5,10&33.2ppm	30
Moyer	Cart	OV-MANU A	Methyl Isocyanate	556	64	25	20	65	20.0	23.0	24.7	28.0	BT@1,5,10&36.7ppm	30
Moyer	Cart	OV/AG-MANU A	Methyl Isocyanate	693	64	25	20		78.0	80.3	83.0		BT@3.3,6.8&10.2ppm	30
Moyer	Cart	OV-MANU A	Methyl Isocyanate	742	64	25	20	63	16.2	19.0	19.1	23.5	BT@1,5,10&16.7ppm	30
Moyer	Cart	OV-MANU A	Methyl Isocyanate	742	64	25	50	66	15.7	20.0	22.2	27.2	BT@1,5,10&38.3ppm	30
Moyer	Cart	OV-MANU B	Methyl Isocyanate	744	64	25	20	97	7.1	14.4	21.5	35.0	BT@1,5,10&27.4ppm	30
Moyer	Cart	OV-MANU B	Methyl Isocyanate	780	64	25	50	66	7.2	23.3	31.1	37.6	BT@1,5,10&27.8ppm	30
Moyer	Cart	OV-MANU B	Methyl Isocyanate	790	64	25	73	91	3.5	7.5	23.3	54.3	BT@0.5,1,2&4ppm	30
Moyer	Cart	OV-MANU A	Methyl Isocyanate	802	64	25	73	93	16.8	22.1	29.9	38.6	BT@1,2,4&8.3ppm	30
Moyer	Cart	OV/AG-MANU A	Methyl Isocyanate	819	64	25	20		72.0	78.0	82.0	86.0	BGT@2,4,6&8.4ppm	30
Moyer	Cart	OV/AG-MANU B	Methyl Isocyanate	841	64	25	20		42.0	44.9	69.2	90.0	BT@0.7,1,2&4.2ppm	30
Moyer	Cart	OV/AG-MANU B	Methyl Isocyanate	861	64	25	20		54.4	63.4	69.9	90.0	BT@0.2,1,2,&5.7ppm	30
Moyer	Cart	OV/AG-MANU A	Methyl Isocyanate	868	64	25	73		1.1	45.1	59.7	70.0	BT@0.7,2,5&8.9ppm	30
Moyer	Cart	OV/AG-MANU B	Methyl Isocyanate	870	64	25	73		0.6	13.9	33.5	56.3	BT@1.2,2.4,&6.0ppm	30
Moyer	Cart	OV-MANU B	Methyl Isocyanate	964	64	25	20	96	5.2	12.6	20.4	30.1	BT@1,5,10&27.4ppm	30
Moyer	Cart	OV-MANU A	Methyl Isocyanate	1004	64	25	20	64	13.4	15.1	16.1	17.7	BT@1,5,10&27.4ppm	30

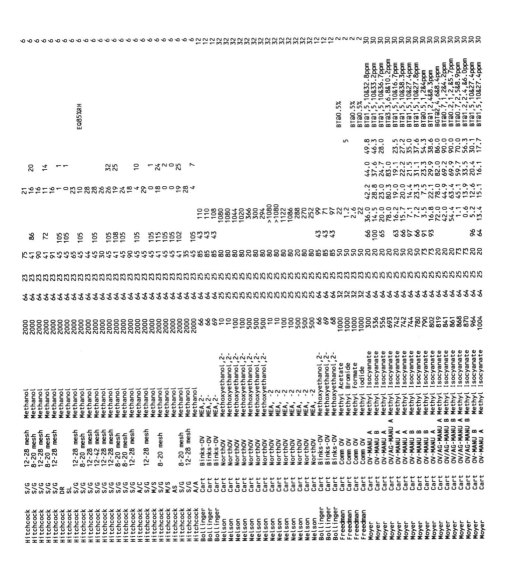

Moyer	Cart	OV/AG-MANU B	Methyl Isocyanate	965	64	25	73		0.7	1.5	24.0	52.1	BT@0.4,1.5,4,&6.8ppm	30
Moyer	Cart	OV/AG-MANU A	Methyl Isocyanate	998	64	25	73		1.5	1.9	51.3	59.7	BT@0.4,2,5&7.1ppm	30
Bollinger	Cart	Binks-OV	Methylene Chloride	15	64	25	85	43	39					12
Bollinger	Cart	Binks-OV	Methylene Chloride	16	64	25	85	43	40					12
Bollinger	Cart	Binks-OV	Methylene Chloride	14	64	25	85	43	43					12
Moyer	Cart	MANU-A	Methylene Chloride	50	64	25	50	72	58.2	69.9	78.6		BT@510&15ppm	31
Moyer	Cart	MANU-A	Methylene Chloride	50	64	25	50	72	65.8	78.6	89.1		BT@510&15ppm	31
Moyer	Cart	MANU-A	Methylene Chloride	50	64	25	80	69	36.3	43.6	49.4		BT@510&15ppm	31
Moyer	Cart	MANU-A	Methylene Chloride	50	64	25	80	69	33.5	39.4	44.9		BT@510&15ppm	31
Moyer	Cart	MSA-OV	Methylene Chloride	113	32	25	50	35	46.3	57.0	62.7	66.4	BT@1%,5,10&15ppm	31
Moyer	Cart	MSA-OV	Methylene Chloride	123	32	25	50	35	49.0	58.6	64.6	68.5	BT@1%,5,10&15ppm	31
Moyer	Cart	MSA-OV	Methylene Chloride	127	32	25	80	36	32.7	40.5	4.4	46.5	BT@1%,5,10&15ppm	31
Moyer	Cart	MSA-OV	Methylene Chloride	114	32	25	80	34	34.8	41.2	45.6	48.0	BT@1%,5,10&15ppm	31
Moyer	Cart	OV-MANU A	Methylene Chloride	125	64	25	50	72	36.0	46.2	52.3	57.5	BT@1%,5,10&15ppm	31
Moyer	Cart	OV-MANU A	Methylene Chloride	127	64	25	50	70	35.0	44.8	50.6	5.6	BT@1%,5,10&15ppm	31
Moyer	Cart	OV-MANU A	Methylene Chloride	127	64	25	50	71	36.5	45.7	52.9	58.1	BT@1%,5,10&15ppm	31
Moyer	Cart	OV-MANU A	Methylene Chloride	127	64	25	50	69	38.0	47.5	35.5	58.1	BT@1%,5,10&15ppm	31
Moyer	Cart	OV-MANU A	Methylene Chloride	125	64	25	80	70	25.5	33.3	37.7	41.4	BT@1%,5,10&15ppm	31
Moyer	Cart	OV-MANU A	Methylene Chloride	125	64	25	80	71	25.0	33.0	37.7	41.6	BT@1%,5,10&15ppm	31
Moyer	Cart	OV-MANU A	Methylene Chloride	125	64	25	80	70	27.5	34.8	39.4	43.3	BT@1%,5,10&15ppm	31
Moyer	Cart	OV-MANU A	Methylene Chloride	123	64	25	80	71	27.0	35.3	40.3	42.8	BT@1%,5,10&15ppm	31
Moyer	Cart	MSA-OV	Methylene Chloride	271	32	25	50	34	34.0	36.0	39.3	41.3	BT@1%,5,10&15ppm	31
Moyer	Cart	MSA-OV	Methylene Chloride	242	32	25	50	35	40.0	42.4	46.2	48.1	BT@1%,5,10&15ppm	31
Moyer	Cart	MSA-OV	Methylene Chloride	242	32	25	80	35	30.9	33.5	36.4	38.0	BT@1%,5,10&15ppm	31
Moyer	Cart	MSA-OV	Methylene Chloride	260	32	25	80	33	26.6	28.5	31.2	32.5	BT@1%,5,10&15ppm	31
Moyer	Cart	OV-MANU A	Methylene Chloride	259	64	25	50	72	28.3	31.5	35.4	37.6	BT@1%,5,10&15ppm	31
Moyer	Cart	OV-MANU A	Methylene Chloride	259	64	25	50	70	32.0	35.0	38.9	41.1	BT@1%,5,10&15ppm	31
Moyer	Cart	OV-MANU A	Methylene Chloride	259	64	25	50	70	30.6	33.5	37.3	39.4	BT@1%,5,10&15ppm	31
Moyer	Cart	OV-MANU A	Methylene Chloride	258	64	25	50	71	32.1	35.4	39.3	41.5	BT@1%,5,10&15ppm	31
Moyer	Cart	OV-MANU A	Methylene Chloride	262	64	25	80	71	24.0	26.7	30.1	32.0	BT@1%,5,10&15ppm	31
Moyer	Cart	OV-MANU A	Methylene Chloride	263	64	25	80	71	23.0	25.5	28.6	30.4	BT@1%,5,10&15ppm	31
Moyer	Cart	OV-MANU A	Methylene Chloride	263	64	25	80	71	22.8	25.3	28.2	29.9	BT@1%,5,10&15ppm	31
Moyer	Cart	OV-MANU A	Methylene Chloride	258	64	25	80	71	21.6	24.2	27.1	28.9	BT@1%,5,10&15ppm	31
Moyer	Cart	MSA-OV	Methylene Chloride	493	32	25	80	35	24.4	26.2	27.2		BT@5,10&15ppm	31
Moyer	Cart	OV-MANU A	Methylene Chloride	499	64	25	50	70	24.9	27.1	28.6		BT@5,10&15ppm	31
Moyer	Cart	OV-MANU A	Methylene Chloride	504	64	25	80	71	15.6	17.5	18.8		BT@5,10&15ppm	31
Moyer	Cart	OV-MANU A	Methylene Chloride	507	64	25	50	71	24.4	26.6	28.2		BT@5,10&15ppm	31
Moyer	Cart	OV-MANU A	Methylene Chloride	507	64	25	80	71	17.6	19.4	20.6		BT@5,10&15ppm	31
Moyer	Cart	OV-MANU A	Methylene Chloride	508	64	25	80	71	14.7	16.5	17.6		BT@5,10&15ppm	31
Moyer	Cart	OV-MANU A	Methylene Chloride	509	64	25	80	70	18.1	20.2	21.3		BT@5,10&15ppm	31
Moyer	Cart	OV-MANU A	Methylene Chloride	510	64	25	50	72	24.0	26.0	27.4		BT@5,10&15ppm	31
Moyer	Cart	OV-MANU A	Methylene Chloride	510	64	25	50	71	25.5	27.8	29.3		BT@5,10&15ppm	31
Moyer	Cart	MSA-OV	Methylene Chloride	523	32	25	50	35	31.0	32.9	34.2		BT@5,10&15ppm	31
Moyer	Cart	MSA-OV	Methylene Chloride	524	32	25	80	34	24.3	26.3	27.2		BT@5,10&15ppm	31
Moyer	Cart	MSA-OV	Methylene Chloride	544	32	25	50	34	36.7	39.6	41.7		BT@5,10&15ppm	31
Moyer	Cart	MSA-OV	Methylene Chloride	1042	32	25	50	35	21.9	23.4	24.4		BT@5,10&15ppm	31
Moyer	Cart	MSA-OV	Methylene Chloride	1000	32	25	50	34	21.1	22.9	24.0		BT@5,10&15ppm	31
Moyer	Cart	MSA-OV	Methylene Chloride	1036	32	25	80	35	18.5	19.9	20.8		BT@5,10&15ppm	31
Moyer	Cart	MSA-OV	Methylene Chloride	1045	32	25	80	36	17.6	19.0	19.8		BT@5,10&15ppm	31
Moyer	Cart	OV-MANU A	Methylene Chloride	985	64	25	50	69	11.6	13.2	13.9		BT@5,10&15ppm	31
Moyer	Cart	OV-MANU A	Methylene Chloride	1012	64	25	50	70	12.2	13.6	14.1		BT@5,10&15ppm	31
Moyer	Cart	OV-MANU A	Methylene Chloride	1026	64	25	50	71	12.2	13.6	14.3		BT@5,10&15ppm	31
Moyer	Cart	OV-MANU A	Methylene Chloride	1026	64	25	50	70	11.1	12.2	12.9		BT@5,10&15ppm	31
Moyer	Cart	OV-MANU A	Methylene Chloride	1030	64	25	50	72	13.3	14.6	15.4		BT@5,10&15ppm	31
Moyer	Cart	OV-MANU A	Methylene Chloride	1034	64	25	50	70	10.6	11.8	12.5		BT@5,10&15ppm	31
Moyer	Cart	OV-MANU A	Methylene Chloride	1002	64	25	80	70	11.0	12.3	13.1		BT@5,10&15ppm	31
Moyer	Cart	OV-MANU A	Methylene Chloride	1026	64	25	80	70	11.1	12.2	12.9		BT@5,10&15ppm	31
Moyer	Cart	OV-MANU A	Methylene Chloride	1034	64	25	80	70	10.6	11.8	12.5		BT@5,10&15ppm	31
Moyer	Cart	OV-MANU A	Methylene Chloride	1038	64	25	80	70	9.8	11.0	11.7		BT@5,10&15ppm	31
Moyer	Cart	OV-MANU A	Methylene Chloride	1081	64	25	80	72	10.5	11.6	12.3		BT@5,10&15ppm	31
Moyer	ChCan	MSA-OV	Methylene Chloride	126	64	25	50	196	182	227	254	268	BT@1%,5,10&15ppm	31
Moyer	ChCan	MSA-OV	Methylene Chloride	129	64	25	50	188	147	186	202	212	BT@1%,5,10&15ppm	31
Moyer	ChCan	MSA-OV	Methylene Chloride	127	64	25	80	197	82	100	109	116	BT@1%,5,10&15ppm	31
Moyer	ChCan	MSA-OV	Methylene Chloride	128	64	25	80	191	87	105	114	121	BT@1%,5,10&15ppm	31
Moyer	ChCan	MSA-OV	Methylene Chloride	257	64	25	50	197	131	143	154	164	BT@1%,5,10&15ppm	31
Moyer	ChCan	MSA-OV	Methylene Chloride	264	64	25	50	195	137	149	159	168	BT@1%,5,10&15ppm	31

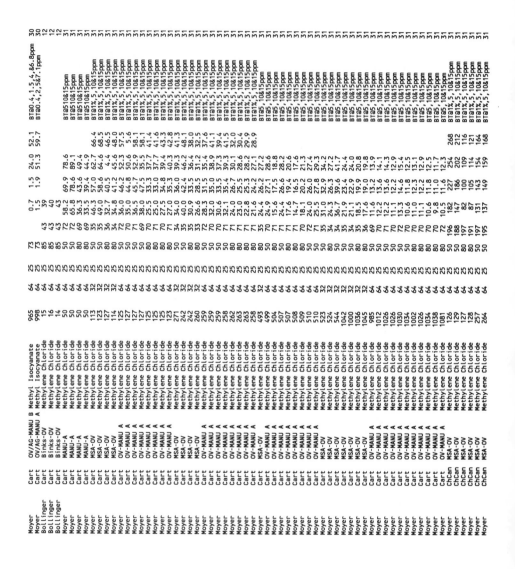

Moyer	ChCan	MSA-OV	Methylene Chloride	268	64	25	80	198	70.5	78.1	85.6	89.4	BT@1%,5,10&15ppm	31
Moyer	ChCan	MSA-OV	Methylene Chloride	272	64	25	80	189	64.5	70.6	77.6	81.4	BT@1%,5,10&15ppm	31
Moyer	ChCan	MSA-OV	Methylene Chloride	999	64	25	50	197	77.1	82.1	84.3		BT@5,10&15ppm	31
Moyer	ChCan	MSA-OV	Methylene Chloride	1000	64	25	50	193	71.4	76.6	79.1		BT@5,10&15ppm	31
Moyer	ChCan	MSA-OV	Methylene Chloride	1029	64	25	80	196	48.4	51.5	53.9		BT@5,10&15ppm	31
Moyer	ChCan	MSA-OV	Methylene Chloride	1039	64	25	80	195	44.4	47.2	49.3		BT@5,10&15ppm	31
Moyer	Can	MSA-OV	Methylene Chloride	123	64	25	50	754	663	744	768		BT@1%,5,10&15ppm	31
Moyer	Can	MSA-OV	Methylene Chloride	129	64	25	50	746	579	648	675		BT@1%,5,10&15ppm	31
Moyer	Can	MSA-OV	Methylene Chloride	120	64	25	80	742	538	634	686		BT@1%,5,10&15ppm	31
Moyer	Can	MSA-OV	Methylene Chloride	125	64	25	80	745	590	679	711		BT@1%,5,10&15ppm	31
Moyer	Can	MSA-OV	Methylene Chloride	495	64	25	50	749	511	542	554		BT@5,10&15ppm	31
Moyer	Can	MSA-OV	Methylene Chloride	483	64	25	80	747	398	413	422		BT@5,10&15ppm	31
Moyer	Can	MSA-OV	Methylene Chloride	487	64	25	80	740	392	407	419		BT@5,10&15ppm	31
Moyer	Can	MSA-OV	Methylene Chloride	1052	64	25	50	738	416	427	433		BT@5,10&15ppm	31
Moyer	Can	MSA-OV	Methylene Chloride	1085	64	25	50	745	406	416	420		BT@5,10&15ppm	31
Moyer	Can	MSA-OV	Methylene Chloride	1004	64	25	80	797	317	327	334		BT@5,10&15ppm	31
Moyer	Can	MSA-OV	Methylene Chloride	1015	64	25	80	745	295	305	315		BT@5,10&15ppm	31
Moyer	SSCan	MSA-OV	Methylene Chloride	272	64	25	50	725	781	795	811	820	multiday run	31
Moyer	SSCan	MSA-OV	Methylene Chloride	255	64	25	80	731	386	411	441	464	BT@1%,5,10&15ppm	31
Moyer	SSCan	MSA-OV	Methylene Chloride	259	64	25	80	734	443	466	487	503	BT@1%,5,10&15ppm	31
Moyer	SSCan	MSA-OV	Methylene Chloride	270	64	25	80	734	473	492	512	523	BT@1%,5,10&15ppm	31
Stampfer	Can	MSA-OV	NDMA	1.0	64	24	<15		2280					7
Stampfer	Can	MSA-OV	NDMA	1.0	64	26	80		1560					7
Cohen	RCT	Willson	Nitroglycerine	0.064	1.5	36	61		234				No BT observed	34
Cohen	RCT	Willson	Nitroglycerine	0.064	2	28	78		858				No BT observed	34
Cohen	RCT	Racal	Nitroglycerine	0.72	3	26	90		774				No BT observed	34
Cohen	RCT	MSA	Nitroglycerine	1.3	3	35	61		474				No BT observed	34
Cohen	RCT	Racal	Nitroglycerine	1.37	3	32	73		222				No BT observed	34
Cohen	RCT	MSA	Nitroglycerine	24.6	3	34	59		246				No BT observed	34
Cohen	RCT	MSA	Nitroglycerine	24.6	2	34	59		276				No BT observed	34
Swearengen	Cart	3M OV	Nitroglycerine	0.01	25	27	90		275				No BT observed	33
Swearengen	Cart	3M OV	Nitroglycerine	0.12	25	27	90		480				No BT observed	33
Swearengen	Cart	3M OV	Nitroglycerine	0.13	25	27	90		480				No BT observed	33
Swearengen	Cart	3M OV	Nitroglycerine	0.13	25	27	90		504				No BT observed	33
Swearengen	Cart	3M OV	Nitroglycerine	0.14	25	27	90		476				No BT observed	33
Swearengen	Cart	3M OV	Nitroglycerine	0.16	25	27	90		480				No BT observed	33
Swearengen	Cart	3M OV	Nitroglycerine	0.19	25	27	90		483				No BT observed	33
Swearengen	Cart	3M OV	Nitroglycerine	0.19	25	27	90		434				No BT observed	33
Swearengen	Cart	3M OV	Nitroglycerine	0.23	25	27	90		270				No BT observed	33
Freedman	Cart	Comm OV	Octane	1000	32	23	50		66				BT @0.5%	2
Freedman	Cart	Comm OV	Pentane	1000	32	23	50		35				BT @0.5%	2
Moyer	Cart	OV	Pentane	275	32	25	50	47	124	129	132		BT @5,10&15 ppm	50
Moyer	Cart	OV	Pentane	275	32	25	50	93	280	286	290		BT @5,10&15 ppm	50
Moyer	Cart	OV	Pentane	275	32	25	50	140	446	452	457		BT @5,10&15 ppm	50
Moyer	Cart	OV	Pentane	275	32	25	80	47	85	89	92		BT @5,10&15 ppm	50
Moyer	Cart	OV	Pentane	275	32	25	80	92	204	208	211		BT @5,10&15 ppm	50
Moyer	Cart	OV	Pentane	275	32	25	80	139	316	320	324		BT @5,10&15 ppm	50
Moyer	Cart	OV	Pentane	500	32	25	50	46	82	86	87		BT @5,10&15 ppm	50
Moyer	Cart	OV	Pentane	500	32	25	50	93	194	198	200		BT @5,10&15 ppm	50
Moyer	Cart	OV	Pentane	500	32	25	50	139	314	318	320		BT @5,10&15 ppm	50
Moyer	Cart	OV	Pentane	500	32	25	50	186	427	432	434		BT @5,10&15 ppm	50
Moyer	Cart	OV	Pentane	500	32	25	80	46	54	56	57		BT @5,10&15 ppm	50
Moyer	Cart	OV	Pentane	500	32	25	80	94	124	126	127		BT @5,10&15 ppm	50
Moyer	Cart	OV	Pentane	500	32	25	80	141	192	194	195		BT @5,10&15 ppm	50
Moyer	Cart	OV	Pentane	500	32	25	80	188	260	263	264		BT @5,10&15 ppm	50
Moyer	Cart	OV	Pentane	750	32	25	50	47	52	54	56		BT @5,10&15 ppm	50
Moyer	Cart	OV	Pentane	750	32	25	50	94	121	124	126		BT @5,10&15 ppm	50
Moyer	Cart	OV	Pentane	750	32	25	50	140	203	205	208		BT @5,10&15 ppm	50
Moyer	Cart	OV	Pentane	750	32	25	50	186	282	283	285		BT @5,10&15 ppm	50
Moyer	Cart	OV	Pentane	750	32	25	80	47	46	48	49		BT @5,10&15 ppm	50
Moyer	Cart	OV	Pentane	750	32	25	80	93	108	110	111		BT @5,10&15 ppm	50
Moyer	Cart	OV	Pentane	750	32	25	80	139	171	173	174		BT @5,10&15 ppm	50
Moyer	Cart	OV	Pentane	750	32	25	80	186	232	234	235		BT @5,10&15 ppm	50
Moyer	Cart	OV	Pentane	1000	32	25	50	46	45	46	47		BT @5,10&15 ppm	50
Moyer	Cart	OV	Pentane	1000	32	25	50	92	106	108	108		BT @5,10&15 ppm	50
Moyer	Cart	OV	Pentane	1000	32	25	50	139	170	172	173		BT @5,10&15 ppm	50

Moyer	Cart	OV	Pentane	1000	32	25	50	185	240	242	243		BT @5,10&15 ppm	50
Moyer	Cart	OV	Pentane	1000	32	25	80	47	35	36	37		BT @5,10&15 ppm	50
Moyer	Cart	OV	Pentane	1000	32	25	80	93	84	85	86		BT @5,10&15 ppm	50
Moyer	Cart	OV	Pentane	1000	32	25	80	141	132	134	135		BT @5,10&15 ppm	50
Moyer	Cart	OV	Pentane	1000	32	25	80	188	184	186	187		BT @5,10&15 ppm	50
Moyer	ChCan	OV	Pentane	1000	32	50	50	206	79	81	83		BT @5,10&15 ppm	50
Moyer	ChCan	OV	Pentane	1000	32	50	80	208	74	77	79		BT @5,10&15 ppm	50
Moyer	ChCan	OV	Pentane	2000	32	50	50	203	57	59	60		BT @5,10&15 ppm	50
Moyer	ChCan	OV	Pentane	2000	32	50	80	215	50	52	53		BT @5,10&15 ppm	50
Moyer	ChCan	OV	Pentane	5000	32	50	50	204	33	34	35		BT @5,10&15 ppm	50
Moyer	ChCan	OV	Pentane	5000	32	50	80	217	30	32	33		BT @5,10&15 ppm	50
Moyer	ChCan	OV	Pentane	7500	32	50	50	214	23	24	25		BT @5,10&15 ppm	50
Moyer	ChCan	OV	Pentane	7500	32	50	80	219	21	22	23		BT @5,10&15 ppm	50
Stampfer	Can	MSA-OV	Proparyl	49	64	26	79		1140					7
Stampfer	Can	MSA-OV	Proparyl	50	64	23	<15		1020					7
Freedman	Cart	Comm OV	Propionaldehyde	1000	32	23	50		8				BT @0.5%BT	2
Freedman	Cart	Comm OV	Proponol	1000	32	23	50		44				BT @0.5%BT	2
Smoot	Cart	Custom	Proponol,2-	1000	32	25	0	34	77	87	105	157		3
Smoot	Cart	Custom	Proponol,2-	1000	32	25	50	34	79	92	114	172		3
Smoot	Cart	Custom	Proponol,2-	1000	32	25	80	34	74	83	98	131		3
Freedman	Cart	Comm OV	Propyl Acetate	1000	32	23	50		62				BT @0.5%	2
Freedman	Cart	Comm OV	Propyl Formate	1000	32	23	50		53				BT @0.5%	2
Henry	Cart	MSA-OVH	Tetraethyl Lead	0.05	30	24	50		>9000				t5%, concn in mg/m3	20
Henry	Cart	North-OV	Tetraethyl Lead	0.12	30	24	50		>18000				t5%, concn in mg/m3	20
Henry	Cart	Surv-OV	Tetraethyl Lead	0.16	30	24	50		15840				t5%, concn in mg/m3	20
Henry	Cart	North-OV	Tetraethyl Lead	0.90	30	24	80		>18000				t5%, concn in mg/m3	20
Henry	Cart	MSA-OVH	Tetraethyl Lead	1.10	30	24	80		>18000				t5%, concn in mg/m3	20
Henry	Cart	Surv-OV	Tetraethyl Lead	1.54	30	24	80		>18000				t5%, concn in mg/m3	20
Smoot	Cart	Custom	Trichloroethylene	1000	32	25	50	34	102	114	131	180		3
Smoot	Cart	Custom	Trichloroethylene	1000	32	25	0	34	105	116	132	174		3
Smoot	Cart	Custom	Trichloroethylene	1000	32	25	80	34	98	108	126	256		3
Werner	Column	Custom OV	Trichloroethylene	50	7.7	23	5	38		100	100			10
Werner	Col	Custom OV	Trichloroethylene	50	7.7	23	25	38		67	90			10
Werner	Col	Custom OV	Trichloroethylene	50	7.7	23	50	38		58	63			10
Werner	Col	Custom OV	Trichloroethylene	50	7.7	23	65	38		32	40			10
Werner	Col	Custom OV	Trichloroethylene	50	7.7	23	85	38		9	9			10
Werner	Col	Custom OV	Trichloroethylene	100	7.7	23	5	38		100	100			10
Werner	Col	Custom OV	Trichloroethylene	100	7.7	23	25	38		82	96			10
Werner	Col	Custom OV	Trichloroethylene	100	7.7	23	50	38		77	85			10
Werner	Col	Custom OV	Trichloroethylene	100	7.7	23	65	38		38	48			10
Werner	Col	Custom OV	Trichloroethylene	100	7.7	23	85	38		15	16			10
Werner	Col	Custom OV	Trichloroethylene	170	7.7	23	5	38		100	100			10
Werner	Col	Custom OV	Trichloroethylene	170	7.7	23	25	38		93	101			10
Werner	Col	Custom OV	Trichloroethylene	170	7.7	23	50	38		80	87			10
Werner	Col	Custom OV	Trichloroethylene	170	7.7	23	65	38		48	55			10
Werner	Col	Custom OV	Trichloroethylene	170	7.7	23	85	38		24	25			10
Werner	Col	Custom OV	Trichloroethylene	225	7.7	23	5	38		100	100			10
Werner	Col	Custom OV	Trichloroethylene	225	7.7	23	25	38		87	99			10
Werner	Col	Custom OV	Trichloroethylene	225	7.7	23	50	38		83	85			10
Werner	Col	Custom OV	Trichloroethylene	225	7.7	23	65	38		51	61			10
Werner	Col	Custom OV	Trichloroethylene	225	7.7	23	85	38		28	27	28		10
Freedman	Cart	Comm OV	Valeraldehyde	1000	32	23	50		57				BT@0.5%	2
Pers Comm	Cart	MSA-OV	Vinyl Bromide	20	55	25	55		240				BT@0.5ppm, 55%EQ	29
Pers Comm	Cart	MSA-OV	Vinyl Bromide	100	55	25	55		125				BT@0.5ppm, 55%EQ	29
Pers Comm	Cart	MSA-OV	Vinyl Bromide	1000	55	25	55		31				BT@0.5ppm, 55%EQ	29
Miller	Cart	Brand G-OV	Vinyl Chloride	9	64	25	70	41			3			15
Miller	Cart	Brand A-OV	Vinyl Chloride	10	64	25	70	41			47			15
Miller	Cart	Brand D-OV	Vinyl Chloride	10	64	25	70	36			29			15
Miller	Cart	Brand B-OV	Vinyl Chloride	10	64	25	70	35			17			15
Miller	Cart	Brand G-OV	Vinyl Chloride	11	64	25	70	41			2		EQ25%RH	15
Miller	Cart	Brand F-OV	Vinyl Chloride	11	64	25	70	30			26		EQ60%RH	15
Miller	Cart	Brand F-OV	Vinyl Chloride	11	64	25	70	30			27		EQ25%RH	15
Miller	Cart	Brand A-OV	Vinyl Chloride	11	64	25	70	41			33		EQ60%RH	15
Miller	Cart	Brand E-OV	Vinyl Chloride	11	64	25	70	69			34		EQ25%RH	15
Miller	Cart	Brand F-OV	Vinyl Chloride	11	64	25	70	30			33			15
Miller	Cart	Brand E-OV	Vinyl Chloride	11	64	25	70	69			34			15

Miller	Cart	Brand H-OV	Vinyl Chloride	11	64	25	70	47			40	EQ60%RH	15
Miller	Cart	Brand B-OV	Vinyl Chloride	12	64	25	70	35			9	EQ60%RH	15
Miller	Cart	Brand H-OV	Vinyl Chloride	12	64	25	70	47			45		15
Miller	Cart	Brand F-OV	Vinyl Chloride	12	64	25	70	69			26	EQ60%RH	15
Miller	Cart	Brand D-OV	Vinyl Chloride	12	64	25	70	36			21	EQ60%RH	15
Miller	Cart	Brand C-OV	Vinyl Chloride	14	64	25	70	41			43		15
Smith	Cart	Manu #4-OV	Vinyl Chloride	50	64	25	85	36	33	48			14
Smith	Cart	Manu #1-OV	Vinyl Chloride	50	64	25	85	50	54	74			14
Smith	Cart	Manu #2-OV	Vinyl Chloride	50	64	25	85	36	28	44			14
Smith	Cart	Manu #3-OV	Vinyl Chloride	50	64	25	85	35	29	48			14
Smith	Cart	Manu #3-OV	Vinyl Chloride	250	64	25	85	35	21	29	41		14
Smith	Cart	Manu #4-OV	Vinyl Chloride	250	64	25	85	36	21	28	39		14
Smith	Cart	Manu #2-OV	Vinyl Chloride	250	64	25	85	36	20	30	43		14
Smith	Cart	Manu #1-OV	Vinyl Chloride	250	64	25	85	50	42	51	67		14
Smith	Cart	Manu #3-OV	Vinyl Chloride	500	64	25	85	35	11	19	27		14
Smith	Cart	Manu #1-OV	Vinyl Chloride	500	64	25	85	50	23	33	43		14
Smith	Cart	Manu #2-OV	Vinyl Chloride	500	64	25	85	36	12	21	30		14
Smith	Cart	Manu #4-OV	Vinyl Chloride	500	64	25	85	36	13	11	27		14
Smith	Cart	Manu #1-OV	Vinyl Chloride	600	64	25	85	50		20			14
Smith	Cart	Manu #1-OV	Vinyl Chloride	600	64	25	85	50		15			14
Smith	Cart	Manu #1-OV	Vinyl Chloride	600	64	25	85	50		18			14
Smith	Cart	Manu #2-OV	Vinyl Chloride	750	64	25	85	36	9	17	24		14
Smith	Cart	Manu #1-OV	Vinyl Chloride	750	64	25	85	50	17	25	33		14
Smith	Cart	Manu #3-OV	Vinyl Chloride	750	64	25	85	35	7	13	20		14
Smith	Cart	Manu #4-OV	Vinyl Chloride	750	64	25	85	36	11	17	22		14
Smith	Can	Manu #1-OV	Vinyl Chloride	100	64	25	85	437	312	356			14
Smith	Can	Manu #4-OV	Vinyl Chloride	100	64	25	85	312	170	209			14
Smith	Can	Manu #2-OV	Vinyl Chloride	100	64	25	85	318	219	284			14
Smith	Can	Manu #3-OV	Vinyl Chloride	100	64	25	85	464	217	277			14
Smith	Can	Manu #2-OV	Vinyl Chloride	1000	64	25	85	318	96	111			14
Smith	Can	Manu #1-OV	Vinyl Chloride	1000	64	25	85	437	144	158			14
Smith	Can	Manu #4-OV	Vinyl Chloride	1000	64	25	85	312	72	84			14
Smith	Can	Manu #3-OV	Vinyl Chloride	1000	64	25	85	464	91	108			14
Smith	Can	Manu #3-OV	Vinyl Chloride	5000	64	25	85	464	31	43			14
Smith	Can	Manu #4-OV	Vinyl Chloride	5000	64	25	85	312	26	33			14
Smith	Can	Manu #2-OV	Vinyl Chloride	5000	64	25	85	318	33	43			14
Smith	Can	Manu #1-OV	Vinyl Chloride	5000	64	25	85	437	55	53			14

Note:

Carbon Tet: carbon tetrachloride
CMME: chlorometyhyl methyl eher
DMMP: dimethylmethyl phosphonate
2-EEA: 2-Ethoxyethanol Acetate
2-MEA: 2-Methoxyethanol Acetate
NDMA: N-nitrososodiummethylamine
Pers Comm: personal communication

A/A: activated alumina
A/C: activated carbon
AG: acid gas
AS: Ascarite
BT: breakthrough time
Can: canister
Cart: cartridge
ChCan: chin cansiter
Col: column
Comm: commercial

DR: Drierite
EQ: equlibration
Manu: manufacturer
M/S: molecular sieve
OV: organic vapor
RCT: RC tubes
RH: relative humidity
S/G: silica gel
SL: sodalime
SS: supersize

Henry[5] also conducted a study to determine the service life of chemical cartridges for protection against formaldehyde. Organic vapor, acid gases, and the combination of organic vapor and acid gases cartridges were selected for testing. The test concentrations chosen were 0.25, 0.5, and 1 time the old OSHA permissible exposure limit for formaldehyde, which was 3 ppm. The flow rate was maintained at 60 l/m, and the temperature and relative humidity was set at 23°C and 50%, respectively. Breakthrough values of 10, 50, and 100% were reported.

The test results are listed in Table 2. The results show that the breakthrough time for organic vapor cartridges are less than those determined for acid gases cartridges, or the combination of acid gases and organic vapor cartridges. It is suggested that the metal oxide catalyst in the acid gases cartridges reacts with formaldehyde and oxidizes it to formate. Another observation is that unlike organic vapor cartridges, the breakthrough time for acid gases cartridges increases with an increase of humidity. The author explains that the reaction of formaldehyde produces a hydrate that eventually reaches equilibrium with formaldehyde vapor before breakthrough occurs.

The breakthrough studies conducted by Freedman and Nelson indicated that activated carbon is a poor sorbent for methanol. Hitchcock et al.[6] made an investigation on other sorbent materials for testing against methanol. In addition to carbon, Ascarite, soda line, Drierite, activated alumina, molecular sieve and silica gel were selected for evaluation. The test concentration was set at 2000 ppm with a temperature of 23 ± 1°C and a relative humidity of 45 ± 3%. The flow rate was set at 64 l/m. Based on the preliminary test results, silica gel was chosen for further study. The second series of study included variations of the size of silica gel, sorbent weight, challenge concentration, and relative humidity. The results of various test conditions are shown in Table 2.

The breakthrough time at 10% of the challenge concentration was reported. The authors conclude that small particle diameter (#12 to #42 Tyler mesh) increases the sorbent service life for methanol without introducing higher breathing resistance. Humidity affects the service life of sorbents because water molecules compete with methanol molecules for available adsorption sites. Even at very humid conditions, silica gels would provide longer service life than active carbon when the carbon is tested at a relatively dry state.

The service life of organic vapor canisters against highly toxic or carcinogenic compounds was investigated by Stampfer of the Los Alamos National Laboratory.[7] Organic vapor canisters were selected for testing against chloroform, benzene, epichlorohydrin, acrylonitrile, propargyl alcohol, 1,2-dibromethane, acrolein, chloromethyl methyl ether (CMME) and N-nitrosodimethylamine (NDMA). The challenge concentration was set at 50 times the assigned TLV except for CMME and NDMA, for which 1 ppm challenge was selected. The tests were performed at ambient temperature with relative humidities set at 15 and 80%. The flow rate was set at 64 l/m. Since Los Alamos has an altitude of 7000 ft, the flow rate was set at 49 l/m when corrected to standard conditions. Six replicate tests were run under each test condition. The duration of the test was set at 16 h or 1% breakthrough, whichever came first.

The test results are tabulated in Table 2. Except for CMME, the service life is at least 4 h for the chemicals tested. The short breakthrough time for CMME suggests that the canister is not safe for protection against CMME. Humidity does have an appreciable effect on the reduction of service life. The effect of humidity on canister service life, however, may be greater than those reported in the literature.

Ackley[8] conducted a study to determine the service life of various types of Scott brand organic cartridges for protection against 1,3-butadiene. The air flow rate was set at 32 and 64 l/m. The challenge concentrations were set at 100 ppm and 1000 ppm. The temperature was set mostly at 25°C and the humidity varied from 0 to 85%. Unlike other studies, he preconditioned several cartridges at 0 and 85% relative humidity. The results of various test conditions are shown in Table 2.

The results indicate that coconut and coal based carbon performed slightly better than petroleum based carbon. The acid gases or combination of organic vapor and acid gases cartridges showed lower service life than the organic vapor cartridges. High moisture conditions reduced cartridge service life for 1,3-butadiene, especially when the cartridges were equilibrated at high humidity. The elevated temperature also reduced the service life of cartridges. The cartridge service life is inversely proportional to the flow rate at a given concentration. There is a good agreement between the test data and the residence time model proposed by the author.

Simon et al.[9] performed a study to determine the service life of chemical cartridges and canisters against chlorine dioxide. A combination of organic vapor, acid gases, organic vapor/acid gases, acid gases/high-efficiency filter cartridges and chin canisters, and an escape mouthpiece acid gas cartridge were selected in this study. The challenge gas contained 450 to 550 ppm of ClO_2 and 10 ppm Cl_2. The air flow was set at 64 l/m. The temperature and relative humidity was set at 25°C and 50%, respectively. Some cartridges were equilibrated at 25°C, 60 to 80% relative humidity and tested at an air flow of 58 l/m for 6 h. A total of 67 tests were performed using 14 different types of cartridges. In 33 of the tests, the cartridges were tested for a minimum of 35 min. For the remaining tests, the cartridges were tested until breakthrough occurs. The breakthrough was reported as the time to reach 0.06 ppm (0.012%) and 0.46 ppm (0.09%) of ClO_2 for as received and equilibrated cartridges and canisters.

The test results are listed in Table 2. It shows that all cartridges are effective for protection against ClO_2 except for the escape cartridge. Based on the removal effectiveness for ClO_2, the acid gases cartridges are more effective than the combination of acid gases and organic vapor cartridges, which in turn are more effective than organic vapor cartridges. The ClO_2 removal capacity is proportional to the weight of sorbent present in the cartridge. Equilibrated cartridges showed a wide range of capacity differences compared to the cartridges tested at *as received* conditions. From a statistical analysis of the data, equilibration does not have a significant effect on the average cartridge capacity.

Werner conducted a study to determine the effect of moisture on the service life of carbon against trichloroethylene (TCE).[10] He used a custom-made column filled with activated carbon. The flow rate was set at 7.7 l/m, and the temperature was set at 23°C. The challenge concentration of TCE were 50, 100, 165, and 220 ppm. The range of relative humidity was set between 5 and 85%. The times at which 10 and 50% breakthrough were measured under these test conditions are presented in Table 2.

In his experimental work to evaluate the Jonas Kinetic model for determining performance of organic vapor cartridges, Moyer selected acetone as the test solvent to generate breakthrough data.[11] The acetone challenge concentration varied between 530 and 1060 ppm. The test results of 1% breakthrough time are shown in Table 2.

In 1987, OSHA requested NIOSH to conduct a study to determine the organic vapor cartridges efficiency for protection against these suspected carcinogens: benzene, methylene chloride, 1,3-butadiene, 2-methyloxyethanol, 2-methyloxyethanol acetate, 2-ethoxyethanol, and 2-ethoxyethanol acetate. In addition, acid gases cartridges were selected to determine the service life against formaldehyde.[12] All cartridges were tested in the "as received" condition. The other testing conditions were: 64 l/m air flow, 25°C, and 85% relative humidity. The challenge concentrations were: 15 ppm for benzene, formaldehyde, and methylene chloride; and 75 ppm for the rest of the chemicals. The duration of the test was set at 8 h or 1% breakthrough, whichever came first. NIOSH used the Binks organic vapor and the MSA acid gases cartridges for testing. The test results are listed in Table 2.

When OSHA was promulgating the standard on ethylene bromide (EDB) in the early 1980s, questions were raised concerning the effectiveness of respirator cartridges and canisters for protection against EDB. Since there was no sorbent service life information available, OSHA contracted Gary Nelson to conduct a study to determine the service life of cartridges and canisters for protection against EDB.[13] Petroleum-based organic vapor cartridges made by Norton, coconut based organic vapor cartridges made by MSA, and coal-based organic vapor canisters made by MSA were selected for testing. The devices were tested in the "as received" condition or preconditioned at 85% relative humidity. In the first series of tests, all tests were conducted at 25°C and 85% relative humidity. Equilibrated devices were tested at 32 l/m while the "as received" cartridges were tested at 64 l/m. The cartridges were equilibrated at 8 l/m, 25°C, and 85% relative humidity for 16 h. Since no breakthrough (i.e., the effluent concentration for EDB was less than the detection limit of the analytical method) was observed after 480 min at a challenge concentration of 5 ppm for the cartridges, a second series of tests were conducted under same test conditions but at higher challenge concentrations. All tests were performed in the "as received" condition. The EDB challenge concentration for cartridges varied from 35 to 200 ppm and the challenge concentrations for the canisters varied between 100 and 2000 ppm. The breakthrough concentrations were measured at 50, 100, and 150 parts per billions (ppb). The test results are listed in Table 2.

When the question on the carcinogenicity of vinyl chloride was raised during the OSHA hearing on the vinyl chloride standard, the only information available was the breakthrough data generated by Nelson at a challenge concentration of 1000 ppm. Since a much lower permissible exposure limit would be promulgated by OSHA, breakthrough data for vinyl chloride was needed. NIOSH conducted a study to determine the service life of organic vapor cartridges and canisters against vinyl chloride.[14] Four types of approved organic vapor cartridges and front or back mounted canisters were selected for testing. A series of preliminary testing was conducted to determine the cartridge service life under different humidities. The cartridges were later tested at 30 l/m, 24°C, and 50% relative humidity. The challenge vinyl chloride concentration were 50, 200, 500, and 750 ppm. All canisters were tested at 60 l/m, 24°C, and 50% relative humidity. The challenge concentrations were 100, 1000, and 5000 ppm vinyl chloride. Breakthrough time was reported as initial, 10%, and 50% penetration. The results were presented in Table 2.

Miller et al. conducted a study to determine the service life of organic vapor cartridges for protection against vinyl chloride.[15] An assortment of organic vapor cartridges from eight manufacturers were tested. Some cartridges were preconditioned at 25 l/m, 25, or 60% relative humidity for 6 h. All conditioned cartridges were tested within 12 h. All cartridges were tested at 25°C and 70% relative humidity. The flow rate was set at 64 or 32 l/m for dual or single cartridge configuration, respectively. A challenge concentration of 10 ppm was selected and the breakthrough concentration at the 50% or 5 ppm was measured in these tests. Test results are contained in Table 2.

The results of both vinyl chloride cartridge and canister service life studies indicate that the activated carbon has a very low adsorption capacity for vinyl chloride. A cartridge can provide a service life of no more than 40 min at 10 ppm challenge of vinyl chloride. Due to higher adsorbent weight, a "super size" canister can provide a service life of about 2 h at a challenge concentration of 100 ppm vinyl chloride. The OSHA standard on vinyl chloride requires that a respirator provides a minimum service life of 2 h for protection against vinyl chloride. Three devices were approved in accordance with the testing schedule prescribed in 30 CFR 11. Since vinyl chloride manufacturers had implemented engineering controls to comply with the 1 ppm standard for vinyl chloride, all approved vinyl chloride respirators were withdrawn from the market.

Beaumont and Garrido conducted a study to determine the service life of organic vapor cartridges for protection against benzene and acrylonitrile.[16] Organic vapor and a combination of organic vapor and acid gases cartridges from three different manufacturers were selected for testing. The cartridges were equilibrated in a desiccator at 55% relative humidity for 16 h. The test conditions were set at 22°C, 50% relative humidity, and 54 l/m. The challenge concentrations varied from 10 to 100 ppm for benzene and from 22 to 1000 ppm for acrylonitrile. The time to reach 1 ppm breakthrough in hours was measured. The test results are listed in Table 2.

The MSHA/NIOSH respirator testing and certification regulations, 30 CFR 11, do not approve cartridges or canisters for protection against substances with poor odor warning properties, except when equipped with an end-of-service-life indicator (ESLI). A canister for protection against carbon monoxide was the first approved device equipped with an ESLI. However, the indicator is a moisture sensor for the hopcalite sorbent used in the CO canister. It may not be directly related to the service life of the canister. With a petition of the 3M Company, NIOSH developed a test schedule for respirators with an ESLI for protection against vinyl chloride (Subpart N of 30 CFR 11) in 1974. A vinyl chloride disposable respirator equipped with an ESLI made by 3M was approved. The indicator did not seem specific for vinyl chloride; it changed color when the respirator was placed in the office environment. Since most employers were in compliance with the vinyl chloride standard, there was a very limited market for vinyl chloride respirators. This situation prevented 3M for making an improvement to the ESLI.

Some sorbents are very effective against inorganic materials. Mercury is an outstanding example. At a challenge concentration of 21.5 mg/m^3 (215 times of OSHA PEL), a cartridge filled with hopcalite tested at a flow of 64 l/m has a service life of 847 min at a breakthrough concentration of 0.05 mg/m^3. MSA markets the hopcalite cartridges with a trade name of Mersorb.[17] The breakthrough data for the

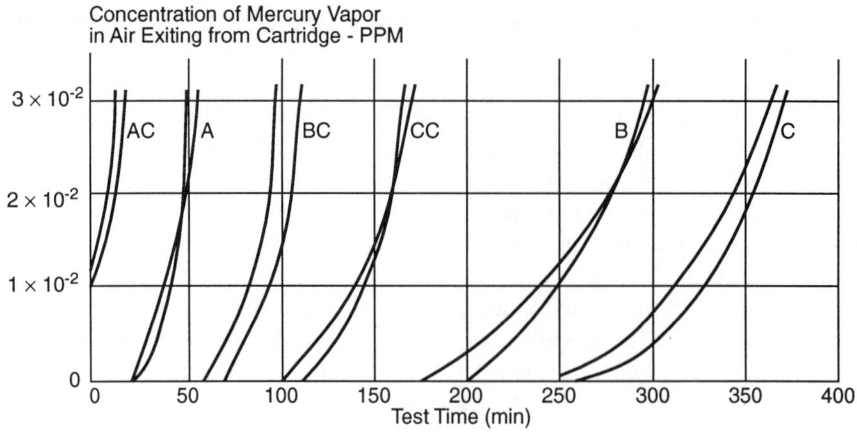

Figure 1 Mercury vapor penetration of cartridges containing activated carbon granules impregnated with iodine, iodine monochloride, and iodine trichloride.

Mersorb cartridges are shown in Table 2. At a challenge concentration of 21.5 mg/m^3, the cartridge provides an average service life of 600 min at a breakthrough concentration of 0.05 mg/m^3 (0.24%). The cartridge is also equipped with a color indicator which change color at 480 min.

Revoir[18] found that activated carbon impregnated with iodine, iodine monochloride (IMC), or iodine trichloride (ITC) would provide even a longer service life for protection against mercury vapor than hopcalite, which is a good sorbent for mercury. At a temperature of 37.8°C, an air flow of 64 l/m, a relative humidity of 50%, and a challenge concentration of 7.5 ppm of mercury, the average service life for hopcalite was 50 min. The iodine impregnated carbon, however, had an average service life of 450 min. Among iodine impregnated carbons, the monochloride impregnated carbon had a 3.5 times longer service life than iodine impregnated carbon, and trichloride impregnated carbon had 5 times more service life than iodine impregnated carbon (see Figure 1).

In 1985, NIOSH developed a test schedule for approving mercury cartridges equipped with an ESLI.[19] The cartridges are tested at both the TLV concentration (0.05 mg/m^3) and the saturation concentration of mercury at 25°C (21.5 mg/m^3 or 420 times the TLV for mercury). Cartridges are tested at both "as received" and "equilibrated" conditions. Other test conditions are: 64 l/m, 25 ± 2.5°C, 25, 50, and 85% relative humidity. The "as received" cartridges are tested at 50% RH. The cartridges equilibrated at 25% RH will be tested at 25% RH. The cartridges equilibrated at 85% RH will be tested at 85% RH. This test regimen will be used to certify all cartridges and canisters equipped with an ESLI. The penetration should not exceed 0.05 mg/m^3 of mercury vapor during the minimum service life of 480 min. The indicator must change color at less than 90% of the actual service life of the cartridge. Four manufacturers have received approvals for a cartridge equipped with an ESLI for mercury.

Henry conducted a study to determine the service life of three different brands of organic vapor cartridges for protection against tetraethyl lead.[20] Cartridges manufactured by MSA, North, and Survivair were selected for testing. The challenge concentrations were 0.075 and 0.75 mg/m^3 respectively, which corresponds to the PEL concentration and 10 times the PEL for tetraethyl lead. The first series of tests were conducted at the PEL concentration, 24°C, and 50% relative humidity. Cartridges were tested in the "as received" condition. The second series of tests were carried out in the "as received" condition, 10 times the PEL of tetraethyl lead and a relative humidity of 80%. Flow rates of 30 and 60 l/m were selected for testing. The upstream and downstream concentrations of tetraethyl lead were analyzed by gas chromatography. The cartridges were tested at the 10 times PEL concentration of tetraethyl lead and a relative humidity of 80% provided a minimum service life of 300 h. The test results are listed in Table 2.

Nelson performed a study to determine the service life of an organic vapor canister for protection against iodine.[21] A Scott combination organic vapor and HEPA filter canister was selected for testing. Canisters were tested in the "as received" and "equilibrated" conditions. The equilibration was carried out at 25% and 85% relative humidity. The test conditions were: 10 ppm iodine, 25°C, 64 l/m for "as received," and 32 l/m for "equilibrated" canisters. The downstream iodine concentration was measured every 30 min. The test results are listed in Table 2. No breakthrough was detected after 120 min.

Canisters and cartridges used for protection against nerve gas or riot control gas such as GA (tabun), GB (sarin), and VX are often tested with stimulants. Cyanogen chloride (CN) and dimethylmethyl phosphonate (DMMP) are used as stimulants. The test conditions for cartridges and canisters are listed in Table 2. The results indicated that cartridges or canisters provide longer service life for protection against DMMP than CK. The TEDA impregnated carbon provided a very long service life against DMMP.

During the OSHA hearing on the proposed rule on 1,3-butadiene, one request asked whether NIOSH would conduct additional respirator cartridge breakthrough testing for 1,3-butadiene. At the request of OSHA, NIOSH conducted a study to determine the service life of organic vapor cartridges against 1,3-butadiene.[22] The test conditions were:

Concentration:	10, 20 and 50 ppm
Temperature:	25°C
Relative Humidity:	50% and 85% (85% for 10 ppm only)
BT Concentration:	2, 3, 5, and 10 ppm

Results of the NIOSH study on 1,3-butadiene are shown in Table 2.

Another approved cartridge with an end-of-service-life indicator for hydrogen sulfide was developed by the Mine Safety Appliance Company.[23] Both acid gas cartridges (GMB) and a combination of organic vapor/acid gas cartridges (GMC) were evaluated. In addition, a super size H_2S canister (GMHS-SSW) was also tested for escape at concentrations up to 10,000 ppm of H_2S. The test conditions were the same as those described for mercury vapor, except that the challenge H_2S concentration was 1000 ppm for the cartridges and up to 10,000 for the canister. The breakthrough concentration was set at 10 ppm for all tests. The test results are shown in Table 2.

The manufacturing of air-purifying respirators is like an automobile assembly line. Except for the facepiece, most components, including filters and sorbents, are provided by vendors. It takes considerable effort to develop a sorbent for protection against a substance for which a standardized sorbent is not available, and such task is only undertaken by a few able manufacturers. Ethylene oxide is an example. Activated carbon is considered as a universal sorbent for organic vapors but carbon is a poor adsorbent for ethylene oxide (EtO). In the late 1970s, many studies indicated that EtO is more toxic than the current OSHA PEL of 50 ppm. Some EtO manufacturers have voluntarily reduced their internal exposure standard for EtO and searched for a sorbent which would be effective against exposure to EtO. This became a joint project for Celanese and Union Carbide, major manufacturers of EtO; and the Mine Safety Appliance Company (MSA), a leading respirator manufacturer.

The preliminary breakthrough information was developed by Nelson.[24] In the first trial, acid gas cartridges and chlorine canisters were selected for test against an ethylene oxide concentration of 5 ppm at relative humidities of 25 and 75%. An immediate breakthrough (BT) was detected for the cartridge at a low humidity. The canister had 1% breakthrough after 15 min.

Additional tests were performed on a variety of sorbents such as silica gel, organic vapor, mercury, ammonia, and ammonia/amine. It was found that the sulfuric acid impregnated carbon cartridges (GMD) used for protection against NH_3/amine was effective against EtO. Further testing was performed on the MSA GMD cartridges and chin style canisters at a preconditioning RH of 85% and a testing RH of 75%. The challenge concentrations were 25 ppm and 100 ppm. It was found that cartridges had immediate breakthrough at such high humidity. The canister showed a 10% breakthrough after 6 h. The test results indicated that humidity has a severe effect on the service life of cartridges, but less pronounced for the canisters. Canisters with large volume should be used for protection against EtO.

The sulfuric acid probably acts as a catalyst to convert the ethylene oxide to ethylene glycol; the reaction proceeds at a slow rate. The presence of a high amount of moisture reduces the active sites for reaction. The sorbent fill volume or the bed depth is critical for the determination of the EtO residence time in the sorbent. Since the reaction takes place at a slow rate, increasing the residence time for EtO would improve the conversion of oxide to glycol. The high carbon volume would increase the reaction sites and minimize the adverse effect of moisture. It was decided to use the largest size canister, supersize, for protection against EtO.

The next phase study was undertaken by MSA. The goal was to develop a canister which could be approved by MSHA/NIOSH. Since ethylene oxide has poor odor warning properties, the canister must be approved with an end-of-service-life (ESLI) indicator. Another question was desorption. The test condition was quite stringent. A 1 ppm breakthrough is permitted for a challenge of 1000 ppm. In addition to the standard test for canisters listed in Subpart J of 30 CFR 11, the canister with an ESLI is equilibrated and tested at 85% relative humidity at an air flow of 64 l/m. The ESLI must change color at less than 90% service life of the canister. An EtO canister equipped with an ESLI was developed. The test results indicated that when the canisters were tested at an equilibration and testing RH of 25%, the minimum service life was 1725 min with an ESLI changes time at 720 min. At an equilibration and testing RH of 85%, the minimum service life was reduced to 653 min with an ESLI changes time at 270 min. The test results for individual test are shown in Table 2.

In the desorption test, the canister was equilibrated at 85% RH. Other test conditions were 25°C, 64 l/m, and 85% RH. The canister was tested at 1000 ppm EtO for 30 min and stood for 4 h; and tested at 1000 ppm EtO for 30 min and stood overnight or over the weekend. This cycle repeats until 1 ppm breakthrough of 1 ppm EtO occurred. For comparison purpose, the supersize organic vapor canister was also tested under the same test conditions. The results are shown on the following table:

Canister type	Service time (ST) (min)	No. cycles	Elapsed time (day)	Window change min/cycles
EtO	660	22	15	300 (10)
EtO	660	22	15	240 (8)
EtO	660	22	15	390 (13)
EtO	720	24	16	240 (8)
OV	31	2	(43 ppm leak occurred during test)	
OV	30	2	(75 ppm leak occurred during test)	
OV	31	2	(59 ppm leak occurred during test)	

It appears that no breakthrough of EtO occurred until the 22nd cycle. Also, the actual service life when the canister was exposed to EtO is not much different from the canister tested normally. This is an indication that desorption of EtO did not occur during storage.

Under a contract with OSHA, Nelson conducted a study to determine the cartridge service life for benzene and acrylonitrile.[25] Two types of Norton organic vapor cartridges with carbon weight of 47 and 87 g were selected for testing. The challenge concentrations for benzene were 100 and 500 ppm. The challenge concentrations for acrylonitrile were 200 and 1000 ppm. The test temperature was set at 25°C, with humidity ranges of 50 and 80%. The breakthrough concentration was measured at 1, 5, and 10%. The results are shown in Table 2.

It is quite common that organizations submitted cartridge or canister breakthrough data to OSHA for supporting the standard development activities. The Dow Chemical Company submitted cartridge breakthrough data on acrylonitrile (AN). The test was performed on the MSA organic vapor cartridges at 20, 100, and 1000 ppm. Breakthrough was measured at 1 and 2 ppm of AN.[26] The test results are shown in Table 2.

NIOSH conducted a study to determine the service life of acrylonitrile (AN) for the MSA and Norton organic vapor cartridges.[27] The test concentration was 141 ppm AN, and the relative humidity was set at 0 and 80%. The breakthrough time was recorded at 1, 5, and 10%. The test results are listed in Table 2.

Revoir of the Norton Company also submitted breakthrough data on acrylonitrile.[28] The first study was performed on two sizes of "deep shell" large Norton organic vapor cartridges. The air flow was set at 64 l/m for a pair of cartridges. The challenge AN concentrations were 20 and 100 ppm. The temperature was 25°C and relative humidity was set at 50 and 80%. Breakthrough concentration was recorded at 1 and 2 ppm. The results indicated that at a challenge concentration of 20 ppm (10 × OSHA PEL), the small cartridge (7500-1) provided a service life of 200 min at a relative humidity of 80%. At a challenge concentration of 100 ppm AN, the large cartridge (7400-1L) provided a service life of 4 h at a relative humidity of 80%. A second study was performed on a smaller cartridge — the "shallow shell" organic vapor cartridges. The test condition was 20 ppm with a relative humidity of 50%. The service life at 1 ppm breakthrough was about 6 h. The test results are shown in Table 2.

The cartridge breakthrough time on vinyl bromide was conducted by the Dow Chemical Company.[29] The test was performed at concentrations of 20, 100, and 1000 ppm. All cartridges were equilibrated at a relative humidity of 55% and the test humidity was 0 and 55%. The air flow was set at 55 l/m. Breakthrough time was reported at 0.3 ppm for a challenge of 20 ppm vinyl bromide, and at 0.5 ppm vinyl bromide for other two challenge concentrations. At a concentration of 20 ppm, breakthrough occurred after 4 h. The breakthrough time was reduced to 30 min when the challenge concentration was increased to 1000 ppm. The test results are listed in Table 2.

After the Bhopal incident, Moyer at NIOSH conducted a study to determine the cartridge service life against methyl isocyanate (MIC).[30] Both organic vapor and acid gas cartridges from two manufacturers were selected for testing. The test concentration of MIC varied from 300 to 1004 ppm. Relative humidity of 0, 50, and 73% were selected. The adsorption capacity of sorbent was also investigated. Breakthrough time at various concentrations are shown in Table 2. The results indicated that under the dry condition, the breakthrough time for the organic vapor cartridge was 30 min for a MIC concentration of 700 ppm. The average breakthrough time for the acid gas cartridges under the dry condition was 50 minutes for the MIC concentration of 700 ppm. When the humidity was increased to 73%, the average service life for the organic vapor and the acid gas cartridges were reduced to 8 min and 1 min, respectively.

In support to OSHA's rulemaking on methylene chloride, NIOSH conducted a study to determine the cartridge and canister service life for methylene chloride.[31] Organic vapor cartridges, chin canisters, and supersize canisters from two manufacturers were selected for testing. The challenge concentrations were 125, 250, 500, and 1000 ppm for cartridges, and 125, 250, and 1000 ppm for canisters. The test relative humidity was set at 50 and 80%. The flow rate was set at 64 l/m for the canister, and the single cartridge configuration (manufacturer A). The flow rate for the dual cartridge configuration (manufacturer B) was set at 32 l/m. Breakthrough time was measured at 5, 10, and 15 ppm. The test results are shown in Table 2. The results indicated that at a concentration of 125 ppm and a relative humidity of 80%, the average 1% breakthrough time is 35 min for the cartridges, 80 min for the chin style canisters, and 9 h for the supersize canisters. The results suggested that cartridges may not provide adequate protection against methylene chloride.

In response to OSHA's proposed standard on glycol ethers, Union Carbide has contracted Gary Nelson to conduct a cartridge breakthrough study for glycol ethers.[32] Organic vapor cartridges made by North (N7500-1) were selected for testing. Tests were conducted at 50 l/m, 20% or 80% RH, and 25°C. The challenge concentrations were 10, 100, and 500 ppm. The tests were performed until 10% breakthrough or 18 h — whichever came first.

The results are presented in Table 2. At a challenge concentration of 500 ppm, the average cartridge service life for glycol ethers is about 4 h. At challenge concentrations of 10 and 100 ppm, service life for organic vapor cartridges is in excess of 18 h.

After OSHA reduced the permissible exposure limit of nitroglycerine from a concentration of 2 to 0.1 mg/m^3, the use of respirators would be necessary to be in compliance with the new requirement. The Naval Ordinance Station in Indian Head, MD has contracted the Lawrence Livermore National Laboratory (LLNL) to conduct a study to determine the service life of nitroglycerine (NG).[33] Due to the potentially explosive nature of NG, it was mixed with 2-nitrodiphenyl amine and triacetin. The tests were performed at 26-28°C, 90% RH, a flow rate of 50 l/m, and at a challenge concentration of 1 to 2 mg/m^3. The test results are shown in Table 2. It indicated that breakthrough was not observed after 300 to 500 min of testing.

Cohen conducted a field study to determine the service of organic vapor cartridges against nitroglycerine.[34] He employed the respirator carbon tubes (RCT) technology which he developed previously.[35] The carbon from organic vapor cartridge made by MSA, Willson, and Racal Airstream PAPR were selected for testing. The RCTs have an inner diameter of 1.1 cm, length of 4 cm, and are packed with 1.8 g of carbon. A total of 10 trials were performed. The ambient temperature varied between 23 and 38°C. The range of relative humidity was from 44 to 100%. Ambient concentrations of nitroglycerine varied between 0.14 and 2.7 mg/m^3. The time-weighted average concentration for nitroglycerine varied between 91 and 685 μg/m^3. The sampling time varied between 7.5 and 215 h with sampling rates between 1.0 and 3.0 l/m. The test results indicated that no breakthrough was observed since all downstream samples of the RCTs showed nondetectable concentrations of nitroglycerine.

Tables 2 and the Nelson Cartridge Service Life Table[36] (Table 1 of Chapter 16) contain the available cartridge or canister service data generated in the last 20 years. Since test conditions and analytical methods vary between different studies, there may not be any correlation between results obtained from same substances by different authors. These data would provide some basis for making a determination whether the breakthrough information for a specific compound would provide adequate service life if some precautions are exercised in applying these data.

II. BREAKTHROUGH TIME FOR BINARY VAPOR MIXTURES

Cartridges or canisters are often approved for protection against a mixture of gas or vapor challenges such as organic vapor, sulfur dioxide, or ammonia. However, the certification test is performed on a single component only. Most breakthrough studies are also concentrated on a single component. There are few studies addressing the issue of service life for mixtures.

Jonas et al. conducted a study to predict the service life for binary vapor mixtures.[37] They selected a small carbon bed which was exposed to binary mixtures of carbon tetrachloride, chloroform, and benzene. The tests were conducted at a flow rate of 285 ml/min and 23°C. The inlet vapor concentrations represented a relative pressure of 0.0936 at 23°C for the single component challenge, and a relative pressure range of 0.0218 to 0.0949 at 23°C for the binary mixture challenge. The breakthrough concentration was set at 1% of the inlet concentration. The authors used the modified Wheeler and the Dubinin-Radushkevich equations outlined in Chapter 16 to determine the kinetic adsorption capacity, W_e, maximum adsorption space, W_0, adsorption potential, ε, and the adsorption rate constant, k_v from experimentally determined breakthrough data. The authors stated that predicting the sorbent's performance for

Table 3 Comparison of Experimentally Determined with Calculated Adsorption Parameters

Vapor	Mole fraction	Adsorption capacity, W_e (g/g) Exptl.	Calcd.	% Dev.	Adsorption rate constant, k_v (min^{-1}) Exptl.	Calcd.	% Dev.
CCl_4	1.0	0.832			356		
C_6H_6	1.0	0.432	0.456	+5.6	411	500	+21.7
$CHCl_3$	1.0	0.645	0.660	+2.3	377	404	+7.2
CCl_4 and	0.541	0.264	0.283	+7.2	508	356	−29.9
C_6H_6	0.459	0.110	0.132	+20.0	715	500	−30.1
CCl_4 and	0.763	0.594	0.637	+7.2	454	356	−21.6
C_6H_6	0.237	0.079	0.081	+2.5	557	500	−10.2
CCl_4 and	0.380	0.290	0.261	−10.0	442	356	−19.5
$CHCl_3$	0.620	0.331	0.298	−10.0	411	404	−1.7
$CHCl_3$ and	0.667	0.440	0.440	0.0	371	404	+8.9
C_6H_6	0.333	0.147	0.153	+4.1	500	500	0.0
$CHCl_3$ and	0.507	0.265	0.247	−6.8	446	404	−9.4
C_6H_6	0.493	0.225	0.226	+0.4	473	500	+5.7
$CHCl_3$ and	0.391	0.146	0.148	+1.4	470	404	−14.0
C_6H_6	0.609	0.263	0.278	+5.7	394	500	+26.9

From Reference 37.

a binary mixture was identical to the single component system but involved one more step. Once the W_v for each constituent vapor was found, it must be multiplied by the mole fraction of the vapor in the mixture.

The results of experimentally determined and calculated adsorption parameters for vapors and vapor mixtures are shown in Table 3. There is a good agreement between these values. The authors concluded that using the mole fraction to predict carbon performance for binary vapor mixtures is appropriate and that there is no interaction among vapors competing for the same adsorption sites. The finding might, however, depend upon the specific vapors tested, the type of carbon selected, the range of mole fractions used experimentally, and/or the fact that both components of the binary vapor mixtures were exposed at the carbon bed simultaneously.

Swearengen and Weaver conducted a study of breakthrough time for binary mixtures of isopropyl alcohol (IPA), methyl ethyl ketone (MEK), n-butyl acetate (BA), ethylbenzene (EB), and hexane (HEX).[38] The cartridges were challenged by a single vapor at 1000 and 2000 ppm at a flow rate of 40 l/m and at 50 and 85% relative humidity. All cartridges were tested in the "as received" condition. The mixtures tested were MEK and IPA; MEK and HEX; EB and BA; MEK and EB; and IPA and EB. The challenge concentration for each component in the mixture was 500 ppm or 1000 ppm, and at two selected humidities. Each test was repeated three times. The test conditions and results of 1 and 10% breakthrough time are listed in Tables 4 and 5.

The compounds involved in this study can be classified into two groups. Group I is made of low molecular weight and low boiling point compounds such as IPA, MEK and HEX. The higher molecular weight and boiling point compounds such as BA and EB belong to Group II. For the single component challenge, the breakthrough time appeared to be affected by the vapor pressure of the challenge solvents. At either concentration and humidity, HEX and MEK in Group I precede the other three compounds in breakthrough time. The separation of the breakthrough time is more pronounced at lower concentration and humidity than at high concentration and humidity.

Table 4 Comparison of 1% Breakthrough Time for Single Components and Mixtures at 50% RH

Solvent	Conc. (ppm)	BT 1%	(min) 10%
Single compound			
Butyl acetate only	2000	76	82
	1000	144	156
Ethyl benzene only	2000	74	84
	1000	154	174
Hexanes only	2000	62	64
	1000	120	130
IPA only	2000	77	92
	1000	136	164
MEK only	2000	62	69
	1000	117	131
Mixtures			
Butyl acetate and ethyl	1000	56	66
benzene	1000	62	78
IPA and ethyl benzene	1000	57	71
	1000	116	141
IPA and MEK	1000	69	78
	1000	74	86
IPA and MEK	1500	77	90
	500	87	95
IPA and MEK	500	67	74
	1500	70	82
MEK and ethyl benzene	1000	60	68
	1000	117	137
MEK and hexanes	1000	50	55
	1000	55	69

From Reference 38.

Table 5 Comparison of 1% Breakthrough Time for Single Components and Mixtures at 85% RH

Solvent	Conc. (ppm)	BT 1%	(min) 10%
Single compound			
Butyl acetate only	2000	72	84
	1000	140	156
Ethyl benzene only	2000	77	83
	1000	145	165
Hexanes only	2000	60	65
	1000	112	124
IPA only	2000	73	81
	1000	125	146
MEK only	2000	55	62
	1000	114	137
Mixtures			
Butyl acetate and ethyl	1000	56	62
benzene	100	60	72
IPA and ethyl benzene	1000	54	66
	1000	106	133
IPA and MEK	1000	62	72
	1000	67	79
IPA and MEK	1500	75	87
	500	82	88
IPA and MEK	500	63	72
	1500	65	75
MEK and ethyl benzene	1000	55	65
	1000	122	140
MEK and hexanes	1000	48	53
	1000	53	67

From Reference 38.

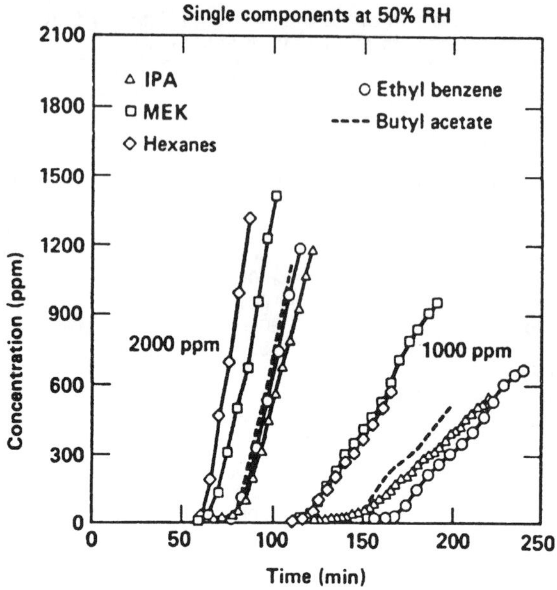

Figure 2 Breakthrough curves for all five compounds at 1000 ppm and at 2000 ppm — 50% RH. (From Reference 38.)

For the binary mixtures of IPA and MEK, and MEK and HEX, the smaller molecule breakthroughs first and was accelerated relative to the single component time in each case. With IPA and MEK, MEK were retarded, but with MEK and HEX, both were accelerated. The larger molecule in the Group I combinations was delayed in breakthrough (when compared with the breakthrough time of the pure compound). For the Group II binary mixture, both compounds broke through ahead of the single-component time (refer to Figures 2 through 4). If the trend proves to be repeatable, a conservative approach may be needed for estimating effective cartridge service life for vapor mixtures. One may anticipate a breakthrough time sooner than would normally be expected from the fastest single compound alone in the total concentration.

In this study of organic vapor mixtures, the smaller, lighter molecules tend to appear through the respirator cartridge sooner than single component challenge would indicate. When two heavy molecules are mixed, both can appear sooner than anticipated from single-component results. These trends are consistent at both moderate and high humidity.

Based on his respirator carbon tube (RCT) study,[35] Cohen applied this technique to examine the performance of RCT tubes for binary mixtures containing carbon tetrachloride.[39] He selected two compounds, pyridine which has a stronger affinity for activated carbon than carbon tetrachloride, and n-hexane which has a weaker affinity; both are soluble in carbon tetrachloride. Another purpose of this investigation was to determine whether intermittent exposure to an adsorbate would affect the adsorption capacity of the carbon.

Organic vapor respirator cartridges made by the MSA Company and RCT tubes having carbon capacities of 1.7 or 3.8 g were selected for testing. The test atmosphere consisted of 1000 ppm of each adsorbate selected for testing. The test chamber had

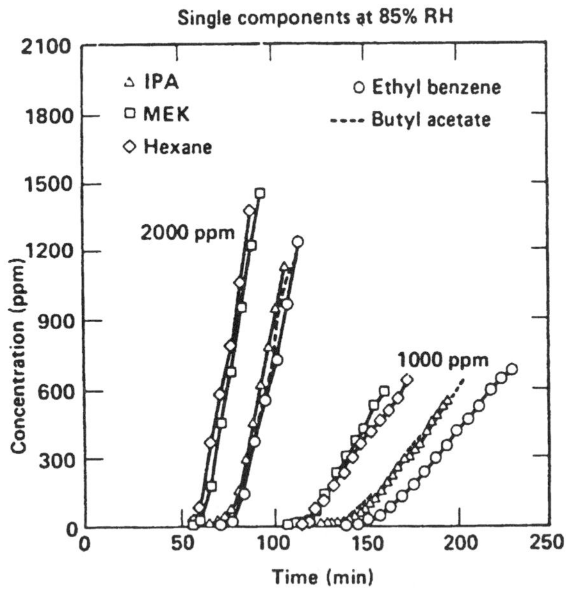

Figure 3 Breakthrough curves for all five compounds at 1000 ppm and at 2000 ppm — 85% RH. (From Reference 38.)

a relative humidity of 50%. The tests were terminated when 10% breakthrough (100 ppm) of the adsorbate was detected. For the binary system, the time needed to reach 10% breakthrough of the first adsorbate was recorded, and the test was terminated when 10% breakthrough of the second adsorbate was detected. For the series of intermittent exposure tests, the tests were conducted with respirator cartridges only. The test procedure was similar to the continuous exposure system except that when approximately half the time to reach the 10% breakthrough had noticed (estimated from the previous exposure data), the cartridge was removed and placed in a sealed plastic bag. After 4 days, the cartridge was removed, reweighed, and tested until 100 ppm of the adsorbate was detected downstream of the cartridge. Each experiment were performed in triplicate.

The mean breakthrough time for the cartridges and the RCTs are presented in Table 6. For the binary system, the first adsorbate listed in the one for which the breakthrough time is given. For example, for carbon tetrachloride and n-hexane, the breakthrough time is for carbon tetrachloride. For n-hexane and carbon tetrachloride mixture, the breakthrough time is for n-hexane. In both cases, the weight was determined only when the second adsorbate was detected downstream of the carbon bed at 100 ppm. The test results for intermittent exposure for the single and binary mixture are presented in Table 7.

The results indicated that for the binary system there was a decrease in the adsorption capacity of the carbon for each component. The pure adsorbate with the highest capacity will be preferentially adsorbed, and the total mass of vapors or gases adsorbed will be somewhere between the two values obtained for each individual adsorbate. Weight gains of the carbon beds for organic vapor mixtures, shown in Table 6, fell within the range of the values for individual components. The weight

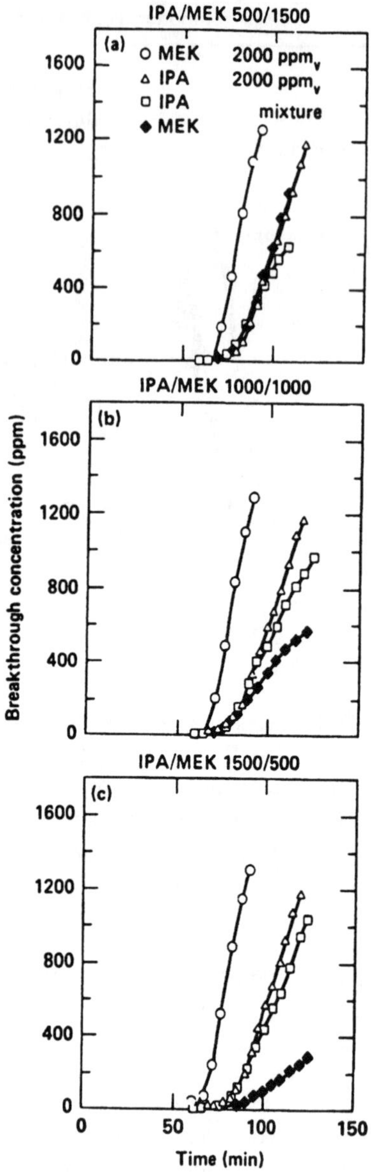

Figure 4 When the breakthrough curves for mixtures are compared with the curves for high concentrations of MEK and IPA alone, the mixture curves fall nearest to the IPA high-concentration curves. (From Reference 38.)

gain of the carbon should be caused by the adsorption of the organic(s) presented and not the water vapor. Reductions in estimates of W_c were observed when binary mixtures were tested, as compared to values obtained with single adsorbates, as expected. Weight gain of carbon beds tested with binary mixtures were within the range observed for individual components. The adsorption capacity for the mixture of n-hexane and carbon tetrachloride approximately equaled to the product of the

Table 6 Results of Breakthrough Studies for Single and Multiple Organics at 50% Relative Humidity[a]

Device[b]	Flow (l/m)	t[c] (min × 10^{-3})	T_b[d] (min)	T_b Pred.[e] (min)	Error (%)	Wt. gain[f] (%)	Wt. gain[f] (mmoles/g)
Carbon tetrachloride							
S-RCT	4.00(0)	0.94(0)	30(3)			42(3)	2.73
S-RCT	2.09(2)	1.81(2)	64(2)			47(2)	3.06
L-RCT	1.99(1)	4.24(1)	162(3)			51(2)	3.32
L-RCT	1.42(1)	5.95(1)	248(1)			56(2)	3.64
Cartridge	28.4(4)	3.32(4)	129(5)	130	0.8	50(2)	3.25
Cartridge	18.8(1)	5.02(1)	205(2)	203	1.0	47(2)	3.06
Carbon tetrachloride and pyridine							
L-RCT	11.90(0)	0.71(0)	26(0)			42(0)	
L-RCT	6.40(0)	1.32(0)	45(4)			37(2)	
L-RCT	2.00(0)	4.22(0)	113(6)			37(0)	
Cartridge	22.6(2)	4.17(2)	114(7)	112	1.8	42(3)	
Carbon tetrachloride and hexane							
L-RCT	5.00(0)	1.69(0)	32(5)			32(0)	
L-RCT	4.00(0)	2.11(0)	40(3)			32(0)	
L-RCT	2.10(0)	4.02(0)	80(2)			32(0)	
L-RCT	1.60(0)	5.28(0)	107(1)			32(0)	
Cartridge	19.1(0)	4.93(0)	105(5)	100	4.8	34(3)	
Hexane							
S-RCT	5.00(0)	0.71(0)	24(2)			14(20)	1.62
L-RCT	4.00(0)	2.11(0)	72(3)			17(7)	1.97
L-RCT	1.60(0)	5.28(0)	169(2)			16(0)	1.86
Cartridge	23.2(1)	4.06(1)	136(2)	132	2.9	19(2)	2.21
Hexane and carbon tetrachloride							
L-RCT	5.00(0)	1.69(0)	40(3)			32(0)	
L-RCT	4.00(0)	2.11(0)	48(4)			32(0)	
L-RCT	2.11(0)	4.02(0)	92(2)			32(0)	
L-RCT	1.60(0)	5.28(0)	125(1)			32(0)	
Cartridge	19.1(0)	4.93(0)	116(5)	115	0.9	34(3)	
Pyridine							
S-RCT	4.90(0)	0.68(0)	47(2)			27(0)	3.41
L-RCT	6.00(0)	1.41(0)	113(4)			32(2)	4.05
L-RCT	2.40(0)	3.52(0)	284(1)			29(12)	3.67
Cartridge	27.1(5)	3.48(5)	301(6)	282	6.3	28(4)	3.54
Pyridine and carbon tetrachloride							
L-RCT	11.90(0)	0.71(0)	54(2)			42(0)	
L-RCT	6.40(0)	1.32(0)	97(1)			37(2)	
L-RCT	2.00(0)	4.22(0)	291(2)			37(0)	
Cartridge	22.6(2)	4.17(2)	294(2)	289	1.8	42(3)	

[a] All data presented are mean values with relative standard deviations in parentheses (n equal to or greater than 3). Relative humidity = 50 ± 5%.

[b] Device: S = short (1.7 g), L = long (3.8 g).

[c] c_t = bed-residence time.

[d] T_b: time to 10% breakthrough corrected to 1000 ppm.

[e] T_b: Pred.: 10% breakthrough time predicted by regression analysis of RCT data (t versus T_b).

[f] mmoles/g = millimoles of adsorbate per gram of carbon.

From Reference 39.

Table 7 Comparison of Breakthrough Data Obtained with Continuous and Intermittent Testing of Respirator Cartridges with Single and Multiple Organics[a]

Exposure	Flow (l/m)	t^b (min × 10^{-3})	T_b[c] (min)	T_b Pred.[d] (min)	Error (%)	Wt. gain[e] (%)	Wt. gain[e] (mmoles/g)
Carbon tetrachloride							
Continuous	28.4(4)	3.32(4)	129(5)			50(1)	3.25
	18.8(1)	5.02(1)	205(2)			47(1)	3.06
Intermittent	23.8(0)	3.96(0)	157(2)	158	0.4	44(1)	2.86
Carbon tetrachloride and pyridine							
Continuous	22.6(2)	4.17(2)	114(7)			31(3)	
Intermittent	22.6(1)	4.17(1)	115(6)	114	0.9	39(5)	
Carbon tetrachloride and hexane							
Continuous	19.1(0)	4.93(0)	105(5)			34(3)	
Intermittent	19.1(0)	4.93(0)	112(4)	105	6.2	36(3)	
Hexane							
Continuous	23.2(1)	4.06(1)	135(2)			19(2)	2.21
Intermittent	22.8(1)	4.13(1)	136(2)	137	0.7	19(2)	2.21
Hexane and carbon tetrachloride							
Continuous	19.1(0)	4.93(0)	116(5)			34(3)	
Intermittent	19.1(0)	4.93(0)	128(5)	116	9.4	36(1)	
Pyridine							
Continuous	27.1(5)	3.47(5)	301(6)			28(4)	3.54
Intermittent	25.0(1)	3.75(1)	278(6)	326	14.7	31(3)	3.92
Pyridine and carbon tetrachloride							
Continuous	22.6(2)	4.17(2)	294(2)			42(3)	
Intermittent	22.7(1)	4.16(1)	273(3)	293	6.8	44(2)	

[a] All data presented are mean values with relative standard deviations in parentheses (n equal to or greater than 3). Relative humidity = 50 ± 5%.

[b] t = bed-residence time.

[c] t_b: time to 10% breakthrough corrected to 1000 ppm.

[d] t_b Pred.: 10% breakthrough time predicted by results of continuous testing of respirator cartridges.

[e] mmoles/g = millimoles of adsorbate per gram of carbon.

From Reference 39.

capacity of each individual component and their molar fraction in the mixture. However, the adsorption capacity for the mixture of carbon tetrachloride and pyridine exhibited only a minor decrease in the adsorption capacity for pyridine when carbon tetrachloride was added.

The results of the intermittent exposure test indicated that bed migration, caused by the desorption and readsorption of contaminants when the respirator cartridges was not in use, was not evident for n-hexane, carbon tetrachloride, and their mixture. A reduction in breakthrough time, although statistically insignificant, was observed for cartridges tested with pyridine and mixtures of pyridine with carbon tetrachloride. It may have been caused by multilayer adsorption. On untested cartridges, some of the pyridine may have desorbed and migrated to an unoccupied site.

A theoretical model developed by Yoon et al.[40,41] to predict the service life of respirator cartridges against a single component has been modified to determine the service life of a binary system of acetone and m-xylene.[42] The new model is based on the observations and assumptions that breakthrough curves can be separated into two regions. In Region 1, the acetone breakthrough increases until it reaches the

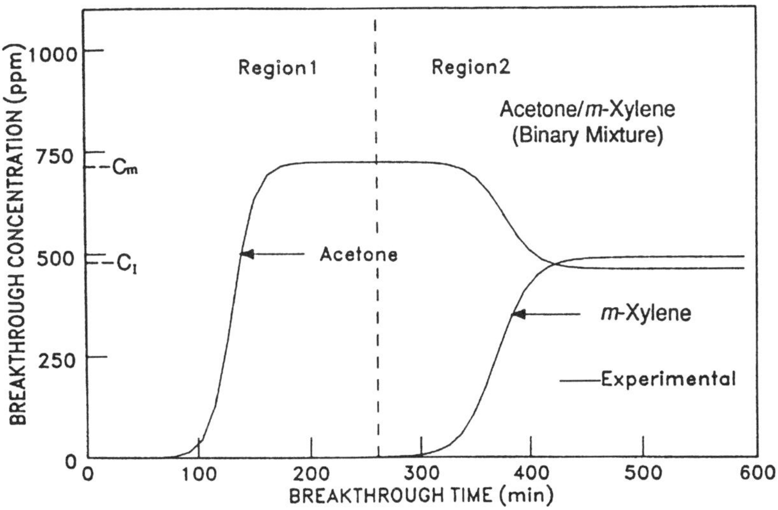

Figure 5 Typical breakthrough curves for a binary assault system. An example system, acetone/m-xylene, is presented. (From Reference 42.)

maximum concentration C_m while in Region 2 the acetone concentration decreases sigmoidally from C_m to C_I, the challenge concentration, as the m-xylene breakthrough concentration increases from zero to C_I. A typical breakthrough curve is shown in Figure 5.

One parameter, A_m has been added to the previously defined parameters τ and k' to predict the service life of components in a binary system. The parameter τ is the time required for 50% breakthrough concentration (min), and k' is the rate constant (min^{-1}). Both τ and k' can be determined experimentally as described by Yoon.[40,41] The breakthrough concentration of acetone (C_b) at any time in either region can be expressed as

$$C_b = C_m P_1 - (C_m - C_1)P_2 \tag{1}$$

where A_m is defined as

$$A_m = \frac{C_m - C_I}{C_I} \tag{2}$$

P is the general term representing the fractional breakthrough of acetone in the presence of m-xylene. The parameter τ_1, associated with P_1, is the time required to observe an acetone breakthrough concentration of 0.5 C_m (or $P_1 = 0.5$). The parameter $\tau_{2,}$ associated with P_2, is the breakthrough time when $P_2 = 0.5$ or 50% breakthrough time of m-xylene in the binary system. When P_2 is negligible in Region 1 of Figure 5, Equation 1 is reduced to

$$C_b = C_m P_1 \tag{3}$$

Therefore, for Region 1:

$$P_1 = C_b/C_m \tag{4}$$

The parameters k_1' and τ_1 may be determined by substituting the value of P_1 for P in the following equation

$$\ln = \frac{P}{1-P} - k'(\tau - t) \tag{5}$$

Note that the value determined for τ_1 corresponds to the condition: $C_b/C_m = 0.5$.

For Region 2 of Figure 5, $P_1 = 1$; therefore, Equation 1 can be rewritten as

$$C_b = C_m - (C_m - C_I)\, P_2 \tag{6}$$

It follows that

$$P_2 = \frac{C_m - C_b}{C_m - C_I} \tag{7}$$

Values of P_2 may be calculated from the experimental data, and the substitution of P_2 for P in Equation 5 permits the determination of k_2' and τ_2.

Experimental breakthrough data on the acetone/m-xylene binary system was performed on organic vapor cartridges made by Scott Aviation at concentrations of 100 to 1100 ppm. The combined concentrations for these two components were 1000 ± 50 ppm. The test temperature was 25°C, and the flow rate through the cartridges was 24 l/m at a relative humidity of 40 ± 1%. Calculated values of various parameters and C_m for the acetone/m-xylene binary system under different concentrations are shown in Table 8. The calculated values for the parameters agree with experimental data.

In general, there was very little difference between the breakthrough curves for m-xylene in the single component system and the corresponding binary system. In both systems, the value of k′ increased with the increase of the challenge concentration while the τ was decreased. In comparing values from the single component and the binary system, there was little difference in the values of k′ and τ for m-xylene since nearly all the acetone adsorbed on the carbon was eventually replaced by incoming m-xylene. When the breakthrough concentration of m-xylene reached the challenge m-xylene concentration (C_I), the carbon bed was saturated with m-xylene, and there was no detectable acetone adsorbed in the bed. This observation was verified by the analysis of acetone content in carbon after completion of the test. One exception was when the challenge concentration (C_I) of acetone was considerably smaller than that of m-xylene. Regions 1 and 2 were not readily distinguishable due to overlapping. Hence, it was difficult to ascertain the value of C_m from experimental results.

The data in Table 8 demonstrates that for the binary system, each data pair of τ_2 and k_2' is consistent with the corresponding data pair of τ and k′. Furthermore, the

Table 8 Values of Theoretical Parameters for Various Acetone/m-Xylene Binary Systems

Conc.[a] (ppm)				Acetone				m-Xylene	
				Region 1		Region 2			
Acetone	m-Xylene	C_m (ppm)	A_m	k'_1 (min^{-1})	τ_1 (min)	k'_2 (min^{-1})	τ_2 (min)	k′ (min^{-1})	τ (min)
813	90	871	0.0713	0.103	117	0.0120	1600	0.0130	1610
749	260	921	0.230	0.107	122	0.0337	652	0.0384	682
474	490	724	0.527	0.0977	132	0.0682	377	0.0626	369
248	730	610	1.46	0.109	154	0.0908	254	0.101	254
109	891	545	4.00	0.0805	163	0.0636	201	0.0694	201

[a] Conc. is the concentration of the pertinent compound in air in ppm (volume/volume).

From Reference 42.

τ values for the single-component m-xylene system (Table 9) are similar to the corresponding τ values for the m-xylene in the binary system (Table 8). Similarly, the value of each k′ for single-component m-xylene is similar to that of the corresponding k′ of the binary system, except for the special case for which the challenge concentration of acetone is considerably smaller than that of m-xylene. In this case, the binary k′ is less than the single-component k′.

In conclusion, in the acetone/m-xylene binary system, respirator cartridge service life is determined by the breakthrough characteristics of acetone, the compound with the shorter breakthrough time. Acetone breakthrough time is determined by the challenge concentrations of both acetone and m-xylene. Acetone has no measurable effect on the breakthrough properties of m-xylene.

Table 9 Values of Theoretical Parameters for Acetone and for m-Xylene Single-Component Systems

Conc.[a] (ppm)	k′ (min^{-1})	τ (min) Calc.[b]	τ (min) Exp.[c]	k
Acetone				
107	0.0414	340	346	14.1
260	0.0531	270	245	14.3
501	0.0760	174	172	13.2
726	0.101	140	139	14.1
1060	0.126	109	110	13.7
Mean ± SD[d]				13.9 ± 0.4
m-Xylene				
93	0.0119	1640	1650	19.5
257	0.0348	702	701	24.4
446	0.0697	386	386	26.9
857	0.110	215	215	23.7
973	0.118	202	206	23.8
Mean ± SD[d]				23.7 ± 2.7

[a] Conc. is the concentration of the pertinent compound in air in ppm (volume/volume).

[b] Calc. is the calculated value of the parameter of interest.

[c] Exp. is the observed experimental value of the parameter of interest.

[d] SD is the standard deviation of the data listed.

From Reference 42.

The authors conducted another series of experiments to investigate the breakthrough characteristics of the acetone/styrene binary system.[43] Organic vapor cartridges made by Scott Aviation were selected for testing. The test conditions were: temperature, 25°C; relative humidity, 40%; and flow rate through the cartridge, 24 l/m. Breakthrough tests were performed under the following conditions:

1. Styrene at concentrations between 250 and 1000 ppm.
2. An acetone/styrene binary system with varied concentration of each component, but the sum of the acetone and styrene concentrations roughly 1000 ppm.
3. An acetone/styrene binary system with a fixed concentration of acetone at 500 ppm, and styrene concentration varied between 250 and 1250 ppm.
4. Acetone/styrene binary system with a fixed concentration of styrene at 500 ppm, and acetone concentration varied between 250 and 1250 ppm.

For binary systems in which Compound 1 (previously adsorbed) was displaced from the carbon by another compound (designed as Compound 2), two types of breakthrough phenomena were considered. The first referred to Compound 1 and the second, Compound 2. In the acetone/styrene binary system, previously adsorbed acetone was displaced by styrene. Therefore, acetone was referred as Compound 1 and styrene was referred as Compound 2.

Prior to the observation of any breakthrough for a given system, acetone and styrene were adsorbed on the cartridge in a ratio equivalent to the ratio of the respective challenge concentration of the two compounds. Acetone breakthrough was observed first. Following the onset of the breakthrough, the acetone breakthrough concentration was enhanced by the displacement of acetone from the bed by styrene. The acetone breakthrough concentration eventually reached a maximum level. When the styrene breakthrough began, a concomitant decrease in the acetone breakthrough was observed. Eventually, all of the adsorbed acetone was displaced by styrene; ultimately, the observed breakthrough concentration of each compound was equivalent to the corresponding challenge concentrations.

The values of parameters such as τs, k′s and A_ms are listed in Table 10. The values of k′ and τ for the styrene single component system are listed in Table 11. The results indicated that the breakthrough characteristics of Compound 2 in a binary system may be described by the expressions developed for the applicable single-component system. Accordingly, the breakthrough curves of styrene in the acetone/styrene binary system are sigmoidal, resembling those of the corresponding single-component system. The experimental data collected in this study were used to determine values of k′ and τ for styrene in the acetone/styrene binary system. These determinations were accomplished by using the approach for single-component systems. At a specified styrene challenge concentration, the values of k′ and τ for single-component styrene are similar to the corresponding values in the binary acetone/styrene system. Applicable theoretical styrene breakthrough curves for various acetone/styrene mixtures may be calculated by using the values of k′ and τ derived for the binary system. Examples of the breakthrough curves of the binary system under different acetone and styrene concentrations are shown in Figures 6 and 7.

Table 10 Values of Theoretical Parameters for Various Acetone/Styrene Binary Systems

Conc.[a] (ppm)		Acetone						Styrene	
Acetone	Styrene	C_m[b] (ppm)	A_m[c]	k'_1 (min^{-1})	τ_1 (min)	k'_2 (min^{-1})	τ_2 (min)	k' (min^{-1})	τ (min)
92	508	258	1.80	0.0618	240	0.0731	374	0.0631	373
99	516	280	1.83	0.0585	239	0.0561	377	0.0571	380
305	503	552	0.810	0.0867	169	0.0829	386	0.0729	386
479	507	763	0.593	0.0894	137	0.0583	372	0.0598	367
779	510	1080	0.390	0.120	115	0.0785	394	0.0580	385
985	496	1340	0.361	0.124	102	0.0652	396	0.0500	405
464	228	582	0.254	0.111	146	0.0358	708	0.0380	726
479	507	763	0.593	0.0894	137	0.0583	372	0.0598	367
492	733	925	0.880	0.0918	124	0.118	274	0.0974	270
494	1036	1110	1.25	0.136	113	0.0903	206	0.102	206
490	1578	1570	2.20	0.158	93	0.169	137	0.130	134
97	892	353	2.64	0.0915	158	0.0921	217	0.0954	217
234	739	492	1.10	0.103	129	0.0852	251	0.0737	257
249	755	513	1.06	0.115	136	0.101	269	0.0994	267
479	507	763	0.593	0.0894	137	0.0583	372	0.0598	367
746	256	882	0.182	0.104	126	0.0393	786	0.0286	762

[a] Concentration of the pertinent compound in air (ppm, volume/volume).

[b] Maximum breakthrough concentration of Compound 1.

[c] Maximum fractional excess breakthrough of Compound 1: $(C_m - C_I)/C_I$.

From Reference 43.

The authors concluded that acetone and styrene are adsorbed on the carbon at a ratio equivalent to that ratio of the respective challenge concentrations of the two compounds. Acetone breakthrough was observed first. Following the onset of this breakthrough, the acetone breakthrough concentration was enhanced by the displacement of acetone from the bed by styrene. The acetone breakthrough concentration eventually reached a maximum level. When the styrene breakthrough began, a concomitant decrease in the acetone breakthrough concentration was observed. Eventually, all of the adsorbed acetone was replaced by styrene; ultimately, the observed breakthrough concentration of each compound was equivalent to the corresponding challenge concentration.

Table 11 Values of Theoretical Parameters for Various Styrene Single-Component Systems

Conc.[a] (ppm)	k' (min^{-1})	τ (min) Calc.[b]	Exp.[c]	k
254	0.0445	728	728	32.4
481	0.0798	418	419	33.4
497	0.0673	393	392	26.4
725	0.118	270	270	31.9
1045	0.138	206	206	27.6
Mean ± SD[d]				30.3 ± 3.1

[a] Concentration of styrene in air in parts per million (volume/volume).

[b] Calculated value of τ.

[c] Observed experimental value of τ.

[d] Standard deviation of the data listed.

From Reference 43.

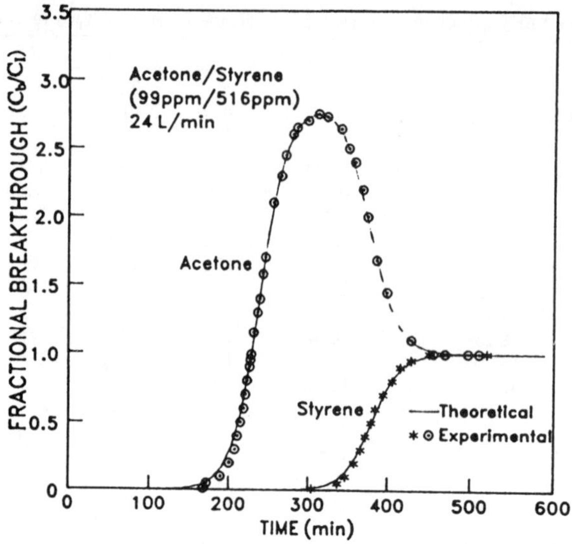

Figure 6 Comparison of theoretical breakthrough curves (solid lines) with experimental data for acetone (99 ppm) and for styrene (516 ppm) in the acetone/styrene binary assault system. (From Reference 43.)

III. CERTIFICATION REQUIREMENTS FOR END-OF-SERVICE LIFE INDICATORS

A major problem of the use of respirators for protection against gases or vapors with poor warning properties is that the wearer has no knowledge regarding the end-of-service life (ESLI) of the sorbent. The most desirable device to detect cartridge or

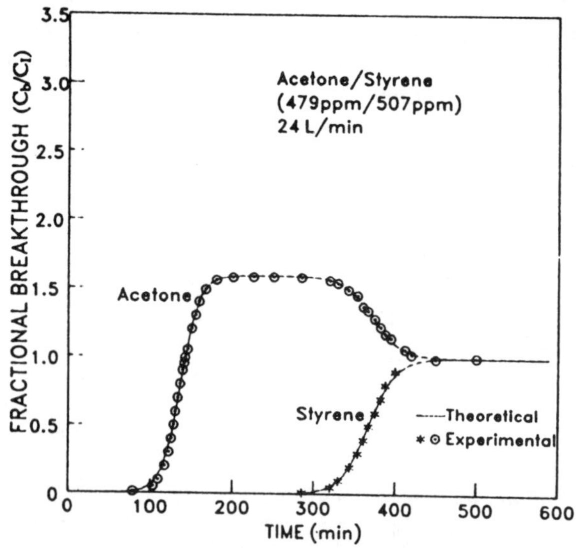

Figure 7 Comparison of theoretical breakthrough curves (solid lines) with experimental data for acetone (479 ppm) and for styrene (507 ppm) in the acetone/styrene binary assault system. (From Reference 43.)

canister breakthrough is an ESLI. In 1984, NIOSH published criteria for certifying respirators equipped with ESLI.[44] The requirements are listed below.

All ESLI must meet the following criteria:

1. The ESLI must provide a reliable indication of sorbent depletion at less than or equal to 90% of its service life.
2. The ESLI must not interfere with the effectiveness of the respirator face seal.
3. The ESLI must not change the weight distribution of the respirator or to the detriment of the facepiece fit.
4. The ESLI must not interfere with the required lines of sight.
5. Any ESLI that is permanently installed in the respirator facepiece shall withstand cleaning and a drop from a 6-ft height. Replaceable ESLI must be able to be easily removed.
6. A respirator with an ESLI must still meet all other applicable requirements set forth in 30 CFR 11.
7. Any electrical components utilized in an ESLI shall conform to the provisions of the National Electric Code and be "intrinsically safe." Where permissibility is required, the respirator shall meet the requirements for permissibility and intrinsic safety set forth in § 15.82 of 30 CFR 11. Also, the electrical system shall include an automatic warning mechanism that indicates a loss of power.
8. (a) Effects of industrial interferences for substances which are commonly found in the workplace where it is anticipated that a given respirator will be used must be determined, and those substances which hinder ESLI performance shall be identified.
   (b) Substances which are commonly found where the respirator will be used must be investigated. Data sufficient to indicate whether the performance is affected must be submitted to NIOSH.
   (c) Manufacturers of respirators equipped with ESLI shall label the respirator to make the user aware of use conditions that could cause false positive and/or negative ESLI responses.
9. The ESLI shall not create any hazard to the wearer's health or safety.
10. Consideration shall be given to the potential impact of common human physical impairments on the effectiveness of the ESLI. A discussion for this problem should be included in the user instruction manual.

In addition, all passive indicators must meet the following criteria:

1. A passive ESLI shall be situated on the respirator so that it is readily visible to the wearer.
2. If the passive indicator utilizes color change, the change shall be detectable to people with physical impairments such as color blindness.
3. If the passive indicator utilizes color change, reference colors for the initial color of the indicator and the final (end point) color of the indicator shall be placed adjacent to the indicator.

Several passive types of ESLIs such as mercury, ethylene oxide, and hydrogen sulfide have been approved.[17,23,24] However, active type ESLIs have not yet been developed. Recently, Stetter et al.[45] developed a prototype active ESLI. This ESLI consists of a sensor which is located within the sorbent bed and connected to a signal-processing module located on the facepiece. The module triggers a LEL alarm that flashes when gas or vapor concentrations reach a preset level, about 90% of the service life of the sorbent.

The sensor used in this study was a chemresistor, a device whose response is a function of the challenge concentration of the gas or vapor. Cyclohexane was selected to test the performance of the sensor at concentrations between 100 and 9000 ppm. In order to determine the effect of temperature on sensor response, tests were conducted at temperatures between 0 and 25°C. The sensor resistance was measured by a Whetstone bridge circuit with a supply voltage of 6 volts. The output voltage of the bridge circuit was used to drive a LED alarm powered by a 9 volt battery.

The test results indicated that the sensor response was rapid. Response time at 90% of the end point was approximately between 30 and 45 s. There was a shift in sensor response due to temperature change. It was corrected by adding a temperature compensation circuit which reduced the change in the base line bridge output from 15% to less than 1%. Additional tests were performed to determine the effect of humidity on sensor response; it was observed that it may be possible to compensate for the effects of humidity by using a reference sensor in the bridge circuit. The screening data for selected hydrocarbons and alcohol suggest that the sensors tested appear to respond more strongly to nonpolar compounds such as benzene than polar compounds such as methanol.

The authors concluded that it is feasible to build a small active type ESLI with a flashing LED alarm which is visible to the respirator wearer. However, the application is limited to full facepiece respirators due to its cost and bulk.

A follow-up study to evaluate the performance of an active type ESLI microsensor was conducted by Moyer et al.[46] The microsensor is fabricated by spraying with a mixture of activated Darco carbon and silicone rubber onto phenolic substrates with two etched copper electrodes. The resistance of the sensor increases when it is exposed to organic vapors, and the resistance is determined with a Wheatstone bridge circuit having a supplied voltage of 6 volts. The output voltage of the bridge circuit is used to drive an LED alarm.

The sensor could be located either inside the facepiece of the respirator or embedded within the cartridge bed. The response of the sensor is fed to a signal-processing unit located on the respirator facepiece. The signal activates a flashing LED alarm when the contaminate concentration reaches the threshold value. The total device and associated electronics takes a volume of approximately 10 to 20 cm^3 and with a weight of 25 g. The estimated cost for the sensor is around $3.00, and the sensor would not affect the fit and function of the respirator.

Tests were performed to determine the accuracy of the sensor in terms of cartridge breakthrough. Based on the test results, calculations were made to determine the best location for the sensor within the sorbent bed. The bed location should be at the NIOSH required 90% of the cartridge service life or at an exit concentration which is much lower than the OSHA permissible exposure limit of the chemical. Ethyl acetate was selected as a challenge at concentrations of 750, 1000, 1500, and 2000 ppm. Tests were performed on Pulmosan organic vapor cartridges at an air flow of 64 l/m and 50% RH. The temperature and relative humidity were 25°C and 50%, respectively. Breakthrough concentration at 1, 5, and 10% were selected. In order to obtain more breakthrough data, four cartridges were stacked together when the tests were performed. This experimental setup was proposed by Moyer.[11] Two infrared spectrophotometers (Miran 1A) were used to monitor the upstream and downstream concentrations of ethyl acetate. The sensor loop was arranged so that each of the four sensors continuously contacted a sample over the entire experimental run, rather than being affected by the switching valves that controlled the IR sampling location. By this arrangement, data from both systems were collected and analyzed in a comparative manner. The experimental setup is shown in Figure 8.

A modified Wheeler equation is used to calculate the sensor location in the sorbent bed. The application of this equation was shown in the previous chapter.

$$t_b = \frac{W_e}{C_o Q}\left[W - \frac{\rho_\beta Q}{k_v} \ln(C_o / C_b)\right] \tag{8}$$

or

$$t_b = \frac{W_e W}{C_o Q} - \frac{W_e \rho_\beta}{k_v C_o} \ln(C_o / C_b) \tag{9}$$

where t_b = Breakthrough time (min)
C_o = Inlet concentration (g/ml)
C_b = Exit concentration (g/ml)
k_v = Rate coefficient (min^{-1})
ρ_β = Bulk density of packed bed (g/ml)
Q = Volumetric flow rate (ml/min)
W = Weight of adsorbent (g)
W_e = Adsorption capacity (g/g).

This equation indicates that a linear relationship exists between the sorbent weight and the breakthrough time. If the breakthrough concentration is plotted against the sorbent weight, the rate constant k_v and the adsorption capacity W_e can be calculated from the slope and the intercept of the plot intercept value.

The Wheeler constants are presented in Table 12. If all sensors have essentially the same response factor, the sensor data would conform to the modified Wheeler equation. The data indicate that a linear relationship does exist between breakthrough time and sorbent weight for the first three sensors at the 1000 ppm challenge. The deviation of Sensor #4 was attributed to its significantly enhanced sensitivity over the other three sensors. If the IR t_b value for the cartridge is replaced with an arbitrary level, i.e., 100 ppm breakthrough into the sensor equation, the carbon bed depth position for the sensor can be determined. The position of the sensor in the first cartridge would be approximately 75% of the way into the carbon bed. The sensor would trigger the alarm when 100 ppm breakthrough has occurred.

The test results indicated that a linear relationship between the breakthrough time and carbon weight also exists for the challenge concentrations of 1550 and 2000 ppm ethyl acetate. The calculated position for the sensor in the first cartridge would be approximately 60% of the way into the carbon at both concentrations.

A series of tests were performed in a three month interval after the first series of 1000 ppm ethyl acetate test runs. The purpose of these tests was to check the stability of the sensor. The results indicated that the sensors had remained stable but the intensity of response had been diminished.

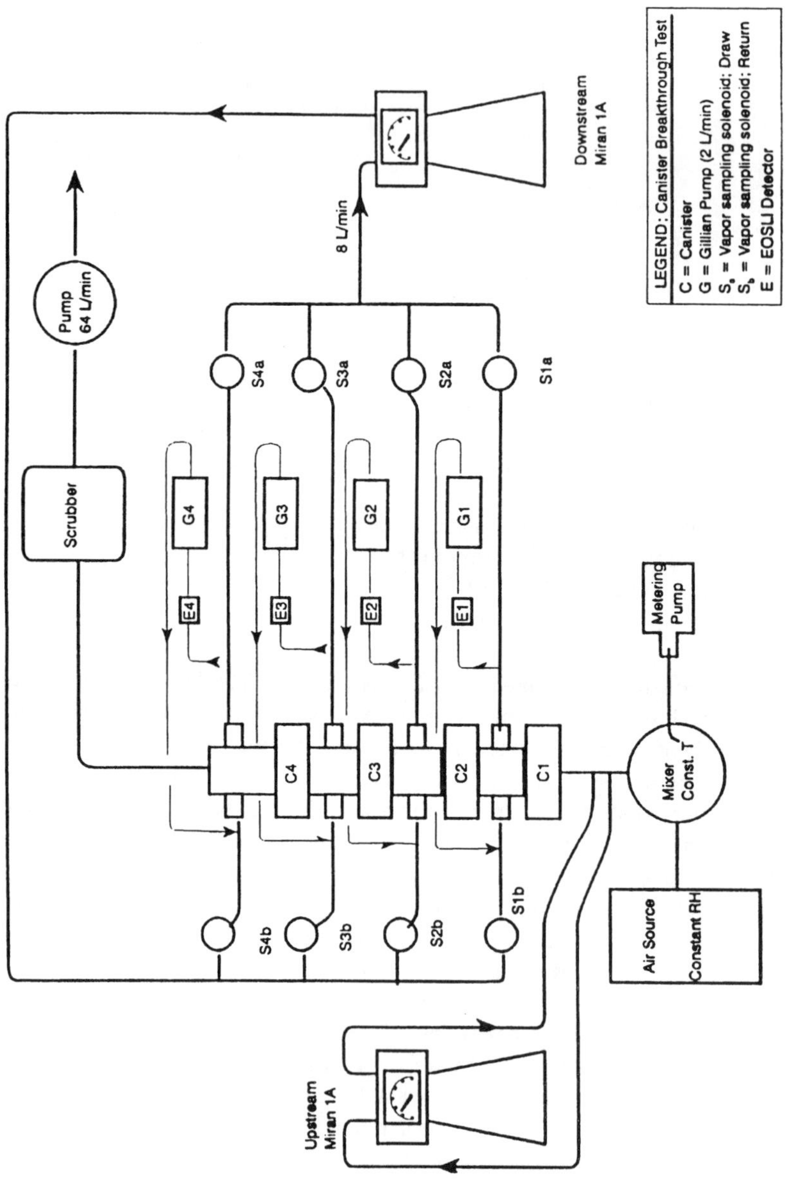

Figure 8 Experimental apparatus for simultaneous monitoring of infrared and microsensor ethyl acetate breakthrough. (From Reference 46.)

Table 12 Wheeler Constants for Ethyl Acetate Challenge Vapor at 1, 5, and 10% Breakthrough for Lot B

Corrected challenge conc. (ppm)	% t_b	# Pts.	% RH	Slope	Y axis intercept	R^2	Average bed bulk density (ρ_β) g/cm^3	W_e kinetic adsorption capacity (g/g)	k_v rate constant (min^{-1})	W_e critical bed weight (g)
750	1	8	50	1.257	–25.05	0.993	0.442	0.218	6540	19.93
	5	8	50	1.260	–17.40	0.992	0.442	0.218	6140	13.81
	10	8	50	1.264	–13.42	0.992	0.442	0.219	6140	10.62
1000	1	12	50	1.018	–16.72	0.999	0.438	0.235	7860	16.42
	1	12	50	0.976	–21.91	0.997	0.449	0.225	5900	22.44
	5	12	50	1.020	–10.71	0.999	0.438	0.235	8000	10.50
	5	12	50	0.983	–16.03	0.997	0.449	0.227	5280	16.31
	10	12	50	1.021	–8.15	0.999	0.438	0.235	8080	7.99
	10	12	50	0.983	–12.99	0.997	0.449	0.227	5010	13.21
1500	1	8	50	0.762	–14.27	0.995	0.437	0.264	6870	18.73
	5	8	50	0.762	–9.82	0.994	0.437	0.264	6500	12.89
	10	8	50	0.765	–8.22	0.994	0.437	0.265	5980	10.75
2000	1	8	50	0.594	–11.72	0.997	0.438	0.274	6540	19.74
	5	8	50	0.596	–8.38	0.997	0.438	0.275	5960	14.08

From Reference 46.

The authors concluded that the chemresistor microsensor is suitable as an active ESLI to detect organic vapor cartridge breakthrough. The preferred sensor location is inside the carbon bed and it is less subject to moisture or environmental pollutant interference.

IV. REPLACEMENT FOR CARBON TETRACHLORIDE

Carbon tetrachloride has been designated as the testing agent for certifying organic vapor cartridges or canisters in 30 CFR 11, the MSHA/NIOSH respirator test and certification regulations. After the International Agency for Research on Cancer (IARC) and the American Conference of Governmental Industrial Hygienists (ACGIH) have announced that carbon tetrachloride is a suspected animal carcinogen, NIOSH was looking for a replacement for this chemical. This process has been accelerated after the Environmental Protection Agency (EPA) proposed regulations to protect the stratosphere ozone from depletion. The rule severely limits the use of many chlorinated hydrocarbons such as carbon tetrachloride.

NIOSH has conducted tests to find a replacement for carbon tetrachloride.[47-49] The most critical testing conditions prescribed in Subpart L of 30 CFR 11 for testing nonpowered negative pressure organic vapor cartridge is in the *"as received"* state. If the *"as received"* organic vapor cartridge is challenged with carbon tetrachloride at 1000 ppm, 64 l/m air flow, 50% relative humidity, and 25°C, a minimum service life of 50 min is required. The correlation testing conducted on pentane to match these critical test conditions resulted in a pentane challenge concentration of 550 ppm, 64 l/m, 80% relative humidity, and 25°C. A minimum service life of 40 minutes at 5 ppm pentane breakthrough was observed for the organic vapor cartridges.

The most critical testing conditions prescribed in Subpart I of 30 CFR 11 for testing the nonpowered negative pressure chin-style organic vapor canister are in the "*preconditioned (equilibrated)*" state. If the chin-style organic vapor canisters are challenged with carbon tetrachloride when the canister has been *"preconditioned"* at 85% RH, and tested at 5000 ppm, 64 l/m air flow, 50% relative humidity, and 25°C, a minimum service life of 27 min at a 5 ppm breakthrough is required. The correlation testing conducted to match these critical test conditions resulted when the *"as received"* chin-style organic vapor canister was tested with a pentane challenge concentration of 4000 ppm, 64 l/m, 50% relative humidity, and 25°C. A minimum service life of 27 min at a 5 ppm pentane breakthrough was observed for the chin-style organic vapor canister.

The most critical testing condition for nonpowered negative pressure organic vapor front or back mounted (FM/BM) canister is in the *"preconditioned"* state. If the FM/BM organic vapor canister is challenged with carbon tetrachloride performed at 20,000 ppm, 32 l/m air flow, 50% relative humidity, 25°C, a minimum breakthrough time of 50 min at 5 ppm is required. Since the lower explosive limit (LEL) for pentane is 1.5%, 10,000 ppm would be the maximum safe test concentration instead of the required 20,000 ppm. The correlation testing was conducted with pentane at 10,000 ppm, 64 l/m, 80% relative humidity, and 25°C. A minimum service life of 50 min at 5 ppm pentane breakthrough was observed for the FM/BM canisters.

The cartridge and canister correlation test results indicated that pentane would be a suitable replacement. NIOSH indicated that it would accept cartridge or canister breakthrough data obtained from pentane for organic vapors certification. Other agents may be used, but the manufacturer must supply data to demonstrate the equivalency between the selected agent and carbon tetrachloride.

Moyer et al. at NIOSH have conducted a comprehensive study to select a suitable replacement for carbon chloride.[50] Four possible replacement chemicals were selected based on their physical properties, toxicity, and carbon adsorption characteristics. These include: ethyl acetate, pentane, n-hexane, and n-heptane. The first phase of this study was screening testing. Cartridges were tested in the *"as received"* state at various concentrations of ethyl acetate, pentane, n-hexane, and n-heptane. Tests were performed 50% or 80% relative humidity, 64 l/m, and 25°C. In order to obtain more information, the "stacked-cartridges" method proposed by Moyer[11] was employed in testing. For comparison purposes, cartridges were also tested with carbon tetrachloride. After the agent that provided the most equivalent test condition to carbon tetrachloride had been identified, side-to-side comparative testing was performed under the following test conditions in the second phase:

1. Preconditioned at 25% RH and tested at 50% and 32 l/m.
2. Preconditioned at 85% RH and tested at 50% and 32 l/m.
3. "As received" tested at 25% RH and 64 l/m.
4. "As received" tested at 50% RH and 64 l/m.
5. "As received" tested at 80% RH and 64 l/m.

Organic vapor cartridges made by three manufacturers were selected for testing. In addition, combination organic vapor/acid gas cartridges made by manufacturer C were also tested. With the exception of manufacturer B, dual cartridges were tested in the second phase. For the stacked cartridge testing, the air was pushed through the cartridges rather than pulled by a vacuum source.

Manufacturer A's organic vapor cartridges were used in the first phase of stacked-cartridges testing. The ethyl acetate challenge was performed at 50% and 80% RH, 64 l/m, 25°C, and at concentrations of 750, 1000, 1500, and 2000 ppm. The breakthrough data at 1000 ppm ethyl acetate challenge are similar to the 1000 ppm carbon tetrachloride breakthrough data. The affinity of ethyl acetate to water makes it unsuitable as a substitute for carbon tetrachloride. The pentane challenges were performed with manufacturer B's cartridges at concentrations of 275, 500, 750, and 1000 ppm. The breakthrough time at a 500 ppm pentane challenge is closest to the carbon tetrachloride data. The stacked-cartridges data from n-hexane were obtained at challenge concentrations of 500, 750, and 1500 ppm. The breakthrough data at 1000 ppm matched the carbon tetrachloride data at 1000 ppm. The n-heptane challenge concentration was set at 800, 1000, 1250, and 1500 ppm with the stacked cartridges. The breakthrough data indicated that the equivalency existed at a minimum concentration of 1250 ppm n-heptane. A summary of test data for carbon tetrachloride, pentane, and n-hexane is shown in Table 13. The test data indicated that the most critical test condition should be "as received" cartridges tested at 64 l/m and 80% RH. Both pentane at a concentration of 550 ppm and n-hexane at 1000 ppm would yield the equivalent cartridge breakthrough concentration of carbon tetrachloride. Since pentane has lower toxicity than n-hexane, it is selected as the replacement for carbon tetrachloride. Selected breakthrough test data for ethyl acetate, pentane, n-hexane, and n-heptane are listed in Table 2.

Table 13 Cartridge Breakthrough Time Comparison Summary

	Breakthrough time t_b (min) against 1000 ppm Carbon tetrachloride			Breakthrough time t_b (min) against 1000 ppm n-Hexane			Breakthrough time t_b (min) against 550 ppm Pentane		
Conditions	**Mfg. B OV**	**Mfg. C OV/AG**	**Mfg. C OV**	**Mfg. B OV**	**Mfg. C OV/AG**	**Mfg. C OV**	**Mfg. B OV**	**Mfg. C OV/AG**	**Mfg. C OV**
Preconditioned 25% RH test: 32 l/m 50% RH	219.6	201.9	210.5	178.9	178.7	191.6	222.0	217.9	232.3
Preconditioned 85% RH test: 32 l/m 50% RH	106.3	80.3	118.3	119.8	83.8	126.6	128.6	87.7	175.2
As received test: 64 l/m 25% RH	106.3	101.3	104.7	87.1	86.7	86.2	120.7	111.1	115.1
As received test: 64 l/m 50% RH	100.9	90.0	97.7	89.8	81.1	84.7	110.6	99.2	110.5
As received test: 64 l/m 80% RH	68.6	68.2	76.6	68.0	70.6	73.2	68.8	63.1	76.1

From Reference 50.

Table 14 Summary of Chin-Style Canister Breakthrough Data at 5 ppm Penetration

Challenge agent	Test relative humidity (%)	t_b at Challenge concentration 1000 ppm	2000 ppm	2500 ppm	5000 ppm	7000 ppm	7500 ppm	10,000 ppm
Carbon	50	140.2	69.7		36.8			21.3
tetrachloride	80	121.7	61.1		33.1			18.1
Ethyl acetate	50	127.4	55.7		24.5		20.1	
	50	125.4						
	80	78.0	44.0		16.6		10.8	
	80	97.7						
	80	98.8						
Pentane	50	78.6	56.7		32.9		23.3	
	80	73.8	49.8		30.4		21.1	
n-Hexane	50	106.0	55.4		28.7	21.2		
	80	97.2	53.7		32.7	21.4		
Heptane	50			59.2	25.9		19.0	
	80			57.1	26.0		15.5	

From Reference 50.

The chin-style canister test was performed at 50% and 80% RH, 64 l/m, and 25°C. The challenge concentrations for ethyl acetate and pentane were 1000, 2000, 5000, and 7000 ppm. The challenge concentrations for n-hexane were 1000, 2000, 5000, and 7500 ppm. The challenge concentrations for n-heptane were 2500, 5000, and 7500 ppm. A summary of test results with comparative breakthrough data of carbon tetrachloride is shown in Table 14. Ethyl acetate was eliminated for further consideration due to its water affinity. Pentane was selected for side-by-side comparative tests with carbon tetrachloride at 50% and 80% RH and 64 l/m. The chin style canisters were tested in the *"as received"* condition. The results indicated that at a challenge concentration of 4000 ppm pentane, equivalent breakthrough concentration to carbon tetrachloride at 5000 ppm was achieved. Selected breakthrough data for ethyl acetate, pentane, n-hexane, and n-heptane are presented in Table 2.

REFERENCES

1. Swab, C. F. and Ferber, B. I., Freon 113 as a test material for chemical cartridge respirators. U.S. Bureau of Mines Report of Investigations 7380, 1970.
2. Freedman, R. W., Ferber, B. I., and Harstein, A. M., Service lives of respirator cartridges versus several classes of organic vapors, *Am. Ind. Hyg. Assoc. J.*, 34, 55, 1973.
3. Smoot, D. M. and Smith, D. L., Development of improved respirator cartridge and canister test methods. NIOSH, DHEW (NIOSH) Publication No. 77-209, 1977.
4. Henry, N. W. and Wilhelme, R. S., An evaluation of respirator canisters to acrylonitrile vapors, *Am. Ind. Hyg. Assoc. J.*, 40, 1017, 1979.
5. Henry, N. W., Respirator cartridge and canister efficiency studies with formaldehyde, *Am. Ind. Hyg. Assoc. J.*, 42, 853, 1981.
6. Hitchcock, R. T., Reist, P. C., and Cooper, S. R., A respirator cartridge for use in removing methanol vapors, *Am. Ind. Hyg. Assoc. J.*, 42, 268, 1981.
7. Stampfer, J. F., Respirator canister evaluation for nine selected organic vapors, Am. Ind. Hyg. Assoc. J., 43, 319, 1982.
8. Ackley, M. W., Chemical cartridge respirator performance: 1,3 Butadiene, *Am. Ind. Hyg. Assoc. J.*, 48, 447, 1987.

9. Simon, C. G., Fisher, R. P., and Davidson, J. D., Evaluation of respirator cartridge for effectiveness of chlorine dioxide removal, *Am. Ind. Hyg. Assoc. J.*, 48, 1, 1987.
10. Werner, M. D., The effects of relative humidity on the vapor phase adsorption of trichloroethylene by activated carbon, *Am. Ind. Hyg. Assoc. J.*, 46, 585, 1985.
11. Moyer, E. S., Organic vapor (OV) respirator cartridge testing — Potential Jonas model applicability, *Am. Ind. Hyg. Assoc. J.*, 48, 791, 1987.
12. Bollinger, N. J. and Coffey, C. C., Report on OSHA requested study. Test and Certification Branch, NIOSH, December 1, 1987.
13. Nelson, G. O., Cartridge and canister service life to ethylene bromide, OSHA Ethylene Dibromide Docket, H-111, 1983.
14. Smith, D. L. and Giesler, W. S., An evaluation of organic vapor respirator cartridges and canister against vinyl chloride, NIOSH, HEW Publication No. (NIOSH) 75-111, 1975.
15. Miller, G. C. and Reist, P. C., Respirator cartridge service lives for exposure to vinyl chloride, *Am. Ind. Hyg. Assoc. J.*, 38, 498, 1977.
16. Beaumont, G. P. and Garrido, C. H., Respirator cartridge test system and test results for benzene and acrylonitrile, *Am. Ind. Hyg. Assoc. J.*, 40, 883, 1979.
17. MSA, Comfo II Mersorb indicator respirator for approval against mercury vapor and chlorine gas, Mine Safety Appliance Co. (MSA), Pittsburgh, PA, 1976.
18. Revoir, W. H. and Jones, J. A., Superior adsorbent for removal of mercury vapor from air, presented at Am. Ind. Hyg. Conference, San Francisco, 1972.
19. Coffey, C., Mercury vapor test system for cartridges — Air purifying respirator section test procedure, NIOSH, January 1985.
20. Henry, N. W. and Chen, K. A., Respirator cartridge breakthrough studies with tetra-ethyl lead (TEL), unpublished.
21. Pentone, Determination of Iodine Mist/Vapor Breakthrough and Pre/Post Test Flow Resistance for Respirator Cartridges: Scott No. 621 OVH, Pentone Corp. and Coors Brewery, Study performed by G. O. Nelson, March 1991.
22. NIOSH, Posthearing Comments of the National Institute for Occupational Safety and Health on the Occupational Safety and Health Administration's Proposed Rule on Occupational Exposure to 1,3 Butadiene. OSHA Docket No. H-041, September 1991.
23. MSA, Breakthrough data on hydrogen sulfide (personal communication). Mine Safety Appliance Co. (MSA), Pittsburgh, PA, 1991.
24. Birkner, L. R., Hanlon, R. G., Nelson, G. O., and Miller, W. B., Ethylene oxide: Air purifying respirator sorbent identification, Unpublished Report, 1981.
25. Nelson, G. O., Service life of respirator cartridges to benzene and acrylonitrile, Miller-Nelson Research, Alamo CA, letter to OSHA on October 27, 1980.
26. Silverstein, L., Data on MSA cartridges versus acrylonitrile concentration, Dow Badische Co., Williamsburg, VA, letter to OSHA on October 3, 1977.
27. Kennedy, E. R. and Smith, D. L., Respirator cartridge test for acrylonitrile, NIOSH, letter to OSHA on March 27, 1978.
28. Revoir, W. H., Test of Norton Company organic vapor cartridges for removal of acrylonitrile, Norton Company, Cranston, RI, Letters to OSHA on June 19, and August 7, 1978.
29. Dow Chemical Company, Cartridge breakthrough data for vinyl bromide (personal communication).
30. Moyer, E. S. and Berardinelli, S. P., Penetration of methyl isocyanates through organic vapor and acid gas respirator cartridges, NIOSH, Report PB87-103057, 1986.
31. Moyer, E. S., Organic vapor (OV) respirator cartridges and canister testing against methylene chloride, NIOSH, March 18, 1992.

32. Nelson, G. O., Respirator cartridge service life for glycol ethers (under contract with Union Carbide), November 17, 1993.
33. Swearengen, P. M., Priante, S. J., and Hannum, A. E., Respirator cartridge service life against nitroglycerine vapor, Lawrence Livermore National Laboratory, UCRL-102428, December 21, 1989.
34. Cohen, H. J., Determining the service life of organic-vapor respirator cartridges for nitroglycerine under workplace conditions, *Am. Ind. Hyg. Assoc. J.*, 54, 432, 1993.
35. Cohen, H. J., Briggs, D. E., and Garrison, R. P., Development of a field method for evaluating the service lives of organic vapor cartridges — Part I: Results of laboratory testing using carbon tetrachloride, *Am. Ind. Hyg. Assoc. J.*, 50, 486, 1989.
36. Nelson, G. O. and Harder, C. A., Respirator cartridge efficient studies. V. Effect of solvent vapor, *Am. Ind. Hyg. Assoc. J.*, 35, 391, 1974.
37. Jonas, L. A., Sansone, E. B., and Farris, T. S., Prediction of activated carbon performance for binary vapor mixtures, *Am. Ind. Hyg. Assoc. J.*, 44, 716, 1983.
38. Swearengen, P. M. and Weaver, S. C., Respirator cartridge study using organic-vapor mixtures, *Am. Ind. Hyg. Assoc. J.*, 49, 70, 1988.
39. Cohen, H. J., Briggs, D. E., and Garrison, R. P., Development of a field method for evaluating the service lives of organic vapor cartridges. Part III: Results of laboratory testing using binary organic vapor mixtures, *Am. Ind. Hyg. Assoc. J.*, 52, 34, 1991.
40. Yoon, H. Y. and Nelson, J. H., Application of gas adsorption kinetics. I. A theoretical modal for respirator cartridge service life, *Am. Ind. Hyg. Assoc. J.*, 45, 509, 1984.
41. Yoon, H. Y. and Nelson, J. H., Application of gas adsorption kinetics. II. A theoretical modal for respirator cartridge service life and its practical application, *Am. Ind. Hyg. Assoc. J.*, 45, 517, 1984.
42. Yoon, Y. H., Nelson, J. H., Lara, J., Kamel, C., and Fregeau, D., A theoretical interpretation of the service life of respirator cartridges for the binary acetone/m-xylene system, *Am. Ind. Hyg. Assoc. J.,* 52, 65, 1991.
43. Yoon, Y. H., Nelson, J. H., Lara, J., Kamel, C., and Fregeau, D., A theoretical interpretation of the service life of respirator cartridge service life for binary systems: Application to acetone/styrene mixtures, *Am. Ind. Hyg. Assoc. J.*, 53, 493, 1992.
44. NIOSH, Notice of acceptance of applications for approvals of air-purifying respirators with end-of-service-life indicators (ESLI). Criteria for Certification of End-of-Service-Life Indicators. NIOSH, *Federal Register*, 49, No. 140, 29270, July 19, 1984.
45. Maclay, G. J., Yue, C., Findlay, M. W., and Stetter, J. R., A prototype active end-of service-life indicator for respirator cartridges, *Appl. Occup. Environ. Hyg.*, 6, 677, 1991.
46. Moyer, E. S., Findlay, M. W., Maclay, G. J., and Stetter, R., Preliminary evaluation of an active end-of-service-life indicator for organic vapor cartridges respirators, *Am. Ind. Hyg. Assoc. J.*, 54, 417, 1993.
47. NIOSH, Letter to all respirator manufacturers, subject: Carbon tetrachloride substitute test agent for nonpowered negative pressure organic vapor cartridges, Division of Safety Research, NIOSH, September 27, 1993.
48. NIOSH, Letter to all respirator manufacturers, subject: Carbon tetrachloride substitute test agent for nonpowered negative pressure organic vapor chin-style canisters, Division of Safety Research, NIOSH, December 29, 1993.
49. NIOSH, Letter to all respirator manufacturers, subject: Carbon tetrachloride substitute test agent for nonpowered negative pressure organic vapor front mounted/back mounted canisters, Division of Safety Research, NIOSH, March 25, 1994.
50. Moyer, E. S., Peterson, J. S., and Calvert, C., Evaluation of carbon tetrachloride replacement agents for testing organic vapor cartridges, NIOSH File Report, February 1993.

18 RESPIRATOR FIT TESTING

I. THE DEVELOPMENT OF ANTHROPOMETRIC FIT TEST PANEL

The fit test requirement prescribed by the Bureau of Mines (BM) respirator test and certification schedules, 30 CFR 11, was retained in the 1972 revision of the regulations when NIOSH took over the BM program.[1] Since the fit test requirement is vague and nonspecific, NIOSH requested the Los Alamos Scientific Laboratory (LASL) to develop detailed anthropometric specifications to replace the requirements in the 30 CFR 11. An anthropometrically selected test panel would be useful for NIOSH to certify respirators, for other governmental agencies to evaluate devices that lack NIOSH certification, and for respirator manufacturers to design facepieces.

During World War II, the Aero Medical Laboratory of the U.S. Army Air Force collected the facial measurements of 1454 flying personnel (men and women),[2] and subsequently measured an additional 1500 individuals as a check on the original findings. The data were used to design an oxygen mask for flying personnel. In 1967, the U.S. Air Force conducted additional anthropometric studies on more than 4000 male and female personnel. The results indicated that females have shorter and narrower measurements in many dimensions, such as face length and lip length.[3,4]

Because anthropometric data on the civilian population were not available, LASL measured the facial dimensions of 200 males to determine whether there was a good correlation in anthropometric measurements between civilian and military populations.[5] The results indicated that of the nine measurements taken in common by both groups, six agreed to within 2 mm. Included in the six measurements were three dimensions which are important in defining test subjects (see Table 1): Bizygomatic Breadth (face width), Lip Length, and Menton-Nasal Root Depression Length (face length). It is interesting to note that more than 43% of the men measured in the LASL were Spanish-American. Although ethnic difference in facial dimensions may be expected, the means for all subjects measured did not differ by more than 2 mm for any measured parameter.

Based on these anthropometric measurements, two panels, one for the full facepiece and one for the half/quarter mask, were constructed. The dimensions of face length and face width were used to develop the full facepiece test panel. Since it is difficult to design a device to cover 100% of the human population, a maximum of 95% is more practical. For a normally distributed population, a ± 2 standard deviation (SD) over the mean of the variable would cover 95% of the population. For the face length, the mean (1967 Air Force data) is 120.3 mm, and the SD is 6.1 mm.

Table 1 Comparison of Male Facial Dimensions Taken by USAF (1967) and LASL (1972) (Listed in Order of Increasing Difference between the Two Surveys)

	Mean			Standard deviation	
Measurement	USAF (mm)	LASL (mm)	Difference (mm)	USAF (mm)	LASL (mm)
Nose length	51.3	51.6	0.3	3.7	4.3
Bitragion breadth	142.5	143	0.5	5.6	6.0
Bizygomatic breadth	142.3	142.9	0.6	5.2	5.7
Lip length	52.3	51.4	–0.9	3.7	3.9
Menton-nasal root depression length	120.3	121.4	1.1	6.1	6.5
Nose breadth	35.4	37.5	2.1	2.9	3.1
Maximum frontal breadth	116.0	112.6	–3.4	4.6	5.0
Bitragion-minimum frontal arc	308.1	302.8	–5.3	10	10.7
Bitragion-submandibular arc	309.8	317.9	8.1	15.8	14.6
Age, yr	30	42.5	12.5	6.3	10
Height, cm	177.3	176.5	–0.8	6.2	6.7
Weight, lb	173.6	170.9	–2.7	21.4	21.6
Size of sample	2420	200			

From Reference 5.

The range for face length which would cover 95% of the population is between 108 and 133 mm. The mean for face width is 142.3 mm with a SD of 5.2, a range between 132 and 153 mm would cover 95% of the population. In order to include females in the test panel, data on the survey of Air Force women[4] were also examined. It was found that males exhibited a longer face length and wider face width, whereas females exhibited a shorter face length and narrowed face width. To accommodate female test subjects, the lower limits were calculated by taking the mean values for face length and face width for the women and subtracting 2 SD. Upper limits were calculated from the male mean values plus 2 SD. Figure 1 shows the limiting face lengths (94 to 133 mm) and face width (118 to 153 mm) for a male and female full facepiece panel. The table is divided into length increments of 10 mm and width increments of 9 mm, thereby creating a 16 category panel. The number and percentage of the females (F) and males (M) are given as part of the total (T). The distribution is the same as for the general population. To simplify subject selection, the six least populated categories were deleted, leaving a 10-category table representing about 91% of the total population. Since NIOSH made a request that the panel size be limited to 25 test subjects, Figure 2 shows the distribution of 25 subjects among the 10 size categories. For testing purposes, it is assumed that a male and female face with the same two key dimensions are equivalent. The sex preferences listed in each box are advisory only, and members of the other sex may be substituted.

Face length and lip length are the key dimensions for selecting a test panel for half- or quarter-masks. The limits for face length are the same for the full facepiece panel. Limits for lip length are 35 to 60 mm, and are derived by adding 2 SD to the male mean value and subtracting 2 SD from the female mean. The limits for lip width are adjusted to allow equal intervals of 9 mm, producing a 12-category sample (see Figure 3). The upper left and lower right boxes containing only about 0.5% of the population are deleted. More than 95% of the population is represented in terms of

Face Length (mm) \ Face Width (mm)	117.5–126.5	126.5–135.5	135.5–144.5	144.5–153.5
133.5–123.5	2 F 0.1% 2 T <0.1%	39 M 1.7% 2 F 0.1% 41 T 1.0%	356 M 15.3% 2 F 0.1% 358 T 8.6%	253 M 10.8% 253 T 6.1%
123.5–113.5	2 M 0.1% 55 F 3.1% 57 T 1.4%	139 M 6.0% 115 F 6.4% 254 T 6.1%	836 M 35.8% 39 F 2.2% 875 T 21.1%	396 M 17.0% 2 F 0.1% 398 T 9.6%
113.5–103.5	1 M <0.1% 295 F 16.3% 296 T 7.1%	46 M 2.0% 612 F 33.8% 658 T 15.9%	190 M 8.1% 132 F 7.3% 322 T 7.8%	68 M 2.9% 5 F 0.3% 73 T 1.8%
103.5–93.5	190 F 10.5% 190 T 4.6%	1 M <0.1% 319 F 17.6% 320 T 7.7%	4 M 0.1% 36 F 2.0% 40 T 1.0%	1 M <0.1% 2 F 0.1% 3 T <0.1%

F = 1808 94.9%
M = 2332 96.4%
T = 4140 95.7%

M = Males
F = Females
T = Total

Figure 1 Male/female panel for testing of full-face masks. (From Reference 5.)

face and lip length in the 10 remaining boxes. Figure 4 gives the distribution of 25 test subjects over these 10 boxes.

The LASL anthropometric study indicated that one facepiece is unlikely to provide satisfactory fit for the entire panel. Later, many respirator manufacturers

Face Length (mm) \ Face Width (mm)	117.5–126.5	126.5–135.5	135.5–144.5	144.5–153.5
133.5–123.5			2 m	2 m
123.5–113.5		1 m 1 f	5 m	2 m
113.5–103.5	2 f	4 f	1 m 2 f	
103.5–93.5	2 f	2 f		

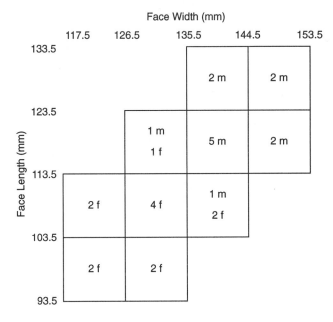

Figure 2 Male-and-female, 25-member panel for testing of full-face masks. (From Reference 5.)

Face Length (mm) \ Lip Length (mm)	34.5–43.5	43.5–52.5	52.5–61.5
123.5–133.5	8 M 0.3% 8 T 0.2%	343 M 14.5% 5 F 0.3% 348 T 8.3%	325 M 13.8% 1 F 326 T 7.7%
113.5–123.5	17 M 0.7% 83 F 4.5% 100 T 2.4%	704 M 29.8% 125 F 6.8% 829 T 19.7%	652 M 27.6% 10 F 0.5% 662 T 15.7%
103.5–113.5	2 M 0.1% 526 F 28.4% 528 T 12.5%	153 M 6.5% 513 F 27.7% 666 T 15.8%	149 M 6.3% 20 F 1.1% 169 T 4.0%
93.5–103.5	306 F 16.5% 306 T 7.3%	4 M 0.2% 248 F 13.4% 252 T 6.0%	2 M 0.1% 12 F 0.6% 14 T 0.3%

Figure 3 Male-and-female panel for testing of half-masks. (From Reference 5.)

developed multisize half-masks. However, Survivair was the first manufacturer to utilize the LASL panel to develop multisize half-mask respirators made of silicone rubber with small, medium, and large sizes. With the emergence of more ethnic groups such as Asians and Hispanics into the work force, it is not clear whether the LASL panel can still be representative of the American work population.

II. RESPIRATOR FIT TESTING METHODS

Two types of respirator fit testing methods are available: the qualitative fit testing (QLFT) and the quantitative fit testing (QNFT). Both methods employ a chemical to determine the individual's response to the odor, taste, or smell of the testing agent as it leaks into the respirator inlet covering, or to perform an actual measurement of the amount of leakage of the chemical inside the respirator inlet covering with a precision instrument. There are many chemicals and methods available for performing the fit testing. Some of these were reported by Kolesar[6] and White.[7]

A. Qualitative Fit Test Method

1. Isoamyl Acetate (Banana Oil) Test

Isoamyl acetate (IAA) has a pleasant and easily detectable odor, with a low odor threshold. It has been used for fit testing before the development of the

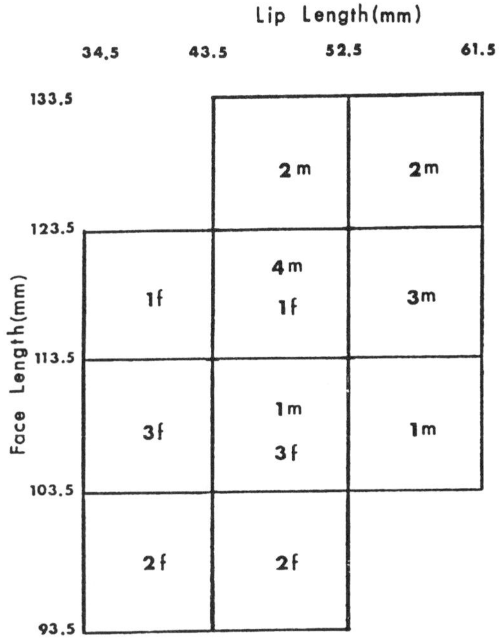

Figure 4 Male-and-female, 25-member panel for testing of half-masks. (From Reference 5.)

quantitative fit testing method. There are at least two versions of the IAA QLFT methods available. The simplest version of this test involves saturating a piece of cotton or cloth with the liquid and passing it close to the respirator face-to-facepiece sealing surface. This method has been improved by the Du Pont Company and has been accepted by the Occupational Safety and Health Administration (OSHA) as an alternative method to QNFT in the amended lead standard promulgated by OSHA.[8]

A more complex version of this test involves the use of a room or small booth. A known concentration of isoamyl acetate vapor, approximately 100 ppm, is generated by evaporating the IAA liquid. The chamber method has been selected for the respirator faceseal test prescribed in the respirator certification regulations, 30 CFR 11.[1]

2. *Irritant Smoke Method*

The irritant smoke QLFT was used by the Los Alamos Scientific Laboratory for respirator fit testing before the development of the QNFT.[7] The irritant smoke is produced by a commercially available smoke tube. These tubes are filled with either stannic chloride or titanium tetrachloride impregnated pumice stone. The irritating smoke consists of hydrochloric acid absorbed on small solid particles. The smoke is directed at the facepiece seal, and leakage is detected by the user's involuntary coughing or sneezing due to irritation of the respiratory tract. The method was refined by the American Iron and Steel Institute and was accepted by OSHA as an alternative method to QNFT in the amended lead standard promulgated by OSHA.[8]

3. *Sodium Saccharin Method*

This method was developed by 3M Company, a major disposable respirator manufacturer, specifically for fit testing disposable respirators because the fine particles of the irritant fume will penetrate readily into the low-efficiency particulate filters and fail the test subject. In spite of issues raised at the OSHA hearing concerning the carcinogenicity of sodium saccharin, an artificial sweetener, and the large size of the saccharin particles which may not penetrate through the respirator faceseal, as well as the lack of validation data developed on the 3M disposable respirators, OSHA accepted this method as an alternative method to QNFT in the amended OSHA lead standard.[8]

4. *Miscellaneous Qualitative Fit Testing Methods*

Other qualitative fit tests have been investigated,[10,11] but have not been as well received as those mentioned above. In one method, a stream of talcum powder or coal dust is directed around the face-to-facepiece seal. The user then removes the respirator and leakage is revealed by any obvious powder or dust streaks. Another test method involves spraying uranin (a fluorescein dye) around the respirator sealing surface. The respirator is then removed and leakage is detected using a fluorescent light source. Finally, the U.S. Army has experimented with a method that used a 1450 mg/m^3 concentration of chloropicrin (trichloronitromethane) as an odor-sensitive, vapor-challenge agent. A leak is considered to have occurred if the user can detect the odor of chloropicrin.

B. Quantitative Fit Test Methods

Both gaseous or aerosol challenge agents can be used for QNFT. Due to complexity of the measuring instrument and the low sensitivity of some measuring instruments for gaseous testing agents, the aerosol methods have gained wide acceptance. The gaseous QNFT systems described below are seldom used anymore.

1. *QNFT with Gaseous Challenge Agents*

Argon. The British Safety in Mines Research Establishment[6,12] has developed a dynamic method to measure the leakage of full facepiece respirators while the subject performs various exercises. Argon is used as the challenge agent because it is physiologically inert, inexpensive, and commercially available in a pure form. In this test, the subject is placed inside a transparent plastic hood which is sealed around the waist. Pure argon is introduced into the top of the hood to maintain hood pressure slightly above atmospheric pressure. To implement this test, the wearer inhales oxygen supplied from a cylinder. The exhaled gas flows through the other breathing tube into a sampling bladder that is fitted with a sampling port. The amount of argon in the exhaled gas is measured with a mass spectrometer. This instrument is capable of measuring the differential amount of argon present in the exhaled breath to a concentration of 10 ppm. This method has been designed to measure respirator leakages of 0.001%.

Ethylene. National Draeger Incorporated has developed two versions of the model 80 facemask fit-test device: One uses a detector tube and the other uses an electronic leak detector instrument. To perform a quantitative respirator fit test, the subject places a transparent plastic hood over the head and seals it around the neck. An air mixture containing ethylene (2% by volume) is used as a challenge gas. The mixture is allowed to flow into the hood from a regulated cylinder supply. The concentration inside the hood is assumed to be the same as the supply challenge gas mixture. The facepiece fit is determined by measuring the concentration of the challenge gas in the exhaled breath. A gas detector tube is used for this measurement; it has a detection range from 0.5 to 10 ppm of ethylene. Respirator fit leakage in the range of 0.0025 to 0.95% may be detected based on a volume challenge gas concentration.

In place of the ethylene detector tube, an electronic leak detector instrument is available for simultaneously measuring the concentration of challenge gas in the hood and exhaled breath. However, when using the electronic leak detector scheme, a 2% mixture of sulfur hexafluoride in air is used as the challenge gas instead of the ethylene mixture. The electrons produce a current between two electrodes. A change in the measured current is proportional to the concentration of the challenge gas. Sampling the exhaled and hood gas are performed simultaneously. Detector sensitivity is approximately the same as the detector tube version.[6]

Dichlorodifluoromethane. Dichlorodifluoromethane (Freon 12), a fluorocarbon with low toxicity has been used by many investigators for performing QNFT.[13-16] Adley[13,14] developed a system using dichlorodifluoromethane as a challenge gas to determine the protection factor for the self-contained breathing apparatus. This system was modified by the Bureau of Mines for determining the protection provided by respirators. The test chamber contains a dynamically generated concentration of 1000 ppm Freon 12. To simulate the actual use condition, the Bureau of Mines developed the following exercises to be performed by the test subject during the testing:

a. Nodding and turning head, 3 min.
b. Smiling, 1.5 min.
c. Frowning, 1.5 min.
d. Reciting alphabet aloud, 3 min.
e. Talking aloud, 3 min.
f. Deep and shallow breathing, 3 min.
g. Pumping air with hand pump into a one cubic foot cylinder, 3 min.

Leakage of the respirator is continuously monitored by a halide meter, and the output is displayed on a strip-chart recorder. With a sensitivity of 2 ppm of Freon 12 as measured by the meter, a maximum leakage of 0.2% may be measured.

Another system was developed by the Dow Chemical Company[15] for training workers who need to wear a mouth piece respirator for escape from a chlorine environment. The chamber concentration was set at 500 ppm. The Bureau of Mines[16] also conducted studies to determine the suitability of using Freon 12 as a fit testing agent.

Helium. The Scott Aviation, a Division of ATI Incorporated, has developed a respirator quantitative fit test procedure that uses helium as a challenge gas. The test chamber contains an air mixture with 10% of helium. Breathing oxygen (medical quality) is supplied to the test subject from a regulated cylinder supply. The wearer's oxygen consumption rate for the leak test is measured by the drop in cylinder pressure with time. The facepiece leakage is measured with a mass spectrometer. A sampling rate of 300 cm^3/min is used. The helium leak detection system can measure penetrations in the range of 1.0 to 0.001%.[12]

n-Pentane. The Federal Aviation Administration, Survival Research Unit, has developed an n-pentane challenge gas respirator quantitative fit test method.[6,17,18] The test chamber has a stable concentration of 120 ppm n-pentane. The leakage measuring probe on the facepiece is made of small stainless steel needles inserted through the facepiece. The respirator facepiece leakage is measured with a gas chromatograph equipped with a hydrogen-flame ionization detector. This system can measure a facepiece leakage of 1%.

Sulfur Hexafluoride. The Los Alamos Scientific Laboratory developed a sulfur hexafluoride respirator quantitative fit test method.[19] To implement this test, the test subject wears a respirator, complete with oxygen breathing tubes, and is placed inside a test chamber with a concentration of 50-ppm sulfur hexafluoride in room air. Samples collected at the breathing zone are corrected through a probe and connecting tube. Samples from the test chamber and respirator facepiece probe are analyzed with a hydrogen-flame detector. In a hydrogen-rich flame, sulfur emits a characteristic luminescence at the 394 nm wavelength, and the intensity of luminescence is a direct function of the sulfur concentration. This system has a lower detection limit of 0.5-ppm sulfur hexafluoride. Since the chamber concentration is maintained at 50 ppm, the sensitivity of the sulfur hexafluoride QNFT system is 1.0%.

2. QNFT with Aerosol Challenge Agents

Biologic Agent Fit Testing System. The U.S. Army has developed an QNFT method that uses a nonpathogenic aerosol of Bacillus subtilis or B. globigii spore-forming bacterium as a challenge agent.[6,20,21] The challenge aerosol concentration inside the test chamber is established by continuously atomizing a water suspension of bacterial spores using a Binks spray nozzle. An aerosol concentration of 300,000 spores/l is maintained by dynamically mixing the test chamber. The chamber concentration is monitored with a Naval Research Laboratory smoke penetration meter. The aerosol can be physically characterized as particles having a mass median diameter of 2.1 μm, and 95% of the particle's diameters fall between 1.0 and 5.0 μm. A specially designed mouth sampler is used to collect the spores penetrating the respirator. The sampler consists of a latex mouthpiece which holds an oval brass cartridge packed with surgical quality cotton. The cotton serves as the sampling medium for leaked spores. During the fit test, the test subject inhales through the sampler and exhales through the nose. After the test, the bacteria spores collected on the cotton is quantified using standard bacteriological plating procedures. Facepiece leakage of 1% has been reported.

Uranin Fit Testing System. The Harvard School of Public Health developed a QNFT method using uranin aerosol as a challenge agent.[6,22,23] Uranin is a commercial dyestuff that is used as a tracer in medical and air pollution studies. The disodium salt of fluorescein is readily soluble in water and excited by blue light ranging from 4400 to 5100 Å, and emits at 5100 to 5900 Å. The test is performed inside a transparent plastic hood which is drawn snugly around the waist of the test subject. The challenge aerosol is generated by a nebulizer using a 2.35% solution of uranin. Two standard Greenburg-Smith impingers remove large particles from the stream. Dilution air is mixed with the uranin particles to reach a challenge aerosol concentration 4 mg/m^3. The geometric mean size of the particles in the challenge aerosol is 0.2 μm with a standard deviation of approximately 2. A Multiplier Fluorescence Meter and a Fluorescence Unit are used to compare the concentrations of the challenge aerosol measured outside to that measured inside the respirator facepiece. This method has the capability of detecting leaks as low as 0.05%.

Coal Dust Fit Testing System. A quantitative fit testing system using polydisperse coal dust was developed by the Bureau of Mines as an alternate to the Freon 12 system described above.[24] The coal dust concentration inside the test chamber has a concentration between 50 and 100 mg/m^3 with a geometric mean of 1.2 μm and a standard deviation of not to exceed 2. The coal dust penetrated into the facepiece is collected on a membrane filter with a sampling rate of 2 l/m. The protection factor is measured by comparing the ambient and inmask weight of coal dust. However, this method was never prescribed as a testing schedule for respirator certification.

The following three aerosol QNFT systems using oil mist, NaCl, and ambient aerosols have gained wider acceptance than the aerosol testing agents mentioned above.

Oil Mist QNFT System. This is a widely used QNFT method.[7] In the early 1960s, the Bureau of Mines used a thermally generated monodisperse di-octyl phthalate (DOP) or di-2-ethylhexyl phthalate (DEHP) aerosol for performing fit testing of respirators. A similar system was developed and used by the Los Alamos Scientific Laboratory (LASL) to determine the protection factors for the SCBAs. The test system consists of a DOP generator, a test chamber, and an analyzing system which contain a forward light-scattering photometer and a data recorder. The concentration of the challenge is between 70 and 100 mg/m^3 with a size of 0.3 μm. The test chamber is 1 × 1 × 2.6 m. The sampling rate for the aerosol is 8 l/m.

In early 1970s, NIOSH asked LASL to develop a system utilizing a cold generated poly disperse DOP as a testing agent since the thermally generated DOP may consist of many contaminants with unknown toxicity. The system consists of a nebulizer and an impactor to remove large particles. The challenge concentration has been reduced to 25 mg/m^3 with a mass median aerodynamic size of 0.6 μm and a geometric standard deviation of 2. The sampling rate was about 1 l/m. A photomultiplier tube (PMT) with a forward scattering light photometer was the aerosol detector. The amount of light scattered is received by the PMT. The output from the PMT is proportional to the aerosol concentration in the sample. Two commercial polydisperse DOP QNFT system, with designs similar to the LASL system have been

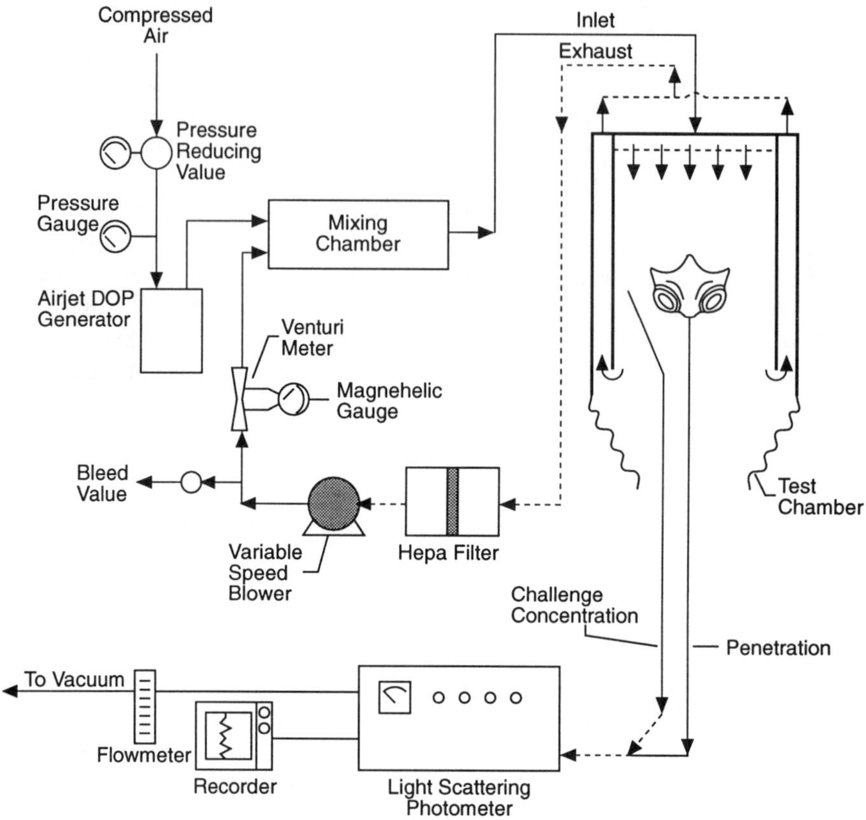

Figure 5 Flow diagram DOP oil mist QNFT system.

developed and widely used by government and industry throughout the 1970s. A schematic diagram of an oil mist QNFT system is shown in Figure 5.

In the late 1970s, an animal toxicological study on DEHP conducted by the National Toxicological Program demonstrated that cancer growth was found on both exposed and control groups.[25] The study concluded that DEHP was a potential animal carcinogen. This action prompted a search for a replacement of DEHP with similar physical characteristics but with a low toxicity.[26,27] Corn oil was found to be a suitable replacement, and became the commonly used substitute for DOP for commercially available portable fit testing instruments. It was also recommended by NIOSH. However, corn oil can become rancid upon prolonged exposure; this requires frequent cleaning of the test chamber to prevent the build up of unpleasant odors and to prevent bacteria growth which could interfere with aerosol generation. Many Department of Energy installations do not use corn oil because of this problem. For example, Los Alamos National Laboratory used DEHS, and the Lawrence Livermore National Laboratory used PEG 400 for performing respirator fit testing. The Department of Energy has authorized the use of these two chemicals for performing the QNFT.

Since the QNFT systems are quite expensive, OSHA asked the aerosol scientists at the National Bureau of Standards (NBS) and the University of Minnesota to determine whether it was feasible to develop a low cost QNFT system.[27] A low cost oil mist QNFT system using inexpensive off-the-shelf components was developed. The aerosol is generated with a clinical nebulizer and diluted with air delivered from a small air pump. The diluted air was then introduced into the test chamber. The aerosol detector is a sensor used in a smoke detector. To simplify operation, the light emitting diode (LED) display was used to indicate the fit factor achieved by the test subject. The test results indicate that the system is capable to measure fit factors up to 450 which is suitable for the widely used negative pressure half-mask respirators. The cost for building up the system is less than $300.

Sodium Chloride Fit Testing System. The NaCl fit testing system has been widely used in Europe because this aerosol does not destroy the charges of electrostatic filter which is dependent upon electric charges to improve the efficiency of the filter. Under a contract with NIOSH, LASL had developed a NaCl fit testing system.[28,29] Commercial sodium chloride QNFT systems based on the LASL design are also available. The system consists of an aerosol generation and an aerosol detection system. The aerosol generation system consists of a compressor, a nebulizer, a dilution air blower, and a mixing chamber. The nebulizer is patterned after the British Wright design and subsequently modified by LASL. A 1% aqueous NaCl solution is used for aerosol generation. The NaCl aerosol generated by the nebulizer is diluted in the mixing and drying chamber with clean dilution air from a blower. NaCl aerosol has a mass median aerodynamic diameter (MMAD) of 0.7 to 0.9 μm, and a wide range of concentrations can be generated from the system. The commonly used concentration is 15 mg/m^3. Because the NaCl aerosol is hygroscopic, a dehumidifier is needed to keep the relative humidity of the dilution air below 50%.

The aerosol detection system consists of a peristaltic pump, a burner assembly, sodium filter and photomultiplier tube assembly, and a solid state amplifier. The peristaltic pump permits the aerosol sample to be drawn out of the QNFT test chamber and fed into the burner assembly without significant loss of aerosol due to the pumping action. The burner assembly consists of a laboratory burner, propane and air feed lines, and a glass chimney with a metal cap. The sample aerosol is mixed with the combustion air and injected into the burner. A PMT with a sodium light filter is placed near the burner flame. Only the yellow sodium emission color is passed through the filter and received by the PMT. The output is proportional to the mass flow rate of sodium in the aerosol sample. A schematic diagram of the NaCl QNFT system is shown in Figure 6.

Ambient Aerosol QNFT System. The aerosol generation and dilution system needs some maintenance work and the testing agent may pose some risk to the test subjects. A system using a continuous flow condensation nuclei counter (CNC) as the aerosol detector was first proposed by Willeke et al. in 1981.[30] The CNC is capable of measuring submicrometer particles that are desirable for fit testing. The CNC uses isopropanol to saturate and enlarge submicrometer particles to supermicrometer size, which can then be counted by the light scattering photometer. The test results indicate that there is a general agreement in measurement between the photometer and the

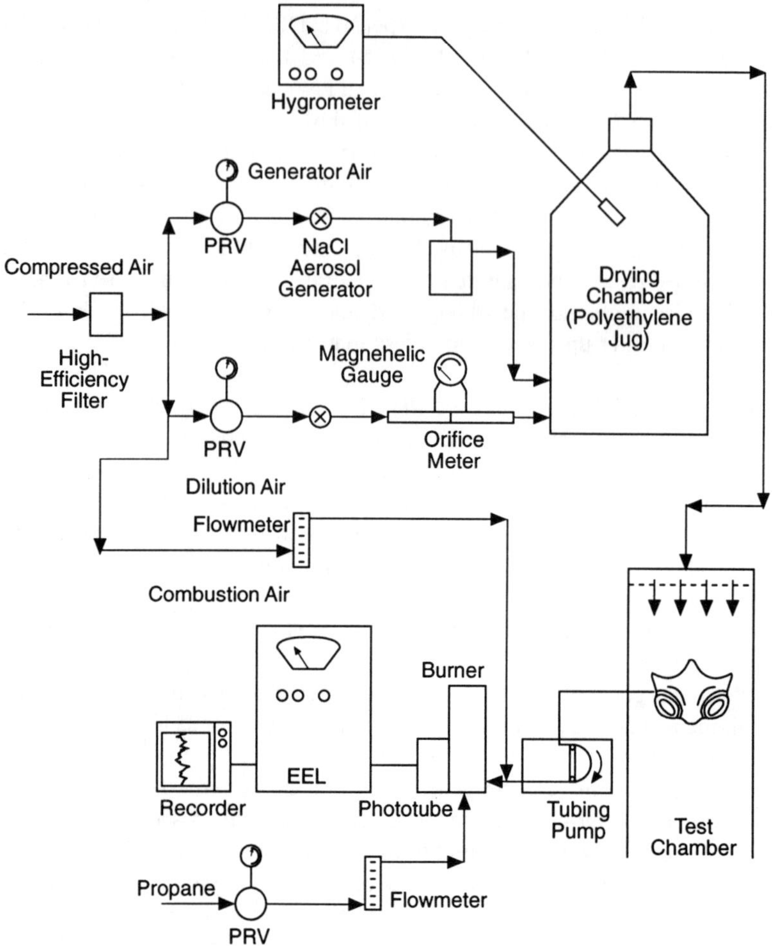

Figure 6 NaCl aerosol QNFT system.

CNC (see Figure 7). Various aerosols such as the room aerosols, cigarette smoke, carbon particles generated from the brushes of a high volume air sampler, or the particles generated from the heated wire of an electric hair dryer can also be used as test agents for performing QNFT.

Later on, Thermo System Inc. (TSI), a manufacturer of the continuous flow CNC, was under contract with the U.S. Army to develop a miniaturized version of the continuous flow CNC used in the Willeke study. A hand-held CNC using the ambient aerosol as the challenge with the trade name "PortaCount®" was introduced in 1988. A schematic diagram of the PortaCount® is shown in Figure 8.

Aerosol is drawn through the instrument by a vacuum pump with switching valves to collect either ambient or inmask samples. The sample then passed a sensor which consists of a saturator, condenser, and optical elements. The alcohol vapor condenses on the particles, causing them to grow into droplets. Each droplet passes through the sensor and scatters light. The light is collected and focused on

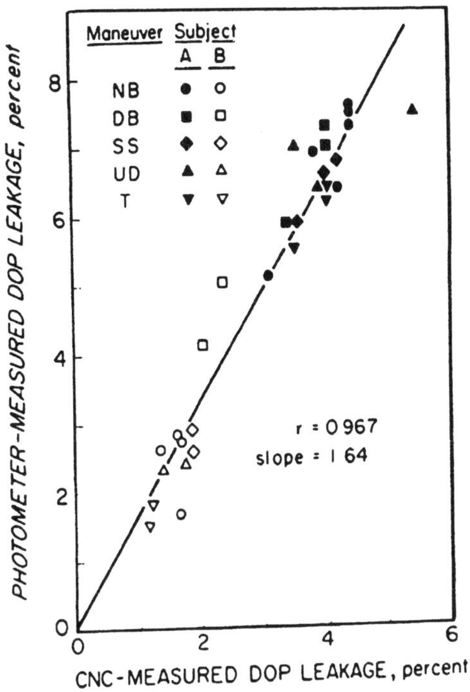

Figure 7 Photometer versus condensation-nuclei-counter response for various fit test maneuvers with DOP aerosols. Three runs for subject A, two runs for subject B. (From Reference 30.)

the photo diode. The photo diode generates an electrical pulse from the scattered light developed by each droplet that passes through the sensing volume. The particle concentration is determined by counting the number of pulses generated. To ease operation, the QNFT procedure is programmed. The instrument can also be used as a particle counter. Due to its simplified operation and without requiring an aerosol generation device, the PortaCount® has gained wide acceptance and is commonly used for preforming fit testing of negative pressure air-purifying respirators. Because it is essentially a "black box" with no user control, the performance of this device cannot be easily evaluated.

Since the beginning of respirator quantitative fit testing, users and manufacturers have expressed concerns that leakage of an aerosol might be significantly different from that of a vapor challenge agent and that the PFs calculated using an aerosol challenge agent might not predict the level of protection for gaseous contaminants. This concern has been studied by two independent laboratories in the U.K.[31,32] One study compared the leakage of a submicrometer aerosol of sodium chloride to Freon 12 or dichlorodifluoromethane and found that the leakages for full facepiece and half-mask respirators were comparable.[33] In another study, the leakage of a submicrometer aerosol of sodium chloride was compared to argon, and the results for a demand self-contained breathing apparatus (SCBA) respirator were similar.[31,32] Finally, a study by the Los Alamos Scientific Laboratory (LASL) compared the leakage of sulfur hexafluoride to a submicrometer aerosol of sodium

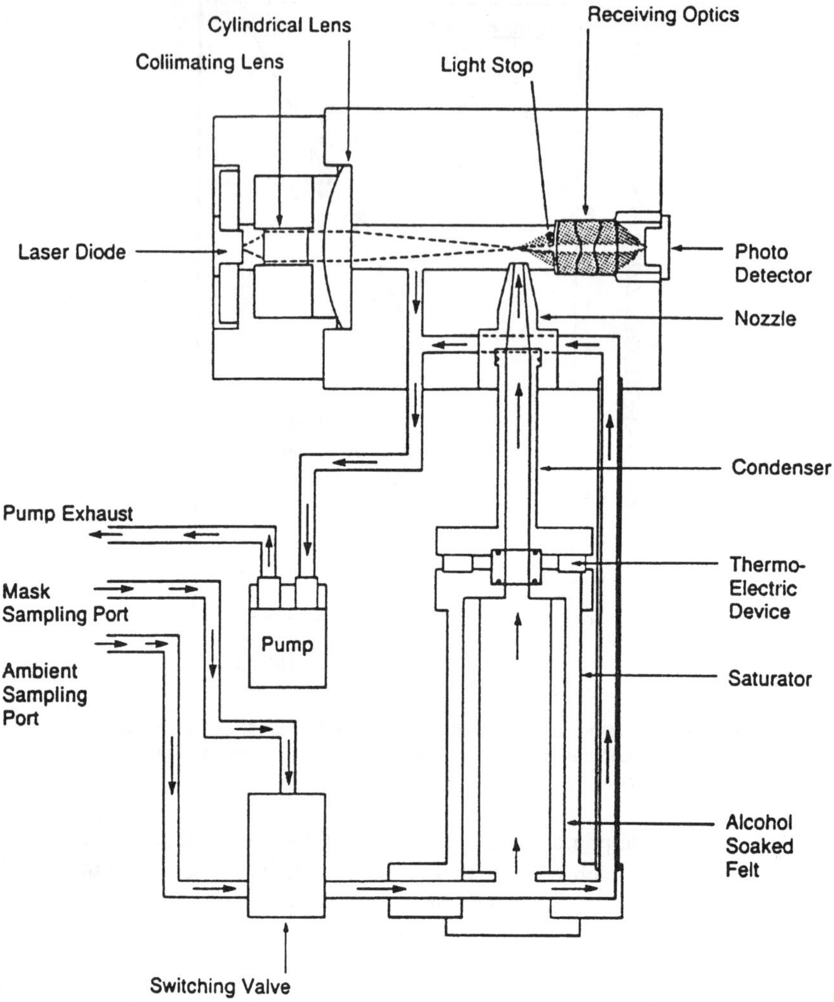

Figure 8 Schematic of PortaCount®.

chloride and di-2-ethylhexyl phthalate (DEHP) (dioctyl phthalate) and found that the results for two types of half-face respirators were similar.[34]

Large Aerosol QNFT System. Because the commonly used QNFT system employs a submicrometer aerosol as a challenge, the small aerosol particles would readily penetrate through the low efficiency disposable dust/mist respirators. Iverson et al.[35] of the 3M Company, developed a system using a large size monodisperse aerosol for performing fit testing. A challenge aerosol having a mass median aerodynamic diameter (MMAD) of 2.5 μm with a geometric standard deviation (GSD) less than 1.2 was selected for performing QNFT. The size was selected so that the filter penetration would be reduced to 0.3%. The MMAD for the silica dust used in the dust/mist filter certification test is about 2 μm with a geometric standard deviation of 2. The commonly used QNFT system employs a polydisperse oil mist or a

Table 2 Particle Size Distribution of Testing Aerosols at Specified Size

Test aerosol	MMAD	% Less than stated size (μm) 4	3	2	1
Polydisperse aerosol	0.6	99.7	98	95	77
Silica dust	2.0	84.5	79	50	16
Monodisperse aerosol	2.5	97.5	73	16	<0.01
Polydisperse aerosol	2.5	75	62	34	9

polydisperse sodium chloride aerosol as the fit testing agent with a MMAD of 0.6 μm and a GSD of 2. If it is assumed that all test aerosols are log-normally distributed, the particle size distribution for these three aerosols are shown in Table 2.

For comparison purposes, the size distribution of a 2.5 μm polydisperse aerosol with a GSD of 2 is also included in the table. Table 2 indicates that 77% of the submicrometer size polydisperse test aerosol have a size less than 1 μm. The submicrometer particle distribution reduces to 16% for the silica dust and 9% for the 2.5 μm polydisperse aerosol, respectively. However, there are very few submicrometer particles having sizes less than 1 μm for the 2.5 μm monodisperse test aerosol.

It is obvious that the monodisperse aerosol contains a dominating fraction of large particles than the other two test aerosols. The facepiece seal leakage studies conducted by Hinds, Willeke and coworkers[36-39] indicated that particles larger than 1 μm do not penetrate through the faceseal readily. The penetration requirement for the dust/mist filter certification test is 1%. The 2.5 μm monodisperse aerosol proposed by 3M has minimized the dust/mist filter penetration to 0.3%. It seems that there should be very little difference in penetration value between 1 and 0.3%. However, the reduced penetration value requires the use of a test aerosol which has very few small particles. This in turn has greatly reduced the possibility for particle penetration through the faceseal. Hence, very high fit factors can be achieved by using this method.

This fact can also be illustrated in the Willeke study[36,37] which compares the fit factors obtained from different sizes of leak holes, test aerosols, and detection methods. The results are shown in Table 3.

The table shows that although the large 2.5 μm monodisperse corn oil aerosol is the least sensitive fit test of the tests shown, it provides the highest fit factor under all test conditions.

The test results indicated that a majority of test subjects achieved a fit factor over 300. The purpose for performing the QNFT is to assess the leakage around the facepiece seal. If the test aerosol contains very few small particles that would

Table 3 Fit Factors Calculated for Each Hole Size Based on Different Detection Methods and Test Aerosols

Test aerosol/ detection method	CMD (μm)	GSD	Fit factors for hole sizes Small	Medium	Large
Corn oil/scatter	0.15	2.0	400	50	19
Corn oil/count	0.15	2.0	830	70	26
Corn oil/scatter	2.5	1.0	1250	250	83
Sodium chloride/mass	0.12	2.0	526	59	20
Silica dust/mass	0.55	2.0	390	83	29

From Reference 37.

penetrate through the facepiece, the large particle QNFT method has little value, except for promoting a high fit factor for disposable dust/mist respirators. Individuals who have been fit tested with the large particle QNFT system could be exposed to toxic air contaminants with particles having a size less than the 2.5 μm monodisperse test aerosol. However, the study did not point to the deficiency associated with the large monodisperse aerosol QNFT method.

QNFT System Using Air as the Challenge Agent. Since the late 1980s, several nonaerosol based quantitative fit test methods have been developed. Carpenter and Willeke[40] proposed a method which measured pressure decay rate caused by respirator leakage. A pressure sensor (transducer) is attached on a modified cartridge and placed on the respirator. The other cartridge is modified to permit capping, so that a negative pressure could be created inside the facepiece. The signal from the transducer is sent to an electronic readout for recording. The pressure test does not require the same initial negative pressure differential between the interior and exterior of the respirator. A flexible diaphragm is attached to the other cartridge in order to conduct the test repeatedly at the same initial pressure. After the test subject took a deep breath the diaphragm was moved outward to decrease the internal pressure, thus initiating a pressure decay record caused by air leakage into the respirator.

The decay pressure differential between the interior and exterior of the respirator is exponential with time. The decay rate is also dependent upon the cavity volume of the respirator. The calculation of fit factor based on the pressure decay rate can be summarized in the following formulae:

$$\text{LCR} = \text{leak to cavity ratio} = D^3/V \tag{1}$$

$$\text{WLS} = c_1(D3) = c_1(D^3/v)^{\alpha} = c_1(\text{LCR})^{\alpha} \tag{2}$$

$$Q_{leak} = 4 \times 10^5\ D^3\ P^{3/4},\ \text{cm/min} \tag{3}$$

$$Q_{clean} = (20{,}000 + 27{,}000)P = 47{,}000\ P\ \text{cm}^3/\text{min} \tag{4}$$

The protection number (fit factor), WPN = Q_{clean}/Q_{leak}, or

$$\text{WPN} = 8.2 \times 10^4\ V^{-1}P^{1/4}\ \text{WLS}^{-3/5} \tag{5}$$

where D is leak hole diameter in cm, V is volume of respirator cavity, and WLS is Willeke's leak slope. Based on limited experimental data, the coefficient c_1 is 5.5 × 10^9/s, and the exponent α is 5/3. P is pressure in cm water gauge. In order to assess the fit of the respirator, the leak flow has to be related to the flow through the clean air-purifying elements. The clean air flow is Q_{clean} which is based on the flow on two filter cartridges,

For example, a half-mask with a cavity volume of 190 ml is worn at an average negative pressure of 1 ml w.g. From the dynamic test, if the leak slope WLS is equal to 1.53/s, then the leak cavity ratio, LCR is calculated from Formula (3):

Table 4 Controlled Negative Pressure Test Variables

Variables	Selection	Values
Inspiratory work rate		**Kcal/h**
	Light	100
	Medium	200
	Heavy	300
	Extreme	350
		Resistance
Cartridge type	Dust	Low
	HEPA/chemical combination	Medium
		High
Mask	Half	
	Full	
		H_2O Pressure
Gender	Male	0.58″H_2O
	Female	0.50 H_2O

$$LCR = 1.85 \times 10^{-6} = D^3/V \tag{6}$$

If D = 0.071 cm and V = 190 cm^3, then the volumetric air flow rate through the leak at a P = 1 cm w.g. can be calculated from Formula 6, and Q_{leak} = 143 cm^3/min. The combined clean air flow through two filter cartridges from Q_{clean} = 47,000 cm^3. WPN can be determined from two air flow rates:

$$WPN = 47{,}000/143 = 329 \tag{7}$$

The air leakage is the inverse of WPN, which is 0.3%.

Another respirator fit test method using air as the challenge is proposed by Crutchfield.[41,42] The controlled negative pressure (CNP) fit test method measures leak rate through the facepiece as a method for determining the fit factor. To perform the test, the subject closes his/her mouth and holds his/her breath. Then, an air pump removes air from the respirator facepiece to a level that is representative of the negative pressure that occurs upon inhalation. The volume of air removed from the facepiece relates to faceseal leakage of the respirator. The fit factor is calculated as the following:

$$FF = \frac{\text{Modeled Breathing Rate}}{\text{Measured Leak Rate}} \tag{8}$$

Based on the selection of parameters listed in Table 4, a modeled breathing rate can be selected. The modeled breathing rate is based on the gender, work rate, type of facepiece, and the type of filters being used. The ratio between the modeled breathing rate and the leak rate is the fit factor for the test subject. For example, if the modeled breathing rate is set at 53.9 LPM and the measured leak rate is 211.9 ml/min, then fit factor = 53.9 × 1000/211.9 = 255.

One important aspect for conducting any respirator fit testing is that the exercises performed by the test subject must be able to simulate actual movements occurring

Table 5 Exercises Performed by Test Subjects

CNP method	OSHA QNFT method
Hold head forward	Normal breathing
Turn head to left and hold	Deep breathing
Turn head to right and hold	Turning head side to side
Hold head up	Moving head up and down
Hold head down	Reading
Hold head forward	Bending over and touch toes
	Jogging in place
	Normal breathing

in the workplace. The fit test protocol prescribed in the OSHA asbestos standard has specified the dynamic body movement exercises. However, the CNP method does not allow the test subject to perform any dynamic exercise during the test and further requires that the test subject must hold the breath when fit factor is being measured. A comparison of exercises between the OSHA and the CNP test protocol is listed in Table 5.

The limited data does not indicate whether exercise has an effect on the fit test results if the test subject performs the same exercises specified in the aerosol based fit test system; there is no degradation of fit when the test subject performs exercises at a higher work rate. Also, there is no human fit test data available to indicate that the CNP method will reject more poor fits than the aerosol based fit testing methods. The author claims that the CNP method yields lower fit factors than the aerosol system and that it is more conservative. However, as 15 mm of water pressure has been selected as the challenge controlled negative pressure this represents only a sedentary work rate.

Both the Willeke and Crutchfield studies proposed a promising alternative method to perform respirator fit testing. However, more studies are needed to prove the virtue of these methods.

III. VALIDATION OF QUALITATIVE FIT TESTING METHODS

In November 1978, OSHA published the lead standard. The standard requires that quantitative fit testing (QNFT) be performed on negative pressure air-purifying respirators and there was also a requirement that mandated the use of high-efficiency particulate air (HEPA) filters for air-purifying respirators. The 3M Company, a disposable respirator manufacturer, which did not produce a HEPA filter which is required for performing a QNFT, petitioned OSHA to reconsider or modify the QNFT requirement for the reasons that there was no standardized protocol for performing QNFT and qualitative fit testing (QLFT) would provide equivalent protection. In the meantime, the National Cancer Institute released a study which stated that the testing agent used for performing QNFT, DEHP, may be a potential carcinogen. Another factor raised by 3M was that QNFT is more expensive than QLFT because it requires sophisticated equipment and trained operator.

On May 19 1981, OSHA published a proposed rule to permit the use of qualitative fit testing (QLFT) method in lieu of the QNFT. Three QLFT methods using different testing agents were submitted to OSHA for acceptance. A method developed by the Du Pont Company used the isoamyl acetate (IAA) as a testing agent. 3M developed a method using sodium saccharin, an artificial sweetener and a suspected carcinogen, as the testing agent. The American Iron and Steel Institute refined a method which used the irritant smoke generated from the ventilation smoke tube as a testing agent. OSHA held a hearing to solicit comments on the acceptability of these QLFT methods. Based on the comments and data received, OSHA accepted all three alternative QLFT methods an alternates to the QNFT method in the amended lead standard.[43]

Both the isoamyl acetate and irritant fume qualitative fit testing methods have been widely used for respirator fit testing before the development of the QNFT. Because these methods are a merely a "pass or fail" screening test, no test protocol was developed. Du Pont made several improvements of the isoamyl acetate method with a detailed test protocol. For example, the protocol requires that the test subject to block the inhalation valves (filter elements) with hands to minimize leakage when a negative pressure leakage test is being performed. Because the design of some filter cartridges does not permit the blocking of all air vents on the cartridge, the Du Pont method calls for the use of a thin latex gloves to cover the whole filter cartridge. The latex gloves would ensure an effective blockage of the air vents on the cartridge.

A. Isoamyl Acetate QLFT Method

Du Pont submitted validation data at the OSHA hearing.[44] The data were based on QNFT results. After the test subject performed the IAA test specified in the protocol proposed by Du Pont, a QNFT was performed immediately to determine whether the minimum fit factor of 100 could be obtained. Du Pont claimed that the test subject who could not smell the odor of IAA could achieve a minimum fit factor of 100 which meets the fit testing requirement prescribed in the ANSI Z88.2-1980 standard on respiratory protection.[9] For test subjects who smelled the odor of IAA during the QLFT test, fit factors less than 100 were reported. One important question is whether the Du Pont method would reject the individual who passed the IAA QLFT and failed to obtain a fit factor of 100 was not addressed adequately. Because Du Pont only used two models of half-mask respirators in the validation study, the straps of the facepieces were purposely loosened to induce leakage to create an artificial "fail" condition. The fit factors obtained from these "failed" tests were generally less than 50. It is not clear whether the method could reject the false positive cases.

Du Pont performed repeated measurements of the isoamyl acetate in the air in the jar which is used to check the ability of a person to sense that the odor of isoamyl acetate vapor has a range from 0.4 to 2.4 ppm and that the concentration of the isoamyl acetate vapor in the air of test enclosure used to perform the QLFT has a range from 71 to 234 ppm. If the concentrations for the odor screening and fit testing are both low, the respirator wearer who "passes" a QLFT may have achieved a fit factor as low as 30.8 (74/2.4). The fit factor claimed by Du Pont that a respirator wearer who "passes" the QLFT achieves a fit factor of 100 may not always be true.

While the IAA test method is quite simple, there is concern about the possible variability in the vapor concentration in the test enclosure. Because the IAA is generated statically inside the test enclosure, the concentration reduces as the test progresses. Since more strenuous exercises are scheduled at the later part of the test regimen, however, the chamber concentration is lower at the time when higher IAA concentration is required, Factors affecting the variability include the ambient temperature, the respiration rate of the individual, the ventilation in the room, and the position and composition of the tissue saturated with IAA.

QLFT methods are generally validated with quantitative fit testing (QNFT). It has been reported that there is a high variability of QNFT validation results. Furthermore, the basis of the IAA QLFT method is to determine whether the test subject could smell the odor of the IAA inside a static chamber which generates a known concentration of IAA. The QNFT method does not validate the reproducibility of the test subject's sensory response to IAA. The dynamic test chamber validation method could be considered as a quantitative validation method since a more accurate and uniform concentration of IAA vapor was generated inside the test chamber. This validation has more direct relationship to the individual's odor perception to isoamyl acetate. The National Bureau of Standards (currently National Institute of Standards and Technology) conducted a study to address these concerns.[45]

The purpose of this study was to compare the fit test results obtained between the static test enclosure and the dynamic test chamber. Exercises performed by the participating test subjects used the Du Pont IAA QLFT test protocol described in the OSHA amended Lead standard with the addition of a jogging-in-place exercise to simulate body movements. This exercise was placed before the second normal breathing test. At first the subject goes through the standard fit testing protocol as described in 29 CFR 1910.1025. Five makes of half-mask respirators (American Optical, MSA, North, Scott, and Survivair) equipped with organic vapor cartridges were available for the subjects to use. Each respirator is available in three sizes. The test subject performed the negative and positive pressure facepiece fit check on the chosen facepiece. Another facepiece was selected and retested if the test subject failed the fit check. In cases where subjects failed the QLFT, that is, they could smell the IAA, they were refitted and/or fitted with a second mask and retested. All individuals passing the first test proceeded immediately to the dynamic test chamber.

A large Dynatech-Frontier quantitative fit testing chamber was used in this study. The IAA vapor is generated from the liquid contained in a gas washing bottle and maintained at 40.0 ± 0.1°C. The air flow through the washing bottle is maintained at a steady flow (±0.2 l/m) at nominally 2 l/m. The lowest concentration during a test was 90 ppm and the highest was 120 ppm. The variability in vapor concentration from floor to ceiling was found to be less than 5 ppm. A portable flame ionization detector (Model OVA-128 Century Organic Vapor Analyzer) was used to monitor the vapor concentration inside the test chamber during the test.

Because there was a significant number of test subjects who passed the Du Pont QLFT test but failed the chamber test, a second series of tests was performed at a lower concentration of 50 ppm of isoamyl acetate. The order of performing the static chamber and dynamic chamber tests was switched for half of the subjects to minimize possible bias resulting from always performing the QLFT first.

A brief summary of the test results is given in Table 6. The laboratory temperature where the qualitative fit tests were performed was 20 ± 2°C.

The results indicate that the isoamyl acetate QLFT method provides a fit factor of at least 50 but less than 100. Adopting a safety factor of 10, which is widely accepted for assigning protection factors for negative pressure respirators, a protec-

Table 6 IAA QLFT Validation Test

Concentration	Test #1	Test #2
100 ppm	QLFT pass: 17	Chamber pass: 11
50 ppm	QLFT pass: 9	Chamber pass: 9
	Chamber pass: 10	QLFT pass: 8

Note: Tally results for test #2 represent the total number of subjects passing test #1 who also passed test #2.

From Reference 46.

tion factor of 5 would be assigned to half-mask respirators where the IAA qualitative fit testing method is utilized. Since three tests are required by OSHA for performing the QNFT and only one test is required for performing the QLFT, the results of this study raises the issue of reproducibility of the IAA method.

B. Sodium Saccharin QLFT Method

The 3M sodium saccharin method is modeled after the IAA method developed by Du Pont. 3M claimed that the saccharin concentration inside testing hood would be 100 times the concentration used for the sensitivity check inside the testing hood. 3M also claimed that this method would reject any respirator having a leakage less than 1%.

In the 3M sodium saccharin QLFT test protocol, the test subject is first tested for his/her ability to detect the characteristic sweet taste of saccharin by exposing the subject to a saccharin aerosol, approximately 1 mg/m^3, which is generated from a saccharin solution, 0.83 g/100 ml (0.83%). If able to detect the saccharin, the participant is then fitted with a disposable dust/mist respirator according to established procedures and a test hood is placed over the subject's head and shoulders. The saccharin aerosol, which is introduced into the hood, is generated from a more concentrated saccharin solution, 83 g/100 ml, by a nebulizer. In order to maintain the required concentration of saccharin inside the hood, continuous squeezes of the nebulizer are needed. The participant then performs each of the following exercises:

1. Normal breathing
2. Deep breathing
3. Turning head from side to side
4. Nodding head up and down
5. Talking
6. Normal breathing

Before each individual exercise, additional saccharin mist is introduced into the test hood to maintain the saccharin concentration at approximately 100 mg/m^3. If the subject detects the taste of saccharin, the participant is fitted with another respirator and the tasks repeated. If the subject does not detect the sodium saccharin, which means the concentration inside the mask is less than 1 mg/m^3, then he/she is considered to be protected with a fit factor of at least 100. The fit factor represents the ratio of the concentration outside the facemask to the value inside the facemask.

3M had submitted validation data at the OSHA hearing. The data were based on QNFT results. After the test subject passed the saccharin test specified in the protocol proposed by 3M, a QNFT was performed immediately to determine whether a minimum fit factor of 100 could be obtained. 3M claimed that the test subject who could not taste the odor of saccharin could achieve a minimum fit factor of 100 which meets the requirement prescribed in the ANSI Z88.2-1980.[9] However, all of the validation tests were performed on a respirator equipped with a elastomeric facepiece and a pair of HEPA cartridges; therefore, it is not clear whether the test results could be applied to the fabric disposable respirators equipped with dust/mist filters.

A potential problem with the saccharin aerosol is the size of the saccharin aerosol particles to which a respirator wearer is exposed to when a respirator fit test is performed in accordance with the protocol developed by the 3M Company. The data submitted by 3M to OSHA during the hearing[46] indicate that using an electronic microscope to measure the "dry" saccharin aerosol particles collected on an electron microscope grid results in a count geometric mean particle size of 2.0 to 2.15 μm with 99% of the particles being smaller than 7.0 μm. These data also indicate that using a LAS-200 particle spectrometer to measure the saccharin particles *in situ* results in a particle size of 2.3 to 2.4 μm. The *in situ* measurement of airborne saccharin particles agrees well with the measurements of "dry" saccharin particles collected on the electron microscope grid.

The respirator wearer is inside a hood-like enclosure approximately 12 in. in diameter by 14 in. tall that fits over the wearer's head and shoulders and this enclosure has a 3/4-in. diameter hole located in front of the wearer's mouth and nose to accommodate the nozzle of the nebulizer which is used to generate the saccharin aerosol. This means that the outlet of the nozzle of the nebulizer will be only a few inches away from the respirator worn by a person inside the hood-like enclosure. It is possible that large size aqueous particles containing dissolved saccharin may be exiting from the outlet of the nozzle of the nebulizer and that these large size droplets are what the respirator wearer is exposed to. The saccharin particle is hygroscopic, i.e., it grows in size under high humidity. Because the respirator wearer's exhaled breath remains inside the hood-like enclosure, the humidity inside is very high. Due to the high humidity, the saccharin mist exiting from the nozzle does not reduce in size and may even increase in size.

Even if a count mean size of 2 μm for the saccharin particle is accepted, converting this count mean size into a mass median aerodynamic size of the saccharin aerosol particle is approximately 4 μm which may not penetrate through the facepiece seal. Saccharin is detected by the tongue of the respirator wearer. The test subject must open the mouth and extend the tongue in order to taste the saccharin. The saccharin QLFT lasts about 5 min. It may not be comfortable for the test subject to open the mouth and extend the tongue during the test. Because it is not possible to verify whether the test subject opened the mouth and extended the tongue during the test, the reliability of the test is questionable.

After the lead hearing, OSHA requested the Los Alamos National Laboratory to conduct a study on the effectiveness of the sodium saccharin QLFT protocol developed by the 3M Company.[47]

The first phase of investigation involved the comparison of the results of a saccharin QLFT with the results of a QNFT. Thirty male and twenty female test subjects participated in this study. Six different models of elastomeric half-mask

Table 7 Results of Saccharin Qualitative and Quantitative Fitting Tests

	Fit factor determined from quantitative fitting test				
	<10	10–100	101–500	>500	Total
Number who did not taste saccharin during qualitative fitting test (passed QLFT)	0	3	3	18	24
Number who tasted saccharin during qualitative fitting test (failed QLFT)	56	33	19	12	120
Total	56	36	22	30	144

From Reference 48.

negative pressure respirators equipped with HEPA filters were selected for testing. The saccharin QLFT which following the 3M protocol was performed first. Upon completion of the QLFT, the test subject was immediately given a QNFT. The di-2-ethyl hexyl sebacate oil mist was used as a test agent with a light scattering photometer as a detector in the QNFT. The exercises performed by the test subjects during the QNFT were the same ones used in the QLFT. Two additional tests were performed after a minimum of 15 min. The test results are shown in Table 7. The 3M "Q" respirators were used in the second phase of study. The "Q" respirator is specifically made for performing QNFT. A total of 40 pairs of saccharin QLFT/QNFT tests were performed. The test results are shown in Table 8.

The purpose of this study is to determine whether the saccharin method would reject poor fit which occurs when the respirator wearer with inadequate fits are not rejects by the QLFT. These statistics are often called as false negative or β error. An inadequate fit is defined as having a measured fit factor (FF) less than 10. The data show that $\beta = 0/56 = 0.0$ with 95% confidence that the statistics would not exceed 0.04. This means 4% of test subjects with inadequate fits (FF < 10) will not be rejected by the saccharin QLFT. If an inadequate fit is defined as having a measured fit factor less than 100, then the estimate of the β error would be $\beta = 3/92 = 0.03$ with 95% upper bound confidence interval of 0.00 to 0.08. This means that 8% of the test subject with inadequate fit (FF < 100) will not be rejected by the saccharin QLFT method.

If the data from repetitive tests on the same individual were not combined, the sample size of inadequate fits is reduced to 34; which means, of 48 subjects tested, 34 were found to have at least one fit factor less than 10. For this sample (FF <10), the estimate of the β error is $\beta = 0/34 = 0.0$ with 95% confidence that the statistics will not exceed 0.07. This means that 7% of the test subjects with an inadequate fit (FF < 10) will not be rejected by the saccharin method. The author concludes that the 3M sodium

Table 8 Results of Saccharin Qualitative and Quantitative Fitting Tests On 3M Respirators

	Fit factor determined from quantitative fitting test				
	<10	10–100	101–500	>500	Total
Number who did not taste saccharin during qualitative fitting test (passed QLFT)	0	1	5	2	8
Number who tasted saccharin during qualitative fitting test (failed QLFT)	5	23	3	1	32
Total	5	24	8	3	40

From Reference 48.

Table 9 Results of Irritant Source Qualitative and Quantitative Fitting Tests

Respirator tested	FF Determined from QNFT				
	<10	10–100	101–500	>500	Total
Full facepiece					
No. passing QLFT	0	0	4	29	33
No. failing QLFT	4	6	5	2	17
Subtotal	4	6	9	21	50
Half-mask respirator A					
No. passing QLFT	0	0	2	14	16
No. failing QLFT	12	18	0	4	34
Subtotal	12	18	2	18	50
Half-mask respirator B					
No. passing QLFT	0	0	0	12	12
No failing QLFT	14	11	6	7	38
Subtotal	14	11	6	19	50
Half-mask respirator C					
No. passing QLFT	0	4	3	0	7
No. failing QLFT	8	31	2	2	43
Subtotal	8	35	5	2	50
All four respirators					
No. passing QLFT	0	4	9	55	68
No. failing QLFT	38	66	13	15	132
Total	38	70	22	70	200

From Reference 49.

saccharin QLFT is effective in rejecting respirator fits with measured fit factors of less than 10 for respirators equipped with a half-mask. This is contrary to 3M's claim that the sodium saccharin method will reject inadequate fits as high as 100.

C. Irritant Fume QLFT Method

OSHA has also requested that the Los Alamos National Laboratory conduct a study on the effectiveness of the irritant fume QLFT protocol developed by the American Iron and Steel Institute.[48] The purpose of the irritant fume QLFT evaluation was to determine whether this method would reject poorly fitted respirators. The evaluation procedure for the irritant fume QLFT method is the same as the saccharin QLFT method. Twenty-nine male and twenty-one female test subjects participated in this study. One full facepiece and two half-mask respirators equipped with HEPA filters were used. In addition, a half-mask "Q" disposable respirator was also used. All four respirators were tested on each test subject with a total of 200 tests performed. The irritant fume QLFT protocol prescribed in the OSHA lead standard was used. Immediately after performing the QLFT, a QNFT was performed on the test subject. The results are tabulated in Table 9, and the estimation of the β error are shown in Table 10.

The results indicated that the number of full facepiece respirator wearers who passed the QLFT and received inadequate fits is much less than the half-mask wearers. The "Q" respirator wearers had the highest number of inadequate fits when they passed the QLFT. The observed β error for the "Q" disposable respirator is much higher than the two elastomeric facepieces. The author concluded that at the 95% confidence level the irritant fume QLFT, following the protocol used, will identify at least 92% of the facepiece fits with FFs less than 10, and at least 92% of the facepiece fits with FFs less than 100.

Table 10 Estimate of Beta Error[a] for Irritant Smoke Qualitative Fitting Test

	Inadequate fit defined as FF < 10	β Calc. basis[b]	Inadequate fit defined as FF < 100	β Calc. basis[b]
Combined data				
Four respirators	0.00 ($0.0 \leq \beta \leq 0.08$)	0/38	0.04 ($0.0 \leq \beta \leq 0.08$)	4/108
Full facepiece				
Respirator	0.00 ($0.0 \leq \beta \leq 0.53$)	0/4	0.00 ($0.0 \leq \beta \leq 0.26$)	0/10
Half-mask				
Respirator A	0.00 ($0.0 \leq \beta \leq 0.22$)	0/12	0.00 ($0.0 \leq \beta \leq 0.10$)	0/30
Respirator B	0.00 ($0.0 \leq \beta \leq 0.19$)	0/14	0.00 ($0.0 \leq \beta \leq 0.11$)	0/25
Respirator C	0.00 ($0.0 \leq \beta \leq 0.31$)	0/8	0.09 ($0.0 \leq \beta \leq 0.20$)	4/43
Combined data				
With no repeat testing on inadequate fits	0.00 ($0.0 \leq \beta \leq 0.12$)		0.04 ($0.0 \leq \beta \leq 0.14$)	

[a] Upper confidence interval = 95%.

[b] Number passing QLFT/number failing QLFT.

From Reference 49.

The developers of these QLFT methods claim that these methods are equivalent to the QNFT. The test results indicated that neither of these three methods would reject poor fits (FF < 100). In other words, there is a fairly high probability that the respirator wearer who passes the QLFT may not achieve an APF of 10 which is required for the half-mask negative pressure air-purifying respirator. Since OSHA health standard requires that only a single QLFT be performed, this further reduces the reliability of the QLFT method. At the OSHA lead hearing, the Du Pont representative stated that there is no time saving between performing a QLFT and a QNFT. Because the QNFT needs to be performed annually, the cost saving between qualitative fit testing and quantitative fit testing equipment may not be justifiable for performing QLFT on workers for protection against exposure to substances with high toxicity or suspected carcinogens such as lead, benzene, or cadmium.

REFERENCES

1. Code of Federal Regulations (CFR), Title 30 Part 11, U.S. Department of the Interior, 1991.
2. Army Air Forces, Material Center, Memorandum Report No. EXP-M-49-695-15, September 9, 1942.
3. United States Air Force Anthropometric Survey — 1967, Unpublished. Wright Patterson AFB, OH, 1968.
4. Clauser, C. E. et al., Anthropometry of Air Force Women, Aerospace Medical Research Laboratory, Wright-Patterson AFB, OH, Report AMRL-TR-70-5, April 1972.
5. Hack, A., Hyatt, E. C., Held, B. J., Moore, T. O., Richards, C. P., and McConville, J. T., Selection of respirator test panels representative of U.S. adult facial sizes, Los Alamos Scientific Laboratory, LA-5488, March 1974.
6. Kolesar, E. S., Respirator qualitative/quantitative fit test method analysis, Review 2-80. USAF School of Aerospace Medicine, Brooks AFB, Texas, 1980.
7. White, J. M., Facepiece leakage and fitting of respirators. Paper presented at the First Canadian Conference on Protective Equipment. AECL-6175, Toronto, Ontario, 1978.
8. Code of Federal Regulations, Title 29 — Part 1910, Section 1025, Appendix D. U.S. Department of the Labor, 1993.

9. ANSI, Practices for respiratory protection, ANSI Z88.2-1980, American National Standards Institute (ANSI), New York, NY, 1980.
10. White, J. M. and Beal, R. J., The measurement of leakage of respirators, *Am. Ind. Hyg. Assoc. J.*, 27, 239, 1966.
11. Billups, N. B. and Oberst, J. C., Chloropicrin leakage test for the M17 protective mask equipped with drinking and resuscitation devices worn by volunteers, U.S. Army Edgewood Arsenal, CRDL Technical Memorandum 2-32, 1965.
12. Cyr, R. R. and Watkins, D. W., Facepiece to face leakage evaluation using a helium leak detection method, *J. Vacuum Sci. Technol.*, 12, 419, 1975.
13. Adley, F. E. and Wisehart, D. E., Method for performance testing of respiratory protective equipment, *Am. Ind. Hyg. Assoc. J.*, 23, 251, 1962.
14. Adley, F. E. and Uhle, R. J., Protection factor of self-contained compressed-air breathing apparatus, *Am. Ind. Hyg. Assoc. J.*, 30, 35, 1968.
15. Packard, L. H. et al., Quantitative fit testing of personnel utilizing a mouthpiece respirators, *Am. Ind. Hyg. Assoc. J.*, 39, 723, 1978.
16. Watson, H. A., Gusey, P. M., and Beckert, A. J., Evaluation of chemical-cartridge respirator face fit. U.S. Bureau Mines Report of Investigation No. 7431, 1970.
17. De Steigurer, D., et al., The use of n-pentane as a tracer gas for the quantitative evaluation of aircrew protective breathing equipment. Proceeding of the Fourteen Annual SAFE symposium, 15, 1976.
18. De Steigurer, D., et al., The objective evaluation of aircrew protective breathing equipment, II, Full-face Masks and Hoods. Proceeding of the Fourteen Annual SAFE symposium, 10, 1976.
19. Douglas, D. D., Lowry, P. L., Hack, A. L., Yasuda, S. K., Wheat, L. D., and Bustos, J. M., Respirator studies for the National Institute for Occupational Safety and Health. Progress Report. July 1, 1975 through December 31, 1976, Los Alamos Scientific laboratory, LA-6722-PR, 1977.
20. Guyton, H. G. and Decker, H. M., Techniques for evaluation biologic penetration of respirator masks on human subjects, *Am. Ind. Hyg. Assoc. J.*, 28, 462, 1967.
21. Guyton, H. G. and Lense, F. T., Method for evaluating biological penetration of respiratory masks on human subjects, *Arch. Ind. Health*, 14, 245, 1956.
22. Burgess, W. A. and Hinds, W. C., Performance and acceptance of respiratory facial seals, *Ergonomics*, 13, 455, 1970.
23. Burgess, W. A., et al., A new technique for evaluating respirator performance, *Am. Ind. Hyg. Assoc. J.*, 22, 422, 1961.
24. Revoir, W. H. and Yurgilas, V. A., Performance characteristics of dust respirators, Bureau of Mines approved and non-approved, *Am. Ind. Hyg. Assoc. J.*, 29, 372, 1968.
25. National Toxicological Program Technical Report on the Carcinogenesis bioassay of Di-(2-Ethylhexyl) Phthalate, U.S. Department of Health and Human Services, National Institutes of Health, DHHS Publication No. (NIH)-82-1771, 1982.
26. Special Occupational Hazard Review — Alternates to Di-2-Ethylhexyl Phthalate (DOP) Respirator Quantitative Fit Testing. National Institute of Occupational Safety and Health, DHHS (NIOSH) Publication No. 83-109, 1983.
27. Mulholland, G. W., Burkowski, R., and Liu, B. Y. H., Application of smoke detector technology to quantitative respirator fit test methodology, National Bureau of Standards NBSIR 86-341, 1986.
28. Douglas, D. D., Lowry, P. L., Pritchard, J. A., Richards, C. P., Wheat, L. D., Geoffrion, L. A., Bustos, J. M., and Hesch, P. R., Respirator studies for the National Institute for Occupational Safety and Health, July 1, 1974 to June 30, 1975. Los Alamos Scientific Laboratory, LA-6386-PR, 1976.

29. Held, B. J., Revoir, W. H., Davis, T. O., Pritchard, J. A., Lowry, P. L., Hack, A. L., Richards, C. P., Geoffrion, L. A., Wheat, L. D., and Hyatt, E. C., Respirator studies for the National Institute for Occupational Safety and Health, July 1, 1973 to June 30, 1974. Los Alamos Scientific Laboratory, LA-5805-PR, 1974.
30. Willeke, K., Ayer, H. E., and Blanchard, J. D., New method for quantitative respirator fit testing with aerosols, *Am. Ind. Hyg. Assoc. J.*, 42, 121, 1981.
31. Dorman, R. G. et al., A comparison of faceseal leakages measured by the argon and sodium flame tests. Test Report, UK Chemical Establishment, Porton Down, U.K., 1970.
32. Hounam, R. F., et al., The evaluation of protection provided by respirators, *Ann. Occup. Hyg.*, 7, 353, 1964.
33. Griffin, O. G. and Longston, D. L., The hazard due to inward leakage of gain to a full face mask, *Ann. Occup. Hyg.*, 13, 147, 1970.
34. Douglas, D. D., Lowry, P. L., Hack, A. L., Yasuda, S. K., Wheat, L. D., and Bustos, J. M., Respirator studies for the National Institute for Occupational Safety and Health. Progress Report. January 1 through December 31, 1977. Los Alamos Scientific Laboratory, LA-7317-PR, 1978.
35. Iverson, S. G., Danisch, S. D., Mullins, H. E., and Rudolph, S. K., Validation of a quantitative fit test for dust/fume/mist respirators. Part I, *App. Occup. Environ. Hyg.*, 7, 161, 1992.
36. Holton, P. M., Tackett, D. Y., and Willeke, K., Particle size-dependent leakage and losses of aerosol in respirators, *Am. Ind. Hyg. Assoc. J.*, 48, 848, 1987.
37. Holton, P. M., Tackett, D. Y., and Willeke, K., The effect of aerosol size distribution and measurement method on respirator fit, *Am. Ind. Hyg. Assoc. J.*, 48, 838, 1987.
38. Hinds, W. C. and Kraske, G., Performance of dust respirators with facial seal leaks. I. Experimental, *Am. Ind. Hyg. Assoc. J.*, 48, 836, 1987.
39. Hinds, W. C. and Bellin, P., Performance of dust respirators with facial seal leaks. II. Predictive model, *Am. Ind. Hyg. Assoc. J.*, 48, 842, 1987.
40. Carpenter, D. R. and Willeke, K., Non-invasive, quantitative respirator fit testing through dynamic pressure measurement, *Am. Ind. Hyg. Assoc. J.*, 49, 485, 1988.
41. Crutchfield, C., Eroh, M., and Van Ert, M., A feasibility study of quantitative respirator fit test by controlled negative pressure, *Am. Ind. Hyg. Assoc. J.*, 52, 172, 1991.
42. Crutchfield, C., Murphy, R. M., and Van Ert, M., A comparison of controlled negative pressure and aerosol quantitative respirator fit test system using fixed leaks, *Am. Ind. Hyg. Assoc. J.*, 52, 349, 1991.
43. Occupational Safety and Health Administration, *Federal Register*, 46, 27359, May 19, 1981.
44. Dickey, J. F., Attorney, Du Pont Environmental Division, Wilmington, WE, Letter of August 4, 1981. OSHA Docket H-049A.
45. Mulholland, G. W. and Brown, J. E., Report of test on qualitative fit testing of half mask respirators using isoamyl acetate, National Bureau of Standards, Report No. FR 3966, June 1987.
46. Schmidt, N. E., Attorney, 3M Company, St. Paul, MN. Letter of July 1, 1981, Occupational Safety and Health Administration Docket, H-049A.
47. Marsh, J. L., Evaluation of saccharin qualitative fitting test for respirators, Los Alamos National Laboratory, LA-UR 82-2389, 1982.
48. Marsh, J. L., Evaluation of irritant smoke qualitative fitting test for respirators, Los Alamos National Laboratory, LA-9778-MS, 1983.

19 RESPIRATOR PERFORMANCE STUDIES I: QUANTITATIVE FIT TESTING

I. INTRODUCTION

The term "protection factor" (PF) was first defined by the Bureau of Mines in the respirator Approval Schedule 21B in 1965.[1] The PF is a measure of the degree of protection provided by the respiratory protective device during use and is defined as the ratio of the concentration of airborne radioactive material outside the respirator to that inside the facepiece. Half-mask air-purifying respirators equipped with high-efficiency particulate air (HEPA) filters were approved for a PF of 10, i.e., it can be used up to 10 times the threshold limit value (TLV) of the air contaminant. The full facepiece air-purifying respirator equipped with HEPA filters were approved with a PF of 100 (100 × TLV). The PFs for these devices were determined from quantitative fit testing (QNFT) using dioctyl phthalate (DOP) or di-2-ethylhexyl phthalate (DEHP) aerosol with six male test subjects performing simulated work exercises.

II. Early Work

In 1967, the Atomic Energy Commission (AEC) Directorate of Regulations published a proposed table of PFs in the Federal Register[2] which included a wide range of respirators. These respirators included the positive pressure self-contained breathing apparatus (SCBA) with a PF of 10,000, and the negative pressure half-mask air-purifying respirator with a PF 10. The table was withdrawn because of the lack of adequate QNFT data on all types of devices. To substantiate or determine universally accepted protection factor for all types of respirators, the AEC Director of Regulation sponsored a respirator testing program at the Los Alamos Scientific Laboratory (LASL) starting in July 1969. Self-contained breathing apparatus was tested first.[3,4]

The LASL respirator quantitative fit testing system consisted of an aerosol generator, a test chamber and an analyzing system consisting of a 5 decade forward light scattering photometer and a recorder. The test was performed inside a 3 × 3 × 8 ft high test chamber. Another connecting chamber having the same size served as an air lock to keep the test chamber concentration stable. Thermally generated monodisperse DOP aerosol with a size of 0.3 μm was used as the test agent. The concentration was maintained between 70 and 100 mg/m^3. A sampling probe was inserted through the facepiece at a point which would not distort the facepiece fit. An

8 liter per minute (l/m) sample of challenge was drawn through a photometer and the recorder was calibrated to 100% on the least sensitive scale (1.0 V). Once a stable 100% was achieved, the photometer was switched to sample "clean" air and a zero base line was established on the 0.1 volt sensitivity scale. A recheck of the 100% concentration was then made and if it had remained stable, the sampling line was connected to the facepiece and the test began.

The exercises performed by the test subjects were: normal breathing, deep breathing, turning the head from side-to-side, moving the head up and down, slowly reciting the alphabet, and finally frowning. It was assumed that these exercises would simulate normal work movements. Exercises were repeated if necessary and additional activities were performed to aid in locating the source of leakage when it occurred. Both demand and pressure demand SCBAs, with a 30-min service life, manufactured by four manufacturers were tested: MSA, Scott, Survivair, and Globe. Except for Globe, the other manufacturers offered two different types of full facepieces and both configurations were tested.

Eighty-four firefighters from the local AEC Fire Department participated in this study. An attempt was made to select a test panel which would relate to the established anthropometric panel of the U.S. Air Force personnel and available statistics on the male labor force in the U.S. It was found that the physical anthropometry of the test panel did not reach agreement with the U.S.A.F. personnel due to age difference and other factors. A meeting was held later regarding the application of anthropometric data and techniques in the determination of a minimum number and type of facial measurements needed for the selection of test panel.

A total of 540 quantitative fit tests were performed. The results are presented in Table 1. The PF was based on the average or integrated penetration determined during exercises performed by the test subject while wearing the device inside the test chamber. The PFs shown in Table 1 were calculated from the average DOP penetration recorded during each of the six exercises. The PF reported in the table indicated the SCBA should provide a fit that would assure no greater than the stated penetration for at least 90% of all the test subjects. The range of PFs from seven demand SCBA varied from 20 to 500. The difference in DOP penetration emphasizes the importance of a good face fit when the SCBA is operated in a negative pressure mode. Both the MSA Ultravue and the Survivair Silicone facepieces provided much better fit than the other type of facepieces. The five pressure demand SCBA evaluated with the same test panel achieved results with 100% of the test subject having DOP penetrations equal or less than 0.01% or a PF of 10,000. This was accomplished by maintaining a positive pressure inside the facepiece regardless of the fitting quality of the facepieces used.

In addition to the SCBA evaluation study, Hyatt et al. conducted QNFT to determine the leakage of half-masks and full facepieces.[4] Five half-mask and six full facepiece respirators with chin style canister or twin cartridge configurations were selected for testing. All respirators were approved by the Bureau of Mines and equipped with HEPA filters.

The test aerosol was a thermally generated monodisperse DOP aerosol having a mass median aerodynamic diameter (MMAD) of 0.3 μm and a concentration of 70 to 100 mg/m^3. Prior to entering the test chamber, each test subject who wore a half-mask respirator was tested for fit by either a negative pressure or a positive pressure fit check.

Table 1 **Comparison of B of M Approved 30-Min Demand SCBA, Representing 4 Manufacturers, and 7 Different Facepieces by a DOP Man Test Panel of 31 Firemen**

		Cumulative % of subjects for whom DOP penetration was value shown						
		Scottoramic		Survivair		MSA 401 Air mask		Globe
Average DOP pen., %	PF	"Old"	"New"	Neoprene	Silicone	Clearvue	Ultravue	Sierra Neoprene
≤ 0.01	10K	16.0	6.5	—	55.0	42.0	81.0	19.0
≤ 0.02	5K	32.0	13.0	—	65.0	48.0	84.0	26.0
≤ 0.05	2K	48.0	26.0	16.0	71.0	65.0	87.0	58.0
≤ 0.10	1K	58.0	42.0	32.0	84.0	74.0	87.0	65.0
≤ 0.20	500	65.0	55.0	42.0	84.0	74.0	94.0	81.0
≤ 0.33	300	68.0	65.0	45.0	84.0	81.0	97.0	87.0
≤ 0.50	200	68.0	71.0	61.0	90.0	94.0	97.0	94.0
≤ 1.00	100	84.0	84.0	77.0	97.0	94.0	97.0	94.0
≤ 2.00	50	87.0	97.0	87.0	97.0	97.0	100.0	94.0
≤ 5.00	20	97.0	100.0	100.0	100.0	97.0	100.0	97.0
≤ 10.00	10	100.0				97.0		100.0

Note: K = 1000.

From Reference 3.

Table 2 Results of Quantitative DOP Tests on Subjects Wearing Half-Mask Particulate Respirators

Leakage average[a] (DOP pen., %)	Percent of subjects for whom penetration was ≤ values shown				
	Resp. A	Resp. B	Resp. C	Resp. D	Resp. E
≤ 0.1	19	33	35	35	6
≤ 0.5	62	44	67	79	28
≤ 1.0	83	53	74	88	50
≤ 2.0	98	72	86	95	59
≤ 5.0	98	84	95	98	81
≤ 10.0	100	97	98	100	100
Average comfort rating[b]	2.2	2.8	2.2	2.7	3.5
Number of men tested	52	36	57	43	32

[a] Average DOP penetration when subject performed specified exercises in man test chamber. The exercises (based on B of M Schedule 21B) include sedentary, turning head side to side and up and down, talking, smiling, and running in place.

[b] Subject makes comfort rating after 10-min DOP man test. The rating is an average for all subjects. The rating system is: 1 — very comfortable, 2 — comfortable (could wear 2 to 4 h), 3 — barely comfortable, 4 — uncomfortable, 5 — intolerable.

From Reference 4.

The fit of the full facepiece respirator wearer was tested with an irritant smoke aerosol. The in-mask sample was drawn from a probe on the facepiece at a flow rate of 8 l/m. The exercises performed by the test subjects consisted of the following:

1. Normal breathing
2. Deep breathing
3. Turning the head from left to right
4. Turning the head up and down
5. Smiling
6. Frowning
7. Reciting the alphabet
8. Reading
9. Running in place
10. Normal breathing

Each exercises was continued from 30 s to 2 min. The average penetration was calculated from the maximum leakage during any exercise. The results are shown in Tables 2 and 3.

The results indicate that for a half-mask respirator, 50 to 88% of the test subjects were able to achieve an overall efficiency of 99% or greater, and 81% to 98% of test subjects could obtain an efficiency of 95%. A properly trained and fitted wearer can obtain an average efficiency of 90% using any of the five respirators tested. These half-mask respirators met the Bureau of Mines performance criteria for protection against up to 10 times the threshold limit value (TLV) of airborne contaminants. For the full facepiece respirator wearer, 93% or more of the test subjects obtained an overall efficiency of 99% or greater using three of the respirators. Penetration in

Table 3 Results of Quantitative DOP Tests on Subjects Wearing Full-Face Particulate Respirators

Leakage average[a] (DOP pen., %)	Percent of subjects for whom penetration was ≤ values shown					
	Mask A	Mask B	Mask C	Mask D	Mask E	Mask F
≤0.01	20	51	77	38	6	27
≤0.05	45	63	95	70	38	61
≤0.10	63	68	100	80	44	61
≤0.50	96	89	—	90	75	73
≤1.0	95	96	—	97	78	76
≤1.0	5	4	0	3	28	24
Average comfort rating[b]	1.9	1.9	1.9	1.9	3.3	2.8
Number of men tested	56	84	84	104	32	33

[a] Average DOP penetration when subject performed specified exercises in man test chamber. The exercises (based on B of M Schedule 21B) include sedentary, turning head side to side and up and down, talking, frowning, and running in place.

[b] Subject makes comfort rating after 10-min DOP man test. The rating is an average for all subjects. The rating system is: 1 — very comfortable, 2 — comfortable (could wear 2 to 4 h), 3 — barely comfortable, 4 — uncomfortable, 5 — intolerable.

From Reference 4.

excess of 1% indicated that the QNFT was terminated when the initial normal breathing exercise resulted in penetration in excess of 1%. In a few tests, the average penetration in excess of 1% was resulted from the frowning exercise. A penetration as high as 18% was recorded in the frowning exercise. A majority of test subjects met the BM full facepiece criterion for protection against up to 100 times the TLV of toxic air contaminants.

The data derived from the above studies were extrapolated in relation to equipment which has not been completely evaluated to determine protection factors. The protection factor for various types of respirators was reported by Hyatt, who pioneered the QNFT work in this country.[5] The protection factors developed by him for various classes of respirators are presented in Table 4. In this publication, the author has provided rationales for setting these PFs, the application and limitation for PFs in the workplace. These PFs have been adopted by OSHA and NIOSH as well as other countries as the maximum use limit for each class of respirators.

In the early 1970s, LASL was asked by NIOSH to develop a system utilizing a cold generated polydisperse DOP as the testing agent since the thermally generated DOP may consist many contaminants with unknown toxicity. The system consists of a nebulizer and an impactor to remove large particles. The challenge concentration was reduced to 25 mg/m^3 with a MMAD of 0.6 μm and a standard deviation of 2. The sampling rate was also reduced to approximately 1 l/m. Two commercial polydisperse DOP QNFT system, with a design similar to the LASL system, were developed and widely used by government and industry throughout the 1970s.

III. SUPPLIED AIR HOODS

Under a contract with the Nuclear Regulatory Commission (NRC), the Los Alamos Scientific Laboratory (LASL) conducted quantitative fit testing studies to determine whether MSHA/NIOSH approved respirators would provide adequate

Table 4 Respirator Protection Factors[a]

Type respirator	Facepiece[b] pressure	Protection factor
Air-purifying		
Particulate[c] removing		
Single-use,[d] dust[e]	–	5
Quarter-mask, dust[f]	–	5
Half-mask, dust[f]	–	10
Half- or quarter-mask, fume[g]	–	10
Half- or quarter-mask, high efficiency[h]	–	10
Full-facepiece, high-efficiency	–	50
Powered, high-efficiency, all enclosures	+	1000
Powered, dust or fume, all enclosures	+	X[i]
Gas- and vapor-removing[j]		
Half-mask	–	10
Full-facepiece	–	50
Atmosphere-supplying		
Supplied-air		
Demand, half-mask	–	10
Demand, full-facepiece	–	50
Hose mask without blower, full-facepiece	–	50
Pressure-demand, half-mask[k]	+	1000
Pressure-demand, full-facepiece[l]	+	2000
Hose mask with blower, full-facepiece	–	50
Continuous-flow, half-mask[k]	+	1000
Continuous-flow, full-facepiece[l]	+	2000
Continuous-flow, hood, helmet, or suit[m]	+	2000
Self-contained breathing apparatus (SCBA)		
Open-circuit, demand, full-facepiece	–	50
Open-circuit, pressure-demand full-facepiece[n]	+	10,000[a]
Closed-circuit, oxygen tank-type, full-facepiece	–	50
Combination respirator		
Any combination of air-purifying and atmosphere-supplying respirator.	Use minimum protection factor listed above for type of mode of operation.	
Any combination of supplied-air respirator and an SCBA.		
Exception: Combination supplied-air respirators, in pressure-demand or other positive pressure mode, with an auxiliary self-contained air supply, and a full facepiece, should use the PF for pressure-demand SCBA.		

[a] The overall protection afforded by a given respirator design (and mode of operation) may be defined in terms of its protection factor (PF). The PF is a measure of the degree of protection afforded by a respirator, defined as the ratio of the concentration of contaminant in the ambient atmosphere to that inside the enclosure (usually inside the facepiece) under conditions of use. Respirators should be selected so that the concentration inhaled by the wearer will not exceed the appropriate limit. The recommended respirator PFs are selection and use guides, and should only be used when the employer has established a minimal acceptable respirator program as defined in Section 3 of the ANSI Z88.2-1969 Standard.

[b] In addition to facepieces, this includes any type of enclosure or covering of the wearer's breathing zone, such as supplied-air hoods, helmets, or suits.

[c] Includes dusts, mists, and fumes only. Does not apply when gases or vapors are absorbed on particulates and may be volatilized or for particulates volatile at room temperature. Example: Coke oven emissions.

[d] Any single-use dust respirator (with or without valve) not specifically tested against a specified contaminant.

[e] Single-use dust respirators have been tested against asbestos and cotton dust and could be assigned a PF of 10 for these particulates.

[f] Dust filter refers to a dust respirator approved by the silica dust test and includes all types of media, that is, both nondegradable mechanical type media and degradable resin-impregnated wool felt or combination wool-synthetic felt media.

[g] Fume filter refers to a fume respirator approved by the lead fume test. All types of media are included.

Table 4 Respirator Protection Factors[a] (*continued*)

[h] High-efficiency filter refers to a high-efficiency particulate respirator. The filter must be at least 99.97% efficiency against 0.3-μm DOP to be approved.

[i] To be assigned, based on dust or fume filter efficiency for specific contaminant.

[j] For gases and vapors, a PF should only be assigned when published test data indicate the cartridge or canister has adequate sorbent efficiency and service life for a specific gas or vapor. In addition, the PF should not be applied in gas or vapor concentrations that are: (1) immediately dangerous to life; (2) above the lower explosive limit; and (3) cause eye irritation when using a half-mask.

[k] A positive pressure supplied-air respirator equipped with a half-mask facepiece may not be as stable on the face as a full facepiece. Therefore, the PF recommended is half that for a similar device equipped with a full facepiece.

[l] A positive pressure supplied-air respirator equipped with a full facepiece provides eye protection but is not approved for use in atmospheres immediately dangerous to life. It is recognized that the facepiece leakage, when a positive pressure is maintained, should be the same as an SCBA operated in the positive pressure mode. However, to emphasize that it basically is not for emergency use, the PF is limited to 2000.

[m] The design of the supplied-air hood, suit, or helmet (with a minimum of 170 l/m of air) may determine its overall efficiency and protection. For example, when working with the arms over the head, some hoods draw the contaminant into the hood breathing zone. This may be overcome by wearing a short hood under a coat or overalls. Other limitations specified by the approval agency must be considered before using in certain types of atmospheres.

[n] The SCBA operated in the positive pressure mode has been tested on a selected 31-man panel and the facepiece leakage recorded as <0.01% penetration. Therefore, a PF of 10,000+ is recommended. At this time, the lower limit of detection 0.01% does not warrant listing a higher number. A positive pressure SCBA for an unknown concentration is recommended. This is consistent with the 10,000+ that is listed. It is essential to have an emergency device for use in unknown concentrations. A combination supplied-air respirator in pressure-demand or other positive pressure mode, with auxiliary self-contained air supply, is also recommended for use in unknown concentrations of contaminants immediately dangerous to life. Other limitations, such as skin absorption of HCN or tritium, must be considered.

From Reference 5.

protection. Since NRC's main interest was atmospheric supplying respirators, air-purifying respirators were not included in these studies.

The first type of respirators studied by the LASL was the supplied air hood.[6] A total of 20 supplied air hoods including Bureau Mines approved, MSHA/NIOSH approved, and nonapproved were selected in this study. The brands and approval numbers of the hoods are presented in Table 5.

Twelve test subjects including four females participated in this study. Test subjects were not chosen for any physical characteristics such as height, weight, or other characteristics. The exercises performed by the test subjects including:

1. Normal breathing
2. Bending forward and touching toes
3. Running in place
4. Raising arms above head repeatedly
5. Deep knee bends
6. Body motion simulating a sandblaster's movements

The air flow rate through each hood at minimum and maximum line pressure recommended by the manufacturer were measured. The approval requirement specifies that the hood should deliver a minimum air flow of 170 l/m. If the air flow measured at minimum specified pressure was below 170 l/m, the hood was tested at that air flow.

Table 5 Hood Names and Approval Numbers

Hood #	Hood name	Approval #
1	"New" MSA Leadfoe	TC-19C-80
2	"Old" MSA Leadfoe	BM-19B-53
3	MSA Blastfoe	BM-19B-52
4	Pulmosan HA-99	BM-19B-44
5	Bullard Abrasive Blasting	BM-19B-57
6	American Optical "ASH"	BM-19B-60
7	Cesco 800-AA	TC-19C-64
8	Cesco 800-DA	TC-19C-64
9	Cesco 800-NA	TC-19C-64
10A	SLY Purair, Before Repairs at LASL	BM-1911
10B	SLY Purair, After Repairs at LASL	BM-1911
11	3M Economy Hood with Vortex Tube	TC-19C-70
12	3M Economy Hood with Air Regulating Valve	TC-19C-69
13	3M General Purpose Supplied-Air Hood	TC-19C-69
14	3M Pesticide Hood	
15	3M Abrasive Blasting Helmet	TC-19C-69
16	"Savannah River Hood"	Not Approved
17	Binks Paint Spray Hood	Not Approved
18	DeVilbiss Model MPH-527 with Clear Plastic Hood	Not Approved
19	Devilbiss Model MPH-527 with Rubberized Fabric Hood	Not Approved
20	Devilbiss Model MPH-527 with Paper Hood	Not Approved

From Reference 6.

The test chamber had a volume of 16 m^3. Air flow inside the chamber was 2.1 m^3/min. The test aerosol was an air-generated polydisperse DOP aerosol with a MMAD of 0.6 μm at a concentration of 25 mg/m^3. A continuous flow sample at a rate of 8 l/m was drawn from the test subject's breathing zone within the hood. The aerosol concentration inside the test chamber and inside the hood was measured with a LASL model 69B forward light-scattering photometer. The concentration of the test aerosol measured inside the hood as compared to the test chamber was expressed as the percentage of penetration.

The percent aerosol penetration inside the hood was recorded continuously and the average penetration was calculated for each exercise. The average penetration for each hood was calculated using the average penetration value for each test, excluding the running-in-place exercise which was not considered representative of physical movements while using the supplied air hood. Performance of the respirator was defined in terms of the PF which was the ratio of the challenge concentration of DOP to the concentration inside the facepiece. The photometer was adjusted to read full scale on the challenge concentration. For PF calculations, the challenge concentration may be considered 100%. If an average penetration of 1% was measured inside the mask, this value divided into 100 (the challenge concentration) to yield a PF of 100. For the air supplied hood, the recommended PF is 2000.

The test results are show in Table 6 and Figure 1. The test results of hood #10 were not used in the calculation of the PF because the hood leaked after receipt and it was subsequently repaired. The results indicated that at a minimum air flow which was not necessarily at or above 170 l/m, the percentage of hoods failed at PFs of 500 to 2000 ranged from 25 to 50%. At a required minimum approval air flow of 170 l/m, there were no failures at a PF of 500, and at a PF of 2000 only 19% of the hoods and 6% of the test subjects failed (omitting the nonapproved devices). At a maximum flow rate recommended by the manufacturer, no failures were noted at PFs up to

Table 6 Summary of Hood PF Results (Hood 10A Omitted)

	Hoods 1 through 20						Hoods 1 through 16 only								
	Hood tested at minimum flow rate[a]		Hood tested at 170 l/m flow rate		Hood tested at maximum flow rate[a]		Hood tested at minimum flow rate[a]			Hood tested at 170 l/m flow rate			Hood tested at maximum flow rate[a]		
	Number failing a PF of						Number failing a PF of								
	2000	500	2000	500	2000	500	2000	1000	500	2000	1000	500	2000	1000	500
Number of hoods	12	8	7	4	3	2	8	6	4	3	1	0	0	0	0
(Percentage)	(60)	(40)	(35)	(20)	(15)	(10)	(50)	(38)	(25)	(19)	(6)	(0)	(0)	(0)	(0)
Number of test subjects	25	14	15	9	8	4	13	7	5	3	1	0	0	0	0
(Percentage)	(40)	(22)	(25)	(15)	(13)	(7)	(26)	(14)	(10)	(6)	(2)	(0)	(0)	(0)	(0)
Total number of test subjects	63		60		60		51			48			48		

Note: Calculations were rounded as follows: 1. PFs below 700 were rounded to the nearest 100; 2. PFs above 700 were rounded to the nearest 1000.

[a] Minimum and maximum flow rates were those resulting from operating the hoods at the manufacturer's minimum and maximum operating pressures.

From Reference 6.

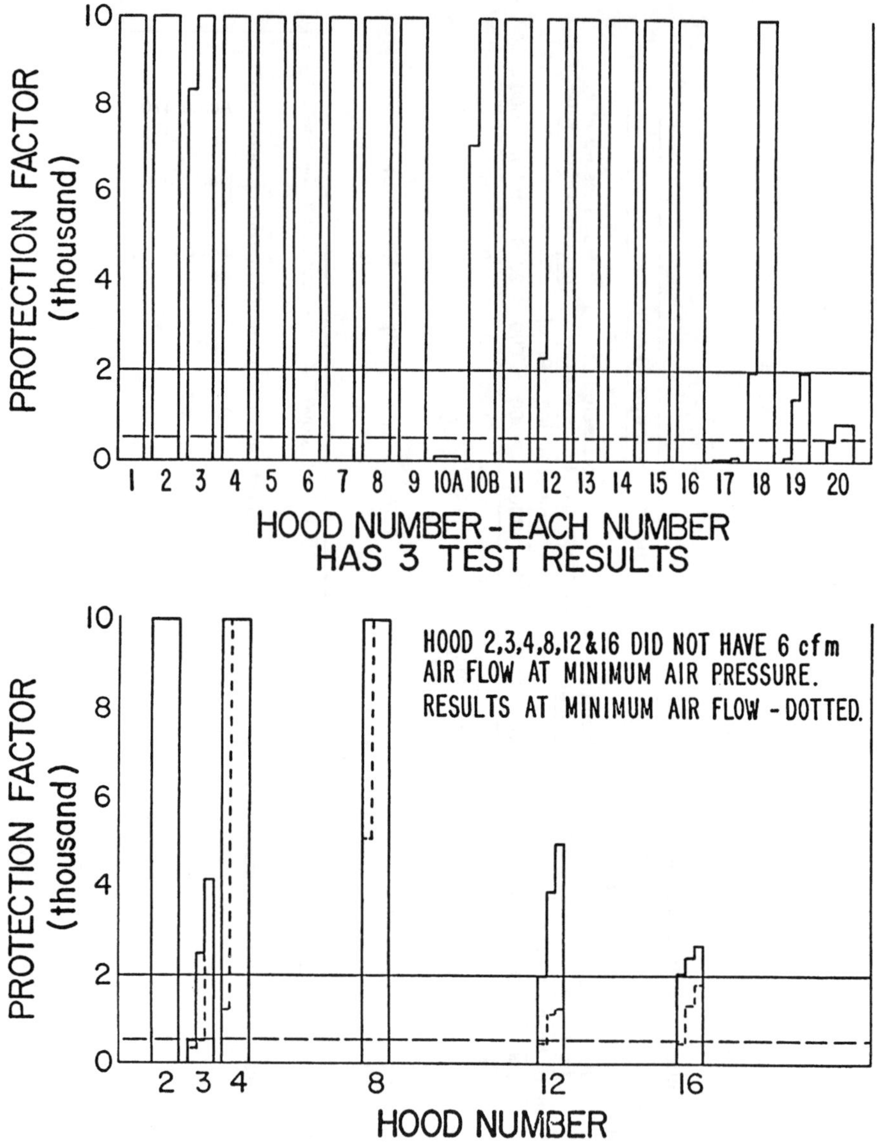

Figure 1 LASL hood test. (From Reference 6.)

2000, even including the nonapproved hoods. The protection provided by the hood is dependent on how the hood is designed; some devices could achieve a PF of 10,000 even with an air flow of 140 l/m.

IV. SUPPLIED AIR RESPIRATORS

The second type of respirators tested was supplied air respirators (SAR).[7,8] Both negative pressure (demand) and positive pressure (continuous flow and pressure

Table 7 Respirators Tested

No.	Manufacturer	TC-19C[a]	Mode[b]	Reg.[c]	Facepiece used[d]
1	Pulmosan Al 180	82	CF	1	Sierra neoprene
2	Scott 4616	72	CF	1	Acme 4704
3	Scott 801450-40	42	CF	1	Scottoramic 742
4	Willson 1810	83	CF	1	Half, valve on bottom
5	Willson 1820	83	CF	1	Half, valve on front
6	American Optical R 6099	86	CF	1	Half, 6000 series
7	Scott 801548-00	74	PD	2	Scottoramic 742
8	Survivair 9811-02	67	PD	2	Silicone
9	Survivair 9011-02	67	PD	2	Neoprene
10	Robertshaw 900002264-01	90	PD	2	Sierra silicone
11	Robertshaw 900002264-11	90	PD	2	Half, Sierra, none elastic straps
12	Survivair 9811-01	65	D/CF	2	Silicone
13	Globe 2185-4	92	D/CF	2	Sierra neoprene
14	Scott 802230	73	D/CF	2	Scottoramic 742
15	Scott 802230	73	D/CF	2	Tite-seal type headband, 742
16	MSA 46063	78	CF	1	Ultravue
17	MSA 460865	78	CF	1	Half, comfo
18	MSA 460862	78	CF	1	Half, welders comfo
19	Binks 40-160	88	CF	1	Half, comfo
20	MSA 457165	75	D	1	Ultravue
21	MSA 46539	75	D	1	Half, comfo
22	MSA 457157	91	D	2	Ultravue
23	MSA 93225	91	D	2	Half, comfo
24	MSA 461669	93	PD	1	Ultravue
25	MSA 89104	93	PD	1	Clearvue

[a] TC-19C: MSHA/NIOSH approval number.

[b] Mode: CF, Continuous Flow; PD, Pressure Demand; D, Demand.

[c] Reg: 1, Regulator, valve, or orifice mounted on belt; 2, Regulator mounted on mask.

[d] Facepieces are full face unless "half" is listed.

From Reference 7.

demand) SARs were selected for this study. A list of respirators being tested is shown in Table 7.

A LASL constructed chamber of 16 m^3 was used for all tests. Air flow into the chamber was 2.1 m^3/min (75 cfm). A polydisperse aerosol of DEHP was generated using air nebulization. A single-stage impactor was used to remove the large particles. The aerosol particle had a MMAD of 0.6 μm, and the chamber concentration was 20 mg/m^3. The sampling rate was 1 l/m. The sample was removed through a probe sealed onto the inlet covering and measured by a forward light-scattering photometer. The facepiece pressure was also measured in this study. The exercises performed by the test subjects included:

1. Normal breathing
2. Deep breathing
3. Turning head from side to side
4. Moving head up and down
5. Talking
6. Smiling (half-mask) or frowning (full facepiece)
7. Normal breathing

Table 8a No. of Persons Achieving Stated Protection Factor with Supplied-Air Respirators (Continuous Flow)

Respirator	Number subjects	Flow[a] (cfm)	Protection factor attained							
			50	100	500	1k	2k	5k	10k	20k
1 Pulmosan	11	4.0	—	—	—	—	—	11	10	10
AL 180	11	5.3	—	—	—	—	—	—	—	11
2 Scott	11	4.3	—	—	11	10	10	8	8	6
4616	11	7.0	—	—	—	—	—	—	—	11
3 Scott	11	4.4	—	—	—	—	—	—	—	11
801450-40	11	7.7	—	—	—	—	—	—	—	11
4 Willson	11	6.5	—	—	—	—	—	—	—	11
1810	11	10.0	—	—	—	—	—	—	11	9
5 Willson	27	6.5	—	—	—	—	—	—	—	27
1820	27	10.0	—	—	—	—	—	—	—	27
6 Am. Opt.	27	4.8	26	25	25	25	24	24	22	21
R6099	27	9.7	—	—	—	—	—	—	—	26
16 MSA	12	6.5	—	—	—	—	—	—	—	12
460863	12	6.5	—	—	—	—	—	—	—	
17 MSA *460865*	12	6.5	—	—	—	—	—	—	12	9
18 MSA *46082*	12	6.5	—	—	—	—	—	—	12	11
19 Binks *40-160*	13	6.5	—	—	—	—	—	—	—	13

Note: Respirator numbers in italics denote half masks, others are full face.

[a] Flow measured at LASL corrected to represent flow at sea level.

From Reference 7.

An exercise, moving blocks from a high to a low shelf, which simulated real work situation, replaced to side-to-side and up to down movements during the later part of the study.

Test subjects were selected according to the anthropometric scheme developed for NIOSH in 1973.[9] The facial dimensions of the panel were taken from measurements of Air Force personnel (male and female) measured in 1967/68. The selection was limited to 90 to 95% of the military population on the assumption that this would represent civilian populations. Male and female test subjects were selected for this study. Originally, 25 test subjects were used to test all SARs. Later, 10 test subjects with one in each grid of the anthropometric panel was selected for testing the positive pressure supplied air respirators because very little leak was detected from these respirators.

The photometer, adjusted to read full scale on the challenge aerosol concentration, provided a direct reading of the percent penetration into the mask. The peak penetrations recorded on the strip chart for each test subject during each exercise were averaged to arrive at a penetration value for the exercise. Penetration values for each exercise was then averaged to arrive at an overall average for the test subject wearing the particular apparatus. Respirator performance was reported as PF, a ratio of the challenge atmosphere (100%) divided by the overall average percent penetration of the challenge aerosol into the mask. For example, a PF of 1000 is 0.1% penetration from the average of all exercises performed.

The test results are shown in Table 8a through 8c. These tables indicate cumulative protection, that is, a subject who achieves a PF of 20,000 is also counted as achieving 10,000, 5000, 200, 50, etc. Test results for the 10 continuous flow supplied

Table 8b No. of Persons Achieving Stated Protection Factor with Supplied-Air Respirators (Pressure-Demand)

Respirator	Number subjects	Line pressure PSIG	Protection factor attained 2k	5k	10k	20k
7 Scott	25	60	—	—	25	24
801548-00	25	100	—	—	25	23
8 Survivair	25	50	—	—	—	25
9811-02	25	100	—	—	25	24
9 Survivair	11	50	—	—	—	11
9011-02	11	100	11	10	10	10
10 Robertshaw	25	50	—	—	—	25
900-002-264-01	25	100	—	—	—	25
11 Robertshaw	11	50	—	—	—	11
900-002-264-11	11	100	—	—	10	11
24 MSA	10	80	—	—	—	10
461669	10	100	—	—	—	10
25 MSA	10	80	—	—	10	9
89104	10	100	—	—	10	9

Note: Respirator numbers in italics denote half masks, others are full face.

From Reference 7.

air respirators are presented on Table 8a. For several SARs, tests were performed for both the minimum and maximum air pressure as specified by the manufacturer. Except for a few operated under a low air pressure, most devices could achieve a protection factor 10,000 or better. At the higher air pressure, all devices could achieve a PF of 10,000.

Table 8c No. of Persons Achieving Stated Protection Factor with Supplied-Air Respirators (Demand) — 25 Subjects Tested

Respirator	Line pres PSIG	Protection factor attained 5	10	20	50	100	500	1k	2k	5k	10k	20k
12 Survivair	50	23	21	20	17	12	6	5	4	—	—	—
9811-01	110	23	22	22	18	12	4	3	2	—	—	—
13 Globe	65	19	13	10	7	5	1	—	—	—	—	—
2185-4												
14 Scott	60	—	—	25	23	17	7	5	2	2	2	1
802230	110	25	24	24	22	16	5	2	1	1	—	—
15 Scott	60	22	20	18	14	10	5	4	3	—	—	—
802230	110	19	19	19	14	13	6	4	3	1	—	—
20 MSA	50	—	—	—	24	22	11	5	4	1	—	—
457165	110	—	—	24	23	22	5	—	—	—	—	—
21 MSA	50	—	—	25	22	18	13	9	4	1	—	—
46539	110	—	25	23	22	20	8	3	1	—	—	—
22 MSA	50	—	24	23	23	23	15	5	3	3	1	—
457157	90	24	23	23	23	23	8	5	3	1	—	
23 MSA	50	22	21	20	18	16	11	9	8	3	2	—
93225	90	23	21	19	19	16	11	8	5	3	1	—
Respirators Nos. 12–15 were also tested in continuous flow												
12 Survivair	50	—	—	—	—	—	25	24	22	21	21	20
13 Globe	65	—	—	—	—	—	—	—	25	24	23	22
14 Scott	60	—	—	—	—	—	25	24	23	22	22	21
15 Scott	60	—	—	—	—	—	—	—	—	—	25	24

Note: Several tests were not completed because of high leakage on respirators Nos. 12, 13, 15, 20, 22, and 23. Respirator numbers in italics denote half masks, others are full face.

From Reference 7.

Table 9 Self-Contained Breathing Apparatus (30-Min) Tested

No.	Device	TC-13F[a]	Facepiece	Type
1	MSA 401 Air Mask, 457152	29	Ultravue	Demand
2	MSA 401 Air Mask, 463861	30	Ultravue	Pressure-demand
3	Scott Air Pak IIa, 900000-00	39	Scottoramic	Demand
4	Scott Pressure Pak IIa, 900014-00	40	Scottoramic	Pressure-demand
5	Globe Guardsman, 2540	43	Sierra Neoprene	Demand
7	Survivair, 9838-02	44	Silicone	Demand
8	Survivair, 9038-22	45	Neoprene	Demand/ PD
11	Scott 4.5, 900450	73	Scottavista	Demand
12	Scott 4.5, 900455	76	Scottavista	Pressure-demand
13	Chubb No. 1	[b]	Chubb	Pressure-demand
14	Auer BD 73	[b]	Auer	Demand
15	Drager PA 54/II	[b]	Panarova S	Pressure-demand

[a] TC-13F-XX is NIOSH/MSHA approval number.

[b] Not approved in the U.S.

From Reference 11.

The test results for the pressure demand SARs are listed in Table 8b. Except for the Survivair (#9), all other devices provide a minimum PF of 10,000. It is interesting to note that the half-mask Robertshaw model also achieves a minimum PF of 10,000.

The test results for the demand SARs are shown in Table 8c. The first four devices are also operated in continuous flow mode. The demand SARs cost more than a continuous flow SAR but offers very low protection. The poorest fitting demand SAR had only 19 out of 25 test subjects achieved a PF of 5. The best fitting MSA demand SAR could only achieve a PF of 50. As a matter of fact, the demand SARs even offer lower protection than the air-purifying respirator equipped with the same type facepiece.

Based on the results of this study, LASL recommended that NRC should not permit demand respirators to be used. The demand SARs should not be used for providing respiratory protection because it is not cost-effective. Currently, very few demand SARS are sold. NIOSH and OSHA, however, are reluctant to prohibit the use of these devices. The American National Standards Institute (ANSI) Z88.2 Subcommittee ignored this study and recommended an assigned protection factor of 100 for the full facepiece demand SAR in the American National Standards Institute (ANSI) Standard "Practices for Respiratory Protection — ANSI Z88.2-1992."[10]

V. SELF-CONTAINED BREATHING APPARATUS

The performance of self-contained breathing apparatus (SCBA) was evaluated by LASL for the second time.[11] Both male and female test subjects were selected for the anthropometric panel instead of a panel consisting all male firefighters used in the previous SCBA study.

Every MSHA/NIOSH approved open-circuit SCBA was selected for testing. Both demand and pressure demand devices were tested. Three foreign made SCBAs were also tested. The service life for the SCBA is 30 min. All tested SCBAs are listed in Table 9.

The equipment used for leak testing was the same used previously to test supplied air respirators. A cold generated polydisperse DOP replaced the monodisperse DOP

Table 10 Protection Factors for SCBAs

		Protection factor achieved by number of subjects									
Respirator	**Mode**	**10**	**20**	**50**	**100**	**500**	**1k**	**2k**	**5k**	**10k**	**20k**
1 MSA 401	D	24	24	22	20	12	9	5	2	1	*
2 MSA 401	P	—	—	—	—	—	—	—	—	—	10
3 Scott	D	—	25	24	24	10	6	1	1	1	*
4 Scott[a]	P	—	—	—	—	—	10	9	9	9	9
5 Globe	D	22	18	14	12	3	1	*	*	*	*
7 Survivair	D	24	22	21	19	7	3	1	1	*	*
8 Survivair[b]	D	23	21	19	17	8	4	1	1	1	*
8 Survivair[b]	P	—	—	—	—	—	—	—	—	—	10
11 Scott 4.5	D	20	19	16	15	7	7	5	3	*	*
11 Scott 4.5[c]	D	23	23	20	19	8	7	5	3	*	*
12 Scott 4.5	P	—	—	—	—	—	—	—	10	7	2
13 Chubb	P	—	—	—	—	—	—	—	—	10	8
14 Auer	D	—	25	24	24	16	12	5	2	*	*
15 Drager	P	—	—	—	—	—	—	—	—	10	8

Note: Mode: D, Demand — 25 subjects tested; P, Pressure Demand — 10 subjects tested.

[a] Scott demand mode not approved, used only for donning.

[b] Survivair is approved for both demand and pressure-demand, and tested in both modes.

[c] The seven subjects who had the lowest PF were retested after donning the mask according to the special instructions of the manufacturer. The results that they achieved during the retest were combined with the results of the 18 subjects not retested and are shown on this line.

* None achieved.

From Reference 11.

used in the previous SCBA study. A 16 m^3 test chamber was filled with approximately 17 mg/m^3 of 0.6 μm polydisperse DOP oil mist generated by air nebulization. The aerosol penetration through the facepiece was by means of a forward light scattering photometer. The test subject performed the following exercises:

1. Normal breathing
2. Deep breathing
3. Side-to-side motion by moving objects from left to right
4. Up-and-down motion by moving objects from a high to a low shelf
5. Talking
6. Frowning
7. Normal breathing

The overall protection factor provided by the device was calculated on the average of the peaks recorded during each of the six exercises, without counting the frowning exercise.

Demand apparatus was tested on 25 subjects that met the criteria of the LANL Anthropometric Test Panel. Only 10 subjects were selected for the pressure-demand apparatus. One test subject filled each block of the test panel.

The test results are shown in Table 10. For pressure-demand SCBA, all subjects wearing the MSA and Survivair SCBAs achieved a PF of at least 20,000, and all subjects wearing the Chubb and Dräger SCBAs achieved a PF of at least 10,000. With the Scott PresurPak, only one subject failed to achieve a PF of at least 20,000. For the Scott 4.5 SCBA, only two subjects achieved a PF of 20,000, while all test subjects achieved a PF of 5000.

The quality of facepiece seal affects the protection provided by demand type SCBA. Twenty-four of 25 test subjects reached a PF of at least 100 wearing the Auer SCBA and the Scott SCBA equipped with the Scottoramic full facepiece. Twenty test subjects also achieved a PF of 100 when wearing the MSA SCBA, and 19 obtained this level of protection wearing the Survivair SCBA and the Scott SCBA equipped with the Vista (4.5) facepiece.

The superior protection shown by pressure-demand devices shown in Table 10 leads to the recommendation that only these breathers should be used. As a result of this study, the use of demand SCBA in the fire services has been phased out. Both the OSHA standard on fire brigade and the ANSI Z88.5 standard on Respiratory Protection for the Fire Services had mandated the use of positive pressure SCBA for firefighting.

To conserve air, most pressure demand SCBA have a switch which permitted donning it in the demand mode. However, the wearer may forget to switch it back to the pressure demand mode. This situation was corrected in the National Fire Protection Association (NFPA) standard on SCBA for firefighting, NFPA 1981, which prohibits donning in the demand mode.[12]

VI. ESCAPE SELF-CONTAINED BREATHING APPARATUS

The escape self-contained breathing apparatus (ESCBA) is a class of respirators developed in the 1970s. It enables the wearer safe exit from atmospheres which may be immediately dangerous to life or health (IDLH). Examples include a sudden release of toxic gas from a chemical reactor or vessel, or perhaps a sudden fire condition. Hack et al.[13] conducted a study to determine the performance of the ESCBA.

ESCBA has a shorter service life than the regular SCBAs: it has a range between 5 and 15 min. ESCBA come in many forms. This can be an air cylinder integrated with a pressure demand supplied air respirator, a self-contained unit like the regular SCBA with a hood, or a mouthpiece.

In terms of use, three different types of devices were tested: (1) devices for escape only, including two half-masks operated in the demand mode, one mouthpiece and two continuous flow hoods; (2) three combination air line/SCBA of less than 15-min duration; and (3) two combination air line/SCBA units of 15 min, or more, duration. A description of various types of ESCBA selected for testing is listed in Table 11.

Since ESCBA is used mainly for escape from IDLH atmospheres, one question of interest is how long it takes the device to deploy. The MSA Air Escape mouthpiece SCBA may be operated within 5 s which provides comparable protection to the other demand half-mask units, but less than the combination pressure demand SAR escape and escape hood systems. Half-masks ESCBA may require 15 to 30 s donning time, while ESCBA equipped with a full facepiece may need a little longer, especially for the wearer to remove safety spectacles. Hood type units require approximately 10 to 20 s to activate. All hoods share the advantage of fitting on all size faces, and over beards and eyeglasses. The service life of the hoods with continuous flow operation is limited by capacity of the stored air supply, about 6.5 min for the Survivair and 5.5 min for the Robertshaw.

The test equipment and penetration measurement methodology were identical to those described in the above studies. A LASL constructed chamber of 16 m^3 was used

Table 11 Escape Respirators Tested

Manufacturer	Approval TC-13F-	Mode	Mask	Weight (kg)	Air source
Scott Ska-Pak	66	Demand (5 min)	Willson Half mask	3.67	SCBA
MSA Air E-Scape	61	Demand (5 min)	Comfo	4.13	SCBA
MSA Air Escape	55	Demand (5 min)	Mouthpiece	3.57	SCBA
Scott Ska-Pak	68	Pressure-demand (5 min)	Scottoramic	4.53	SCBA+Air line
Robertshaw		Pressure-demand	Sierra		
Ram 15 (min)	63		Full face	11.11	SCBA+Air line
Ram 15 (min)	63		Half mask	10.71	
Robertshaw		Pressure-demand	Sierra		
Ram 5 (min)	64		Full face	4.39	SCBA+Air line
Ram 5 (min)	64		Half mask	3.96	
Robertshaw 5000	28 (5 Min, air stored in tubing)	Continuous flow	Hood	2.46	SCBA
Survivair 2878	86 (5 Min, 28% O_2 in two disposable cylinders)	Continuous flow	Hood	2.88	SCBA

From Reference 13.

for all tests. A polydisperse aerosol of DEHP was generated using air nebulization. The MMAD for the test aerosol was 0.6 μm, and the concentration was 20 mg/m^3. A sample of 1 l/m was removed through a probe sealed onto the inlet covering of the ESCBA and measured by a forward light-scattering photometer.

Test subjects were selected according to a LASL anthropometric test panels consisted of male and female test subjects. This was developed for NIOSH in 1973.[5] Originally, this test protocol called for selection of test subjects for half masks by face length and lip width. More recent work had indicated that lip width was not as important as previously thought and accordingly the full face panel was used for testing both full face masks and half masks.

The test panel called for different numbers of subjects (male and female) in 10 different size box categories. The two demand mode half mask respirators were tested on 25 persons as shown. For pressure-demand equipment, only 10 persons, one from each size category, were chosen. Face size was of less importance with positive pressure devices because a good face seal is maintained by an outward flow of air if leak occurs. Five test subjects participated in the escape hood test. There was no anthropometric criteria for selecting subjects to test mouthpieces or hoods ESCBAs.

The test exercises performed by the test subjects were:

1. Normal breathing
2. Deep breathing
3. Moving small discs from side to side on a frame
4. Moving blocks from a high to a low shelf
5. Talking, running in place
6. Normal breathing

For the two hoods and mouthpiece unit, facial movements are of less importance, so the following exercises were used instead:

1. Normal breathing
2. Bending over and touching toes

Table 12 Escape Type SCBA Number of Persons Achieving Stated Protection Factor

	Protection factor attained										
Less than	**10**	**20**	**50**	**100**	**200**	**500**	**1k**	**2k**	**5k**	**10k**	**20k**
Demand mode											
Scott Ska-Pak Half mask 25 subjects	25	17	14	9	9	6	6	6	3	2	1
MSA Air E-Scape Half mask 25 subjects	25	19	13	5	—	—	—	—	—	—	—
MSA Air Escape Mouthpiece 5 subjects	—	5	4	4	4	3	—	—	—	—	—
Pressure-demand mode 10 subjects tested											
Scott Ska-Pak (full face)	—	—	—	—	10	9	9	9	9	7	6
RobertShaw Ram 15 (full face)	—	—	—	—	—	—	—	—	—	10	8
RobertShaw Ram 15 (half mask)	—	—	—	—	—	—	—	—	10	9	5
RobertShaw Ram 5 (full face)	—	—	—	—	—	—	—	—	—	—	10
RobertShaw Ram 5 (half mask)	—	—	—	—	—	—	—	10	8	7	4
Continuous flow hood 5 subjects tested											
Robert Shaw Air Capsule	—	—	—	—	5	3	3	1	1	—	—
Survivair	—	—	—	—	—	—	—	5	4	4	3

From Reference 13.

3. Running in place
4. Normal breathing

The penetration determination method was the same as used in other PF studies. The peak penetration recorded on the strip chart for each exercise was used to calculate protection factors. Then the penetration recorded for each exercise were averaged to arrive at an overall average penetration for the test. Results for smile and frown were not used to calculate the overall penetration.

Table 12 lists the PF achieved by each test subject wearing each mask. The table indicates cumulative protection factors, that is, a subject who achieved a PF of 1000 is also counted as achieving 500, 200, etc. The two demand half-mask ESCBA provided a PF of at least 10 to all test subjects. This agreed with published results for the half-mask respirator used in the air-purifying mode, but it is not acceptable for protection against the potentially high concentrations of toxic materials created in an emergency escape situation. The mouthpiece respirator tested (also a demand device) provided a PF of 200 for four of the these subjects tested, with the remaining subject achieving a PF of 20.

For the continuous flow ESCBAs, all test subjects achieved a PF of 200 and 2000, respectively, for the Robertshaw and the Survivair models. For pressure demand devices, all test subjects achieved a PF of 2000 for the Robertshaw models. However, only a PF of 200 was achieved by all test subjects on the Scott unit. The results of this study agrees with other LASL protection factor studies for atmosphere supplying respirators that the superior protection of pressure demand or continuous flow over demand devices. However, the PFs achieved by devices tested in this study are much lower than the regular use atmospheric supplying respirators. The probable cause is the low air flow provided by these devices. For example, the air flow for the continuous flow hood is only 1 cfm.

Table 13 Closed-Circuit SCBA Tested by LANL

Brand and model	Service life (min)	Facepiece pressure	Breathing gas
Aerolox	180	–	Liquid O_2
Biomarine 45	45	–	Compressed O_2
Biomarine 60	60	+	Compressed O_2
Chemox w/Ultravue	60	–	Chemical O_2
Chemox w/Dual Vision	60	–	Chemical O_2
Dräger BG174	240	–	Compressed O_2
McCaa w/Ultravue	120	–	Compressed O_2
McCaa w/Dual Vision	120	–	Compressed O_2
Scott 900050	240	–	Compressed O_2

From Reference 14.

VII. CLOSED-CIRCUIT SELF-CONTAINED BREATHING APPARATUS

Closed-circuit self-contained breathing apparatus (CCSCBA) is the last series of respirators evaluated by the Los Alamos National Laboratory (LANL former LASL).[14] Devices evaluated are shown in Table 13.

The aerosol generation and the leak detection system are identical to those used in the previous protection factor studies. Test subjects were selected according to the Los Alamos anthropometric panel.[9] The panel consisted of 13 male and 12 female test subjects. The distribution of these panel members are listed in Figure 2. The following exercises were performed by the test subjects:

1. Normal breathing
2. Deep breathing
3. Moving small discs from side to side on a frame
4. Moving blocks from a high to a low shelf
5. Talking

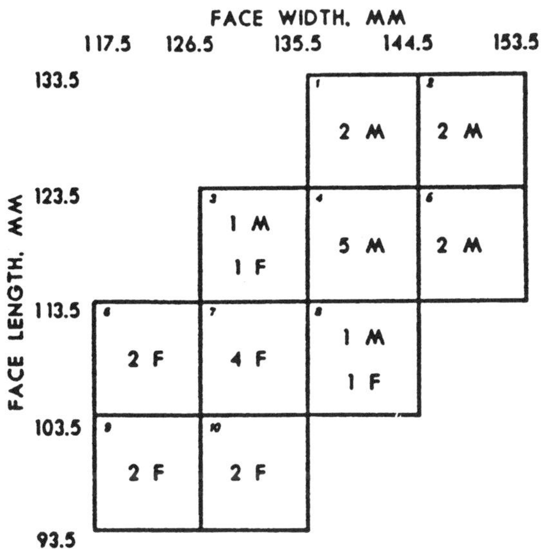

Figure 2 Los Alamos panel of test subjects of different size faces, including 13 males and 12 females.

Table 14 Number of Subjects Achieving Stated Protection Factor Wearing Rebreather SCBA

	10	20	50	100	500	1k	2k	5k	10k	20k
Aerorlox (5)	—	20	18	17	—	—	15	13	12	6
Biomarine 45	25	23	22	21	9	7	5	—	—	—
Biomarine 60	—	25	—	24	—	22	21	20	—	15
Chemox Ultravue (1)	—	—	—	24	23	21	13	1	—	—
Chemox DualVision (11)	14	—	—	12	—	9	7	1	—	—
Draeger BG174 (2)	23	21	—	—	19	17	13	10	9	3
McCaa Ultravue	—	25	—	23	7	1	—	—	—	—
McCaa DualVision	24	22	18	17	9	4	1	—	—	—
Scott 900050	—	—	25	24	12	5	2	—	—	—

Note: 25 subjects tested, numbers in parentheses are the terminated tests.

From Reference 14.

6. Running in place
7. Normal breathing

Each exercise was continued for at least 1 min. Peak penetrations measured by the photometer for each test subject during each exercise were averaged to arrive at a penetration value for the exercise. Penetration values for each exercise were then averaged to arrive at an overall average for the test subject wearing the particular rebreather.

The test results expressed as the PF achieved by each test subject are listed in Table 14. The table indicates cumulative protection factors. The median PF for all 25 test subjects for each device is shown in Figure 3. The median value is calculated by

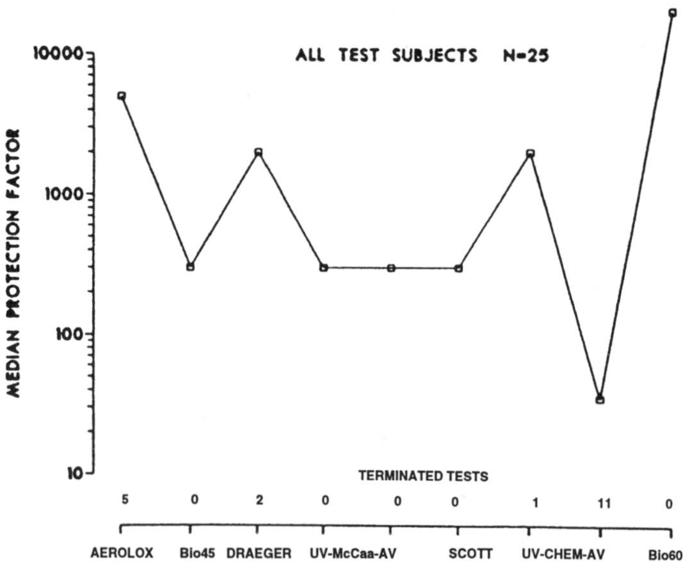

Figure 3 Median protection factor for those test subjects who completed wearing each respirator system. Subjects who did not complete tests because of excessive leakage are listed at the bottom.

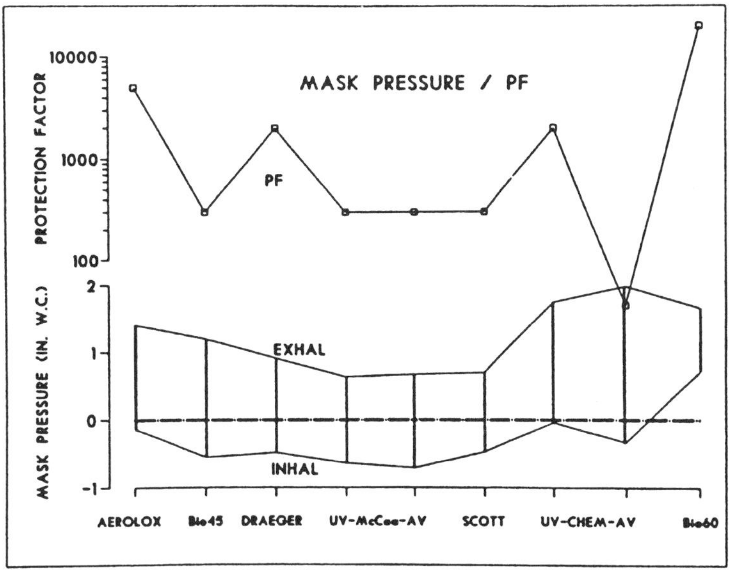

Figure 4 Relationship of facepiece pressure (minimum during inhalation, maximum during exhalation) to protection provided.

ranking all test subjects in the order of PFs achieved by each test subject. The test was terminated when the test subject developed a peak leakage above 20% (PF < 5); however, these subjects were included in determining the median value.

The results indicate a widespread of PFs among different devices. The Scott, Biomarine 45, and the two McCaas achieved a median PF of 250. The Dräger device obtained a higher median PF of 2000 which may be contributed to the design of the facepiece. The Chemox SCBA equipped with a Ultravue facepiece provided much higher protection than the one with a Dual Vision facepiece. The liquid oxygen filled Aerolux demonstrated a median PF of 5000. The Biomarine 60P achieved a median PF well above 10,000 because it is a positive pressure device.

There is a possible link between the facepiece pressure and the protection achieved by the test subjects. Figure 4 shows that most rebreathers exhibits a negative pressure during inhalation except the Biomarine 60 P which exhibits an average facepiece pressure of +0.5 inch water pressure since it is a positive pressure device. Both the Chemox and the Aerolox show a slight negative pressure which contributes to the higher PFs achieved by these rebreathers.

All the rebreathers tested have only one size available for each line of full facepiece. For long duration mine rescue work, the choice is between the Aerolox and the Dräger. However, both these facepieces fit large faces better than small faces. Persons with small facepiece need to be fitted with additional facepieces. Since the CCSCBA recycles the breathing air, the operating temperature is much higher than the open-circuit devices due to heat generated by the carbon dioxide absorber. The liquid oxygen filled Aerolox provides cooling effect when the liquid oxygen is vaporized. However, the filling of liquid oxygen immediately before each use may not be convenient for some applications.

Table 15 Comparison of Single Use Respirator Man Tests; Air Generated DOP Vs. NaCl Aerosol

Type of single use respirator	1.2 µm DOP # Tests	1.2 µm DOP Facepiece leakage, % Range	1.2 µm DOP Facepiece leakage, % Average	0.8 µm NaCl # Tests	0.8 µm NaCl Facepiece leakage % Range	0.8 µm NaCl Facepiece leakage % Average
Mfg. A. w/valve (Open Cell foam seal as approved)	3	13.0–22.5	16.8	2	10.2–11.4	10.8
Mfg. A. w/valve (Exp. Closed cell foam seal)	4	4.6–7.0	5.9	5	5.3–6.4	5.7
Mfg. A. valveless (Experimental)	3	2.7–12.7	7.1	4	4.0–12.8	7.6
Mfg. M valveless (Experimental)	5	8.5–16.0	12.4	4	12.0–14.0	13.1
Mfg. W w/valve (Experimental)[a]	—	—	—	3	4.7–6.0	5.2
Sepestok 200[a]	—	—	—	7	1.0–3.8	2.1

[a] Filter media degradable.

From Reference 15.

VIII. SINGLE USE RESPIRATORS

There are very few QNFT studies on single use (disposable) respirators. The first study was conducted by the Los Alamos Scientific Laboratory in 1971 on several disposable respirators.[15] Both a polydisperse DEHP aerosol with a MAAD of 1.2 µm, and a polydisperse NaCl aerosol with a MAAD of 0.8 µm were used. The results are shown in Table 15. The results indicate that the DOP aerosol is more penetrating than the sodium chloride aerosol in general. However, there is no significant difference in facepiece leakage with two different size aerosols tested on the same respirator.

The second QNFT study on single use (disposable) respirators was also conducted by LASL by Lowry et al.[16] Six single use respirators were selected for testing. These respirators were manufactured by AO, 3M., Welsh, and Willson. The respirators were evaluated by the LASL anthropometric panel consisted of a total of 10 male and female test subjects in each block of the panel. Since the filter media of single use respirators would be degraded by the oil mist aerosol challenge, a sodium chloride aerosol with a MMAD of 0.7 µm and a standard deviation of 2.15 was selected for testing. The challenge concentration of the NaCl aerosol was 15 mg/m^3. The aerosol penetration was measured by a sodium flame photometer.

The test subject performed the following exercises:

1. Normal breathing
2. Deep breathing
3. Turning head from side to side
4. Moving head up and down
5. Talking
6. Normal breathing

The average inspiration peak penetration for all six exercises was used to calculate the protection factor.

The test results are shown in Figures 5 and 6. Protection factors were expressed as 1, 2, 5, 10, 20, 100, and 1000. If a measured protection factor fell between two

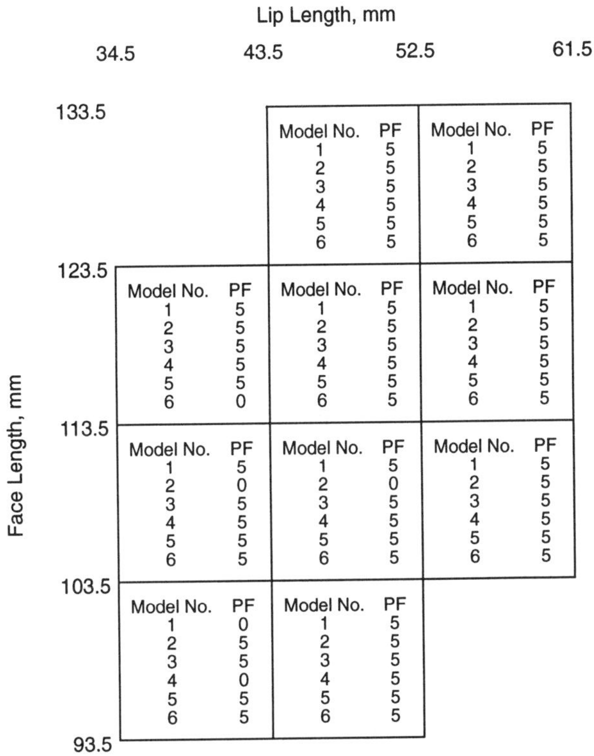

Figure 5 Test subject — respirator model combinations obtaining protection factors of at least 5.

values, the lower protection factor was assigned. Figure 5 shows that only two of the six models provided a protection factor of 5 to all 10 test subjects. However, all models provided a protection factor of 5 for at least 8 of 10 test subjects. Figure 6 indicates the maximum PF obtained by the six single use respirators. From a total of 60 data points, only one test subject achieved a PF in excess of 1000. Almost one half of the test subjects achieved a PF of 5 or less. The protection factor provided by the half-mask equipped with an elastomeric facepiece is usually in excess of 1000. Even taking account of the increased leakage of the dust/mist filters as compared to the HEPA filters, the single use respirators still provide less protection than the elastomeric facepiece respirators.

In his workplace protection factor study for single use respirators for protection against cotton dust, Revoir[17] conducted a QNFT study using coal dust as a challenge. Three single use respirators were tested. The results are shown in Table 16.

IX. POWERED AIR-PURIFYING RESPIRATORS

The performance of the powered air-purifying respirators (PAPR) was evaluated by Lowry et al. of the LASL in the mid-1970s.[18] The initial and final flow rates of the devices evaluated are shown in Table 17. Both the Burgess and the Lawrence

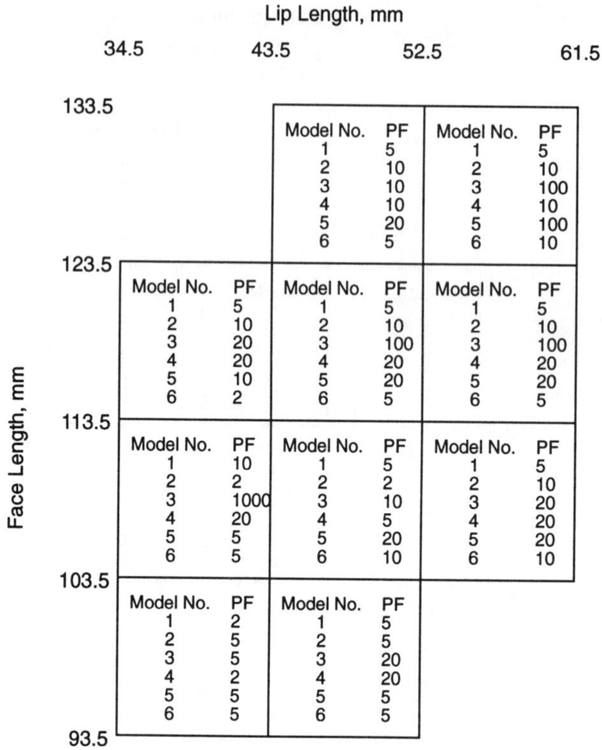

Figure 6 Protection factor obtained by each of the 6 models of singe-use respirators on the 10 test subjects.

Table 16 Quantitative Coal Dust Aerosol Respirator-Fitting Test Results

Respirator	Human test subject	Coal dust concentration (mg/m³)	Penetration of coal dust into interior of respirator (% of ambient concentration)
A	M	42	8.1
	N	85	8.6
	O	74	4.5
	P	81	4.4
	Q	85	5.6
	R	77	4.8
B	M	75	1.0
	N	74	2.9
	O	84	5.2
	P	74	0.6
	Q	82	0.6
	R	77	0.1
C	M	76	1.3
	N	84	1.6
	O	80	3.1
	P	72	0.7
	Q	88	2.9
	R	72	0.2

From Reference 17.

Table 17 Initial and Final Airflow Rates Delivered by Powered Air-Purifying Respirators

Device	Initial flow (l/m)	Final flow time[a] (l/m)
Burgess with 6-V battery	91	51 (1.4 h)
Burgess with 6-V power supply	125	—
Daisy Pak (low speed)	122	116 (8 h)
Daisy Pak (high speed)	204	108 (8 h)
3M White Cap and Economy Hood	244	—

[a] For devices operated on batteries.

From Reference 18.

Livermore Laboratory (LLL) developed Daisy Pak PAPRs were not approved. The approved 3M White Cap PAPR was tested with the shirt and shroud configurations.

The QNFT was performed on the LASL anthropometric test panel with one test subject in each of the 10 grids. The aerosol generation and detection system was the same as described in the supplied air respirator testing.[7] The test results are summarized in Figure 7. The PF for the power off mode for the Burgess and the Daisy Pak was also measured. The results indicated that most devices achieved a PF of at least

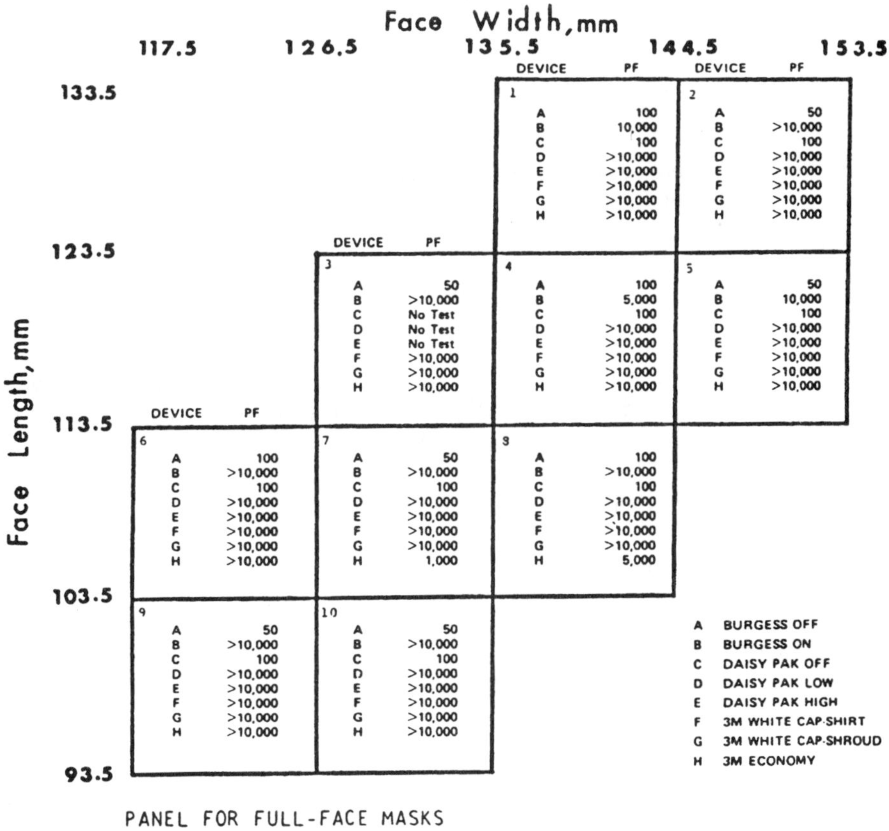

Figure 7 Summary of protection factors provided to 10-person test panel by powered air-purifying respirators. (From Reference 18.)

Table 18 Results of Protection Factor Determinations by Subject and Time Interval

Subject	Week 0	Week 2	Week 4	Week 6	Week 8
A	>20,000	>20,000	3133	517	661
	>20,000	12,857	10,937	180	416
	>20,000	15,167	>20,000	2647	514
B	>20,000	6536	186	95	72
	>20,000	13,000	308	346	70
	>20,000	8227	150	348	117
C	>20,000	2065	278	152	46
	>20,000	5000	1542	620	292
	—	4842	455	344	330
D	>20,000	1175	569	940	414
	>20,000	13,071	3067	650	575
	7038	9050	1264	650	723
E	18,900	5531	377	186	65
	>20,000	17,600	450	344	218
	>20,000	12,714	775	556	135
F	>20,000	486	368	80	14
	4889	3480	149	59	25
	>20,000	2557	229	33	40
G	>20,000	1156	92	38	25
	>20,000	430	75	50	41
	>20,000	669	54	49	29
H	>20,000	6133	617	421	525
	>20,000	>20,000	2350	1162	965
	>20,000	2676	3854	2618	1067

From Reference 19.

10,000. The PFs obtained by the Burgess and the Daisy Pak in the blower off position are between 50 and 100.

X. PROTECTION FACTOR STUDY ON A CLOSED-CIRCUIT SCBA

The protection factor for the Biomarine 60-P, a positive pressure closed-circuit SCBA, was reported by McKee[19] in a study to determine the effect of beard growth on the performance of the respirator. Eight test subjects participated in this study. A Dynatech Frontier high flow DEHP QNFT instrument was used in this study. The test aerosol had a MMAD of 0.6 μm, and a challenge concentration of 20 mg/m³. The exercises performed by the test subjects were:

1. Normal breathing
2. Deep breathing
3. Turning the head from side to side
4. Moving the head up and down
5. Talking
6. Normal breathing

The protection factor was first measured when all test subjects were clean shaven. Then the PF was measured every 2 weeks for a period of 8 weeks. Three replicates were measured for each person at each test period. The test results are shown in Table 18.

The results indicate that there is a definite deterioration of the fit factors with the beard growth. The range of fit factor for the clean shaven test subject is similar to the one obtained by Hack.[14] The test results may be of great concern for the bearded person when he wears a tight-fitting positive pressure SCBA since this device is often used under IDLH conditions.

XI. FACEPIECE SELECTION

It is generally believed that a silicone rubber facepiece would provide a better fit than a natural rubber facepiece due to its pliability. Oestenstad and Zwissler[20] conducted a comparative study on the fit obtained between silicone and rubber facepieces. One purpose of this study was to evaluate the intrasubject variability in making the comparison. Medium size silicone and natural rubber half-masks made by U.S. Safety Service Co. were selected for testing. Forty-five test subjects, 23 males and 22 females, participated in this study. Respirator faceseal leakages were measured with a portable condensation nuclei counter, the PortaCount®. A series of six tests were performed on each test subject:three with the silicone rubber facepiece and three with the natural rubber facepiece. The type of facepiece worn in each test was randomly selected. A positive pressure fit check was performed prior to each test. Each test subject performed the following exercises for 1.5 min:

1. Normal breathing
2. Deep breathing
3. Moving the head from side to side
4. Moving the head up and down
5. Talking
6. Normal breathing

The test results are presented in Table 19, and a summary of facepiece fit factors appears in Table 20. Table 20 shows that the maximum and minimum FFs for the natural rubber facepiece are higher than those of the silicone rubber facepiece. However, the median FF for the silicone rubber facepiece is higher than for the natural facepiece. Both independent and nonparametric statistical tests found FFs for the silicone rubber facepiece to be significantly greater than those for the natural rubber facepiece. The distribution of the mean log fit factor for the natural rubber facepiece followed an approximately straight line indicating a lognormal distribution. However, the values for the silicone facepiece deviated substantially from a lognormal distribution. When the mean log FFs for the silicone rubber facepiece were sorted by gender, it was found that the distribution for the males was lognormal, whereas it was not for females.

The authors concluded that although these differences are statistically significant, they do not confirm the order of magnitude differences reported by Hyatt[5] who conducted QNFT studies on silicone and natural rubber full facepieces. Hyatt also reported lognormally distributed fit factors. One possible explanation for the discrepancy in findings between these two studies is that FFs reported by Hyatt were obtained from a variety of full facepieces tested on an anthropometric test panel of

Table 19 Subject Fit Factors

Subject no.	Natural rubber facepiece					Silicone rubber facepiece				
	Test 1	Test 2	Test 3	GM[a]	GSD[b]	Test 1	Test 2	Test 3	GM[a]	GSD[b]
1	713	53	311	228	3.76	7038	8749	5385	6921	1.28
2	32233	29321	29127	30194	1.06	33477	33575	30214	32384	1.06
3	3790	1975	1875	2412	1.48	24135	19985	16081	19795	1.23
4	3626	5548	5657	4845	1.29	63786	14703	10965	21746	2.57
5	17031	10173	9760	11914	1.36	29970	19698	20527	22969	1.26
6	15912	30059	4998	13371	2.48	6008	18340	14413	11667	1.80
7	457	2055	5349	1713	3.45	35350	17279	32827	27167	1.48
8	57829	9952	32024	26415	2.45	31324	27143	24127	27374	1.14
9	12740	12704	7870	10840	1.32	61	36	13	31	2.18
10	8211	4135	2827	4578	1.72	8	2159	50	96	17.09
11	8270	8521	6509	7712	1.16	264	4019	1997	1284	4.11
12	188	5128	2472	1335	5.68	4277	7107	6688	5880	1.32
13	14333	8768	8213	10106	1.36	3531	3881	2681	3324	1.21
14	4404	1208	1025	1760	2.22	15755	32777	12322	18531	1.66
15	2287	2399	1944	2201	1.12	24380	27654	2697	12205	3.70
16	12944	10697	8959	10745	1.20	2332	2972	3001	2750	1.15
17	27116	48835	37763	36841	1.34	2086	309	2921	1235	3.36
18	2891	2058	1919	2252	1.25	35840	16564	17685	21897	1.53
19	6505	5197	3072	4700	1.47	29209	5522	9360	11472	2.34
20	389	463	1727	678	2.26	11362	5164	7729	7683	1.48
21	3324	1834	12628	4254	2.69	26368	29543	18277	24237	1.29
22	9982	7714	7665	8388	1.16	15794	16571	14246	15507	1.08
23	2097	2357	1362	1888	1.33	2106	18935	218	2055	9.33
24	1385	4059	3663	2741	1.81	7741	7957	7981	7892	1.02
25	922	338	2440	913	2.68	1120	120	19	136	7.78
26	11221	15036	28220	16824	1.60	18319	22415	18607	19695	1.12
27	46813	42726	18582	33376	1.66	21581	15610	13552	16589	1.27
28	23148	18120	22412	21105	1.14	27425	40086	17577	26834	1.51
29	6389	7224	5125	6184	1.19	6458	5844	9400	7079	1.28
30	16050	10141	10358	11902	1.30	2451	13557	646	2779	4.60
31	6650	7790	10666	8206	1.27	543	10832	10650	3971	5.60
32	2329	5494	4981	3995	1.60	42913	35653	13470	27417	1.86
33	12083	9248	6348	8933	1.38	2466	13622	12329	7454	2.61
34	4083	517	9463	2714	4.46	218	210	1022	361	2.47
35	1104	1975	8445	2641	2.85	27022	10236	16426	16563	1.62
36	513	554	2186	853	2.26	7391	2202	6292	4678	1.93
37	3097	481	1687	1360	2.58	10770	7877	1887	5430	2.53
38	10637	7390	18999	11431	1.61	13143	20439	16194	16324	1.25
39	2903	568	596	994	2.53	9622	9663	563	3742	5.15
40	10475	8830	2337	6001	2.27	10193	13631	8747	10672	1.25
41	11442	6834	7087	8214	1.33	18968	12869	9810	133778	1.39
42	6933	5539	5894	6094	1.12	4265	15808	16585	10380	2.16
43	6649	6499	5634	6244	1.09	6828	6620	6877	6774	1.02
44	12160	12985	2425	7262	2.59	1212	13172	15268	6247	4.14
45	15349	5289	3931	6833	2.05	10730	2990	9160	6648	2.01

[a] GM, geometric mean.

[b] GSD, geometric standard deviation n = 3 replicates.

From Reference 21.

experienced test subjects who tended to show a much less variation in fit factors. Hyatt selected a variety of full facepiece in his study. However, only one brand of half-masks was selected in the Oestenstad study, and their pliability may not have been the same as those tested by Hyatt.

Many OSHA substance specific health standards such as Asbestos, 29 CFR 1910.1001,[21] require that an employer must provide a selection of various sizes of respirators from different manufacturers. The selection shall include at least five

Table 20 Summary of Facepiece Fit Factors

	Natural rubber facepiece		Silicone rubber facepiece	
	GM[a]	GSD[b]	GM[a]	GSD[b]
Maximum	36845	5.68	32386	9.33
Median	6140	1.54	9136	1.60
Mean	8401	1.94	11800	2.34
Minimum	228	1.06	31	1.02

[a] GM, geometric mean of three replicates with the same respirator on the same subject.

[b] GSD, geometric standard deviation of replicates with the same respirator on the same subject.

From Reference 20.

sizes of elastomeric half-masks from at least two manufacturers. Gross and Hortsman[22] conducted a study to determine whether women could receive comparable protection as men when using currently available facepieces at the Hanford Environmental Health Foundation.

Facial parameters of the test subject such as face length, face width, lip length, menton-nasal bridge length, nose breath, bizygomatic breath, etc, were anthropometrically measured on each test subject. The results of anthropometric measurements showed that a comparison was made on anthropometric results of this study with the U.S. Air Force data collected in 1966 and 1968.[9] It was found that parameters such as face length, face width, nose breadth, and lip length were within one standard deviation between this study and the Air Force data. The female facial anthropometric measurements generated in this study (Table 21) averaged 91.5% of comparable male measurements. This agrees with the often stated rule of usage that female body measurements average 92% of male body measurements. A bivariate frequency distribution scattergram of differences in face length and face width between sexes was plotted as a 95% eclipse. It was found that while the mean values of male and female data are often quite different, the degree of overlap of the female and male eclipse ranged between 20 and 50% (see Figure 8).

The orientation of the major and minor axes of the ellipse differ considerably, depending on the facial parameters plotted. A continuity of respirator sizes, ranging from the smallest female to the largest male, is appropriate. The degree of mask fit, however, will be highly influenced by the parameters used while the facepiece was designed and engineered. The differing bivariate ellipses also help explain why the respirator brand did not extrapolate sized up and down, but rather designed in a separate small-, medium-, and large-shaped mask, obtained fit factor ratings similar to the extrapolated brands. The most important measurements for determining respirator fit have not been determined. But, the most often stated combination of more than one of the parameters such as face length, mouth width, and face width provides the critical factor.

A specific critical facial measurement, the smile lip length, has not been measured in previous anthropometric studies. The anthropometric data indicate that the "smiling lip length" is the only female facial parameter which is larger than the males. The authors noted that more women lost their respirator faceseal when they smiled, or counted the numbers "5" and "10." Tables 23 and 24 illustrate the differences in the major parameters between the test subjects who failed the fit test

Table 21 Facial Attributes of the Study Group

Measurement	Males (n = 61)	Females (n = 60)	Average (all subjects)	% Difference female/male
Menton — Nasal root	12.21 cm	11.09 cm	11.65 cm	90.9
Depression (face length)	±0.71	±0.65	±0.88	
Menton — Nasal bridge	10.55 cm	9.52 cm	10.04 cm	90.2
Length	±0.65	±0.62	±0.82	
Lip length	5.33 cm	5.16 cm	5.26 cm	96.4
	±0.45	±0.39	±0.43	
Lip length smiling	6.22 cm	16.23 cm	6.22 cm	100.2
	±0.58	±0.44	±0.52	
Nasal bridge	±3.13 cm	2.75 cm	2.94 cm	87.9
Breadth, maximum	±0.26	±0.25	±0.32	
Nasal bridge	1.23 cm	1.08 cm	1.16 cm	87.8
Breadth, minimum	±0.16	±0.14	±0.17	
Nose breadth	3.53 cm	3.13 cm	3.33 cm	88.7
	±0.35	±0.29	±0.39	
Nose length	5.16 cm	4.76 cm	4.96 cm	92.2
	±0.35	±0.38	±0.41	
Horizontal nose	3.55 cm	3.13 cm	3.34 cm	88.2
Protrusion	±0.31	±0.25	±0.35	
Bizygomatic breadth	14.06 cm	13.01 cm	13.54 cm	92.5
(face width)	±0.64	±0.57	±0.80	
Age	37.9 yrs	37.1 yrs	37.5 yrs	
	±11.3	±12.8	±12.0	

From Reference 22.

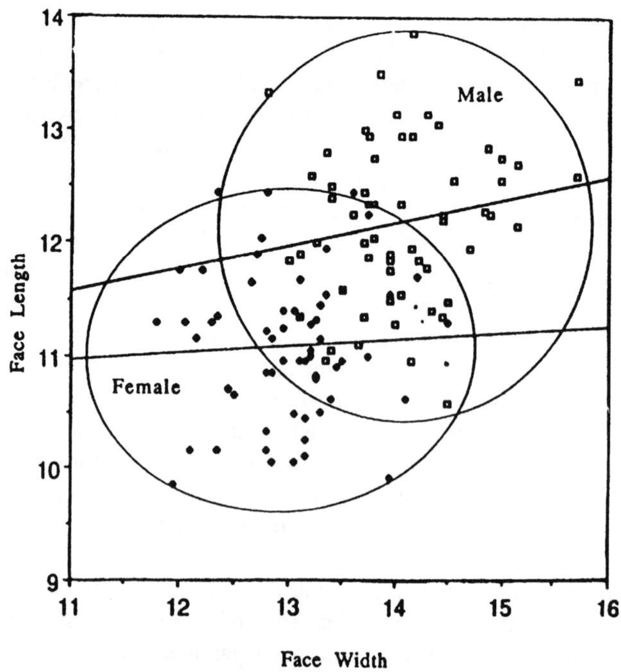

Figure 8 Ellipse plot of face length vs. face width for men and women. (From Reference 22.)

Table 22 Differences in Major Anthropometric Measurements for Subjects Failing and Passing Fit Test[a]

		Mean of outlier data	Mean of group excluding outliers	% Difference
	Males:	N = 9	N = 52	
	Females:	N = 10	N = 50	
Age	Males:	42.00 yr	37.15 yr	
	Females:	32.30 yr	38.10 yr	
Menton — Nasal root depression (face length)	Males:	12.27 cm	12.19 cm	100.0
	Females:	10.87 cm	11.13 cm	97.7
Menton — Nasal bridge length	Males:	10.42 cm	10.57 cm	98.6
	Females:	9.34 cm	9.56 cm	97.7
Lip length	Males:	5.46 cm	5.33 cm	102.4
	Females:	5.19 cm	5.15 cm	100.8
Lip length smiling	Males:	6.37 cm	6.19 cm	102.9
	Females:	6.29 cm	6.22 cm	101.1
Nasal bridge breadth, maximum	Males:	3.14 cm	3.12 cm	100.6
	Females:	2.78 cm	2.75 cm	101.1
Nasal bridge breadth, minimum	Males:	1.21 cm	1.23 cm	98.4
	Females:	1.07 cm	1.09 cm	98.2
Nose breadth	Males:	3.58 cm	3.52 cm	101.7
	Females:	3.05 cm	3.14 cm	97.1
Nose length	Males:	5.23 cm	5.15 cm	101.6
	Females:	4.74 cm	4.76 cm	99.6
Horizontal nose protrusion	Males:	3.66 cm	3.53 cm	103.7
	Females:	3.18 cm	3.12 cm	101.9
Bizygomatic breadth (face width)	Males:	14.31 cm	14.01 cm	102.1
	Females:	12.98 cm	13.02 cm	99.7

[a] Fit failures (outliers) were defined as having greater than 10% leakage in mask number 1.

From Reference 22.

and those who passed, as well as fit failure by brand and size of respirator. The outlier was defined as having greater than 10% mask penetration. In general, the men who failed the test had larger facial dimensions than the others. The women who failed the test mainly had smaller noses, shorter face lengths, or wider lips than the women who passed. Of the 9 men failed, none were tested with small size masks. For the women, only 2 of the 10 that failed wore medium masks; 8 wore small masks.

Quantitative fit testing was performed on 61 male and female test subjects from Hanford. Small medium and large size half-masks made by MSA (Comfo II), Norton (7500 series), and Survivair (Blue 1) were selected for testing. The fit factor was measured by a photometer using DEHP as the test aerosol. Exercises performed by

Table 23 Number of Fit Failures by Brand and Size of Respirator

	Male	Female
By brand		
MSA	3	3
Norton	3	4
Survivair	3	3
By size		
Small	0	8
Medium	5	2
Large	4	0

From Reference 22.

Table 24 Failure Rates for Subjects Tested with Two Different Sizes of Respirators

% of Subjects failing both fit tests	Dual Failure Rates		
	MSA/Norton	Norton/Survivair	MSA/Survivair
Male (n = 61) (4.9% failure)	1	1	1
Female (n = 60) (5.0% failure)	2	0	1

Note: Brands fit in either order A/B or B/A.

From Reference 22.

each test subject consisted of normal breathing, deep breathing, moving the head from side-to-side, moving the head up-and-down, talking, and normal breathing. The test period for each exercise was 2 min. After completing the first QNFT, a second QNFT was performed on a second respirator. The passing fit factor (FF) was 10. After each quantitative fit test, the test subject was asked to rate the comfort of the respirator on a scale from 1 to 5. The Comfort rating scale was as follows:

1. Very comfortable. Mask could be worn for an indefinite period without becoming unbearable, bothersome, or painful. No pain points.
2. Comfortable. Mask could be worn for 2 to 4 h without undue discomfort. Some pressure points with slight discomfort.
3. Barely comfortable. Mask could be worn for approximately 0.5 to 1.0 h without intolerable comfort. Some discomfort from pressure.
4. Uncomfortable. Mask could be tolerated for the period of the test only.
5. Intolerable. Mask could not be worn at all without discomfort.

The quantitative fit test results indicated that 5% of the men and women did not receive a fit factor of 10 or greater when tested with two random brands of half-mask respirators (Table 24). The percentage of men and women obtaining acceptable fits (FF 10 or greater) by brand is shown in Figure 9. Between 83 and 85% of male test

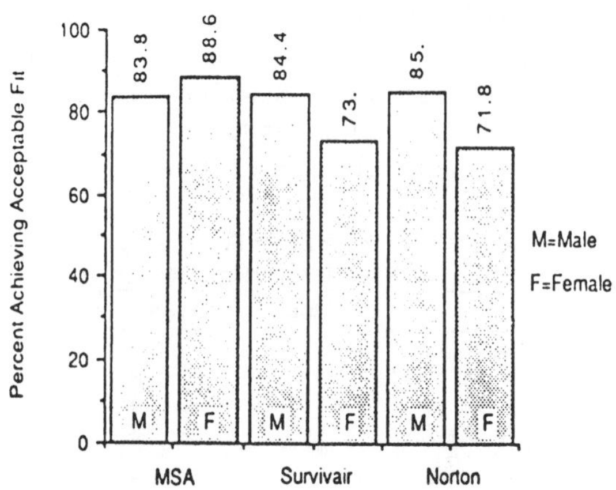

Figure 9 Single brand results — Acceptable fit (%); M = male, F = female. (From Reference 22.)

Table 25 Comfort Ratings of Respirator by Brand

	Comfort rating				
Brand	**1 (most)**	**2**	**3**	**4**	**5 (least)**
MSA n = 80	13.8%	43.8%	38.8% (96.4%)	2.3%	1.3%
Survivair n = 81	25.9%	50.6%	22.3% (98.8%)	1.2%	0
Norton n = 78	19.2%	41.0%	29.5% (89.7%)	9.0%	1.3%

From Reference 22.

subjects obtained an acceptable fit given any one brand of respirator. Acceptable fits for women were found to vary considerably depending on respirator brand. The range of acceptable fit by brand was 71 to 89%.

The comfort ratings of the respirators are shown in Table 25. Only 1.2% of the test subjects thought the silicone facepiece were uncomfortable or intolerable, whereas 3.6 to 10.3% of the subjects thought that the hard rubber facepieces were unacceptable for working conditions. The distribution of respirator size to the test population is shown in Table 26. The results indicated that men were found to wear all three sizes of the Norton series, but only the medium and large in MSA and Survivair. Women were found to fit only the small and medium sizes of all three brands. When any two brands (in multiple sizes) of respirators were available, 95% of both men and women could be safely fitted. When only one brand of respirator was available, between 71 and 89% of the test subjects achieved acceptable fits.

The authors concluded that in order to achieve a passing fit factor of 10, it would be necessary for an employer to offer a variety (two or more) of half-mask respirators brands. Each brand must be available in a range of sizes and/or shapes, either extrapolated or manufactured in a unique design. For the 5% of individuals who cannot obtain an acceptable half-mask fit from either available brand, a full facepiece may then be required. If a full facepiece fit test fails, a positive pressure respirator would then be appropriate. The respirator manufacturers have developed a variety of facepiece designs and/or sizes, and these offer both men and women an equal chance of obtaining a good fit. It is up to each employer to ensure that appropriate devices are available for their workers.

The limitation of this study is that racial characteristics were not investigated. Ethnic and racial groups may have a significantly more difficult time fitting any given brand of respirators. Since a passing fit factor of 100 is usually required in OSHA standards, two types of respirators may not be sufficient to fit 95% of the work population when a passing FF of 100 is required.

Table 26 Distribution of Respirator Sizes Fitting the Test Population

	Male			Female		
Brand	**Small (%)**	**Medium (%)**	**Large (%)**	**Small (%)**	**Medium (%)**	**Large (%)**
MSA	0	64.5	35.5	48.7	51.3	0
Survivair	0	68.4	31.6	66.7	33.3	0
Norton	11.8	44.1	44.1	75.0	25.0	0

From Reference 22.

REFERENCES

1. U.S. Bureau of Mines, Filter-type dust, fume, and mist respirators. Schedule 21 B, January 19, 1965.
2. U.S. Atomic Energy Commission, Proposed Rule Making, Standard for Protection Against Radiation, 10 CFR Part 20, 32, 215, 1967.
3. Hyatt, E. C. and Richards, C. P., A study of facepiece leakage of self-contained breathing apparatus by DOP man tests, Los Alamos Scientific Laboratory, LA-4927-PR, 1972.
4. Hyatt, E. C., Pritchard, J. A., and Richards, C. P., Respirator efficiency measuring using quantitative DOP man tests, *Am. Ind. Hyg. Assoc. J.*, 33, 635, 1972).
5. Hyatt, E. C., Respirator protection factors, Los Alamos Scientific Laboratory, UC-41, 1976.
6. Douglas, D. D., Hesch, P. R., and Lowry, P. L., Supplied air hood, Los Alamos Scientific Laboratory, LA-NUREG-6612-MS, 1976.
7. Hack, A. L., Bradley, O. D., and Trujillo, A., Respirator studies for the Nuclear Regulatory Commission, October 1, 1976 to September 30, 1977, Protection Factors for Supplied Air Respirators. Los Alamos Scientific Laboratory, 1978.
8. Hack, A. L., Bradley, O. D., and Trujillo, A., Respirator protection factors. Part II: Protection factor of supplied air respirators, *Am. Ind. Hyg. Assoc. J.*, 376, 1980.
9. Hack, A. L. and McConville, J. T., Respirator protection factors. Part I: Development of an anthropometric test panel, *Am. Ind. Hyg. Assoc. J.*, 39, 970, 1978.
10. ANSI, Practices for Respiratory Protection, ANSI Z88.2-1992. Amercian National Standards Institute (ANSI), New York, 1992.
11. Hack, A. L., Trujillo, A., and Bradley, O. D., Respirator studies for the Nuclear Regulatory Commission, October 1, 1977–September 30, 1978: Evaluation of open-circuit self-contained breathing apparatus, Los Alamos Scientific Laboratory, NUREG/CR-1235, LA-8188-PR, 1980.
12. NFPA, Open-Circuit Self-Contained Breathing Apparatus for Fire Fighting, NFPA 1981. National Fire Protection Association (NFPA), Quincy, MA, 1987.
13. Hack, A., Trujillo, A., and Bradley, O. D., Respirator studies for the Nuclear Regulatory Commission, October 1, 1978–September 30, 1979: Evaluation and performance of escape-type self-contained breathing apparatus, Los Alamos National Laboratory, NUREG/CR-1586, LA-8188-PR, 1980.
14. Hack, A. L., Trujillo, A., Bradley, O D., and Carter, K., Evaluation and performance of closed-circuit self-contained breathing apparatus, Los Alamos National Laboratory, NUREG/CR-2652, LA-9266-MS, 1982.
15. Hyatt, E. C., Pritchard, J. A., Richards, C. P., Geoffrion, L. A., and Kressin, I. K., Respirator research and development related to quality control. LASL Project R-037, Quarterly Report, July 1 through September 30, 1971. Los Alamos Scientific Laboratory, LA-4908-PR, March 1972.
16. Lowry, P. L., Hesch, P. R., and Revoir, W. H., Performance of single use respirators, *Am. Ind. Hyg. Assoc. J.*, 38, 462, 1977.
17. Revoir, W. H., Respirators for protection against cotton dust, *Am. Ind. Hyg. Assoc. J.*, 35, 322, 1974.
18. Lowry, P. L., Wheat, L. D., and Bustos, J. D., Quantitative fit-test method for powered air-purifying respirators, *Am. Ind. Hyg. Assoc. J.*, 40, 291, 1979.
19. McGee, M. K. and Oestenstad, R. K., The effect of growth of facial hair on protection factors for one model of closed-circuit pressure-demand self-contained breathing apparatus, *Am. Ind. Hyg. Assoc. J.*, 44, 480, 1983.

20. Oestenstad, R. K. and Zwissler, A. M., Comparison of fit provided by natural and silicone rubber facepieces of the of the same brand of half-mask respirator, *Appl. Occup. Environ. Hyg.*, 6, 785, 1991.
21. Code of Federal Regulations, Title 29, Part 1910.1001. Government Printing Office, Washington, DC, 1992.
22. Gross, S. F. and Hortsman, S. W., Half-mask respirator selection for a mixed worker group, *Appl. Occup. Environ. Hyg.*, 5, 229, 1990.

20 RESPIRATOR PERFORMANCE STUDIES II: SIMULATED WORKPLACE STUDIES

The exercises for the quantitative fit testing are performed under sedentary conditions with little or no body movements. The activities performed by the respirator wearer in the workplace involve many body movements. In order to simulate real work situations, exercises for the QNFT should be performed with body movements, or under a high work rate, to induce leakage at the respirator faceseal.

I. LANL CONTROLLED ENVIRONMENTAL CHAMBER STUDY

When the NIOSH was conducting workplace protection factor (WPF) studies to determine assigned protection factors for respirator certification, the technical staff of the Occupational Safety and Health Administration believed that in order to assign protection factors for different types of respirators the tests should be performed under the same test conditions for better correlation. Since respirators are often used under stressful working conditions, such as high temperature, high humidity, and heavy work rate, the test conditions should reflect real use situations. The controlled environmental chamber study would be the method of choice, and since all respirators are evaluated under the same test conditions, reproducible results could be obtained. OSHA and the Nuclear Regulatory Commission (NRC) jointly sponsored a controlled environmental chamber study which was conducted by the Los Alamos National Laboratory (LANL).[1]

The study had three objectives:

1. Measure the protection factors of respirators under different work conditions.
2. Determine the difference between protection factors obtained from the test subjects during quantitative fit and under simulated work conditions measured on the test subject in the environmental chamber.
3. Determine the effect of temperature and humidity on respirator facepiece leakage.

Seven respirators were selected for this study, divided into two series:

Series I — Positive Pressure Respirators

1. MSA Comfo half-mask powered air-purifying respirator
2. Racal AH-3 loose fitting powered air-purifying respirator

3. 3M Airhat loose fitting powered air-purifying respirator
4. Survivair full facepiece pressure demand supplied air respirator
5. Bullard supplied air hood

Series II — Negative Pressure Respirators
6. MSA Comfo II half-mask air-purifying respirator
7. MSA Ultravue full facepiece air-purifying respirator

All air-purifying respirators were equipped with high-efficiency particulate air (HEPA) filters.

The study was performed in a large controlled environmental chamber which had enough space for two test subjects to perform, at the same time, a variety of exercises that simulate actual work conditions. The test chamber consisted of a 28.3-m^3 (1000 ft^3) room with an attached 5.5-m^3 (200 ft^3) antechamber. The di-2-ethylhexyl sebacate (DEHS) test aerosol was generated by Laskin nozzles. The test aerosol concentration in the chamber had an average concentration of 25 mg/m^3 with an average mass median aerodynamic diameter (MMAD) of 0.67 μm. Another small chamber was used for performing other quantitative fit testing. The test temperature and humidity in the test chamber varied at the following:

Temperature (°C)	Humidity (%)
32	85
0	15
32	15
0	85
21	85
21	15

The tests were performed on the above sequence to alleviate possible acclimatization by test subjects. Each test subject performed exercises in the following regimen:

Exercise	Time (min)
Step up and back down a two-step platform	5
Rest	5
Move oiled gravel between two bins	10
Rest	5
Pound nails into an overhead board	10
Rest	5
Move and lay cinder blocks	10
Rest	5
Pound a board with a sledge hammer	5
Rest	5
Total time:	65

Twenty-two male and female test subjects participated in this study, but only 10 test subjects were tested on each respirator.

A qualifying quantitative fit testing was performed on each prospective test subject wearing negative pressure respirators (Series II). The test subject must obtain a minimum fit factor of 1000 and 3000 respectively for the half-mask and the full facepiece before being selected. Since not all test panel members available at LANL would meet this requirement, it was decided that the facial dimensions of test subjects would not be required to meet the anthropometric requirement of the LLNL test panel.

The test was performed on each test subject at the following sequence:

1. Prefit
2. Prework
3. Simulated work
4. Postwork

At first, a quantitative fit test was performed on the test subject in the small fit test chamber. After the test, the test subject entered the environmental chamber to perform the prework fit test. The simulated work test followed the prework test. A postwork test was performed after the completion of the simulated work test. In order to permit two test subjects to be tested in the environmental chamber simultaneously, a time lag of 5 min was set for performing the required tests. Except for the simulated work test, all test exercises were those prescribed in the American National Standard Institute Practice for Respiratory Protection, ANSI A-88.2, 1980.[2]

The ambient and in-mask aerosol concentrations were measured by two photometers, which permitted the simultaneous measurement of the fit factors of two test subjects. The limit of detection of the photometer was set at 20,000, so that any value reported above this value was reported as 20,000. The fit factors for the test subjects were measured for the prefit, prework, simulated work, and postwork stages. Ten sets of tests under each of the six environmental conditions were performed on each respirator, except for the Bullard hood for which only two sets of tests were performed because there was very little change in the fit factors obtained from different test conditions. The geometric mean fit factors obtained from prefit, prework, simulated work and postwork are listed in Table 1. The simulated work fit factors obtained by each test subject are tabulated in Table 2.

Based on the test results, the following conclusions were made:

1. The tight-fitting half-mask PAPR provided a much higher protection than the loose fitting helmet type PAPRs. Different protection factors should be assigned.
2. The fit test exercises prescribed by ANSI Z-88.2 for tight or loose fitting respirators do not adequately simulate real work situations. More dynamic full-body exercises should be used for QNFT to better duplicate work situations, especially those motions in which the individual bends over and stands up repeatedly.
3. Tests at all environmental conditions and exercises confirm a protection factor of 1000 proposed by Hyatt[3] for both the supplied air respirator and the PAPR.
4. The environmental conditions and exercises had no obvious effect on the performance of the Bullard continuous flow supplied air hood.
5. The performance of the Racal and 3M PAPRs degraded during simulated exercises at 32°C and 85% relative humidity. However, these conditions had nonsignificant effects on the performance of the MSA PAPR and the Survivair pressure demand SAR.
6. The performance of negative pressure air-purifying respirators (Series II) degraded in the environmental chamber tests. The simulated work exercises are more

Table 1 LANL Environmental Chamber Study — Geometric Mean Fit Factors and Simulated Work Protection Factors (× 1000)

Brand	Temp. (C)	RH (%)	PreFit FF	PreWork FF	Sim Work FF	PostWork FF
3M PAPR	0	15	14.3	7.1	5.6	8.3
W-344	0	85	18.1	14.1	4.3	11.8
	21	15	12.4	14.3	3.4	11.5
	21	85	17.1	13.2	3.3	7.9
	32	15	18.8	18.2	4.3	8.6
	32	85	13.9	11.0	1.9	4.5
RACAL	0	15	8.6	8.7	2.4	1.9
PAPR	0	85	7.1	6.2	3.5	3.5
AH-3	21	15	6.6	4.9	2.4	2.5
	21	85	8.2	3.6	2.4	1.7
	32	15	5.2	2.4	2.0	2.3
	32	85	4.4	1.9	1.2	0.9
MSA	0	15	20.0	17.1	14.2	13.8
PAPR	0	85	18.4	18.7	15.3	12.7
COMFO	21	15	19.3	20.0	20.0	13.2
	21	85	20.0	18.2	19.0	14.6
	32	15	19.8	20.0	20.0	18.9
	32	85	19.7	19.8	19.1	17.4
SURVIVAIR	0	15	20.0	20.0	19.7	18.9
PD-SAR	0	85	20.0	20.0	19.1	13.6
FF	21	15	19.6	19.8	15.8	19.4
	21	85	17.7	19.9	20.0	20.0
	32	15	13.3	13.6	12.4	12.8
	32	85	19.8	19.8	8.5	7.9
BULLARD	0	15	20.0	20.0	20.0	20.0
#999	0	85	19.5	20.0	16.3	20.0
CF Hood	21	15	20.0	20.0	13.7	20.0
	21	85	18.7	20.0	15.8	20.0
	32	15	20.0	20.0	7.5	20.0
	32	85	20.0	20.0	19.1	20.0
MSA	0	15	1.3	1.5	1.6	1.3
Comfo	0	85	4.8	8.4	5.7	2.5
	21	15	1.4	3.2	2.9	0.9
	21	85	2.8	2.1	1.5	0.5
	32	15	2.6	3.2	2.0	1.5
	32	85	1.6	1.5	0.8	0.6
MSA	0	15	8.1	5.6	5.3	3.1
Ultra	0	85	7.2	5.2	1.0	1.5
FF	21	15	4.1	3.7	5.0	5.8
	21	85	2.9	4.2	2.6	5.3
	32	15	3.6	3.7	3.5	4.4
	32	85	6.5	4.9	2.6	2.6

From Reference 1.

effective in degrading fit factors for the half-mask and full facepiece respirators than the prework or postwork exercises.

7. The fit factors for the half-mask negative pressure respirators degraded at high humidity (85% RH) at both 21 and 32°C because sweat would cause the facepiece to slide down the nose during simulated work and postwork exercises.
8. The fit factors for the negative pressure half-mask respirator degraded to 50 during simulated work at high temperature and humidity, and each test subject was preselected to achieve a minimum fit factor of 1000. The commonly used fit factor "pass criterion of 100" for the half-mask may not provide an adequate margin of safety under adverse weather conditions and a moderately heavy workload.

Table 2 LANL Environmental Chamber Study — Simulated Work Protection Factors for Individual Test Subjects

Test sub #	Temp (C)	RH (%)	Brand	Mean PF	Brand	Mean PF	Brand	Mean PF	Brand	Mean PF	Brand	Mean PF	Brand	Mean PF	Brand	Mean PF
1	0	15	3M	4350	RACAL	600	MSA	9000	SUVIV	20000	BULL		MSA	15100	MSA	20000
2	0	15	PAPR	6000	PAPR	210	PAPR	20000	PDSAR	20000	Hood		Comfo	13100	ULTRA	20000
4	0	15		6300		500		18000								
5	0	15		430		1140		20000		20000						
9	0	15		5520		5260		11800		20000		20000		20000		4700
10	0	15		6900		9800		20000		17400				90		840
11	0	15		6560		400		18200								
15	0	15		12500		14300		20000		20000		20000		260		8200
16	0	15		5670												
17	0	15		16700		6000		2950		20000						
18	0	15		8500		4500		20000		20000				150		20000
19	0	15												10700		6900
20	0	15								20000				290		6600
21	0	15								20000				15400		19500
22	0	15												170		80
1	0	85	3M	9200	RACAL	1270	MSA	20000	SUVIV	20000	BULL		MSA	18200	MSA	1990
2	0	85	PAPR	10300	PAPR	4500	PAPR	20000	PDSAR	20000	Hood		Comfo	3650	ULTRA	620
4	0	85		8500		980		3620								
5	0	85		960		8400		20000		20000						
9	0	85		1900		3600		8300		20000		20000		20000		5160
10	0	85		770		3900		20000		20000				20000		210
11	0	85		2520		4500		17800								
15	0	85		16000		8600		20000		20000		13300		60		20
16	0	85														
17	0	85		5300		6250		20000		20000						
18	0	85		9400		2150		20000		20000				390		4900
19	0	85												9500		470
20	0	85								20000				20000		4200
21	0	85								12700				20000		20000
22	0	85												17000		140

Table 2 LANL Environmental Chamber Study — Simulated Work Protection Factors for Individual Test Subjects (*continued*)

Test sub #	Temp (C)	RH (%)	Brand	Mean PF	Brand	Mean PF	Brand	Mean PF	Brand	Mean PF	Brand	Mean PF	Brand	Mean PF	Brand	Mean PF
1	20	15	3M	550	RACAL	1350	MSA	20000	SUVIV	20000	BULL		MSA	18200	MSA	5930
2	20	15	PAPR	5000	PAPR	6250	PAPR	20000	PDSAR	20000	Hood		Comfo	14300	ULTRA	2560
4	20	15		4680		1550		20000								
5	20	15		1100		20000		20000		20000						
9	20	15		1530		2270		20000		20000		20000		100		7400
10	20	15		3760		2600		20000		20000				50		500
11	20	15		1540		340		20000								
15	20	15		5300		7690		20000		20000		9400		50		4200
16	20	15														
17	20	15		11800		230		20000		18200						
18	20	15		2400		6250		20000		20000				6100		16000
19	20	15												20000		2820
20	20	15								20000				20000		20000
21	20	15								20000				19000		20000
22	20	15												16000		2260
1	20	85	3M	5400	RACAL	3600	MSA	13800	SUVIV	20000	BULL		MSA	16300	MSA	20000
2	20	85	PAPR	2180	PAPR	5230	PAPR	20000	PDSAR	20000	Hood		Comfo	16300	ULTRA	290
4	20	85		11100		500		20000								
5	20	85		2900		17000		20000		20000						
9	20	85		2600		1200		17800		20000		20000		3800		5100
10	20	85		1200		9000		20000		20000				18100		190
11	20	85		2300		810		19000								
15	20	85		8900		650		20000		20000		12500		50		2400
16	20	85														
17	20	85		2830		5000		20000		20000						
18	20	85		2230		1220		20000		20000				2770		400
19	20	85												310		3500
20	20	85								20000				43		7200
21	20	85								20000				18100		10800
22														65		7800

1	32	15	3M	6200	RACAL	2320	MSA	20000	SURVIV	20000	BULL		MSA	5630	MSA	11940
2	32	15	PAPR	8300	PAPR	6100	PAPR	20000	PAPR	20000	Hood		Comfo	20000	Ultra	8700
4	32	15		5840		2330		20000								
5	32	15		6670		9300		20000		20000						
9	32	15		3000		500		20000		20000		20000		90		20000
10	32	15		6780		2500		20000		250				17000		20
11	32	15		400		1550		20000								
15	32	15		9090		19000		20000		20000		2778		16670		13300
16	32	15		1650												
17	32	15		19000		300		20000		20000						
18	32	15		690		250		20000		12900				6350		1170
19	32	15												45		20000
20	32	15								20000				3570		20000
21	32	15								20000				20000		5560
22	32	15												16		200
1	32	85	3M	4520	RACAL	460	MSA	20000	SUVIV	6300	BULL		MSA	19500	MSA	10390
2	32	85	PAPR	1760	PAPR	370	PAPR	20000	PDSAR	20000	Hood		Comfo	9100	ULTRA	3540
4	32	85		2220		4730		20000								
5	32	85		1310		19000		20000		20000						
9	32	85		280		110		15400		13300		20000		560		19500
10	32	85		7140		150		20000		700				50		190
11	32	85		1500		370		20000								
15	32	85		8200		15700		20000		20000		18182		140		40
16	32	85		970												
17	32	85		1840		3300		20000		20000						
18	32	85		630		1800		16700		520				70		1080
19	32	85												110		17400
20	32	85								19500				60		20000
21	32	85								20000				17800		10800
22	32	85												20000		580

From Reference 1.

II. LLNL POWERED AIR-PURIFYING RESPIRATORS (PAPR) STUDY

The Lawrence Livermore National Laboratory (LLNL)[4] conducted another simulated work place study for OSHA. The study was performed on three powered air-purifying respirators. The purpose of the study was to determine:

1. Whether these positive pressure respirators would maintain positive pressure under a heavy work load.
2. The effect of air flow on the performance of the PAPR.
3. The effect of beard growth on the seal of the inlet covering of the PAPR.

The tight-fitting half-mask PAPR made by MSA, the loose fitting facepiece Racal Breathe-Easy 1 PAPR, and the 3M W-344 Airhat PAPR were selected for testing. All PAPRs were equipped with high-efficiency particulate air filters. The approval air flow requirement is 115 liters per min (l/m) (4 cfm) for the MSA PAPR, and 170 l/m (6 cfm) for the Racal and 3M PAPRs. To assess the effect of air flow, the flow to the PAPR was controlled by replacing the battery pack with a dc power supply. When the test subject's work rate reached 80% of the maximum, the air flow varied for each PAPR in the following sequence:

PAPR	Flow (l/m)	Flow (cfm)
MSA	170	6
	198	7
	142	5
	114	4
	86	3
	170	6
RACAL	202	7
	226	8
	170	6
	142	5
	114	4
	202	7
3M	255	9
	283	10
	198	7
	170	6
	114	4
	255	9

Six test subjects participated in this study. The work rate was controlled by the treadmill grade. At first, the subject walked on the treadmill at 5.3 km/h (3.3 mph) at a grade of 2.5%, the grade of the treadmill was increased until the subject reached 80% of his cardiac reserve which usually took 10 min. Then the penetration of the inlet covering of each type of PAPR was measured at the air flow rate listed above. The flow rate changed at a two minute interval.

The treadmill was placed in a large quantitative fit testing chamber. The challenge aerosol was polyethylene glycol 400 (PEG 400) with a MMAD of 0.78 μm and a concentration of approximately 20 mg/m³. The pressure inside the inlet covering of each PAPR was measured to determine whether a positive pressure could be maintained at a heavy work rate.

Table 3 Physiological Response at Various Grades

Grade (%)	Heart rate (min)	Breathing rate (l/min)	Minute volume (l)
2.5	116	23	30
5.0	124	23	36
10.0	155	30	53
15.0	165	33	72
20.0	177	41	103

From Reference 4.

The effect of beard growth was determined by measuring the penetration of the inlet covering of the PAPRs when the test subjects were clean shaven, with 3 days of stubble and a full beard of 2 to 3 months growth.

The physiological response of the test subject on the treadmill is shown in Table 3. The effect of facial hair growth at 80% work rate for these PAPRs is shown in Figures 1 through 3. The effect of air flow on the facepiece pressure of the MSA PAPR is shown in Figure 4.

The following conclusions can be drawn from the test results.

1. For the tight-fitting PAPR, a positive pressure cannot be maintained inside the facepiece at the approved flow rate of 115 l/m at a high work rate. A minimum flow rate of 170 l/m (6 cfm) is required to maintain a positive pressure inside the facepiece.
2. The beard growth degrades the performance of the tight-fitting PAPR.
3. The pressure inside the two loose fitting PAPRs does not change much with air flow.
4. The loose fitting helmet type PAPRs provide very low protection when the test subjects perform heavy work.

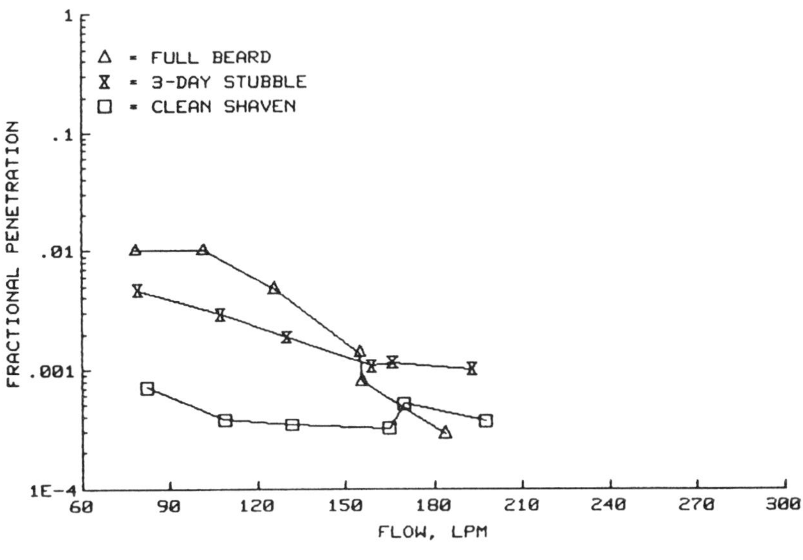

Figure 1 Effect of facial hair at 80% work rate, average subject, MSA half-mask PAPR.

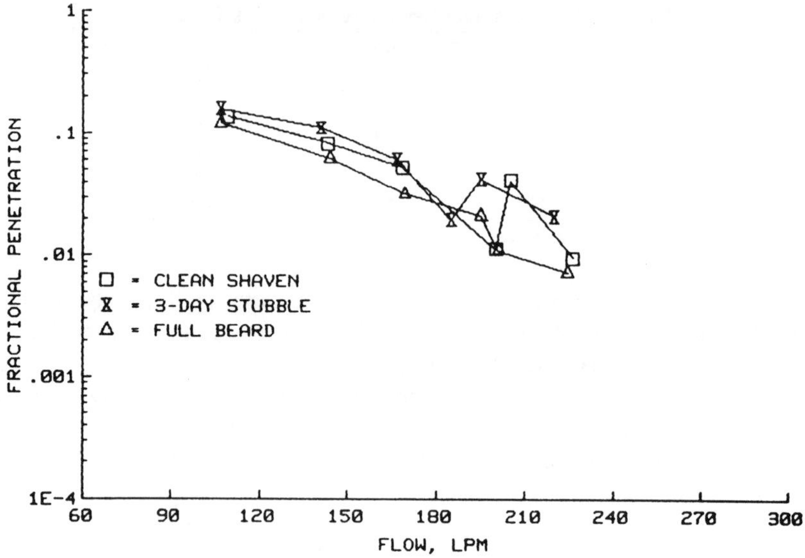

Figure 2 Effect of facial hair at 80% work rate, average subject, Racal helmet PAPR.

5. The loose fitting PAPRs could not consistently maintain a positive pressure inside the inlet covering, even at a flow rate as high as 280 l/m (10 cfm).

Based on the results of this study, at least two PAPR manufacturers presently sell their PAPRs with a minimum flow rate of 170 l/m (6 cfm) for the tight-fitting facepiece, with a minimum flow rate of 230 l/m (8 cfm) for the loose fitting facepiece.

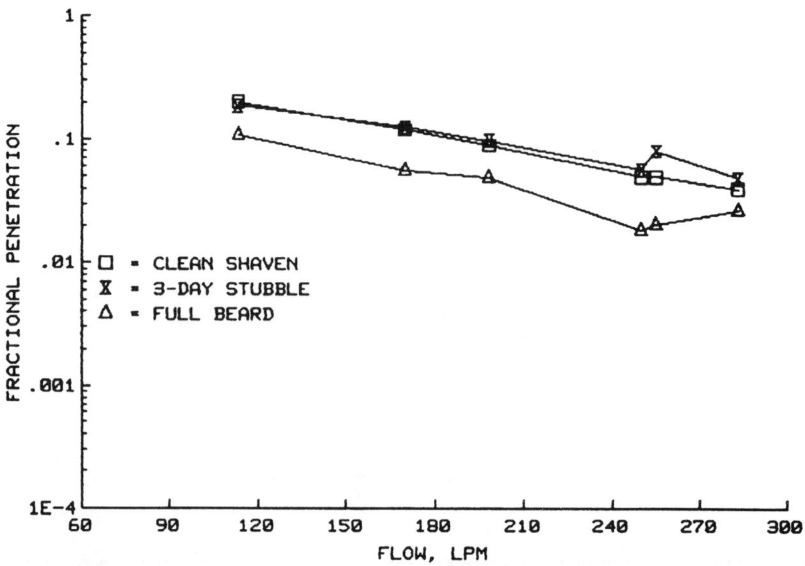

Figure 3 Effect of facial hair at 80% work rate, average subject, 3M helmet PAPR.

III. MANUFACTURER SPONSORED SIMULATED WORKPLACE STUDIES

A. Bullard Abrasive Blasting Hood

The NIOSH Respirator Decision Logic (RDL)[5] assigned the loose fitting supplied air hood a protection factor (PF) of 25. A part of the Assigned Protection Factors (APF) for respirators listed in the RDL are based on workplace protection factor studies. NIOSH had conducted WPF studies for the loose fitting facepiece PAPRs such as the 3M Airhat and the Racal AH-3. Both devices showed poor performance.[6,7] NIOSH assigned these respirators with a PF of 25. NIOSH also assigned the loose fitting supplied air hood the same PF as the PAPR. NIOSH's rationale is that both devices have a loose fitting inlet covering and the same approval flow rate of 170 l/m. However, NIOSH has no workplace testing data for the supplied air hood.

The Bullard Company is a major manufacturer of the supplied air hood. In order to demonstrate the performance of the hood, Bullard conducted a simulated workplace study on a hard shell abrasive blasting hood (Model 77).[8] The test was performed in a test chamber with the dimensions of 12' wide, 15' long, and 8' high. Two test subjects performed in the blast chamber. Both subjects wore a blasting hood. One person performed the blasting and the other one was a helper and observer. A third person stayed outside the chamber and acted as a helper and monitor of the air compressor.

The work performed by the blasters lasted approximately 1 to 2 min. The blasting operation was continued for a 2 h time period. The blasting operation was stopped for chamber cleaning over a 15-min period. The total ambient dust concentration varied between 70 and 33,000 mg/m^3, with a geometric mean concentration of 3200 mg/m^3. The ambient respirable dust ranged between 5 and 1000 mg/m^3, with a geometric mean concentration of 250 mg/m^3. The inward leakage for total and respirable dust was 0.06 and 0.02 mg/m^3, respectively. The measured PFs for ambient and respirable dust concentrations were 53,000 and 12,500, respectively. This study agreed with the LANL study that the performance of the supplied air hood is greatly superior to the loose fitting facepiece PAPR. These two devices should not have the same assigned protection factor of 25, as designated in the NIOSH RDL.

B. MSA PAPR

In 1981, NIOSH received two user complaints concerning the poor performance of the MSA PAPR. A workplace study on a silica bagging operation performed by NIOSH as a part of field investigation[9] showed that the MSA PAPR achieved a protection factor which was significantly less than the expected 1000. MSA contracted Ayer of the University of Cincinnati to conduct a simulated workplace study to evaluate the performance of its PAPR.[10]

Four test subjects participated in this study. The exercises consisted of lifting a 50 lb sandbag from a low stool to a high stool on a signal, and back to the low stool on the second signal. The signals were set at a 30-s interval. These tests were further modified to increase the lift height and body movements during bending and turning. In one test, the lifting interval was reduced to 20 s. A total of 14 tests were performed

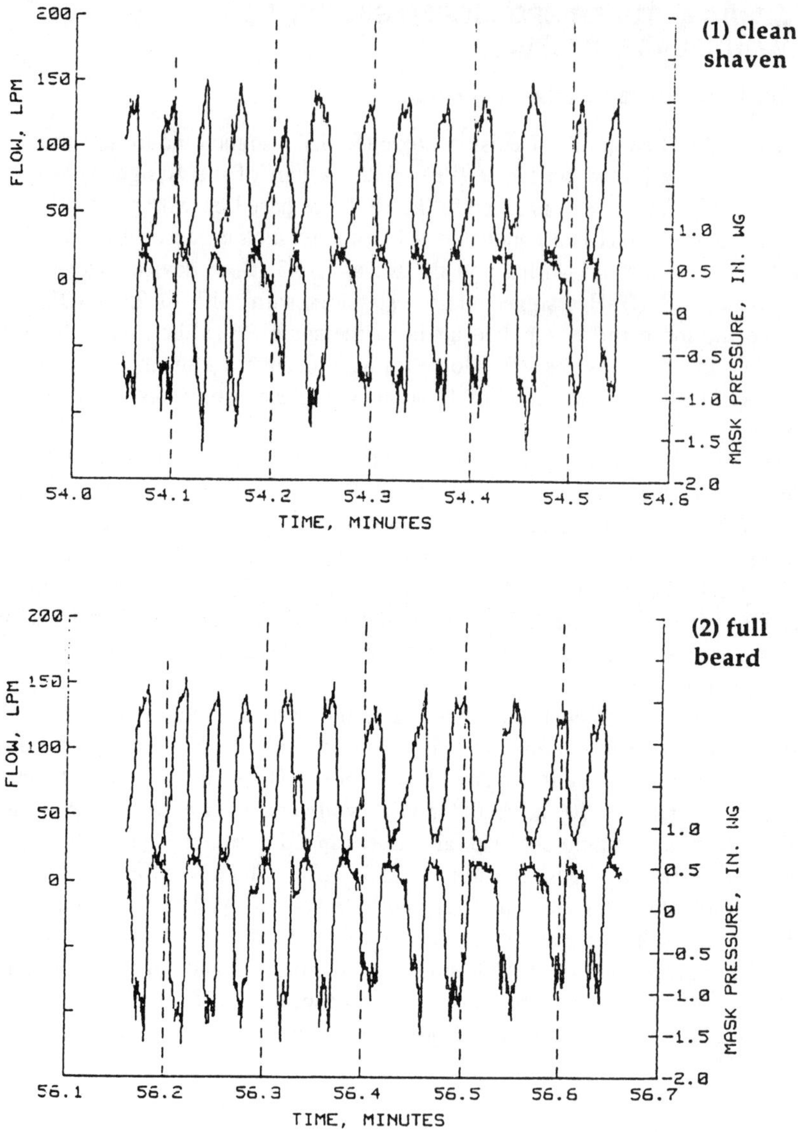

Figure 4 (a) Pressure and flow vs. time, MSA half-mask PAPR, flow of 3 cfm; Subject BGH. (From Reference 4.)

during a 5 day period. Eight of the 14 samples had a minimum sampling time of 3 h. The remaining lasted for more than 2 h in all but one test.

The challenge aerosol was an unclassified finely ground silica dust (<325 mesh) and test chamber dust concentration varied from 20 to 200$^+$ mg/m^3. The particle size of the chamber aerosol was measured by an Andersen impactor which showed a geometric mean diameter between 5.1 and 7.2 μm and a geometric standard deviation from 2.2 to 2.5. Approximately 35 to 40% of the airborne aerosol had an aerodynamic

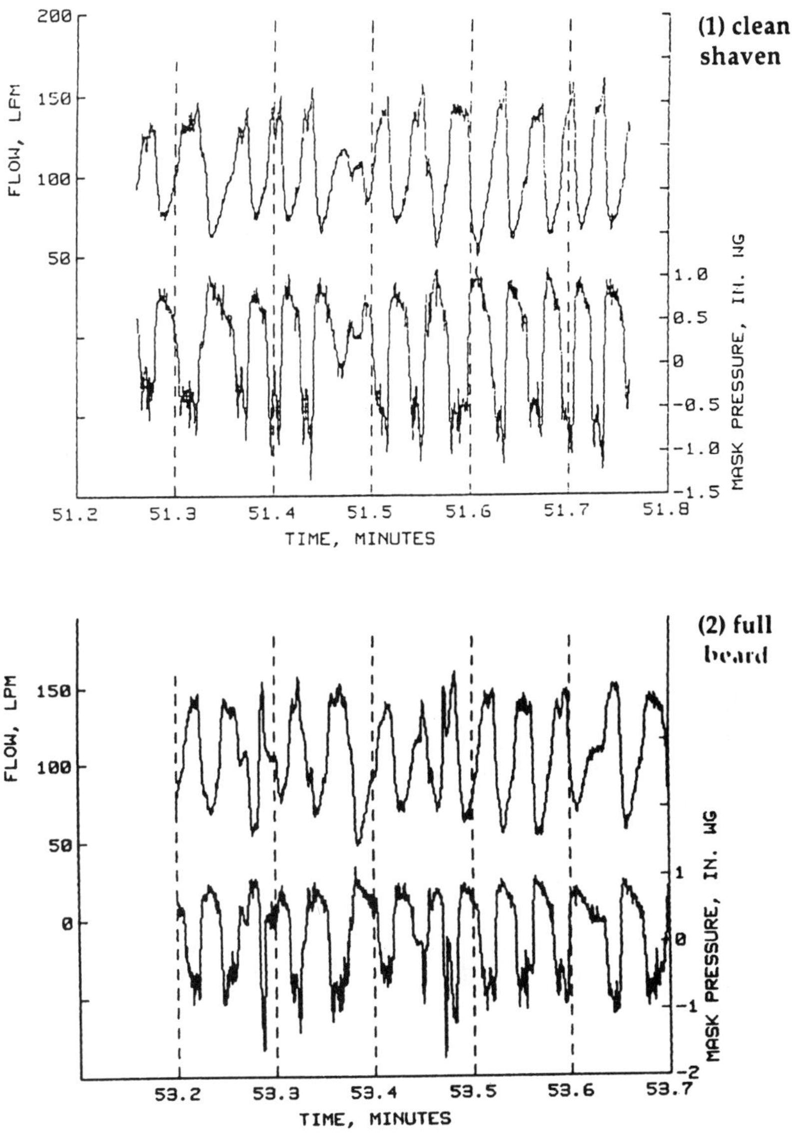

Figure 4 (b) Pressure and flow vs time, MSA half-mask PAPR, flow of 4 cfm; Subject BGH. (From Reference 4.)

size of less than 5 μm. Ambient samples were collected at both lapel and belt positions to determine the dust exposure without the use of a respirator, and the exposure of the air-purifying element. The lapel filter sample of test No. 14 was collected through a 10 mm nylon cyclone.

The test results are shown in Table 4. All ambient and in-mask samples were weighed gravimetrically for calculating protection factors. The weights of in-mask samples were less than 0.01 mg. Except for test No.10, all other samples showed a

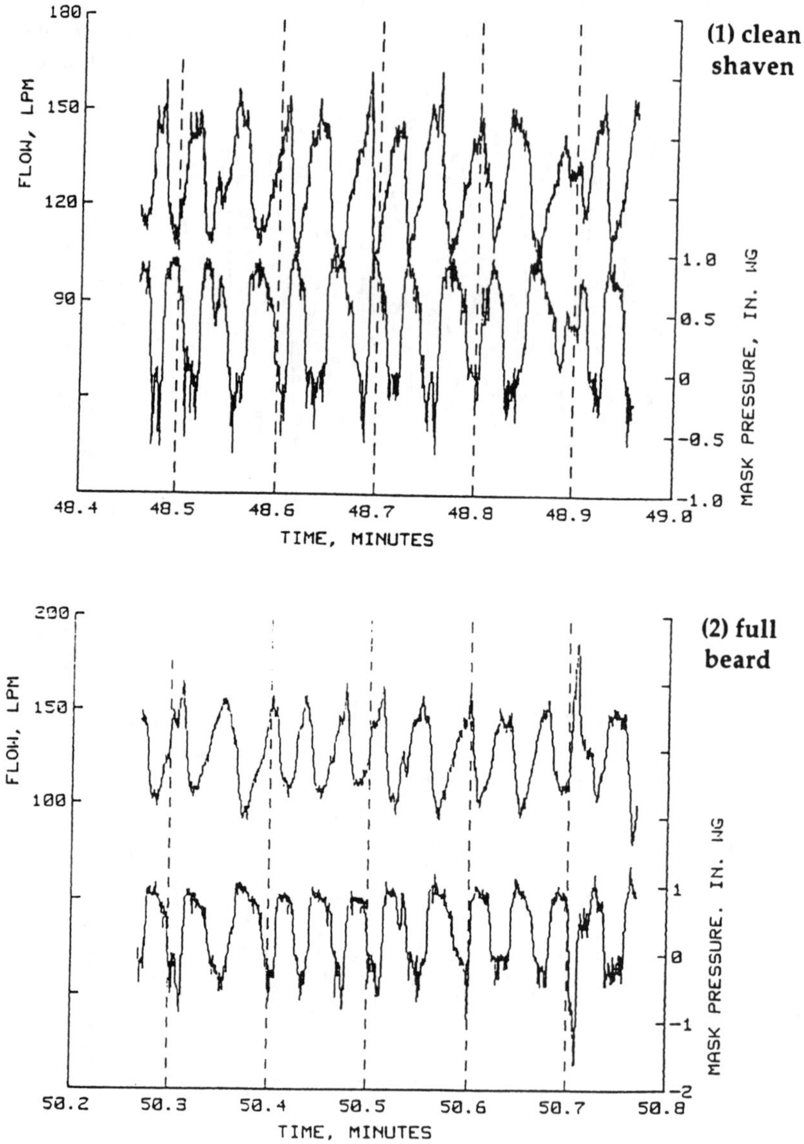

Figure 4 (c) Pressure and flow vs time, MSA half-mask PAPR, flow of 5 cfm; Subject of BGH. (From Reference 4.)

minimum protection factor of 1600 and a maximum protection factor in excess of 10,000, for the belt mounted samples. The lapel samples showed a minimum protection factor of 1700 and a maximum protection factor of in excess of 7800. After an investigation, it was found that the No. 10 facepiece sample was probably contaminated. The test results indicated that under the simulated workplace test, the MSA PAPR provided much higher protection than tests conducted at the workplace.

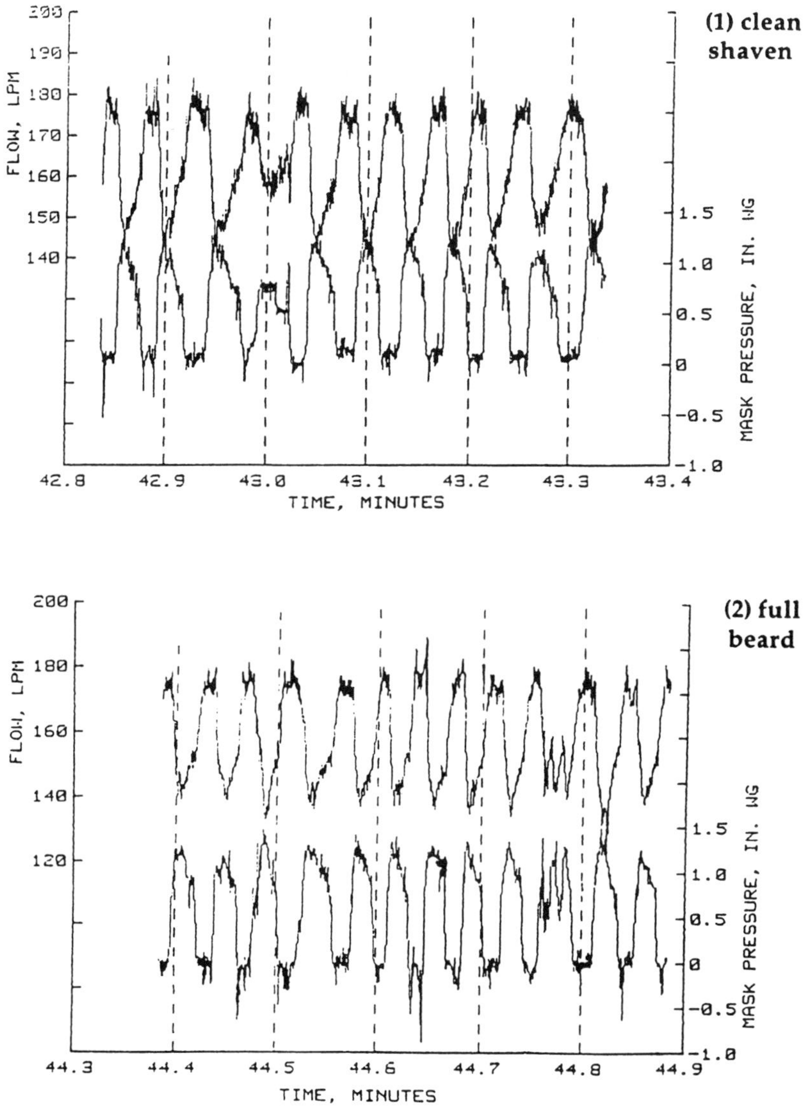

Figure 4 (d) Pressure and flow vs time, MSA half-mask PAPR, flow of 6 cfm; Subject BGH. (From Reference 4.)

IV. CONTINUOUS FLOW ESCAPE SELF-CONTAINED BREATHING APPARATUS

Continuous Flow Escape Self-Contained Breathing Apparatus (CFESCBA) is used for escape from atmospheres which are immediately dangerous to life or health (IDLH). These devices should provide the wearer with adequate protection against escape from a sudden surge in contaminant concentration, or from an oxygen deficient environment. The device usually consists of a compressed air cylinder, a

Table 4 Simulated Workplace Test of MSA PAPR

Test no.	Chamber test concentrations (mg/m³) Potential exposure Shoulder	Belt	Facepiece concentration	"Protection factor" Shoulder	Belt
1	24.2	27.6	<0.006	>4000	>4000
2	18.4	21.5	0.008	2300	2700
3	20.2	21.6	0.012	1700	1800
4	31.8	35.6	<0.009	>3500	>3900
5	16.8	19.4	<0.004	>4200	>4800
6	21.1	22.7	0.014	1500	1600
7	31.5	37.4	<0.004	>7800	>9300
8	18.2	16.1	<0.005	>3600	>3200
9	19.1	21.2	0.010	1900	2100
10	91.6	103.4+	~0.79	~120	~130
(10[a])			(<0.01)	(>9000)	
11	82.7+	106.1+	0.012	~6900	~8800
12	122.0+	128.7+	0.039 (resp.)	~3100	~3300
13	79.6+	96.5+	0.016 (resp.)	~5000	~6000
14	27.2 (resp.)	235+	<0.005	>5400	>10,000
Blank 1	0.02 mg		–0.011 mg		
Blank 2	0.01 mg		–0.012 mg		

Note: Blanks for potential exposure represent ~0.07 mg/m³ and ~0.03 mg/m³ respectively. Facepiece blanks represent ~ –0.01 mg/m³.

[a] Sample weighed by MSA technicians; see discussion.

From Reference 10.

control valve, and a hood. Breathing air is fed to the hood continuously. The Environmental Protection Agency requires that workers who work at hazardous waste sites must carry a CFESCBA for emergency escape if a regular SCBA is not worn. Since 30 CFR 11 has no performance criteria for testing these devices, devices with air flow varying from 28 to 70 l/m have been approved. Questions have been raised as to whether these devices could maintain a positive pressure, and whether the concentrations of oxygen is not low enough, and the concentration of CO_2 is not high enough, to impede escape. If the air flow rate inside the hood does not meet the respiratory demand under a moderate work rate, the CFESCBA wearer may rebreathe the air inside the hood having an increase in CO_2 level and a decrease in the level of oxygen. Several investigations were made on these issues.

A. Kennedy Space Center

The Kennedy Space Center conducted a test to determine the performance of two nonapproved CFESCBAs.[11] One had a yellow color and another one had an orange color. These devices had a higher air flow (45 l/m) than the approved device (34 l/m) and an increase in service life from 5 to 10 min. The test consisted of three series. The first series was a treadmill test. Three test subjects walked on a treadmill at 4 kph and a 10% grade. The second series was a quantitative fit testing. Three test subjects participated in this series of test which was carried out on a treadmill in the same configuration as the first series. The third series was to measure the air flow rate inside the hood. Three devices from each type were selected for evaluation.

The test results indicated that the oxygen concentration was 17.5% at the 8th min after deployment. The CO_2 concentration increased to 2% within 2 min. The maximum CO_2 concentration was less than 2.5%. The QNFT was performed with a challenge aerosol having a 0.6 μm MMAD at a concentration of 26 mg/m³. Fit factors exceeding 10,000 were achieved from 6 tests. The flow rate was not constant throughout the test period. For the yellow color unit, the initial flow rate of 46.4 l/m was decreased to 40.2 l/m after 5 min, and further reduced to 7.4 l/m after 7 min. The orange color unit has an initial flow of 46.4 l/m. The flow rate was reduced to 43.3 l/m after 3 min and further reduced to 25.5 l/m after 7 min. The pressure inside the hood remained positive during the deployment.

The noise level inside each device was measured by a dosimeter with the sensor located in the hood near the ear of the test subject. The yellow device had a noise range from 103.1 dB to 111.3 dB. The orange device had a range from 96.3 dB to 101.6 dB.

B. National Dräger

National Dräger, a major respirator manufacturer, contracted Penn State University to conduct a study to determine the acceptable performance of ESCBAs.[12] Five-minute service life ESCBAs operating at 40, 55, and 70 l/m were selected for testing. These devices are available with slot or hole type diffusers. The test subject who wore the ESCBA walked on a treadmill at different work rates. The oxygen consumption, CO_2, and oxygen concentrations inside the hood were measured. The test results led to the following conclusions:

1. At an oxygen consumption rate (VO_2) of 1.5 l/m, the average CO_2 concentration inside the hood varied from 3.9 to 4.3% with an air flow of 40 l/m. The test subjects experienced some difficulty during the test due to the increase of CO_2 concentration inside the hood. The CO_2 level varied from 2 to 2.8% at an air flow rate of 55 l/m. The CO_2 level varied from 1.0 to 1.6% with an air flow of 70 l/m.
2. At an oxygen consumption rate of 2.5 l/m, the CO_2 concentration inside the hood reached 4.3% within 3 min, with an air flow of 55 l/m. The test was terminated after 4 min. The test subject experienced some disorientation due to the increase of CO_2 concentration. The CO_2 level reached a maximum of 3.85% at an air flow rate of 70 l/m after 5 min.
3. At an oxygen consumption rate of 3.0 l/m, the CO_2 concentration inside the hood reached 5.7% after 3 min with an air flow of 40 l/m. The CO_2 concentration inside the hood reached 5.7% after 3 min with an air flow of 40 l/m. The test subject experienced extreme discomfort. The CO_2 concentration inside the hood reached 4.25% after 4 min with an air flow of 55 l/m. The test for the two lower flow rates was terminated due to the high CO_2 concentration inside the hood. The CO_2 level reached a maximum 4.2% at an air flow rate of 70 l/m after 5 min. The test subject did not experience disorientation despite the high CO_2 concentration inside the hood.
4. At a constant flow of 70 l/m, the device provides adequate protection at VO_2 of 2.5 l/m, but provides marginal protection at a VO_2 of 3.0 l/m. The other two devices provide inadequate protection at a VO_2 of 2.5 l/m.

Table 5 Characteristics of Commercially Available Emergency Escape Respirators

Unit name	Weight	Duration	Flow rate
ISI			
ELSA 5	6-3/4 lb	5 min	36–39 l/m
ELSA 5XF	10-1/4 lb	5 min	70–72 l/m
ELSA 7Xf	10-1/2 lb	7 min	55–60 l/m
National Draeger			
MAX	6-1/4 lb	5 min	70 l/m
ERMA	15-1/2 lb	10 min	70 l/m
North			
Model 845	8 lb	5 min	42 l/m
Model 850	10 lb	10 min	42 l/m
Model 855	10 lb	5 min	75 l/m
MSA			
Custom Air 5 Escape Hood	7.3 lb	5 min	72 l/m
Scott			
Skat-Pak	9.2 lb	5 min	65 l/m
Survivair			
Emergency Escape Apparatus	6.5 lb	5 min	35 l/m
Emergency Escape Apparatus	8.5 lb	10 min	32 l/m
Respirator Systems, Inc.			
LIFEAIR 5	7 lb	5 min	40 l/m
LIFEAIR 10	7 lb	10 min	33 l/m

5. The air diffuser located in front of the hood near the breathing zone provides the best performance.
6. The hole diffuser resulted in less fogging than the slot diffuser.

C. Lawrence Livermore National Laboratory

A more comprehensive study on the ESCBA performance was conducted by the Lawrence Livermore National Laboratory.[13] LLNL selected 11 models from seven manufacturers for testing. The service life varied from 5 to 10 min. The flow rate varied between 33 to 75 l/m. A summary of characteristics of the models tested is presented in Table 5.

At first, the air flow with respect to time was determined for each device. The test models were separated into "low flow" and "high flow" groups. The high flow models had a minimum air flow of 60 l/m. The air flow measurements are shown in Figures 5 and 6. The simulated workplace protection factor study was performed on a treadmill at a "moderate" and a "high" work rate. The treadmill setting for the moderate work rate was set at 4 kph and 10% grade, which corresponds to an average breathing rate of 45 l/m. The treadmill setting for the high work rate was 5.3 kph and 10% grade, which corresponds to an average breathing rate of 69 l/m. Both the oxygen and CO_2 concentrations, pressure inside the hood, and the protection factors for each device were measured for the duration of the test. The protection factor was measured with a forward scattering photometer. The test aerosol was polyethylene glycol 400 with a MMAD of 0.6 μm and a concentration of 15 mg/m³.

The results of oxygen concentration inside the hood for the moderate and the high work rate tests are shown in Figures 7 and 8. The results of CO_2 concentration inside the hood for the moderate and the high work rate tests are shown in Figures

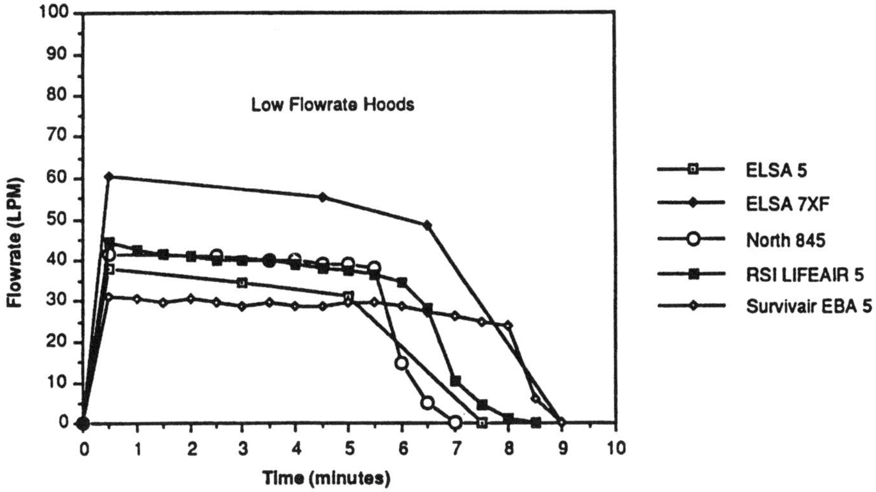

Figure 5 Flow rate vs. time for moderate to low flow emergency escape hoods.

9 and 10. The results of the protection factor measurement of selected model of ESCBAs are shown in Table 6.

The test results indicated that the flow rate showed some fluctuations. At the moderate work rate, some low air flow models have more significant drops in oxygen level than the high air flow models. After 2 min of deployment, the oxygen content of two low flow rate models (<40 l/m) reduced to 16%, which is considered as immediately dangerous to life or health. At the high work rate, the oxygen concen-

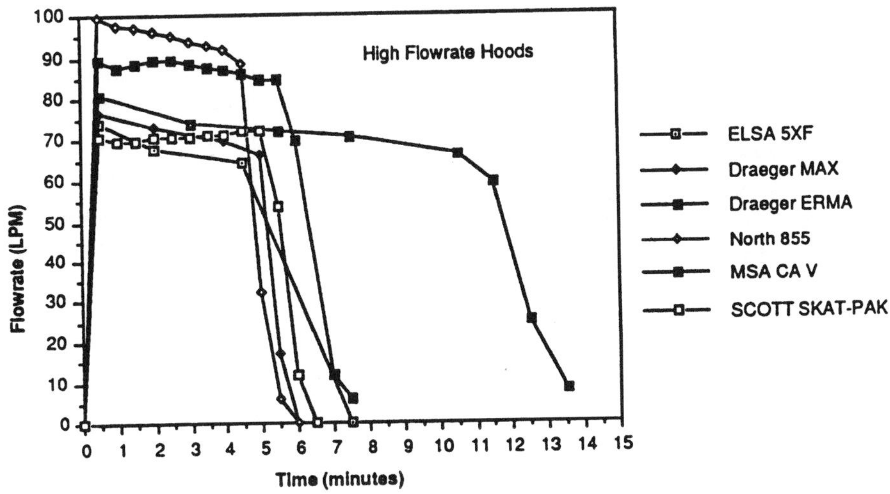

Figure 6 Flow rate vs. time for high flow emergency escape hoods.

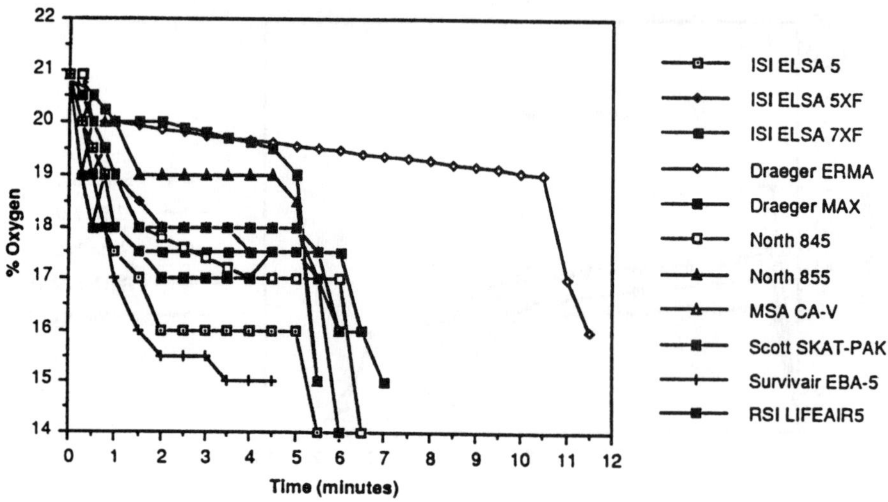

Figure 7 Percent oxygen concentration at moderate work rates.

tration of some devices with a low flow rate reduced to 16% within 1 min of deployment. All other devices maintained an oxygen level around 17%.

The results of the CO_2 measurements were similar to the oxygen measurement. The models with air flow less than 40 l/m showed a rapid increase of CO_2 concentration to 5% at a moderate work rate. At the high work rate, there was a further increase of CO_2 concentration to 7% for the low flow models. The pressure inside the hood reduced to negative pressure for the low flow rate models at the high work rate. The pressure differential (ΔP) inside the hood varied from 0 to 1.5 in. water pressure between inhalation and exhalation. Some models reached zero ΔP inside the hood faster than other models. This may be due to hood design since some high flow rate

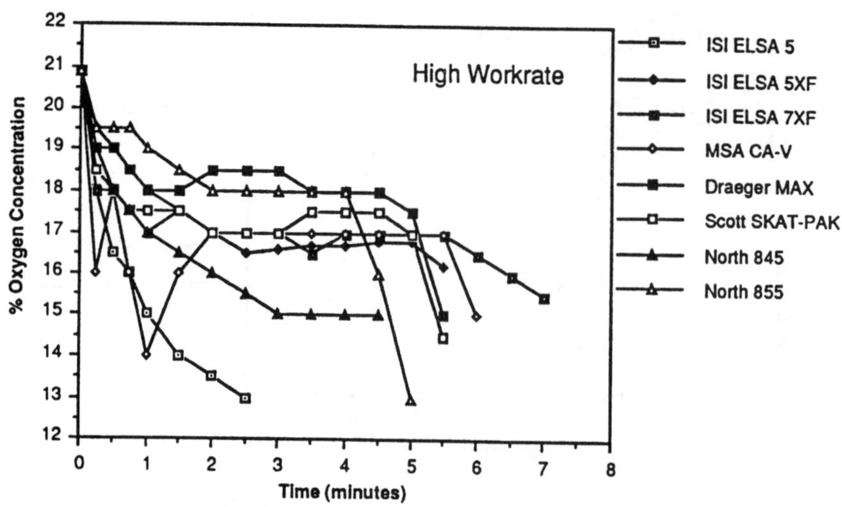

Figure 8 Percent oxygen concentration at high work rates.

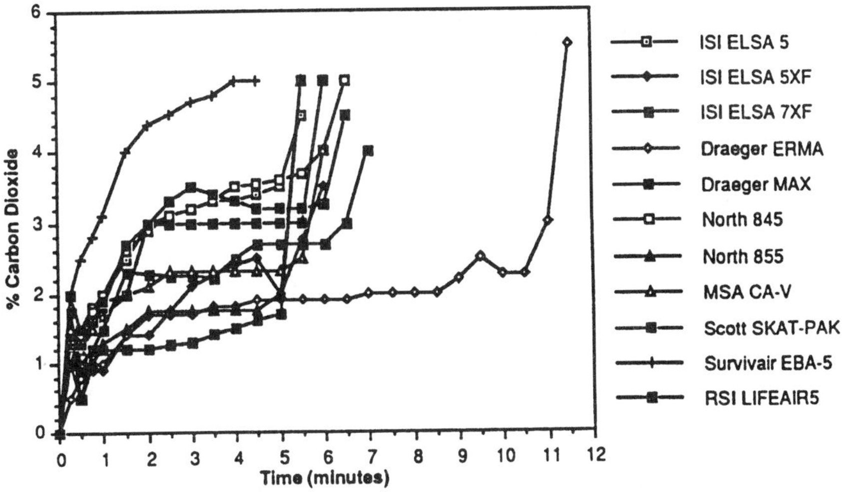

Figure 9 Percent carbon dioxide at moderate work rates.

models also showed a fast zero ΔP. The time reached zero ΔP also corresponds to the protection factor obtained in the QNFT test. The low flow model with a low PF reached zero ΔP faster than other models. The test results concluded that a minimum flow rate of 60 l/m is required to provide adequate protection for the ESCBA wearer.

D. Occupational Safety and Health Administration

The compliance safety and health officers (CSHOs) of OSHA routinely inspect workplaces. The air contaminant concentrations at these workplaces may reach IDLH concentrations during abnormal operating conditions. An ESCBA would be

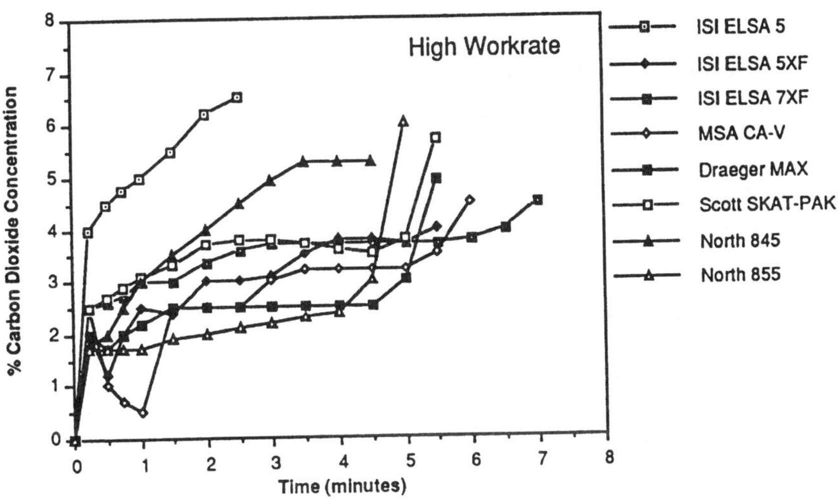

Figure 10 Percent carbon dioxide concentration at high work rates.

Table 6 Average Hood Penetration of Moderate and High Work Rates

		Average protection factor	
	Flow rate >60 l/m	Moderate work rate	High work rate
Draeger MAX	X	>1000	~1000
ELSA-5		>200	~100
ELSA-5XF	X	>2000	~10,000
MSA CA-V	X	>10,000	1000
North 855	X	>10,000	>10,000
Scott SKAT PAK	X	>10,000	>10,000

suitable for the safe exit from IDLH environments. Since there are no performance criteria prescribed in 30 CFR 11, OSHA developed the following procurement performance test criteria for the 5-min service life ESCBA:[14]

1. Air Flow. Since the ESCBA may be worn over a wide range of temperatures, the device should provide adequate air flow under these temperature extremes. The air flow is measured when the device is cold soaked at –18°C for at least 4 h, and also measured when the device is hot soaked at 62°C for at least 4 h. The air flow is measured immediately after each conditioning. The device must deliver an average air flow of 70 l/m for the 5 min service life. The minimum air flow is 65 l/m during the test.
2. Rapid Deployment. In order for the wearer to escape to a safe environment in a timely fashion, the device should be deployed within a reasonable time period. Five test subjects participated in this test. Each person was permitted to practice donning three times. The deployment time for the fourth donning is timed. The maximum permitted donning time is 20 s.
3. Simulated workplace test. Five test subjects walk on a treadmill at a speed of 5.1 kph and a slope of 9%. The ESCBA is deployed after the test subject has walked on the treadmill for 3 min. The facepiece pressure, oxygen, and CO_2 concentration inside the hood is measured continuously for the duration of the service life of the breather.

The pressure inside the hood must maintain positive at all times. The oxygen content inside the hood must be higher than 17% during the test. The average CO_2 concentration inside the hood must not be exceed 3% with a maximum of 4% during any time of the test.

The devices received from the contractor (Mine Safety Appliance Co.) met all the above requirements. A summary of the test results is presented in Table 7.

V. SELF-CONTAINED BREATHING APPARATUS

OSHA has also procured SCBAs which meet the performance requirements specified in the National Fire Protection Association (NFPA) standard 1981.[15] OSHA has set an additional requirement for protection factor measurement. The Lawrence Livermore National Laboratory was retained by the contractor to conduct this test. The PF determination is performed inside an aerosol test chamber. The PFs are measured by a forward light scattering photometer when the test subject walks on a

Table 7 Test Results: MSA ESCBA

High temperature air flow (l/m)	74.3, 72.3, 73.6
Low temperature air flow (l/m)	82.4, 82.3, 82.7
Rapid deployment (s)	11, 7.5, 12.7, 6.8, 7.1
Treadmill	
Facepiece pressure	>0
Oxygen (%)	
Avg:	19.6, 19.3, 19.4, 19.4, 19.1
Min:	19.2, 18.7, 19.2, 19.1, 18.5
Carbon dioxide (%)	
Avg:	1.8, 1.9, 1.4, 1.6, 1.7
Max:	2.0, 2.1, 1.6, 1.7, 1.9

treadmill at a very high work rate. The measurement starts when the heart rate of the test subject reaches 80% of his/her cardiac reserve. The test terminates when the low air-flow alarm of the air cylinder is activated. The average PF for the test subjects must be higher than 10,000. The test results indicated that SCBAs delivered by the contractor (Survivair) met the requirement.[16]

VI. CONCLUSIONS

The following conclusions can be made from the above studies concerning adopting simulated workplace protection factor studies for respirator certification.

1. The simulated workplace study is preferred to the workplace protection factor (WPF) study for respirator certification since all respirators can be tested under the same controlled test conditions.
2. High temperature and humidity could degrade the performance of the respirator, especially the negative pressure type. There should be a test requirement for certifying negative pressure respirators. However, reproducible adverse environmental conditions cannot be easily established in the workplace to determine such effects.
3. The simulated workplace study can provide real time measurement for any desired parameter. In general, the WPF can only provide an integrated reading over the test period which may not be desirable for many applications.
4. For positive pressure respirators, maintaining a positive pressure under sustained heavy work rate should be a requirement for certification. Work at 80% of the maximum cardiac rate of the test subject should be able to separate a good performing respirator from a poor one.
5. The simulated workplace study is able to measure many performance parameters simultaneously such as facepiece pressure, oxygen, and CO_2 concentrations.
6. As far as assigning protection factors for different classes of respirators for regulatory purpose is concerned, the simulated workplace protection factor study would be the only method of choice. Because the protection factor for each class of respirator is interrelated, they must be tested under the same test conditions. The test results of the simulated workplace study would indicate that a positive pressure SCBA would provide a much higher protection than a disposable respirator equipped with a dust/mist filter. However, by tailoring the test conditions in a WPF study, one can demonstrate that the disposable respirator provides a higher protection than the positive pressure SCBA.

REFERENCES

1. Skaggs, B. J., Loibl, J. M., Carter, K. D., and Hyatt, E. C., Effect of temperature and humidity on respirator fit under simulated work conditions, Los National Laboratory, NUREG/CR-5090, LA-11236, 1988.
2. ANSI, Practices for Respiratory Protection, ANSI Z88.2-1980. American National Standards Institute (ANSI), New York, 1980.
3. Hyatt, E. C., Protection factors, Los Alamo Scientific Laboratory Report LA-6084-MS, 1975.
4. da Roza, R. A., Cadena-Fix, C. A., and Kramer, J. E., Powered air-purifying respirator study final report, Lawrence Livermore National Laboratory, UCRL-53757, 1986.
5. NIOSH, Respirator decision logic, National Institute for Occupational Safety and Health. DHHS (NIOSH) Publication No. 87-108, May 1987.
6. Myers, W. M., Peach, M. J., III, Cutright, K., and Iskander, W., Workplace protection factor measurements on powered air-purifying respirators at a secondary lead smelter-results and discussion, *Am. Ind. Hyg. Assoc. J.*, 45, 681, 1984.
7. Myers, W. R., Peach, M. J., III, Cutright, X., and Iskander, W., Field test of powered air-purifying respirators at a battery manufacturing facility, *J. Int. Soc. Respir. Prot.*, 4, 62, 1984.
8. King, J., Simulated work place testing in a sandblasting environment, E. D. Bullard Company (unpublished), November 1990.
9. Myers, W. R. and Peach, M. J., Performance measurement on a powered air-purifying respirator made during actual field use in a silica bagging operation, *Ann. Occup. Hyg.*, 3, 251, 1983.
10. Ayer, H. E., Report on testing of powered air purifying respirators, University of Cincinnati, August 24, 1981.
11. Doerr, D., Emergency life support apparatus-extra flow, Kennedy Space Center, National Aeronautics and Space Administration, 1988.
12. Penn State University, Test Report on Constant Flow Escape Sets. Penn State Physiology Laboratory, Project ED1008, 1984.
13. Johnson, J. S., da Roza, R. A., Foote, K. L., and Held, K., An evaluation of emergency escape respirators for use in a space launch environment, Lawrence Livermore National Laboratory, Livermore, CA, 1991.
14. Bien, C. T., Performance testing criteria for continuous flow escape self-contained breathing apparatus, Presented at the American Industrial Hygiene Conference and Exposition, Boston, MA, 1992.
15. NFPA, Open-circuit self-contained breathing apparatus for fire fighting, NFPA 1981. National Fire Protection Association, Quincy, MA, 1992.
16. Johnson, J. S., da Roza, R., and McCormack, C. E., Evaluation of a commercial SCBA's compliance to the NFPA 1981 standard for fire fighter and measurement of simulated workplace protection factors at high work rates, Lawrence Livermore National Laboratory, Presented at the American Industrial Hygiene Conference, Boston, MA, 1992.

21 RESPIRATOR PERFORMANCE STUDIES III: WORKPLACE PROTECTION FACTOR STUDIES

The workplace respirator performance studies were begun during the 1970s. These studies can be classified into two phases. The first phase consisted of studies sponsored by NIOSH that included several comprehensive evaluations of respirator effectiveness in coal mines, abrasive blasting operations, and paint spray operations. These studies covered different operations over a large geographic area. The second phase represents the workplace respirator performance studies conducted by the NIOSH staff, as well as private sectors. The first NIOSH study was conducted in the early 1980s after a silica processing plant operator notified NIOSH that the powered air-purifying respirators (PAPR) purchased by the plant did not provide adequate protection. NIOSH conducted a workplace study to determine the performance of the PAPR. The results suggested that the PAPR did not provide the required protection as prescribed in the 30 CFR 11. In order to determine whether other approved respirators also performed adequately, NIOSH initiated additional workplace studies for other types of respirators. Subsequently, other respirator users and a few respirator manufacturers also conducted workplace respirator performance studies for a variety of respirators.

I. REVIEW OF WORKPLACE PROTECTION FACTOR STUDIES

Harris et al.[1] conducted a respirator performance study in coal mines. The test was carried out in four different mines. At each of the four mines, tests were performed concurrently on two mining sections for a 5-day period. Five miners from each mine section participated in the study. The test subjects were involved in 37 different job classifications. Five different types of half-mask air-purifying respirators equipped with dust/mist filters were selected for this study. The 10-mm nylon cyclones were used throughout to collect respirable ambient and in-mask dust samples. Samples were collected for the work shift except during the lunch period. The ambient dust concentration varied between 1.3 and 2.1 mg/m^3. A time-of-wear device equipped with a thermistor was employed to monitor the respirator wearing time. A total of 187 samples were collected in this study.

The protection provided by the respirator is expressed as the effective protection factor (EPF), which is the ratio of ambient and in-mask concentrations of the challenge. The results indicated that a mean workplace protection factor (EPF) of 5.7

Table 1 Quantitative Cotton Dust Aerosol Respirator Evaluation Test Results

Respirator	Human test subject	Operation	Time period (hours)	Cotton dust concentration (mg/m³)	Penetration of cotton dust into interior of respirator (% of ambient concentration)
A	V	Picking	4	2.73	9.1
	W	Picking	4	2.08	8.2
	X	Carding	2	6.73	12.5
	Y	Floating[a,b]	2	22.9	5.7
	Z	Floating[a]	2	1.42	5.3
B	V	Picking	2	1.57	1.2
	V	Carding	2	4.73	1.8
	W	Picking	2	1.28	3.6
	X	Picking	2	1.89	4.5
	Y	Floating[a]	2	1.33	1.9
	Z	Floating[a]	2	1.09	2.3
C	V	Carding	2	5.22	1.7
	W	Carding	2	3.69	7.3
	X	Carding	2	3.43	5.7
	Y	Floating[a]	2	2.27	1.9
	Z	Floating[a]	2	2.56	2.2

[a] Floating included picking, carding, spinning.

[b] Test subject carried out a machine-cleaning operation.

From Reference 2.

with a median value of 3.2 was obtained for all mines. The mean and median EPF values for mine 1 was 12.2 and 3.9 respectively. The mean and median WPF values for mines 2, 3, and 4 were 4.4 and 3.2, 4.5 and 2.6, and 5.6 and 3.6, respectively. Since this paper was a progress report, no conclusion was provided by the authors concerning the test results.

Revoir[2] conducted a study to determine the performance of single-use respirators for protection against cotton dust in a textile mill. One Bureau of Mines approved, one nonapproved single-use, and one prototype nonapproved single-use respirator were selected for testing. Prior to conducting the field test, all respirators were evaluated by a silica dust test prescribed in 30 CFR 11 and a coal dust quantitative fit test. The test results indicated that except for the nonapproved single-use respirator, all other respirators passed the above mentioned tests. Six test subjects participated in this study. The operations of the test were picking, carding, spinning, and floating operations. The test area was equipped with air conditioning. The sampling period varied from 2 to 4 h, and the ambient cotton dust concentration varied between 1.09 and 22.9 mg/m³.

The test results are shown in Table 1. The penetration of cotton dust into the interior of the respirator varied between 1.2 and 12.5%. The corresponding WPF varied between 5.5 and 83. The results indicated that the approved respirator and the prototype single-use respirator designed to meet the certification requirements provided adequate protection against cotton dust. Since cotton dust has a relatively large size as compared to other types of industrial dust, the protection factors obtained from this study may not be representative of other operations.

Almost all WPF studies were performed on a particulate challenge. One published WPF study performed on a gaseous challenge was conducted by Moore and Smith[3] in a copper smelter. The study was designed to measure the SO_2 protection factors of three chemical cartridge respirators. Six workers at the feeder operation

Table 2 Average SO_2 Concentrations Inside and Outside Respirator Masks and Average Protection Factors

		SO_2 (mg/m^3)				Protection factor	
		Outside mask		Inside mask			
Respirator	n	Mean	SD	Mean	SD	Mean	SD
A	26	61.1	40.2	5.0	4.0	22.1	22.6
B	25	53.0	25.6	4.6	3.8	18.4	14.2
C	25	53.0	35.6	6.2	4.5	12.9	11.0

From Reference 3.

participated in this study. Each worker wore all three types of respirators in turn. All respirator evaluations were made during the furnace charging operation, where the highest SO_2 concentration was expected. The SO_2 was collected in two impingers filled with hydrogen peroxide and subsequently analyzed by atomic absorption spectrophotometer. The sampling time was approximately 80 min. A total of 81 sampling sets were collected. The authors used the term effective protection factor (EPF), which is the ratio between the ambient and in-mask concentrations. The results are shown in Table 2.

The results indicated that the ambient SO_2 concentration averaged approximately 21 ppm and varied from 6.2 to 75.4 ppm. The in-mask concentration averaged approximately 1.9 ppm and varied from 0.3 to 7.0 ppm. The average and median EPF for type A respirators was 22.1 and 15.3, respectively. The EPF exhibited a similarly wide range, between 2.6 and 83.1. The distribution of average and median WPFs for respirator B and C were 18.4, 13.7 and 12.9, 9.6 respectively. The type A mask had 38.5% of its EPF less than 10, the type B mask had 30.4% of its EPF less than 10, and the type C respirator had 56.0% of its EPF less than 10. The authors found that the overall in-use SO_2 protection provided by these three respirators was poor. The best mask had 30.4% of its tests with an EPF <10, and the worst respirator had 56% of its tests with an EPF <10 . Since the half-mask respirator is permissible at ten times the OSHA SO_2 permissible exposure limit of 5 ppm (which is 50 ppm), the worker who wears this type of respirator would be overexposed in a substantial amount of time to a SO_2 concentration of 50 ppm.

A WPF study on worker exposure to cadmium was conducted by Smith et al.[4] Nine workers participated in this study. Samples were collected over a whole shift and included lunch and break periods. The primary respirator tested was the Welsh 7500-8 with the Protex 7500-43 combination acid gas and fume cartridges. The geometric mean lapel concentration was 107 μg/m^3 with a range of 19.3 to 3600 μg/m^3. The in-mask concentration had a geometric mean of 15.4 μg/m^3 and a range of 3.4 to 67 μg/m^3. The geometric mean EPF was 5.6 with a range of 1.12 to 146, and a 95% confidence limit of 2.9 to 10.9 about the mean. Only 22.2% of the EPFs exceeded 10, and 33% of the EPFs were less than 2. All of the low WPFs occurred for low ambient cadmium concentrations. Workers in low exposure areas were observed wearing their respirators less frequently than workers in posted high exposure areas where respirator use was mandatory. The test results concerning EPFs are shown in Table 3.

The results indicated that intermittent respirator usage provides a highly variable degree of effective protection factor between individuals, and also for the same individual on a day-to-day basis. The protection factor ranged from 1.12 to 146, and had a pooled geometric standard deviation of 2.62 for day-to-day variability. It is

Table 3 Geometric Mean Lapel Cadmium Concentration and Effective Protective Factors for Each Subject's Three Days of Observations

Subject	Geometric mean lapel cadmium exposure (μg/m³)	SDg[a]	Geometric mean effective protection factor	SDg[a]
1	22	1.16	1.5	1.56
2	37	2.03	4.0	1.33
3	54	2.80	2.6	2.82
4	58	1.29	2.1	1.66
5	64	2.07	6.6	1.24
6	84	1.35	4.9	3.44
7	143	2.71	3.3	2.93
8	352	6.15	15.4	6.76
9	2530	1.33	103.0	1.78
Pooled day to day SDg				
All subjects		2.36		2.62
Without subject no. 9		2.47		2.72

[a] Geometric standard deviation.

From Reference 4.

important to distinguish the studies that measured effective protection factors during a period when the respirator was worn continuously from those that measured protection factors during intermittent respirator usage. The former evaluates the accumulated effects of the worker's repeated decisions to use the respirator, and the variable leakage produced with each "*fitting*" during the course of a work shift, whereas the latter evaluates the leakage associated with a single "*fitting*" of the worker to his face. The authors concluded that it is not surprising that the effective protection factor during intermittent use is less than the protection factor measured during a single continuous use of the respirator.

Que Hee and Lawrence[5] conducted a study on the Racal PAPRs on lead and brass foundry workers. Seven ladle and furnace room attendants participated in this study. The sampling duration varied from 4 to 8 h and included breaks and lunch. Two models of the Racal AH-3 PAPRs equipped with high-efficiency particulate air filters were selected for this study. The model AH3-1 is identical to the AH3 except that the former is equipped with a Tyvek mask side seal to allow a better fit at the side of the visor. The measured ambient concentration varied from 36 to 2316 μg/m³ and the in-mask concentration varied between <5 and 230 μg/m³. For the ladle operator, the measured effective protection factor for the Racal AH3 varied from 6 to 30 and the WPF for the AH3-1 was 67. Effective protection factors for the furnace attendants varied from 1.1 to 4 for the AH3 and varied between 3 and 5 for the AH3-1 PAPR. The activity of the furnace attendants caused the PAPR to hang vertically from the face, and hot and dusty conditions often caused the attendants to raise their visors. These factors may contribute to the poor protection provided by the PAPR.

Toney and Barnhart[6] conducted a survey to determine the efficiency of respiratory protection used in paint spray operations. Originally 253 companies were contacted but only 159 companies were visited. The respirators used at the sites varied from half-mask respirators to supplied air respirators. One nonapproved particulate respirator, the nonapproved 3M 8500 single use respirator, was used in 30 operations. Both particulate and gas samples collected over a 10-min period were used to calculate the protection provided by the respirators. The results of selected samples are shown in Table 4. The protection factor is expressed as the following:

Table 4 On-Site Tests — All Areas Protection Factors by Respirator Fit and Condition

				Protection factors	
Test no.	Respirator type	Fit	Condition	Solvent	Pigment
2	MSA 8556	Poor	Poor	56.8%	71.4%
3	Willson 841CP	Good	Poor	28.6%	81.4%
4	DeVilbiss MSC-505	Good	Good	72.2%	95.9%
5	Willson 841CP	Good	Poor	46.7%	94.0%
8	MS 8556	Good	Good	94.7%	95.0%
9	Welsh 7511	Good	Good	66.3%	72.0%
10	Welsh 7511	Good	Good	—	88.2%
11	MSA 85556	Good	Good	87.1%	100.0%
12	MSA 85556	Good	Good	82.0%	100.0%
13	MSA 85556	Good	Good	87.2%	100.0%
14	MSA 85556	Good	Good	94.4%	100.0%
18	Pulmosan C-251	Good	Poor	38.4%	62.5%
19	DeVilbiss MSC-506	Poor	Poor	67.7%	66.6%
21	A. O. 5051P	Poor	—	55.6%	75.0%
22	MSA 85556	Good	Good	70.8%	71.4%
23	Pulmosan C-259	Good	Good	83.4%	96.4%
26	MSA 85556	Poor	Good	75.2%	50.0%
28	Willson 941CP	Poor	—	68.4%	100.0%
29	MSA 85556	Good	Good	78.3%	96.2%
30	MSA 85556	Poor	Good	96.8%	75.0%
31	Willson GA-2	Poor	Good	69.0%	83.6%
32	Willson 1221-14	Poor	Good	65.3%	96.2%
33	MSA 8556	Poor	Good	67.0%	69.1%
34	DeVilbiss MPH-529	Good	Good	90.8%	96.3%
35	A. O. 5091	Poor	Poor	48.2%	64.9%
39	DeVilbiss MSC-505	Poor	Good	40.6%	80.0%
40	Welsh 7531	Good	Poor	54.5%	100.0%
41	Cesco P/N 70-440	Poor	Good	36.8%	55.7%
42	Binks Unk	Good	Good	81.3%	100.0%
43	A. O. 5051	Good	Poor	74.1%	24.0%
45	A. O. 5051	Good	Poor	45.5%	68.8%
47	DeVilbiss MSC-506	Poor	Good	56.2%	95.1%
48	DeVilbiss MSC-506	Poor	Good	76.6%	80.8%
50	MSC-506	Good	Good	14.5%	84.2%
51	Welsh 7531	Good	Good	93.4%	97.8%
52	Welsh 7531	Good	Good	89.4%	98.8%
53	Safeline 5215	Good	Poor	8.1%	51.1%
54	Willson 122114	Good	Good	75.0%	94.4%
55	Willson 122114	Good	Good	91.8%	90.0%
56	Safeline 5211	Good	Poor	35.9%	54.2%

$$\text{Protection Factor (PF)} = 1 - \left[\frac{\text{Components analyzed in respirator sample}}{\text{Component analyzed in ambient sample}}\right] \times 100 \qquad (1)$$

In comparing this term to the commonly used workplace protection factor (WPF), a WPF of 5 equals 80% PF, a WPF of 10 equals 90% PF, and a WPF of 100 equals a PF of 99%. The results indicated that protection provided by these respirators is generally poor. But, the cartridge seems to perform better in removing vapors rather than particulates. This may be due to factors such as improper selection, poor training, inadequate fit, and poor maintenance.

The respiratory protective practices in the abrasive blasting operations were evaluated by Blair under a contract with NIOSH.[7] Locations selected for this survey included Houston (TX), Philadelphia (PA), Seattle (WA), Mobile (AL), Portland (ME), and Wichita (KS). About 3800 firms in these areas were contacted and only 903 returns (23%) were received. The survey started with a questionnaire and was followed by on-site visits to firms that responded. Abrasive blasting operations visited included the monument industry, shipyard, painting, and primary metal industries. Sand was the primary abrasive material used for blasting, from 115 of 257 (44%) operations responding, followed by steel shot (16.7%), steel grit (9.7%), alumina (9.3%), and flint/garnet (7%). Dry blast consisted of 147 of 192 processes surveyed. Thirty processes used airless blasting, and wet blasting only accounted for 11 processes. Supplied air respirators were used in a majority of the operations surveyed. The ricochet hood, alone or equipped with a respirator, was also observed in many operations. In some operations where respiratory protection was required, however, it was not worn. More than 80% of blasters wore the respirator for more than 4 h/day.

The mean and range of ambient and in-mask concentrations, protection factors, exposure time, assigned TLV, time-weighted average exposure and "times TLV" used to evaluate the efficiency of the particular dust exposure situations found in the monument industry, shipyard, painting/sandblasting operations, and primary metal industries are presented in Table 5.

The highest ambient concentrations of dust exposure and longer work hours were observed in the shipyard and painting sandblasting operations. Protection factors between 1.9 and 3,750 were found in the painting/sandblasting operations. However, PFs between 1.6 and 5.6 were obtained when the ricochet hood was used. PFs between 1.7 and 122 were obtained when the ricochet hood was used in combination with other air-purifying respirators. The poor protection provided by the supplied air helmets may have resulted from poor design, inadequate fit, and poor maintenance.

The first workplace protection factor (WPF) study conducted by the NIOSH staff was performed by Myers and Peach.[8] The study was performed at a silica bagging operation on a tight-fitting PAPR manufactured by the Mine Safety Appliance Company (MSA). Both the half-mask and the full facepiece configurations were selected for this study. The results indicated that the ambient concentration varied between 2 and 3.6 mg/m^3. The in-mask concentration varied between 0.06 and 0.17 mg/m^3. The WPFs were calculated as free silica analyzed from the ambient and in-mask samples. The free silica, as determined from x-ray diffraction, showed a geometric mean (GM) of 54, a geometric standard deviation (GSD) of 2.24, and a range of 16 to 215. The particle mass median aerodynamic diameter (MMAD) as determined from impactor samples collected on 2 different days showed a GM of 5.8 and 5.5 μm and a GSD of 7.1 and 7.7, respectively. All other pertinent information is listed in Table 6.

The second NIOSH study of the MSA PAPR was conducted by Lenhart and Campbell[9] at a sinter plant and blast furnace area of a primary lead smelter. The study was performed on a MSA Comfo II half-mask in the powered and the nonpowered configurations. Both respirators are equipped with high-efficiency particulate air (HEPA) filters. Particle size distribution for the ambient aerosol samples was determined from impactor samples. The samples showed that geometric mean mass

Table 5 Abrasive Blasting Respiratory Protective Devices

Respirator code (1)[a]	Ci (2)[b]	Co (3)[c]	PF (4)[d]	Contaminant type (5)[e]	Time h (6)[f]	Assn. TLV (7)[g]	Exposure (8)[h]	X TLV (9)[i]
Monument industry								
	1.08			Si-Al-Fe	6	0.40	0.81	2.00
	0.19			Si-Al-Fe	4	0.36	0.09	0.25
66	0.9	35	3.9	Al	6	5.00	0.07	0.01
	1.2			Al	2	5.00	0.30	0.06
66	1.43	3.73	2.6	Si-Al	6	0.60	1.07	1.80
66	0.12	0.36	3	Si-Al	8	3.90	0.12	0.31
	0.33			Si-Al	8	1.50	0.33	0.22
Momo/Ric	0.14	2.45	17.5	Si	6	0.10	0.10	1.00
	0.05			Al	4	5.00	0.03	0.01
2160	0.48	1.44	3	Si-Al-Fe	3	0.45	0.18	0.40
	1.73			Si-Al-Fe	2	0.27	0.43	1.60
	2.55			Si-Al-Fe	2	1.00	0.64	0.64
3M	1.75	3.5	2	Si-Al-Fe	6	0.33	1.32	4.00
Shipyard								
19B-57	0.1	16.9	169	Al	7	5.00	0.08	0.02
19B-57	0.1	28.7	955	Al	7	5.00	0.03	0.01
19B-57	0.08	49.6	633	Al	7	5.00	0.07	0.01
LB	0.68	48.9	72	Al	7	5.00	0.60	0.12
HA-99	0.19	16	85	Al	7	5.00	0.17	0.03
Blastfoe	0.13	35.7	275	Al	7	5.00	0.11	0.02
Clem/Ric	0.69	2.5	3.6	Si	4	0.11	0.34	3.10
Clem/Ric	0.67	6.73	10	Si-Pb	5	0.20	0.42	2.10
19B-57	0.5	5.1	10	Si-Pb	6	0.46	0.37	0.80
6901C	0.54	2.7	5	Si-Pb	7	0.38	0.47	1.20
MSA Tight	1.4	14	10	Al-Si-Fe	7	0.33	1.23	3.70
19B-57	4.16	73.65	17.7	Si-Pb-Fe	7	0.20	3.66	18.30
19B-34	0.08	11.3	114	Si-Pb-Fe	6	0.20	0.06	0.30
MSA Tight	0.08	13.1	255	Pb-Fe	6	0.40	0.06	0.15
MSA Tight	0.21	13.1	63	Si-Fe	6	0.50	0.16	0.28
19B-34	0.02	21.9	1095	Si-Pb-Fe	6	0.20	0.02	0.10

Table 5 Abrasive Blasting Respiratory Protective Devices (*continued*)

Respirator code (1)[a]	Ci (2)[b]	Co (3)[c]	PF (4)[d]	Contaminant type (5)[e]	Time h (6)[f]	Assn. TLV (7)[g]	Exposure (8)[h]	X TLV (9)[i]
Painting/sandblasting								
Ric/2301	8.87	15.32	1.7	Si-Fe	5	4.70	5.56	1.18
Pul/Ric	3.37	5.37	1.6	Si-Fe	3	0.20	1.26	6.29
HM	2.15	7.24	3.3	Si	6	0.10	1.62	16.20
Clem/Ric	5.23	8.88	1.7	Si	8	0.10	5.23	52.30
SAnst/Ric	4.35	8.28	1.9	Si-Al	5	0.10	2.72	27.20
HA-99	0.19	2.5	13.1	Si	8	0.26	1.33	5.22
MSA Tight	0.07	11	158	Al	8	5.00	0.70	0.11
Ric/Cus C	0.35	1.25	3.6	Si	8	0.11	0.35	3.17
Clem Ric	0.57	3.2	5.6	Si	5	0.10	0.36	3.60
Blastfoe	0.05	3.25	65	Al	7	5.00	0.04	0.01
Blastfoe	0.63	7.71	12.1	Si-Fe	4	0.68	0.32	0.47
Cesco 691	2.28	9.98	4.2	Si	8	0.12	2.28	18.90
Cesco 691	6.3	11.2	1.9	Si	9	0.12	6.30	52.20
19B-53	0.26	29.4	113	Si	2	0.13	0.05	0.38
19B-53	1.7	49.4	28.9	Si	5	0.10	0.16	1.60
19B-53/3M	0.17	34.7	203	Si	5	0.10	1.06	10.60
19B-53	0.62	43.4	70	Si	5	0.26	0.39	1.49
19B-53/3M	0.08	9.4	117	Si	5	0.10	0.05	0.50
19B-40	0.04	3.77	94	Si	5	0.10	0.03	0.30
Bul/No BM	0.53	2	3.8	Si-Fe	4	0.59	0.26	0.24
Blastfoe	0.07	6.2	89	Si	6	0.10	0.05	0.50
19B-57	0.05	4.8	96	Si	3	0.10	0.02	0.20
Ric/Air Ln	0.02	7.2	360	Si	3	0.10	0.01	0.10
Pul/Ric	0.88	4.88	5.5	Si-Pb	3	0.20	0.32	0.16
Ric/Air Ln	0.49	7.94	16.2	Si-Fe	2	0.13	0.12	0.93
19B-40	0.05	3.75	75	Si-Fe	7	0.19	0.05	0.26
Cesco 690C	0.1	1.55	15.5	Si-Fe	6	0.17	0.07	0.41
SBH30/Wal	0.12	14.6	122	Si	6	0.23	0.09	0.39
Home-air	0.17	10	58	Si-Pb	6	0.22	0.13	0.65
19B-57	0.12	268	2220	Si-Pb	6	0.12	0.09	0.75
Home Air/3M	0.79	5.6	7.1	Si	5	0.24	0.49	2.10
Shield/3M	0.18	9	50	Si-Pb	6	0.10	0.13	1.34
MSA Tight	0.21	31.8	151	Si	6	0.10	0.15	1.50
19B-57	0.04	150	3750	Si-Fe	6	0.12	0.03	0.25

Primary metal industries								
988	0.43	3.97	9	Fe	4	5.00	2.10	0.42
Breathy	79.24	142.91	1.8	Si-Fe-Al	2	0.22	19.90	90.50
Pang HD	0.06	1.25	21	Fe	4	5.00	0.03	0.01
Pang HD	0.6	33.1	55	Fe	4	5.00	0.30	0.06
Pang HD	0.1	1.87	18.7	Fe	4	5.00	0.05	0.01
Pul/Ric	2.73	8.45	3.1	Si	2	0.11	0.68	6.20
Welsh 7200	0.05	1.9	1.9	Al	2	5.00	0.01	0.01
Welsh 7200	0.54			Al	6	5.00	0.41	0.08
Welsh 7200	1.85			Al	6	5.00	1.39	0.28
Welsh 7200	0.26			Al	6	5.00	0.19	0.05
Pul ND	2.03	4.28	2.1	Al	6	5.00	1.52	0.03
Pang HD	0.24	5.6	23.2	Si	6	0.12	0.18	1.51
Pulmo AF	1.23	5.99	5.3	Si	6	0.11	1.23	11.20
Rag/Ric	4.9	42.5	8.7	Si-Cu	6	0.11	3.68	33.30
Wil 52	0.32	5.6	17.5	Fe	6	5.00	0.24	0.05
Wil 52	0.88			Si-Al	2	1.26	0.22	0.17
19B-57	0.05	71.7	1430	Al	6	5.00	0.04	0.01
Face Shield	0.55	1.55	2.8	Al	1	5.00	0.07	0.02
19B-57	1.1	2.5	2.3	Si	4	0.10	0.55	5.50
19B-57	0.17			Al	2	5.00	0.04	0.01
Pang HD	0.22	24	1.9	Si	4	0.10	0.11	1.10
Pang HD	42.5			Al-Fe	1	5.00	5.31	1.06
Ric/Saf	1.22	lost	—	Si-Mg	6	0.23	0.92	4.00
19B-57	1.68	31.2	18.6	Al-Si	5	0.90	1.26	1.40
Blastfoe	0.36	14.4	40	Pb-Fe	5	0.40	0.22	0.56
Blastfoe/Sa	0.13	82	630	Si-Fe	5	0.10	0.08	0.80
Primary metal industries airless process								
	9.29			Fe	6	5.00	7.00	1.40
	27.7			Si-Fe	4	1.80	13.90	7.75
	7.06			Si-Fe	3	0.24	2.70	1.12
	nil			Fe	2			0.00
	0.85			Glass	2	5.00	2.10	0.42
	0.18			Fe	1	5.00	0.24	0.01
	1.21			Si-Fe	4	0.41	0.60	1.47
	0.15			Fe	4	5.00	0.07	0.01
	0.3			Fe	8	5.00	0.30	0.06

Table 5 Abrasive Blasting Respiratory Protective Devices (*continued*)

Respirator code (1)[a]	Ci (2)[b]	Co (3)[c]	PF (4)[d]	Contaminant type (5)[e]	Time h (6)[f]	Assn. TLV (7)[g]	Exposure (8)[h]	X TLV (9)[i]
Primary metal industries airless process (*continued*)								
	0.25			Fe	2	5.00	0.06	0.01
	0.22			Fe	2	5.00	0.05	0.01
Wil 52	0.39	84	215	Fe	6	5.00	0.29	0.06
	0.04			Fe	6	5.00	0.03	0.01
	2.1			Fe	8	5.00	2.10	0.42
	0.06			Fe	6	5.00	0.05	0.01
	0.09			Fe	3	5.00	0.03	0.01
	0.76			Fe	2	5.00	0.19	0.04
	2.87			Fe	1	5.00	0.36	0.07
	1.44			Fe-Al	1	5.00	0.18	0.03
	0.86			Fe	2	5.00	0.21	0.04
	0.34			Fe	2	5.00	0.08	0.17

Note: Tables will now be presented showing the protection factors and exposures vis-a-vis the calculated TLV's. Each table will be divided into nine columns. 66, MSA Dustfoe 66 nuisance dust; Mono/Ric, Welsh Monomask under Pulmosan ricochet hood; 2160, Pulmosan 2160 nuisance dust respirator; 3M, 3M mask (nuisance dust); 19B-57, Bullard 19B-57 air fed helmet; LB, Leather covered Bullard (no BM approval); HA-99, Pulmosan HA-99 air fed helmet; Blastfoe, MSA Blastfoe air fed helmet; Clem/Ric, Clemco ricochet hood only; Clem/Met, Clemco metal air fed helmet (no BM approval number obvious); 6901C, Guardian 6901C air fed helmet; MSA Tight, MSA BM approved with tight full face air line respirator under apron; Ric/2301, Clemco ricochet hood over MSA 2301 organic vapor cartridge half mask; Pul/Ric, Pulmosan ricochet hood only; HM, Home made ricochet hood only; Sanst/Ric, Sanstorm ricochet hood only; Ric/77, Pulmosan ricochet hood over MSA Dustfoe 77 nuisance dust respirator; Ric/Cus C, Empire ricochet hood over MSA Custome Comfo nuisance dust respirator; Cesco 691, Cesco #691 air supplied helmet; 19B-53, MSA Leadfoe (not CE approved); 19B-53/3M, 19B-53 with 3M underneath; 19B-40, Bullard 19B-40 air supplied helmet; Bull/no BM, Bullard (no BM approval); Ric/Air Ln, Pulmosan ricochet hood over Scott full face air line respirator; Cesco 690C, Cesco 690C air supplied helmet (no apparent BM approval); 988, Whitecap 988 air supplied helmet; Breathzy, Whitecap Breatheasy air supplied hood; Pang HD, Pangborn heavy duty air supplied helmet; Welsh 7200, Welsh Bantam 7200 nuisance dust respirator; Pulm ND, Pulmosan nuisance dust respirator; Pulmo AF, Pulmosan air fed helmet; Rag/Ric, Dirty undershirt wrapped bandit fashion over nose and mouth (Figure 29) and covered with a worn out Pulmosan ricochet hood; Wil 52, Wilson #52 air fed helmet; Face Shield, Home made 5-mil face shield only; SBH30/Wel, Kelco SBH-30 ricochet helmet over Welsh 7100 nuisance dust respirator; Home air, Homemade air supplied helmet; Home Air/3M, Homemade air supplied helmet over 3M nuisance dust respirator; Shield/3M, 3M under face shield; Ric/Saf, Ricochet hood over Safeline nuisance dust respirator; Blastfoe/Saf, MSA Blastfoe over Safeline nuisance dust respirator; and 19B-34, Air line respirator plus sweat shirt hood.

[a] Column 1 is the respirator code given at the end of the table series. Several marketers sell the same helmet manufactured by the same firm under different trade names. Others sell several of their own approved helmets under the same trade name. Where possible USBM approval numbers will be used.

[b] Column 2 is the breathing zone respirable dust measured. Where no entry appeared in Column 1 only breathing zone (BZ) tests were made as no respirator was worn.

[c] Column 3 gives the ambient respirable dust concentration. Where no notation occurs no respirator was worn.

[d] Column 4 gives the protection factor calculated where a respirator was worn using the formula:

$$PF = \frac{\text{Ambient Respirable Dust}}{\text{BZ (in mask) Respirable Dust}}$$

[e] Column 5 lists symbols for the predominant respirable dust contaminants as measured chemically. Calculations were made on the basis of the most likely oxide.

[f] Column 6 lists the exposure hours per day for each monitored workman. Some of these are startling.

[g] Column 7 is the assigned TLV based upon the chemical and x-ray diffraction analyses of the dust collected.

[h] Column 8 is an 8-h day exposure factor based on the working hours and assigned TLV.

[i] Column 9 is the "Times TLV" factor to better evaluate the efficacy of the particular dust exposure situation.

From Reference 6.

Table 6 Summary of WPF Studies

Ref #	Author	Affiln	Year	Published	Type	Respirator Brand	Model	Operation
8	Myers	NIOSH	83	Yes	PAPR	MSA	FF& Half	Bagging
8	Myers	NIOSH	83	Yes	PAPR	MSA	FF& Half	Bagging
9	Lenhart	NIOSH	84	Yes	PAPR	MSA	Half	Blaster Furnace
9	Lenhart	NIOSH	84	Yes	PAPR	MSA	Half	Sinter
9	Lenhart	NIOSH	84	Yes	AP	MSA	Half	Sinter
9	Lenhart	NIOSH	84	Yes	AP	MSA	Half	Blaster Furnace
10	Myers	NIOSH	84	Yes	PAPR	3M	W-344	Smelter
10	Myers	NIOSH	84	Yes	PAPR	Racal	AH-3	Smelter
11	Myers	NIOSH	86	Yes	PAPR	3M	W-316,DM	Battery
11	Myers	NIOSH	86	Yes	PAPR	Racal	AH-5,DM	Battery
12	Gaboury	Alcan	93	Yes	PAPR	Racal	BE-10	Smelter
12	Gaboury	Alcan	93	Yes	AP	AO-WIL-SU	Half,OV-DFM	Smelter
13	Keys	Syntex	90	No	PAPR	3M	White Cap II	Pharmaceutical
13	Keys	Syntex	90	No	PAPR	Racal	BE-10	Pharmaceutical
13	Keys	Syntex	90	No	PAPR	Bullard	Quantum	Pharmaceutical
14	Stokes	3M	87	No	PAPR	3M	W-344	Roofing
14	Stokes	3M	87	No	PAPR	3M	W-316,DM	Roofing
15	Colton	3M	90	No	PAPR	3M	White Cap 3205	Smelter
16	Dixon	DuPont	84	No	PAPR	3M	W-316,DM	Pigment
17	Johnston	3M	87	No	SAR	3M	W-8100	Blasting
18	Dixon	DuPont	84	Yes	AP	Survivair	2000-OV/H	Pigment
19	Colton	3M	90	No	AP	3M	EA,FF	Smelter
20	Wilmes	3M	84	No	Disp	3M	8710,DM	Brake
20	Wilmes	3M	84	No	Disp	3M	9920,DM	Brake
20	Wilmes	3M	84	No	PAPR	3M	W-344	Brake
20	Wilmes	3M	84	No	AP	3M	EasiAir	Brake
20	Wilmes	3M	84	No	Disp	3M	9910,DM	Brake
20	Wilmes	3M	84	No	AP	3M	EA-DM	Brake
21	Dixon	DuPont	85	No	Disp	AO	R-1050,DM	Asbestos Removal
21	Dixon	DuPont	85	No	Disp	3M	8710,DM	Asbestos Removal
21	Dixon	DuPont	85	No	AP	MSA	Comfo II	Asbestos Removal
21	Dixon	DuPont	85	No	Disp	3M	9910,DM	Asbestos Removal
21	Dixon	DuPont	85	No	AP	North	7700-H	Asbestos Removal
21	Dixon	DuPont	85	No	AP	Survivair	2000-DFM	Asbestos Removal
22	Mullins	3M	87	No	Disp	3M	8715,DM	Polish & Grinding
22	Mullins	3M	87	No	Disp	3M	8715,DM	Polish & Grinding
22	Mullins	3M	87	No	Disp	3M	8715,DM	Polish & Grinding
23	Colton	3M	90	No	Disp	3M	9906,DM	Carbon Change
24	Colton	3M	90	No	Disp	3M	9970	Foundry,Brass
24	Colton	3M	90	No	Disp	3M	9970	Foundry,Brass
25	Nelson	DuPont	NA	No	Disp	3M	9900,DM	Pigment
25	Nelson	DuPont	NA	No	AP	Survivair	2000	Pigment
26	Reed	NIOSH	87	Yes	Disp	3M	9910,DM	Pcakaging
27	Cohen	Olin	84	Yes	Disp	3M	8716,DM/Hg	Chlorine Cell
28	Galvin	UC	90	Yes	AP	North	75000V	Fiberglass
29	Johnston	3M	89	No	SAR	3M	W-8000	Grinding
29	Johnston	3M	89	No	SAR	3M	W-8000	Grinding
30	Wallis	Duracell	93	Yes	Disp	3M	8710	MnO2 processing
31	Colton	3M	93	No	Hood	3M	W-3258	Foundry
31	Colton	3M	93	No	Hood	3M	W-3258	Foundry
32	Colton	3M	94	No	Half	3M	6000	Ship Wrecking
32	Colton	3M	94	No	Half	3M	6000	Pigment

Note: 1. Except as noted, all air-purifying respirators tested are equipped with HEPA filters
2. For fit testing methods, the number after QNFT is the fit factor achieved.
For QLFT method, IAA: isoamyl acetate, IS: irritant smoke, SA: saccharin.
3. NR: not reported. PIXEA: Proton Induced X-ray Emission Activation. Imp: Impactor.
Atom Abs: Atomic absorption. Grav: Gravimetric. Grap: Graphic furnace. GC: Gas chromatography.
L: lapel sample. I: In-mask sample. nd: not detected.

Conta-minant	Number of Workers	Days of Sampling	Sampling time, min	Particle Method	Size Distribution MMAD,um	Std Dev	Fit Testing	Author	Ref #
Silica	4	3	84-320	Imp	5.5	7.7		Myers	8
Silica	4	3	84-320	Imp	5.8	7.1		Myers	8
Lead	18		480	Imp	1-8	9.5-28.5		Lenhart	9
Lead	7		480	Imp	9-16	2.5-5.1		Lenhart	9
Lead	7		480	Imp	9-16	2.5-5.1	QNFT-250	Lenhart	9
Lead	18		480	Imp	1-8	9.5-28.5	QNFT-250	Lenhart	9
Lead	12	4	330-420	Imp	0.68-17	35%>17u	QNFT-5100	Myers	10
Lead	12	4	330-420	Imp	0.68-17	30%<0.68u	QNFT-7900	Myers	10
Lead	12	4	330-420	Imp	17			Myers	11
Lead	12	4	330-420	Imp	17			Myers	11
BaP	22		160-260	Imp	0.5			Gaboury	12
BaP	22		160-260	Imp	0.5			Gaboury	12
Steroids	NR							Keys	13
Steroids	NR		30-180					Keys	13
Steroids	NR							Keys	13
Silica	5	4	30-60	NR				Stokes	14
Silica	5	4	30-60					Stokes	14
Lead	20	5 Shifts	60-240	Imp	>10		QNFT-500	Colton	15
Lead	7		30-120		NR		No	Dixon	16
Silica	4	3	10-60					Johnston	17
Lead	11		NR	Micro	1.8	2.7	IAA	Dixon	18
Lead	13	4	30-180	Imp	15%<0.9,65%>17		QNFT-500	Colton	19
Asbestos	7		30					Wilmes	20
Asbestos	6		30					Wilmes	20
Asbestos	5		30					Wilmes	20
Asbestos	3		30					Wilmes	20
Asbestos	8		30					Wilmes	20
Asbestos	5		30					Wilmes	20
Asbestos	17							Dixon	21
Asbestos	17		30-120					Dixon	21
Asbestos	17		30-120					Dixon	21
Asbestos	17		30-120					Dixon	21
Asbestos	17		30-120					Dixon	21
Asbestos	17		30-120					Dixon	21
Si	5		35-235					Mullins	22
Al	5		35-235					Mullins	22
Ti	5		35-235					Mullins	22
Al	5	5	67-315	Imp	>5		QLFT	Colton	23
Zinc	17		67-315	NR	NR			Colton	24
Lead	17	5	30-270	NR	NR		QLFT	Colton	24
Cadmium	10	5	NR	NR	NR		QLFT-Sa	Nelson	25
Cadmium	10		NR					Nelson	25
Concrete	7	3	240	Imp	3.7-20		QNFT-200	Reed	26
Mercury	7		10-30	-	-		None	Cohen	27
Styrene	13		60	-	-		QLFT-IS	Galvin	20
Si	6							Johnston	29
Fe	6			Optical	1.27-2.55			Johnston	29
MnO2	10		30-40	Imp	0.52->9.8	2-2-6.2		Wallis	30
Si	4							Colton	31
Si	4							Colton	31
Pb	18			Imp	2.24-11.1	1.2-1.9	QLFT-Sa	Colton	32
Cd	18			Imp	1.3-10	1.3-1.7	QLFT-Sa	Colton	32

Table 6 *(continued)*

Ref #	Author	Analytical Method	Method Sensitvty, ug	Sample Sets	PEL ug/m3 or f/c	Range	Co ug/m3 GM	GSD
8	Myers	Grav	300	11		4400-41700		
8	Myers	X-ray Diff		11		2000-36800		
9	Lenhart	Grap Fur(I)	0.2	18	50	95-1700	480	2.2
9	Lenhart	Atom Abs(L)	10	7	50	460-2500	480	2.2
9	Lenhart	Atom Abs(L)	10	7	50	570-28000	580	3.4
9	Lenhart	Grap Fur(I)	0.2	18	50	92-2900	580	3.4
10	Myers	Grap Furnace(I)	0.3	22	50	439-5464		
10	Myers	Atom Abs(L)	3	22	50	366-3621		
11	Myers	Grap Furnace(I)	0.3	23	50	112-534	236	1.45
11	Myers	Atom Abs(L)	3	24	50	125-387	237	1.42
12	Gaboury	Alcan#1223-84	0.001	20	0.2	2.5-111	16.7	2.32
12	Gaboury	Alcan#1223-84	0.001	18	0.2	2.5-23	7.9	1.88
13	Keys	RIA	0.00005	22	0.15	8.7-1330	NR	
13	Keys	RIA	0.00005	29	0.15	8.7-1330	NR	
13	Keys	RIA	0.00005	9	0.15	8.7-1330	NR	
14	Stokes	PIXEA	0.001-0.01	12		NR	NR	
14	Stokes	PIXEA	0.001-0.01	90		NR	NR	
15	Colton	PIXEA	0.01	55	50	162-4421	NR	
16	Dixon	PIXEA	0.01-0.1	NR	50	NR		
17	Johnston	PIXEA	0.001-0.01	68		10-1332	0.13-24.7	
18	Dixon	PIXEA	0.002	37	50	NR	225	
19	Colton	PIXEA	0.01	20	50	150-3380	NR	
20	Wilmes	Counting		13	2	0.5-7.4	2.3	
20	Wilmes	Counting		10	2	1.8-19	6.5	
20	Wilmes	Counting		9	2	1.2-8	4.1	
20	Wilmes	Counting		7	2	0.3-2	1.1	
20	Wilmes	Counting		13	2	0.6-9.4	2.9	
20	Wilmes	Counting		9	2	0.4-2.3	1.8	
21	Dixon	Counting		18	2	0.01-10.5	NR	
21	Dixon	Counting		18	2	0.01-10.5	NR	
21	Dixon	Counting		17	2	0.01-10.5	NR	
21	Dixon	Counting		14	2	0.01-10.5	NR	
21	Dixon	Counting		14	2	0.01-10.5	NR	
21	Dixon	Counting		15	2	0.01-10.5	NR	
22	Mullins	PIXEA	0.009-0.035	14	15000	10-300	7.4	
22	Mullins	PIXEA	0.009-0.035	10	15000	10-300	2.5	
22	Mullins	PIXEA	0.009-0.035	14	15000	10-300	18	
23	Colton	PIXEA	0.027	23	10000	NR	NR	
24	Colton	PIXEA	0.003	62	10000	90-17,000	NR	
24	Colton	PIXEA	0.009	43	50	3.4-708	NR	
25	Nelson	PIXEA	0.008	18	200	NR	80	
25	Nelson	PIXEA	0.008	28	200	NR	80	
26	Reed	Grav	10	22	15000	2600-53000	15000	2.3
27	Cohen	Atom Abs	3	26	100	55-1097	NR	
20	Galvin	GC-NIOSH1501		63	215000	11900-762000	NR	
29	Johnston	PIXEA	0.008	37	15000	<100-1500	1000	
29	Johnston	PIXEA	0.008	37	15000	100-2800	1500	
30	Wallis	Atom Abs	0.25	70	5000	150-77400		
31	Colton	ICP/PIXEA	0.006	14	15000	18400-209000	83.3	1.76
31	Colton	ICP/PIXEA	0.008	15	15000	18400-209000	83.3	1.76
32	Colton	ICP/PIXEA	0.008	47	50	14-1630	183	3.44
32	Colton	ICP/PIXEA	0.008	59	5	3.5-423	25.7	2.39

Data Presented	Ci, ug/m3 Range	GM	GSD	WPF Range	GM	GSD	5th %	Author	Ref #
Yes	50-1500			8-181	35	2.8		Myers	8
Yes	30-560			16-215	54	2.24		Myers	8
Yes	<0.23-5.5			94-1600	380	2.6		Lenhart	9
Yes	1.2-20			23-930	380	2.6	79	Lenhart	9
Yes	1.2-13			110-2200	180	4.1	18	Lenhart	9
Yes	<0.22-30			10-1700	180	4.1		Lenhart	9
Yes	0.2-66	5.5	3.2	28-5500	165	3.57	28.1	Myers	10
Yes	0.6-33.6	4.9	2.89	42-2323	205	2.83	30.5	Myers	10
Yes	0.5-5.4	1.8	1.8	31-392	135	1.89	48.2	Myers	11
Yes	0.3-12.4	2	2.51	24-1063	120	2.66	26.6	Myers	11
Yes	0.006-0.072	0.012	1.94	371-8658	1414	2.51	275	Gaboury	12
Yes	0.008-1.056	0.169	3.32	13.1-410	47	2.52	10	Gaboury	12
No	0.0006-0.162	NR		1810-470000	42260	9.8	997	Keys	13
No	0.0006-0.162	NR		1150-304000	11137	3.9	1197	Keys	13
No	0.0006-0.162	NR		1230-62700	9574	3.1	1470	Keys	13
No	NR	NR		NR	5370	3.0	762	Stokes	14
No	NR	NR		NR	2480	7.0	95	Stokes	14
No	nd-1.45	NR		NR	8843	3.2	1335	Colton	15
No	NR	NR		37-1500	230	3		Dixon	16
Yes	<0.0001-0.0033	NR		39-28600				Johnston	17
No	NR	NR		94-27000	3400	3.8	512	Dixon	18
No	0.03-3	NR		NR	3929	9.6	95	Colton	19
No	NR	NR		29-484	81	1.99	25	Wilmes	20
No	NR	NR		71-1050	223	2.38	45	Wilmes	20
No	NR	NR		66-603	199	2.36	42	Wilmes	20
No	NR	NR		34-82	56	1.35	31	Wilmes	20
No	NR	NR		24-584	107	2.5	20	Wilmes	20
No	NR	NR		27-119	68	1.66	28	Wilmes	20
No	0.0003-1.4	NR		10-970	52	4.2	5	Dixon	21
No	0.0003-1.4	NR		7-3200	310	5.3	20	Dixon	21
No	0.0003-1.4	NR		12-7900	94	3	0	Dixon	21
No	0.0003-1.4	NR		94-5600	580	4.2	55	Dixon	21
No	0.0003-1.4	NR		12-3100	250	6.9	11	Dixon	21
No	0.0003-1.4	NR		15-4200	240	6.3	12	Dixon	21
No	0.5-2	NR		NR	172	3.1	24	Mullins	22
No	0.5-2	NR		NR	145	2.3	32	Mullins	22
No	0.5-2	NR		NR	59	1.7	24	Mullins	22
No	5.4-60	NR		NR	27	1.5	13	Colton	23
No	0.01-27	NR		NR	681	5.6	40	Colton	24
No	nd-4.3	NR		NR	310	4.3	28	Colton	24
No	NR	0.2		11-2260	316	4.3	25	Nelson	25
No	NR	0.2		11-8800	337	5.3	22	Nelson	25
Yes	200-4100	800	2.7	1.6-150	18	3.1	3	Reed	26
Yes	2-20.9	NR		8.6-63	26			Cohen	27
Yes	400-215000	NR		3.4-982	79			Galvin	20
No	1-5			220-1417				Johnston	29
No	0.2-0.8			273-1012				Johnston	29
Yes	8-160			2.8-848	50		7.5	Wallis	30
No	8-200	13	3.28		7883	3.01	1284	Colton	31
No	1-89	13	3.28		8718	3.02	1411	Colton	31
No	0.4-39.3	1.3	3.07		205	3.56	25	Colton	32
No	0.03-1.47	0.07	1.54		558	3.85	61	Colton	32

median aerodynamic diameter (GM MMAD) for lead in the sinter plant varied from 9 to 16 μm and the GSD varied from 2.5 to 5.1. The geometric mean of the MMAD for lead in the blast furnace area varied from 1 to 8 μm and the GSD varied between 9.5 and 28.5. The ambient concentrations for the PAPR varied between 460 to 2500 μg/m³, and 95 to 1700 μg/m³ in the sinter plant and in the blast furnace area, respectively. The in-mask concentrations for the PAPR varied between 1.2 and 20 μg/m³ and <0.23 and 5.5 μg/m³ in the sinter plant and in the blaster furnace area, respectively. The ambient concentrations for the half-mask respirator varied between 570 and 28,000 μg/m³ and 92 to 2900 μg/m³ in the sinter plant and in the blast furnace area, respectively. The in-mask concentrations for the half-mask varied between 1.2 and 13 μg/m³ and <0.22 to 30 μg/m³ in the sinter plant and in the blaster furnace area, respectively. A summary of test results and test conditions are shown in Table 6.

The calculated WPF varied between 23 to 1600 for the PAPR, with a GM of 380 and a GSD of 2.6. The calculated WPF for the nonpowered respirator varied between 10 and 2200 for the PAPR with a GM of 180 and a GSD of 4.1. This is the first time a confidence limit at 95% (or a fifth percentile) has been applied to all WPF values determined from a WPF study. It is:

$$WPF_a = \frac{GM}{(GSD)^{1.64}} \tag{2}$$

The WPF_a was calculated as 79 and 18 for the PAPR and half-mask, respectively. In the respirator selection table of OSHA health standards, this type of PAPR has a protection factor of 1000 based on the quantitative fit test results conducted by the Los Alamos Scientific Laboratory. As a result of this study, NIOSH reduced the assigned protection factor (APF) for the tight-fitting PAPR from 1000 to 50. It should be pointed out that the mean ambient concentration in this study was less than 50 times the OSHA permissible exposure limit (PEL) of lead, and the in-mask concentration was much less than the OSHA PEL. There is no indication that a higher APF would not be achieved by the PAPR if the test was conducted at a workplace with higher ambient concentrations.

The performance of loose-fitting helmet type PAPRs made by Racal and 3M have been evaluated by Myers, Peach and coworkers of NIOSH in two studies.[10,11] PAPRs used in the first study were equipped with HEPA filters, and the dust/mist filter was selected for the second study. The first study was performed in a secondary lead smelter. The ambient concentration varied between 366 and 5464 μg/m³, and the in-mask concentrations varied between 0.2 and 66 μg/m³. The fifth percentile WPF for the Racal and the 3M PAPR are 30.5 and 28, respectively. The second study was performed in a lead acid battery manufacturing plant. The ambient concentration varied between 112 and 534 μg/m³, and the in-mask concentration varied between 0.3 and 12.4 μg/m³. The fifth percentile WPF for the dust/mist filter equipped Racal and 3M PAPRs are 26 and 48, respectively. The sampling time covered the whole shift in both studies. Details on WPF and test conditions are shown in Table 6.

Grabory of Alcan and Burd of Racal Airstream[12] conducted a study to determine the performance of the Racal PAPR and several half-mask negative pressure air-purifying respirators (AP). Information from this study is shown in Table 6. The study was performed at an aluminum reduction plant. Benzo(α)pyrene was the parameter used to calculate WPFs. The ambient concentration varied between 2.5 and 111 μg/m^3 and the in-mask concentration varied between 0.006 and 1.05. The WPFs varied between 385 and 8658 for the Racal PAPR and between 13 and 410 for the air-purifying respirators. The fifth percentile WPFs for the Racal and other respirators were 275 and 9, respectively.

Keys et al.[13] of Syntax conducted a study to determine the performance of three loose-fitting helmet type PAPRs — the Bullard Quantum, the Racal Breath Easy 10 and the 3M White Cap II — at a pharmaceutical manufacturing plant. Steroids were manufactured in that operation and the WPFs were determined from these substances. The analytical method for the steroids is very sensitive, with a sensitivity of 50 pg. The sampling time was about 2 h. The ambient concentration varied between 8.7 and 13,300 μg/m^3, and the in-mask concentration varied between 0.0006 and 0.162 μg/m^3. The fifth percentile WPFs were 1197, 1470, and 997 for the Racal, Bullard, and 3M PAPRs, respectively. A summary of WPF and test conditions is shown in Table 6. Details on sampling results were not reported.

Stokes et al.[14] of the 3M Company conducted a study to determine the performance of the 3M loose-fitting facepiece Airhat PAPR in the dust/mist filter and the HEPA filter configurations. The study was performed in a roofing operation and silica was measured for the WPF determination. The sampling periods ranged from 30 to 60 min. The proton induced x-ray emission analysis (PIXEA) was used to analyze samples. This method is quite sensitive, with a sensitivity in the nanogram range. Both the ambient and the in-mask concentration, as well as the range and standard deviation of WPFs were not reported in this study. The geometric mean WPFs were 2480 and 5370 for the dust/mist and the HEPA filters configurations, respectively. The fifth percentile WPFs achieved by the PAPR were 95 for the dust/mist filter and 762 for the HEPA filter configuration. A summary of data presented in this study is shown in Table 6. Details on sampling results were not reported.

Colton[15] of the 3M Company conducted a WPF study for the 3M Whitecap full facepiece PAPR with the HEPA filter at a secondary lead smelter. The contaminant lead was measured by the PIXEA method. Samples were collected over five shifts. However, the sampling time was not reported. The ambient concentration varied between 162 and 4421 μg/m^3, and the in-mask concentration varied between nondetectible and 1.45 μg/m^3. The geometric mean and the geometric standard deviation for the WPF were 8843 and 3.2. The fifth percentile WPF was reported as 1335. The mean and standard deviation of the ambient and in-mask concentrations and details of sampling results were not reported in this study. Data presented in this study are summarized in Table 6.

Dixon and Nelson[16] of the DuPont company conducted a "program" study to determine the performance of the 3M W316 Airhat PAPR at a workplace using lead. The information presented in this study is summarized in Table 6. The sampling period varied between 30 and 120 min. Samples were analyzed by the PIXEA method. The WPFs had a range between 37 and >1500. However, no other details on test conditions and test results were reported.

Johnston et al.[17] of the 3M Company conducted a study to determine the performance of a helmet type abrasive blasting supplied air respirator, the 3M White Cap II. Sand and grit were used to strip paint off a barge. Both total and respirable dust samples were collected for the ambient aerosol. Samples were analyzed by the

PIXEA method. Since the PIXEA method cannot be used with a high dust load, the sampling time was reduced to very short periods of 10 to 60 min. The ambient respirable dust concentration varied between 10 and 1332 μg/m³, and the in-mask concentration varied between <0.02 and 0.04 μg/m³. The WPFs varied between 39 and 28,000. Although higher WPF values were reported in this study, the low ambient concentration and short sampling time is hardly representative of abrasive blasting operations. Information presented in this study is summarized in Table 6.

Dixon and Nelson[18] of the DuPont Company conducted a study to measure the WPF for a negative pressure half-mask respirator. The study was performed in a pigment operation containing lead. The ambient concentration varied between 40 and 1410 μg/m³. The in-mask concentration varied between <0.01 and 1.1 μg/m³. Sampling time was not reported. The WPF had a range between 126 to 36,000. The geometric mean and standard deviation was 35 and 2.8, respectively. The particle size was determined by microscope with a mean size of 1.8 μm reported. The PIXEA method was used to analyze the samples for WPF determination. This method has a detection limit of 2 ng. Almost all WPF values in excess of 10,000 were reported with in-mask concentrations less than the detection limit of the PIXEA method. Since the aerodynamic size of the lead pigment was unknown, it is likely that a majority of the airborne lead pigment particles were in the nonrespirable size range that does not penetrate through the respirator facepiece. This may explain the presence of a large number of high WPFs in this study. Information presented in this study is summarized in Table 6.

Colton et al.[19] of the 3M Company conducted a study on the performance of a 3M #7800 full facepiece negative pressure respirator at a secondary lead smelter. Both lead dust and fume were presented at the workplace. The sampling time varied between 30 and 180 min. More than 65% of the ambient aerosols were larger than 10 μm. The geometric mean and standard deviation were 3929 and 9.6, respectively. The fifth percentile WPF was 95. No other details of the study were reported. Information presented in this study is summarized in Table 6.

A number of unpublished WPF studies were performed on disposable respirators (filtering facepieces), and a majority of these studies were conducted by the 3M company, a major disposable respirator manufacturer.

Wilmes et al.[20] of the 3M Company conducted a study to determine the performance of 3M disposable and elastomeric facepiece air-purifying respirators for asbestos dust. The study was performed at a brake manufacturing operation. The ambient and in-mask concentrations for each respirator being studied were not reported; however, concentrations of asbestos in each different operation were measured and reported. The ambient asbestos had a range between 0.28 and 18.9 fibers/ml, with a geometric mean between 1.00 and 4.86 and a geometric standard deviation between 1.28 and 3.60. No in-mask concentrations were reported. The sampling time was about 30 min. Individual ambient and in-mask concentrations for this WPF study were not reported. Information presented in this study is summarized in Table 6.

The WPFs for the disposable respirators had a range of 24 to 1050. The geometric means varied between 81 and 233, and the geometric standard deviation varied between 1.99 and 2.38. The fifth percentile WPFs for disposable respirators had a range of 25 to 45. The WPFs for elastomeric facepiece respirators had a range of 27 to 603. The geometric means varied between 56 and 199, and the geometric standard deviation varied between 1.35 and 2.36. The fifth percentile WPFs for elastomeric facepiece respirators had a range between 28 and 42.

Another WPF study for asbestos was conducted by Dixon and Nelson[21] of the Du Pont Company. Both disposable and elastomeric facepiece air-purifying respirators were selected for this study. The study was performed on asbestos removal operations. The ambient and in-mask concentrations for each respirator were not reported. Details of test results were not presented, but were submitted to the OSHA docket on asbestos. The demolition operation was performed at two sites. The ambient asbestos concentrations at site #1 varied between 2.4 and 10.5 fibers/ml, and the concentration at site #2 varied between 0.001 and 2.61 fibers/ml. The in-mask asbestos concentrations varied between 0.0013 and 1.2, and 0.00032 to 1.37 fibers/ml, respectively at these sites. Because in-mask asbestos concentrations were low, the field of counting was increased from 100 to 500 to increase fiber count. However, the standard NIOSH asbestos counting method only permits counting of 100 fields. The increase in counting fields extended the sensitivity of the analytical method and artificially boosted the WPF value.

The WPFs for disposable respirators varied between 9.7 and 5600, and the WPFs for the elastomeric facepiece respirators varied between 12 and 7900. Even with such a high field count, very few fibers were found in many in-mask samples, especially for those with WPFs higher than 1000. The geometric mean of WPFs for disposable respirators varied between 52 and 580, and the fifth percentile WPFs varied between 5 and 55. The geometric mean of WPFs for the elastomeric facepiece respirator varied between 94 and 250, while the range of the fifth percentile WPFs varied between 12 and 16. Information presented in this study is summarized in Table 6.

The 3M staff conducted additional unpublished WPF studies to determine the performance of various types of 3M respirators. The PIXEA method was used for WPF determination in all studies. The first WPF study using the PIXEA method was conducted by Johnston and Mullins[22] for a polish and grinding operation. The WPFs were determined from the values of ambient and in-mask concentrations of aluminum, titanium, and silicon dust, and had a range of 100 to 300. The ambient concentrations varied between 100 and 300 μg/m^3 and the in-mask concentrations varied between 0.5 and 2 μg/m^3. The fifth percentile WPFs for aluminum, titanium, and tungsten were 32, 24, and 24, respectively. Individual ambient and in-mask concentrations, and particle size distribution of the ambient aerosols, were not reported. It should be noted that ambient concentrations were so low that respiratory protection was not required. Information presented in this study is summarized in Table 6.

A 3M disposable respirator WPF study was conducted by Colton and Mullins[23] against aluminum dust in a carbon change operation. Information presented in this study is summarized in Table 6. The respirator selected in this study was the 3M 9906 equipped with a dust/mist filter. Ambient dust was collected by a cyclone, but the concentration was not reported. The in-mask concentration had a range from 5.44 to 60.1 μg/m^3. The reported WPF had a geometric mean of 27 with a standard deviation of 1.5. The fifth percentile WPF was 13. Few details concerning particle size distribution, ambient, and in-mask concentrations were available.

The performance of a 3M disposable respirator equipped with a HEPA filter, 3M 9970, was examined by Colton[24] of 3M. Information presented in this study is summarized in Table 6. The ambient lead concentration varied between 3.4 and 708 μg/m³ and the in-mask lead concentration varied between nondetectable and 4.3 μg/m³. The ambient zinc concentration varied between 0.09 and 17,000 μg/m³, and the in-mask lead concentration varied between nondetectable and 4.3 μg/m³. The study was performed in a zinc and lead environment. The WPF determined for lead had a geometric mean of 310 and a standard deviation of 4.3. The WPF determined for zinc had a geometric mean of 681 and a standard deviation of 5.6. The fifth percentile WPFs for lead and zinc were 28 and 40, respectively. Details on particle size distribution, ambient and in-mask concentrations were not reported.

Nelson[25] of the Chemical Manufacturers Association (CMA) conducted an unpublished study to determine the performance of a 3M 9910 disposable dust mist respirator in a pigment manufacturing operation. The PIXEA method was used to measure the concentration of cadmium. A Survivair elastomeric facepiece negative pressure respirator was also selected for this study. The ambient concentration for cadmium was 80 μg/m³, and the in-mask cadmium concentration was 0.2 μg/m³. For the disposable respirator, the geometric mean of the WPF was 316 with a standard deviation of 4.3. For the elastomeric facepiece respirator, the geometric mean of the WPF was 681 with a standard deviation of 5.6. The fifth percentile WPF for the disposable and the elastomeric facepiece respirator was 24.8 and 21.9, respectively. Except for the above, no other details, such as ambient concentration, particle size, and sampling period, were reported. It should be noted that this study was performed in a workplace with a very low ambient concentration, and respiratory protection was not required. Information presented by the author of this study is shown in Table 6.

Reed et al.[26] of NIOSH conducted a WPF study to determine the performance of a 3M 9910 disposable dust/mist respirator at a cement product packaging operation. Seven workers participated in this study. The sampling was conducted in 3 days, with a sampling period of approximately 4 h. The particle size distribution was determined by an impactor. The aerodynamic size varied between 3.7 and 20 μg/m³. The gravimetric method with a detection limit of 10 μg was used for WPF determination. The geometric mean ambient concentration was 15,000 μg/m³, with a standard deviation of 2.3. The ambient concentrations varied between 2600 and 53,000 μg/m³. The geometric mean in-mask concentration was 800 μg/m³, with a standard deviation of 2.7. The in-mask concentrations varied between 200 and 4100 μg/m³. The geometric mean WPF was 18, with a standard deviation of 3.1. The fifth percentile WPF was 3. Information presented in this study is summarized in Table 6.

Very few studies have examined the performance of respirators for protection against vapors. Cohen[27] of Olin conducted a study to determine the performance of a nonapproved disposable respirator for protection against mercury vapor. The study was performed at a chloro-alkali plant for workers who were exposed to mercury vapor during cell maintenance. Mercury vapor was analyzed by atomic absorption spectrophotometry. The sampling time ranged between 10 and 30 min. The ambient mercury concentrations varied between 55 and 1097 μg/m³. The in-mask mercury concentrations varied between 2 and 20.9 μg/m³. The geometric mean WPF was 25.8, with a range between 8.6 and 62.9. Information presented from this study is summarized in Table 6.

Table 7 C_i (mg/m³)/C_o (mg/m³) = Penetration for 13 Workers[a]

Worker	Job		Sample no. 1	2	3	4	5	6
1	Nonsprayer	C_o	324.95	265.35	281.39	251.10	210.91	
		C_i	1.22	1.79	1.72	2.28	2.52	
		p	0.0037	0.0068	0.0061	0.0091	0.0119	
2	Nonsprayer	C_o	350.43	223.20	192.79	285.57	243.27	283.31
		C_i	1.69	1.57	8.12	1.36	1.48	2.09
		p	0.0048	0.0070	0.0421	0.0048	0.0061	0.0088
3	Nonsprayer	C_o	311.14	195.11	187.39	527.27		
		C_i	2.28	1.51	1.54	2.28		
		p	0.0073	0.0077	0.0082	0.0043		
4	Nonsprayer	C_o	253.78	488.56	382.64	475.26	666.87	
		C_i	3.15	49.32	0.39	4.03	7.39	
		p	0.0124	0.1010	0.0010	0.0085	0.0111	
5	Nonsprayer	C_o	224.87	499.89	470.68	562.56	255.83	
		C_i	2.45	22.86	40.38	84.25	7.94	
		p	0.0109	0.0457	0.0858	0.1498	0.0310	
6	Nonsprayer	C_o	281.30	550.56	368.53	535.57	545.12	
		C_i	5.05	1.19	2.32	11.59	3.30	
		p	0.0180	0.0022	0.0063	0.0216	0.0061	
7	Nonsprayer	C_o	189.92	242.41	330.49	366.99	327.96	
		C_i	2.11	1.11	1.89	17.44	8.43	
		p	0.0111	0.0046	0.0057	0.0475	0.0257	
8	Sprayer	C_o	456.25	569.29	689.86			
		C_i	0.80	11.27	4.77			
		p	0.0017	0.0198	0.0069			
9	Sprayer	C_o	11.90	456.55	761.78	611.27	650.39	
		C_i	3.53	23.21	4.27	1.43	14.94	
		p	0.2965	0.0508	0.0056	0.0023	0.0230	
10	Sprayer	C_o	640.15	436.12	204.48	585.44	585.98	
		C_i	16.09	3.04	1.20	13.19	16.98	
		p	0.0251	0.0070	0.0059	0.0225	0.0290	
11	Sprayer	C_o	453.57	392.06	360.01	672.16	450.33	
		C_i	2.60	2.17	4.79	3.70	2.34	
		p	0.0057	0.0055	0.0133	0.0055	0.0052	
12	Sprayer	C_o	497.27	502.62	560.17	680.94	568.44	
		C_i	3.51	14.94	10.72	39.48	9.31	
		p	0.0071	0.0298	0.0191	0.0580	0.0164	
13	Sprayer	C_o	389.89	437.73	597.45	631.65	556.29	
		C_i	6.57	10.60	215.17	4.86	12.73	
		p	0.0168	0.0242	0.3601	0.0077	0.0229	

[a] C_i values corrected for pulmonary retention as discussed in text.

From Reference 28.

A NIOSH sponsored WPF study for styrene was conducted by Galvin et al.[28] of the University of California at Berkeley. The study was performed with a chemical cartridge respirator equipped with organic vapor cartridges, at a reinforced fiberglass bath tub manufacturing operation that uses styrene as a resin. Styrene was measured for WPF determination. Thirteen workers participated in this study. The ambient concentrations varied between 11,900 and 762,000 μg/m³, and the in-mask concentrations varied between 400 and 215,000 μg/m³. In order to avoid the breakthrough of styrene, the sampling time was limited to 1 h. Samples were collected between styrene sprayers and nonsprayers, and it was found that no difference in penetration between job classifications existed, despite the different range of body movements observed during work. The range of WPF was between 2.8 and 588 for sprayers, and between 6.7 and 1000 for nonsprayers. The geometric mean WPF for the sprayers and nonsprayers was 67 and 91, respectively. Information presented in this study is summarized in Table 7.

The results indicated that there were variations of protection between the single worker and a group of workers. The geometric standard deviation (GSD) between workers was 1.92, the common GSD within workers was 2.93, and the total GSD was 3.51. One half of a population of wearers with similar protection would be expected to obtain an average protection factor in excess of 44 and one half below that value. In order for the in-mask concentration to be less than the short-term excursion level (STEL) for styrene (100 ppm), the ambient styrene concentration should be less than 1000 ppm for 95% of a population of workers. This does not mean that each individual will receive the same level of protection because the calculation does not account for individual differences. The similar ambient concentration which would prevent overexposure for 99% of a population is reduced to 488 ppm styrene. The authors pointed out that the difference between individuals, as well between different wearings, must be considered in setting maximum use limits for a given class of respirators.

Johnston et al.[29] of the 3M Company conducted a workplace study to evaluate the performance of a supplied air respirator. The study was conducted in a foundry with 6 workers. The test subjects wore a 3M W-8000 Whitecap II supplied air helmet equipped with a vortex cooling device or an air regulating valve. The flow rate was set at 6.7 cfm. The line pressure was set for 60 psi with a vortex tube, and 25 psi with the regulating valve.

The workers were grinding iron parts. The ambient air samples were collected with a cyclone and filter assembly. Samples were analyzed by the PIXEA method for iron and silicon. The ambient concentration for iron dust varied between 100 and 2800 $\mu g/m^3$, with an average of 1500 $\mu g/m^3$. The ambient respirable dust concentration for silicon dust varied between <100 and 1500 $\mu g/m^3$. The OSHA PEL for iron and silicon is 15,000 $\mu g/m^3$. The in-mask concentration was at, or near, the detection limit for iron (0.2 to 8 $\mu g/m^3$), and silicon (1 to 5 $\mu g/m^3$). A total of 30 samples were collected. The particle size for area samples was analyzed by optical microscopy, and mean particle sizes varied from 1.27 to 2.55 μm. The aerodynamic diameter for the dust samples should be larger than those reported. The sampling time was not reported. The GMWPF for iron dust was 273 to 1012, grouped as ambient filter weight expressed as multiples of filter blank. The GMWPF for silicon dust, calculated by the same method, varied from 220 to 1417. Since the authors did not report the individual ambient and in-mask concentrations for each sample, it is difficult to estimate the mean and fifth percentile WPFs for this study. Information presented in this study is summarized in Table 6.

Wallis et al.[30] of Dura-Cell Corp. conducted a WPF study at alkaline battery manufacturing facilities involving the handling of manganese oxide. The 3M 8710 dust/mist filtering facepiece was selected for testing. Test subjects were involved in ten job classifications. A total of 60 samples were collected in powder transfer, processing and packaging. The sampling time varied between 30 and 40 min. Samples were analyzed by atomic absorption spectroscopy, with a detection limit of 0.25 μg/filter. The ambient concentration varied from 0.14 to 77.4 mg/m^3. The in-mask concentration varied from 0.008 to 0.16 mg/m^3. The measured workplace protection factors varied from 2.8 to 847.9.

The WPFs values are plotted in three groups and shown in Figure 1. Group A consists of all WPF values which have ambient concentrations between 0.1 and 77.4

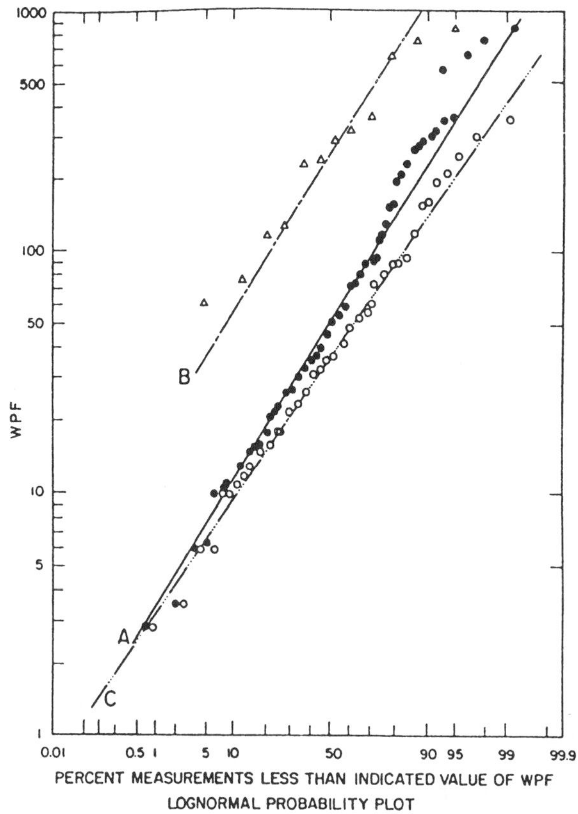

Figure 1 Lognormal probability plot of WPFs. WPFs represented by Line A were measured in the range $0.14 < C_o < 77.4$ mg/m^3. Line B represents WPFs in the range $5 < C_o < 77.4$ mg/m^3. Line C represents WPFs in the range $0.14 < C_o < 5$ mg/m^3. (From Reference 30.)

mg/m^3. The geometric mean (GM) WPF for this group is 50 and the fifth percentile WPF was 7.5. The ambient concentrations for the second group varied between 5 and 77.4 mg/m^3, for which respiratory protection is required. The GMWPF for this group is 250 with a fifth percentile WPF of 35. The third group consists of ambient concentration values from 0.14 to 5 mg/m^3. The GMWPF for this group is 37 with a fifth percentile WPF of 6.8. These plots show that WPF values are proportional to the ambient concentration. By examining the particle size distribution for samples collected at powder drop and bag slitting operations, it is evident that at least 60% of the particles are larger than 6.0 µm. It is obvious that the large particles do not penetrate through the facepiece, and the WPF becomes a function of ambient concentration. All relevant data are summarized in Table 6.

Colton et al.[31] of 3M Company conducted a study to evaluate the performance of a 3M W-3258 supplied air hood. The test respirator was equipped with a loose-fitting hood and a hard hat. The hood was operated at 75 psig with an air flow of 8–9 cfm, which is higher than the approved flow of 6 cfm. The study was performed at the foundry during the furnace tear down. The test subjects used pneumatic chippers to cut the furnace wall, allowing the pieces of the wall to fall into a small barrel to be removed. The operation took most of the 8-h shift. The workers worked in pairs and would work for about 1 h at a time before switching with other workers due to the physical and hot nature of work. Sampling was performed for 2 days. Four workers participated in this study. The samples were analyzed for silicon by PIXEA and x-ray emission. A total of 37 samples were collected. The OSH PEL for silicon is 15 mg/m^3.

The silicon content of ambient samples varied from 18.4 to 209 mg/m^3, with a geometric mean of 83.3 mg/m^3 and a geometric standard deviation of 1.76. The silicon content of in-mask samples varied from 0.001 to 0.089 mg/m^3, with a geometric mean of 0.013 mg/m^3 and a geometric standard deviation of 3.28. The authors used two methods to calculate the WPF. The first method used ICP to analyze the ambient samples and used the PIXEA method to analyze in-mask samples. The WPF obtained from 14 samples had a geometric mean of 7883, with a geometric standard deviation of 3.01 and a fifth percentile WPF of 1284. The second method used the regressional equation to predict the PIXEA equivalent of the ambient concentration. The WPF obtained from 15 samples had a geometric mean of 8718, with a geometric standard deviation of 3.02 and a fifth percentile of 1411.

Samples collected for particle size analysis showed that over 50% of the mass was greater than 10 μm. Since the majority of ambient dust are in the nonrespirable size, and the analytical method for silicon is very sensitive, it is expected that high WPFs would be achieved from this combination. Furthermore, the higher air flow rate of the hood also improves the performance since it is operated at the high end of air flow.

Colton et al.[32] also conducted a WPF study on the 3M 6000 series respirator equipped with a disk type HEPA filter. The study was performed at two locations. The first was a plastic colorant manufacturing operation using cadmium containing pigment. The second location was a ship wrecking facility where the worker used an oxygen-acetylene torch to cut metals containing lead. The sampling time for each operation was not specified.

Eighteen workers participated in the study at each site. Each worker had to pass a saccharin qualitative fit test. Samples were analyzed with inductively coupled plasma (ICP) spectroscopy. Fifty nine samples were collected at the pigment operation and 47 samples were collected at the ship wrecking operation. The ambient cadmium concentration varied between 3.45 and 423 μg/m^3 with a geometric mean of 25.7 μg/m^3 and a geometric standard deviation of 2.39. The in-mask cadmium concentration varied from no detection to 1.47 μg/m^3 with a geometric mean of 0.07 μg/m^3 and a geometric standard deviation of 1.54. The ambient lead concentration varied from 14 to 1,630 μg/m^3 with a geometric mean of 183 μg/m^3 and a geometric standard deviation of 3.44. The in-mask lead concentration varied from no detection to 39.3 μg/m^3 with a geometric mean of 1.3 μg/m^3 and a geometric standard deviation of 3.07.

The 59 cadmium samples resulted in a geometric mean of 558, a geometric standard deviation of 3.85, and a fifth percentile WPF of 61. The 47 lead samples resulted in a geometric mean of 205, a geometric standard deviation of 3.56, and a fifth percentile WPF of 25. Samples of cadmium collected for particle size analysis showed a bimodal distribution with geometric means of 1.33 and 9.77 μm at the blending operation, and geometric means of 1.33 and 9.77 μm at the mixing operation. The two samples collected for particle size distribution analysis for the ship wrecking operation showed that lead also had bimodal distributions with geometric means of 0.24, 11.1, 0.43, and 7.5 μm. The test results are summarized in Table 6.

The authors proposed a method for rejecting outlier ambient samples. The criterion is 100 times higher than the blank concentration. The detection limits for cadmium and lead are 0.01 and 0.1 μg, respectively. In order to declare that the respirator would achieve a PF of 10, the ambient concentration must be higher than 10 times the PEL of cadmium and lead which is 50 and 500 $\mu g/m^3$, respectively. Assume that the filter blank is at the detection limit, 100 times the blank concentration may still be less than the PEL concentration.

II. Factors Affecting WPF Determination

The results obtained from workplace protection factors (WPF) studies are often proposed as the OSHA assigned protection factor (APF) for this type of respirator. In order to apply the results of WPF study as the APF, the study must be representative of other workplaces. There are many factors which would affect the application of WPF data.

A. Particle Size

Quite often the total ambient particulate concentration is used to calculate the WPF values. This tends to exaggerate the WPF values, especially when the ambient aerosol contains a significant fraction of large particles. Another consideration is that gas or vapor molecules are always respirable; therefore, WPF values based on ambient aerosols may not be applicable to gases or vapors. Since the respirator facepiece is a size selective device, nonrespirable particles should not be present inside the facepiece unless the particles penetrate through the gaps in the facepiece seal. Recent studies on facepiece leakage conducted by Willeke and Hinds indicated that the upper limit for the most penetrating size is 1 μm.[33-36] If large particles exist in the workplace, then the WPF becomes a function of ambient concentration.

B. Work Time

Except as specified in the OSHA lead standard, there is no requirement governing how long a worker is required to wear a respirator. It is quite common that respirators are worn for most of a shift. OSHA even issued variances allowing employers to require employees to wear respirators for the full shift during lead exposure. Since wearing a respirator may cause physiological as well as psychological stress, performance of the respirator degrades with an increase in wearing time. A short sampling time of less than 4 h is not representative of many work situations.

C. Environmental Conditions

It is understandable that adverse environmental conditions such as high temperature and humidity impose additional stress to the respirator wearer. This has been demonstrated from a simulated workplace study conducted by the Los Alamos National Laboratory.[37] The test subjects performed tasks inside the environmental chamber at moderately heavy to heavy work rates under high temperature and humidity. The test results indicated that for a negative pressure half-mask air-purifying respirator, there was a 20 times reduction in the protection factor when the test subjects performed exercises inside the environmental chamber. It was observed that facepieces slipped due to moisture accumulated on the face. This may not have occurred if the respirator was tested under normal environmental conditions. Since summer is often hot and humid in most parts of the country, it is necessary to perform the WPF study under high temperature and humidity in order for it to be representative of many workplaces.

D. Work Rate

The Lawrence Livermore National Laboratory conducted an environmental chamber study[38] to measure the performance of positive pressure respirators, such as the powered air-purifying respirators (PAPR). The test subject walked on a treadmill at a work rate which was 80% of the maximum cardiac capacity of that individual. The results indicated that at the certified air flow of 115 l/m, a tight-fitting PAPR does not maintain positive pressure inside the facepiece until the air flow has been increased to 170 l/m.

Another Lawrence Livermore treadmill study[39] on the performance of the escape self-contained breathing apparatus (ESCBA) conducted by the Lawrence Livermore National Laboratory indicated that at the rated air flow of 40 l/m, the concentration of oxygen and carbon dioxide inside the hood of the ESCBA reached a dangerous level a few minutes after donning. This situation could be corrected by increasing the air flow to at least 65 l/m.

These examples illustrate that work rate is critical to the performance of a respirator, especially for the commonly used negative-pressure air-purifying respirators. It is important that the task performed by the test subjects in the WPF study be high enough to impose considerable physiological stress on the respirator wearer.

E. Analytical Method

The analytical method should yield reproducible results. For example, asbestos counting by optical microscopy is subject to many errors such as the variation between counters, the field to be counted, and the distribution of fibers on the filter. For these reasons, asbestos counting should not be used in a WPF study. Furthermore, the method being used in a WPF study should be validated. For a validated method, variables such as accuracy, precision, and recovery rate have been validated through round robin tests. All errors associated with this method would be identified and their magnitudes determined.

F. Sensitivity of Analytical Methods

The sensitivity or detection limit of the analytical method should be appropriate for the protection factor being determined. Since carbon monoxide (CO) is a major source of contaminant during a fire, it may be used to measure the performance of the self-contained breathing apparatus (SCBA) in the fire environment. Infrared spectrophotometry (IR) can detect CO at a concentration of 5 ppm. It is generally recognized that positive-pressure SCBAs would provide a PF of 20,000 or higher, even at a high work rate. If the IR method is used in the WPF study to measure CO, no WPFs higher than 1000 would be measured when the CO concentration in the ambient is less than 5000 ppm.

On the other hand, a low detection limit tends to boost the WPFs of a respirator. This is especially true for respirators with low assigned protection factors. A photometer used in quantitative fit testing is considered a very sensitive instrument. For a challenge concentration of 20 mg/m^3, a fit factor of 20,000 is the limit of detection that corresponds to a concentration of 1 µg/m^3. A fit factor of 3000 was reported for this respirator. However, WPFs as high as 15,000 have been reported for a negative pressure half-mask respirator when the sensitivity of the analytical method of the WPF study was in the nanogram range.

Recently, more sensitive analytical instruments have been used in the WPF measurement. Several nonvalidated methods have sensitivity in the nanogram to the picogram range. It may be justifiable to use a very sensitive method in a WPF study for positive pressure respirators with APFs of 1000 or more. However, very sensitive methods have been used in WPF studies for half-mask negative pressure air-purifying respirators and disposable respirators. If the WPF study is performed at very low ambient concentrations with the presence of large particles, very few particles penetrate through the respirator facepiece. The WPF then becomes a function of sensitivity of the analytical method, in which the lower the detection limit of the analytical method, the higher the WPF that would be achieved.

III. APPLICATION OF WPF STUDY DATA

At a NIOSH sponsored prerulemaking technical conference on the Assessment of Performance Levels for Industrial Respirators, Howie[40] presented a discussion of WPF data interpretation. The WPF data are often interpreted by deriving the "95th percentile" protection factor from the data, including any replicates being added together and analyzed under the assumption of log-normality. This is a fundamentally wrong assumption since 3 tests from each of 10 wearers is not 30 points from a single distribution, but 10 points from each of 10 potentially independent distributions. At best, the WPF obtained predicts 95% donning rather than, generally assumed, 95% of the population.

Both inter- and intrasubject variation should be considered in setting the WPF. The WPF assigned should be relevant to the risk. Greater emphasis should be placed on acute, rather than chronic, situations. For chronic situations, it may be adequate to set WPF at the minimum protection given to 95% of the population. For acute situations it might be necessary to set WPF at the minimum PF, i.e., 95% of the population for 95% of donning by each member of the population, perhaps with a minimum limit set for those individuals and donning out with 95th percentiles. For acute situations, sufficient replicates will be required to ensure that adequate information is obtained about the distribution of results.

The following example was given to show how WPF should be assigned. The results of replicate WPF measurements from five workers are shown in the following table.

Wearer	GM	GSD	95% Percentile
1	165	5.5	9.85
2	165	5.2	10.8
3	200	2.4	45.5
4	641	3.1	97
5	671	1.1	573

If all data are pooled together, the geometric mean (GM) WPF is 298 and the 95% percentile WPF is 88. For acute exposure, the WPF based on 95% of the population for 95% of donning by each member of the population should be 3 — the value which would be achieved by all wearers.

At the NIOSH public meeting, Howie also discussed the selection of the right class of respirator to achieve the required level of protection.[40] The current workplace protection factor (WPF) is derived from either the 95th percentile or worst result from field or laboratory measurements of respirators. Under certain circumstances, an arbitrary "safety factor" is applied. For example, the fit test results for a negative pressure full facepiece respirator obtained from the laboratory is 1000, whereas current German practice is to limit such a device to use in concentrations up to 200 times the Maximum Allowable Concentration. In the U.K., no such safety factor is applied, therefore the full facepiece could be used in concentrations up to 1000 times the relevant British Occupational Exposure Level.

The technique may have the benefit of simplicity, but may result in an unduly "high" or "low" protection factor being assigned. The wearer may be required to use unnecessarily uncomfortable and expensive equipment for the conservatively assigned "low" PFs. On the other hand, workers may not be adequately protected if a "high" PF is assigned, i.e., higher than the equipment merits.

Howie proposed that the following factors should be taken into consideration when selecting RPE:

- The health consequences of inadequate protection
- Any warning characteristics of the airborne contaminant
- Duration of required wear
- Availability of in-facepiece monitoring data for equipment and/or wearers
- Quality of training and supervision

The consequence of equipment failure would be more severe for substances with lower PELs. Lower APFs should also be set for substances with "poor warning" properties. The duration of wear should also be a factor since the quality of fit would be reduced for long periods of wearing a tight-fitting facepiece. The quality of training and supervision is another factor of consideration. A well-trained and well-supervised wearer is likely to achieve a high level of protection.

Using the above criteria, the overall APF should be the product of three individual APFs based on the toxicity and warning properties: duration of wear, exposure concentration (IDLH or not), and level of supervision and training, or:

$$APF = APF_1 \times APF_2 \times APF_3 \tag{3}$$

where APF_1 addresses toxicity, concentration, and warning properties, APF_2 addresses the duration of wear, and APF_3 addresses training and supervision.

A term standard protection factors (SPF) is proposed. The SPF value is used for respirator selection. The SPF values under different exposure conditions are shown on the following tables.

APF-1: Toxicity, Concentration, and Warning Properties

Subject detection level	Low toxicity		Medium toxicity		High toxicity	
	Concentration		Concentration		Concentration	
	<IDLH	>IDLH	<IDLH	>IDLH	<IDLH	>IDLH
<<TLV	SPF	SPF/2	SPF/2	SPF/4	SPF/4	SPF/8
≈TLV	SPF/2	SPF/4	SPF/4	SPF/8	SPF/8	SPF/16
>TLV	SPF/2	SPF/4	SPF/8	SPF/16	SPF/16	SPF/32

APF-2: Duration of Wear,
APF-3: Training and Supervision

Duration of wear			Training and supervision	
< 30 min	30–60 min	> 60 min	Good	Poor
×4	×1	×1/2	×2	×1

Howie gave an example of how PFs should be assigned for three toxic substances with different toxicity. These are: NaOH, xylene, and crocidolite. The ambient concentration is 50 times the PEL for each substance. The respirator wearing time is 2 h. The following table summarizes the information on these three substances.

Toxic substance	NaOH	Xylene	Crocidolite
Ambient Concn.	100 mg/m³	5000 ppm	10 fibers/ml
Toxicity	Low	Medium	High
TLV	2 mg/m³	100 ppm	0.2 fibers/ml
IDLH Concn.	200 mg/m³	1000 ppm	Carcinogen
50 × TLV	< IDLH	> IDLH	> IDLH
Detection level	at TLV	at TLV	
APF_1	SPF/2	SPF/8	SPF/32

To calculate the required protection factor, Howie used a 2-h exposure time for the three substances mentioned above, and training on the use of the respirator is poor. Then the APF for NaOH is:

$$\begin{aligned} APF &= APF_1 \times APF_2 \times APF_3 \\ &= SPF/2 \times 1/2 \times 1 = SPF/4 \end{aligned} \tag{4}$$

The APF for xylene and crocidolite is SPF/16 and SPF/64, respectively.

At a concentration of 50 times the TLV, the required SPFs for the three substances are:

For NaOH:	50 = SPF/4,	or SPF = 200
For Xylene:	50 = SPF/16,	or SPF = 800
For Crocidolite:	50 = SPF/64,	or SPF = 3200

Howie indicated that the most protective respirator, such as a positive pressure device, would be required for protection against highly toxic substances.

IV. INTERPRETATION OF WPF DATA

The WPF is usually expressed as the ratio of the ambient and the in-mask concentration of the challenge:

$$WPF = C_o/C_i \tag{5}$$

The Occupational Safety and Health Administration (OSHA) health standards prescribe the use limit for each class of respirators. It is expressed as the maximum use limit (MUL), which is the product of protection factor (PF) and the permissible exposure limit (PEL):

$$MUL = PF \times PEL \tag{6}$$

For example, if the PF for the half-mask negative pressure respirator is 10 and the PEL is 10 parts per million (ppm), then the MUL for this air contaminant is 100 ppm.

A WPF value of 10 means that the respirator would provide a ten-fold reduction in ambient concentration under the prescribed test conditions. A PF of 10 means that the respirator wearer would be exposed to the air contaminant at the PEL level for an ambient concentration up to 10 times the PEL. By definition, the PF and the WPF are not the same. The WPF is equal to PF only when the WPF study is conducted at 10 times the PEL concentration of the challenge.

If a WPF study was conducted in a workplace with a contaminant concentration of only three times the PEL and a WPF of 10 was obtained for the respirator, by the OSHA definition of MUL, this respirator should be certified as having a PF of 3 instead of 10. There is no study to indicate that WPF values are independent of ambient concentrations.

Many investigators ignored the difference between MUL and WPF and made the misleading statement that the protection provided by a respirator is expressed as the OSHA MUL, even when the study was performed in the workplace with very low ambient concentrations. In most cases a respirator was not even needed.

The sensitivity of an analytical method has a dramatic impact on the WPF study. The sensitivity for the gravimetric, atomic absorption (AA), graphic furnace (GF), and the PIXEA method is 10, 3, 0.3, and 0.003 μg, respectively. For an ambient concentration of 300 μg/m^3, the maximum attainable WPF, when the in-mask concentration is at or less than the detection limit, under the different analytical methods is shown in the following table.

Analytical method	C_o µg/m³	C_i µg/m³	WPF
Gravimetric	300	10	30
AA	300	3	100
GF	300	0.3	1000
PIXEA	300	0.003	100000
(QNFT)	20000	1	20000

The table also lists the detection limit for the commonly used oil mist quantitative fit testing instrument, which has a detection limit of 20,000 µg/m³. Using the data obtained from a lead pigment WPF study[18] as an example, the WPF calculated by different analytical methods is shown in Table 8.

Table 8 Comparison of WPF Obtained by Different Analytical Method

	C_i			WPF		
C_o	PIXEA	Gr Fur	AA	PIXEA	Gr Fun	AA
261	0.216	0.3	3	1210	870	87
231	<0.014	0.3	3	16500	770	77
306	0.017	0.3	3	18000	1020	102
298	0.111	0.3	3	2680	993	99
798	0.596	0.6	3	1340	1339	266
1079	0.144	0.3	3	7490	1810	360
192	0.22	0.3	3	872	640	64
1410	0.20	0.3	3	7040	4700	470
221	0.049	0.3	3	4520	737	74
160	<0.024	0.3	3	6660	533	53
145	0.012	0.3	3	12100	483	48
154	0.091	0.3	3	1690	513	51
149	0.021	0.3	3	7110	497	50
366	<0.012	0.3	3	30500	1220	122
97	0.038	0.3	3	2540	323	32
97	<0.010	0.3	3	9650	323	32
342	0.056	0.3	3	6110	1140	114
1088	1.10	1.1	3	990	989	363
381	1.12	1.1	3	339	340	127
1088	0.43	0.43	3	2520	2519	363
120	0.06	0.3	3	1930	400	40
45.9	0.01	0.3	3	4590	153	15
39	<0.015	0.3	3	2600	130	13
69	0.032	0.3	3	2140	230	23
389	0.091	0.3	3	4270	1297	130
134	1.063	1.06	3	126	126	45
326	0.081	0.3	3	4020	1087	109
397	<0.011	0.3	3	36100	1323	132
40	<0.022	0.3	3	1800	133	13
389	<0.011	0.3	3	35400	1297	130
160	0.057	0.3	3	2810	533	53
400	<0.018	0.3	3	22200	1333	133
735	<0.044	0.3	3	16700	2450	245
458	<0.016	0.3	3	28700	1527	153
474	<0.022	0.3	3	21500	1580	158
42	<0.039	0.3	3	1070	140	14
469	<0.035	0.3	3	13400	1563	156
GM:				4553	702	82
GSD:				3.79	2.49	2.61
95% Percentile:				512	157	17

The PIXEA method was used to analyze the lead from samples collected in this study. The results indicated that the geometric mean (GM) WPF is 4553 and the fifth percentile (95%) WPF is 512. If samples are analyzed by the graphic furnace method with an analytical sensitivity of 0.3 μg, the GMWPF and the 95 percentile WPF is reduced to 702 and 157, respectively. If a less sensitive atomic absorption method is used (analytical sensitivity of 3 μg), the GMWPF and the 95 percentile WPF is further reduced to 82 and 17, respectively. Table 8 shows that the majority of in-mask samples with WPF in excess of 10,000 has nondetectable concentrations of lead. It is interesting to note that the average fit factor obtained from a quantitative fit testing with a submicrometer aerosol is 3000. The GMWPF obtained from the PIXEA method; however, is 4500. The author stated that the test subjects performed heavy work in this study. The data indicated that less facepiece leakage was found in the WPF study when test subjects performed heavy work, rather than when the test subjects performed sedentary exercises. The explanation is that large particles that existed in the workplace may not penetrate the facepiece. Then the WPF becomes a function of the sensitivity of the analytical method and the ambient concentration.

A low detection limit tends to boost the WPFs of a respirator. One example is a WPF study on asbestos abatement.[21] Water spray was applied to the asbestos insulation before removal. Fibers saturated with water droplets had increased the size of asbestos from respirable to nonrespirable and very little of the wet fibers penetrated through the respirator facepieces. However, fiber counting of ambient samples was performed in the dry state which counted only the fibers with evaporated water vapors. The NIOSH method 7400 for asbestos measurement requires that counting be terminated at 100 fields or 100 fibers, whichever occurs first. Since there were very few fibers found in the in-mask samples, counting fields were increased to 500 in order to search for fibers. The increased counting fields artificially boosted the WPF values for many samples, especially for those reported WPFs in excess of 1000. For example, a WPF of 7860 was reported in one sample. However, only one fiber was found from the in-mask sample. If two more fibers had been counted, the WPF would have been reduced from 7860 to 2620.

To illustrate the effect of concentration and particle size on WPF, a study conducted by Myers for the National Paint and Coatings Association is selected as an example.[41] The performance of the same respirators was evaluated in a foundry and a spray painting operation. Cyclones were used to collect ambient respirable samples in the foundry. Total particulate sample was collected for the spray painting operation because the high concentration of the paint droplets would have clogged the cyclone. The results on WPF obtained are shown in the following table.

Respirator	Foundry	Spray painting
#1	119	2211
#2	94	5218
#3	108	6629

The WPF obtained from the spray painting operation appears to be much higher than the WPF obtained from the foundry. The reason for this discrepancy is that both the concentration and particle size of the air contaminant found in the spray painting operations were much greater than those in the foundry.

By comparing WPF studies conducted by NIOSH, respirator users, and respirator manufacturers, as shown in Table 6, it is found that WPF values vary widely for the same type of respirators in different studies. It is interesting to note that studies conducted by NIOSH staff or sponsored by NIOSH reported the lowest values, and studies conducted by the respirator manufacturers reported the highest WPF values. It is also found that a study which reported a high WPF resulted from using a very sensitive analytical method. WPF values are proportional to ambient concentration and particle size, but inversely proportional to work time, work rate, and the sensitivity of the analytical method. When higher ambient concentration is associated with the larger particle size, the more sensitive analytical method, the shorter work time, or the lighter work rate would yield higher workplace protection factors. The reproducibility is poor for the same WPF study when it is conducted on different days. It is apparent that artificially high WPFs could be obtained when the study is performed in a workplace with large aerosols and using a very sensitive analytical method is used. If the test conditions at the workplace are carefully selected, any WPF value can be achieved.

NIOSH sponsored a prerulemaking technical conference on the Assessment of Performance Levels for Industrial Respirators from January 9–11, 1991. Many papers were presented at the meeting. The consensus opinion was that the reproducibility of the WPF study is poor and it is not appropriate to use the WPF values for certifying respirators. The controlled environmental chamber study would be a choice for respirator certification.

Opinions expressed by the attendees can be summarized as the following:

- It is not possible to identify a representative workplace.
- There is no standard test protocol for performing WPF studies.
- Field testing would introduce uncontrolled test conditions and a large number of poorly defined variables.
- The reproducibility of test results are poor.
- There is insufficient information to implement a test procedure of this type into a formal certification program.
- Test protocols are different in most cases; environmental conditions are poorly defined; particle size information is not always measured, even through particle size and contaminant concentration seem to influence the measured WPF. This raises questions regarding the interpretation of any single study and the comparison of different studies.
- The validity of extrapolating the results of any study to general use situation is in question.
- The validity of extrapolating test results between different work situations is in question.

REFERENCES

1. Harris, H. E., De Sieghardt, W. C., Burgess, W. A., and Reist, P. C., Respirator usage and effectiveness in bituminous coal mining operations, *Am. Ind. Hyg. Assoc. J.*, 35, 159, 1974.
2. Revoir, W. H., Respirators for protection against cotton dust, *Am. Ind. Hyg. Assoc. J.*, 35, 503, 1974.

3. Moore, D. E. and Smith, T. J., Measurement of protection factors of chemical cartridge, half-mask respirators under working conditions in a copper smelter, *Am. Ind. Hyg. Assoc. J.*, 37, 453, 1976.
4. Smith, T. J., Ferrell, W. C., Varner, M. O., and Putnam, R. D., Inhalation of cadmium workers: Effects of respirator usage, *Am. Ind. Hyg. Assoc. J.*, 41, 624, 1980.
5. Que Hee, S. S. and Lawrence, P., Inhalation exposure of lead in brass foundry workers: The evaluation of the effectiveness of a powered air-purifying respirator and engineering controls, *Am. Ind. Hyg. Assoc. J.*, 44, 746, 1983.
6. Toney, C. R. and Barnhart, W. L., Performance evaluation of respiratory protective equipment used in paint spray operations, HEW Publication No. (NIOSH) 76-177, 1976.
7. Blair, A., Abrasive blasting respiratory protective practices, HEW Publication (NIOSH) 74-104, 1974.
8. Myers, W. R. and Peach, M. J., III, Performance measurements on a powered air-purifying respirator made during actual field use in a silica bagging operation, *Ann. Occup. Hyg.*, 27, 251, 1983.
9. Lenhart, S. W. and Campbell, D. L., Assigned protection factors for two respirator types based upon workplace performance testing, *Ann. Occup. Hyg.*, 28, 173, 1984.
10. Myers, W. M., Peach, M. J., III, Cutright, K., and Iskander, W., Workplace protection factor measurements on powered air-purifying respirators at a secondary lead smelter — results and discussion, *Am. Ind. Hyg. Assoc. J.*, 45, 681, 1984.
11. Myers, W. R., Peach, M. J., III, Cutright, X., and Iskander, W., Field test of powered air-purifying respirators at a battery manufacturing facility, *J. Int. Soc. Respir. Prot.*, 4, 628, 1984.
12. Graboury, A. and Burd, D. H., Workplace protection factor evaluation of respiratory protective equipment in a primary aluminum smelter, *Appl. Occup. Environ. Hyg.*, 8, 19, 1993.
13. Keys, D. R., Guy, H. P., and Axon, M., Workplace protection factors of powered air-purifying respirators. Presented at the American Industrial Hygiene Conference, May, Orlando, FL, 1990.
14. Stokes, D. W., Johnston, A. R., and Mullins, H. E., Respirator workplace protection factor studies — powered air loose fitting helmet. Presented at the American Industrial Hygiene Conference, Montreal, Canada, June, 1987.
15. Colton, C. E. and Mullins, H. E., Respirator protection factor studies — Whitecap powered air purifying respirator. Presented at the American Industrial Hygiene Conference, Orlando, FL, May, 1990.
16. Dixon, S. W., Nelson, T. J., and Wright, J., Program protection factor study on the 3M W-316 Airhat. Presented at the American Industrial Hygiene Conference, Detroit, MI, May 22, 1984.
17. Johnston, A. R., Stokes, D. W., Mullins, H. E., and Rhoe, C. R., Workplace protection factor study on a supplied air abrasive blasting respirator. Presented at the American Industrial Hygiene Conference, Montreal, Canada, June, 1987.
18. Dixon, S. W. and Nelson, T. J., Workplace protection factors for negative pressure half-mask facepiece respirators, *J. Int. Soc. Respir. Prot.*, 2, 347, 1984.
19. Colton, C. E., Johnston, A. R., Mullins, H. E., and Rhoe, C. R., Respirator workplace protection factor studies — Full facepiece respirator. Presented at the American Industrial Hygiene Conference, Orlando, FL, May 1990.
20. Gosselink, D. W., Wilmes, D. P., and Mullins, H. E., Workplace protection factor study for airborne asbestos. Presented at the American Industrial Hygiene Conference, Dallas, TX, May 1986.

21. Nelson, T. J. and Dixon, S. W., Respirator protection factors for asbestos. Parts I and II. Presented at the American Industrial Hygiene Conference, Los Vegas, NV, May 1985.
22. Johnston, A. R. and Mullins, H. E., Workplace protection factor study for airborne metal dusts. Presented at the American Industrial Hygiene Conference, Montreal, Canada, June 4, 1987.
23. Colton, C. E., Johnston, A. R., Mullins, H. E., Rhoe, C. R., and Myers, W. R., Workplace protection factor study on a half-mask dust/mist respirator. Presented at the American Industrial Hygiene Conference, Orlando, FL, May 17, 1990.
24. Colton, C. E. and Mullins, H. E., Workplace protection factor tests — Brass foundry. Presented at the American Industrial Hygiene Conference, Orlando, FL, May 1990.
25. Chemical Manufacturers' Association, CMA/Cadmium pigments study. Unpublished. Undated. Washington D.C.
26. Reed, L. D., Lenhart, S. W., Stephenson, R. L., and Allender, J. R., Workplace evaluation of a disposable respirator in a dusty environment, *Appl. Ind. Hyg.*, 2, 53, 1987.
27. Cohen, H. J., Determining and validating the adequacy of air-purifying respirators used in industry. Part I. Evaluating the performance of a disposable respirator for protection against mercury vapor, *J. Int. Soc. Respir. Prot.*, 2, 296, 1984.
28. Galvin, K., Selvin, S., and Spear, R. C., Variability in protection afforded by half-mask respirators against styrene exposure in the field, *Am. Ind. Hyg. Assoc. J.*, 51, 625, 1990.
29. Johnston, A. R., Colton, C. E., Stokes, D. W., Mullins, H. E., and Rhoe, C. R., Workplace protection factor study on a supplied air respirator. Presented at the American Industrial Hygiene Conference, St. Louis, Mo, May 1989.
30. Wallis, G., Menke, R., and Chelton, C., Workplace field testing of a disposable negative pressure half-mask dust respirator, *Am. Ind. Hyg. Assoc. J.*, 54, 576, 1993.
31. Colton, C. E., Mullins, H. E., and Bidwell, J. O., Workplace protection factor study on air-line respirator with a loose fitting hood during furnace teardown. 3M Company. Presented at the American Industrial Hygiene Conference and Exposition, New Orleans, May 1993.
32. Colton, C. E., Mullins, H. E., and Bidwell, J. M., Workplace protection factors of a half-facepiece high-efficiency filter respirator in different environments. 3M Company. Presented at the 1994, *Am. Ind. Hyg. Conference and Exposition*, Anaheim, CA, 1994.
33. Holton, P. M., Tackett, D. Y., and Willeke, K., Particle size-dependent leakage and losses of aerosol in respirators, *Am. Ind. Hyg. Assoc. J.*, 48, 848, 1987.
34. Holton, P. M., Tackett, D. Y., and Willeke, K., The effect of aerosol size distribution and measurement method on respirator fit, *Am. Ind. Hyg. Assoc. J.*, 48, 838, 1987.
35. Hinds, W. C. and Kraske, G., Performance of dust respirators with facial seal leaks. I. Experimental, *Am. Ind. Hyg. Assoc. J.*, 48, 836, 1987.
36. Hinds, W. C. and Bellin, P., Performance of dust respirators with facial seal leaks. II. Predictive model, *Am. Ind. Hyg. Assoc. J.*, 48, 842, 1987.
37. Skaggs, B. J., Loibl, J. M., Carter, K. D., and Hyatt, E. C., Effect of temperature and humidity on respirator fit under simulated work conditions. Los National Laboratory, NUREG/CR-5090, LA-11236, 1988.
38. da Roza, R. A., Cadena-Fix, C. A., and Kramer, J. E., Powered air-purifying respirator study final report. Lawrence Livermore National Laboratory, UCRL-53757, 1986.
39. Johnson, J. S., da Roza, R. A., Foote, K. L., and Held, K., An evaluation of emergency escape respirators for use in a space launch environment. Lawrence Livermore National Laboratory, Livermore, CA, 1991.

40. Howie, R. M., Interpretation of workplace protection factor data. Presented at the NIOSH Assessment of Performance Levels For Industrial Respirators: Prerulemaking Technical Conference. January 9-11, 1991.
41. Myers, W. R., Determining workplace protection factors for negative-pressure elastomeric and disposable half-mask air-purifying respirators. Presented at the NIOSH Assessment of Performance Levels for Industrial Respirators: Prerulemaking Technical Conference. January 9-11, 1991.

22 RESPIRATOR PERFORMANCE STUDIES: OTHER RELATED ISSUES

I. FACEPIECE LEAKAGE

It is commonly acknowledged that facepiece leakage is dependent on the particle size. However, the relationship between particle size and facepiece leakage was not reported until the last few years. Holton, Willeke and coworkers[1] conducted a study to measure particle size-dependent leakage into and losses inside a respirator, the mechanism occurring at the leak site and the flow dynamics inside the respirator. Particles with sizes between 0.07 and 4.4 μm were selected for testing. The fine test aerosol was generated from a mixture of smoke from burning incense and a nebulized corn oil. The two larger aerosols were generated from limestone dust and corn oil. Three types of aerosol measurement systems were used in this study: an electrostatic aerosol classifier (EAC) and a condensation nuclei counter measured particles 0.1 μm or less; an active-scattering aerosol spectrometer (ASAS) measured particles between 0.1 and 3.0 μm; and an aerodynamic particle size (APS) measured aerodynamic diameters between 0.5 and 16 μm.

The parameters measured in this study were:

Hole shapes: slit and circular with a depth of 4 mm.
Probe locations: at the center line between the mouth and the nose, and along the side of the facepiece near the leak site.
Hole locations: the first one located at the side of nose and the second one located at the chin between the inhalation valve and the exhalation valve.
Hole sizes: 0.57, 1.07, and 1.68 mm in diameter and 4 mm deep.
Slit size: 11 mm long, 0.076 mm wide, and 4 mm deep.
Hole length: 4 and 10 mm.

The midpoint diameters for the size ranges for the three test aerosols were:

Fine aerosol:	0.07, 0.10, and 0.22 μm.
Corn oil:	0.16, 0.25, 0.55, and 1.09 μm.
Limestone:	0.72, 1.11, 2.3, and 4.4 μm.

A negative pressure air-purifying respirator equipped with organic vapor/HEPA filter cartridges was selected for testing.

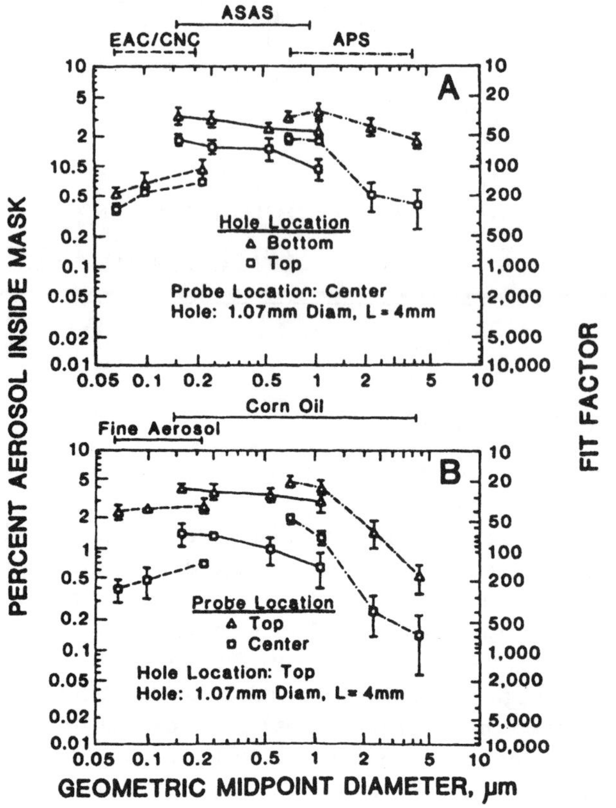

Figure 1 Percent aerosol measured inside negative-pressure half-mask respirator during nose breathing (A) Two hole locations; (B) two probe locations. (From Reference 1.)

During the test, the test subjects breathed through the nose and sat quietly inside the test chamber without any movements. To ensure that all leakages would occur only at the leak site, the test subjects did not perform any head movements and the facepiece leakage was minimized by applying petroleum jelly around the faceseal. The test results are shown in Figures 1 through 4. In the hole location test, the leak measured with the bottom hole open ranged from 1.2 to 4.9 times larger than the leak measured with the top hole open (Figure 1A). For 0.5 µm particles, this translates to fit factors of 64 for the top hole open and 41 for the bottom hole open. From the same figure, the ratio for the bottom hole to the top hole of the percent aerosol inside the mask is between 0.7 and 4.4 µm and increases with particle size. The air flow through the filter cartridges that carried a greater number of all particle sizes into the sampling zone, when the bottom hole was open, also may be carrying more particles with sizes greater than 1.0 µm.

The author postulated that for particles greater than 1 µm entering through the top holes, more were lost inside the mask from settling and inertial impaction. Therefore, a relatively large measured leakage at the midpoint diameters greater than 1.0 µm for the bottom hole compared to the top hole was seen. Leak site at the chin area of a half-mask seems to pose a more serious hazard than the leak sites at the nose position because of the ability of the clean air entering through the filter cartridges to carry the leaked aerosol into the breathing zone.

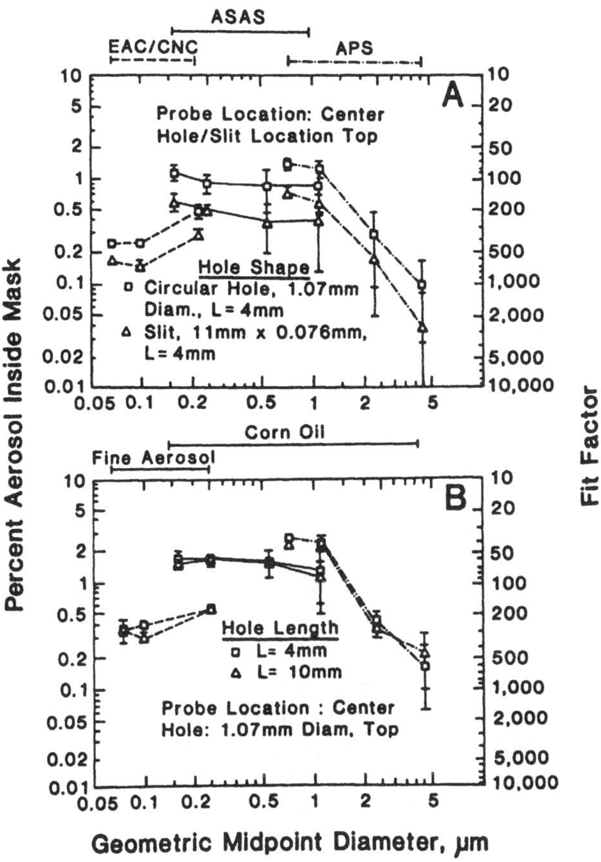

Figure 2 Percent aerosol measured inside a negative pressure half-mask respirator during nose breathing. (A) Two hole shapes; (B) two hole lengths. (From Reference 1.)

In the probe location test the top probe measured aerosol concentrations inside the mask that were from 2.3 to 6 times as high as the aerosol concentrations measured with the center probe (Figure 1B). At 0.5 μm this translates to a fit factor of 100 when measuring with the center probe and 30 when measuring with the top probe. The center probe measured relatively fewer particles between 0.07 and 0.1 μm, and between 2.3 to 4.4 μm, compared to the top probe. Part of the reason may be that the center probe, which is nearer to the nose, samples for exhaled air. Therefore, because of increased lung losses, fewer small particles and fewer large particles are being counted. The center probe, further from the leak site, may be measuring decreasing numbers of particles smaller than 0.1 μm because of some diffusional losses within the mask cavity, and measuring decreasing numbers of particles in the 2.3 to 4.4 μm size range because of settling and inertial losses within the mask cavity.

There may be some settling losses of particles larger than 1 μm and diffusional losses of particles smaller than 0.2 μm within the mask. The locations of the leak and the sampling probe may have a significant effect on the measured leakage. The further the sampling probe is from the leak site, and the closer it is to the clean air supply, the smaller the overall leakage that is measured.

In the hole-shape test, the percent aerosol inside the mask for all particle sizes was smaller for the slit as compared to the circular hole (Figure 2A). The percent aerosol inside the mask when the circular hole was open was from 1.4 to 2.5 times higher that when the slit was open. Although the areas of the two leak sites were within 10%, there probably was an increased air flow resistance for the slit, as compared to the circular hole, that would cause both the flow into the mask and the particle count in the mask to decrease. A slit or narrow gap in the faceseal, compared to a circular hole, not only decreases the total aerosol leakage but also diminishes the entry of the larger particles.

Hole length made little difference in the percent aerosol measured inside the facepiece (Figure 2B). This would indicate that there was no essential difference in the losses due to settling or diffusion within the leak sites because of length, but that entry into the hole was the primary cause for particle losses.

In the hole size tests, the measured aerosol inside the mask increases as the hole size increases (Figure 3). As the particle size increases from 1 to 4.4 μm for all three hole sizes, the percent aerosol inside the mask decreases. Likewise, as the particle size decreases from 0.22 to 0.07 μm, the percent aerosol measured inside the mask once again decreases.

For the fine aerosol test, for all three hole sizes, particle size had a significant effect on amount of leakage into the mask. Similarly, for the corn oil and limestone aerosol tests using the APS as the measurement device, particle size had a significant effect on the aerosol concentration inside the mask. For the corn oil measured by the ASAS, only for the large hole size did particle sizes between 0.16 and 1.1 μm have a significant effect on the aerosol concentration inside the mask. For the limestone aerosol measured by the ASAS for particles between 0.33 and 1.9 μm; however, particle size had a significant effect on particle concentrations inside the mask for all three hole sizes.

In general, for all three hole sizes, there is little difference in the percent aerosol inside the mask for particles between 0.2 and 1.0 μm in size. For particle sizes smaller than 0.2 μm and larger than 1.0 μm, however, size-dependent leakage into the mask does appear. Between 0.2 and 1.0 μm, the measured percent aerosol inside the mask shows little size dependence. However, because of inertial entry losses of particles greater than 1.0 μm and diffusional losses of particles smaller than 0.2 μm, size dependent particle leakage does occur. As the hole size decreases, the total leakage decreases, but a greater percentage of larger particles enters through the leak.

Figure 4 shows smooth curves with corrections for coagulation and humidity effects. The fit factors for the small, medium, and large hole sizes were 500, 63, and 21 at 0.5 μm and show the increasing leakages as the hole size increases. The average flow rates through the three hole sizes decreases as the hole size decreased. The small hole size would allow the penetration of more particles larger than 0.9 μm than the medium and large hole sizes because of inertial losses that would occur at the entry of the medium and large hole sizes. The curves shown in Figure 4 can be used to calculate the leakage into a respirator for different aerosol size distributions, and for different detection methods in a quantitative test. The mass leakage into a mask of an exposure aerosol with any size distributions can be calculated.

In the second phase of this study,[2] the test was also performed on human test subjects. To ensure that all aerosol leakage into the respirator was through the leak

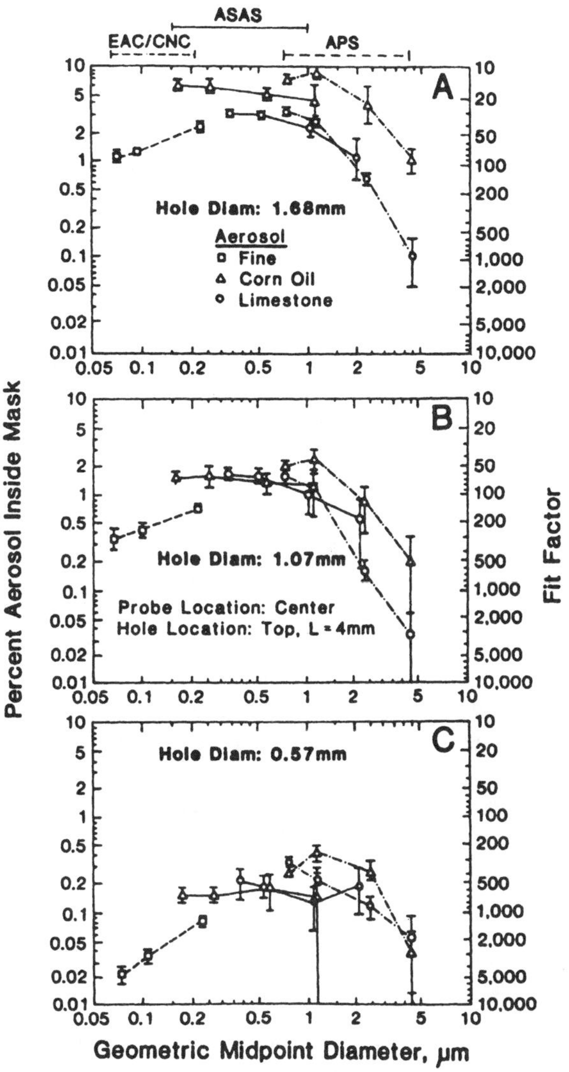

Figure 3 Percent aerosol measured inside a negative-pressure half-mask respirator during nose breathing for three hole sizes. (From Reference 1.)

sites, the test subject breathed through the nose and sat quietly during the testing without performing any exercise. Petroleum jelly was applied around the facepiece seal to further minimize facepiece leakage.

The dashed line curves in Figure 5 show the average percent aerosol measured inside the mask for particle sizes between 0.07 and 4.4 μm. Curves corrected for lung deposition are also shown (solid line curves). When the leak holes on the facepieces were closed, the leakage of aerosol into the respirator was very low. The maximum measured fit factors are 85,000 for the APS at 1.1 μm, 67,000 for the ASAS at 0.16 μm, and 87,000 for the EAC/CNC at 0.07 μm. Figure 6 gives the percent leakage curves from Figure 5. Measurement of the flow rates through the three hole sizes

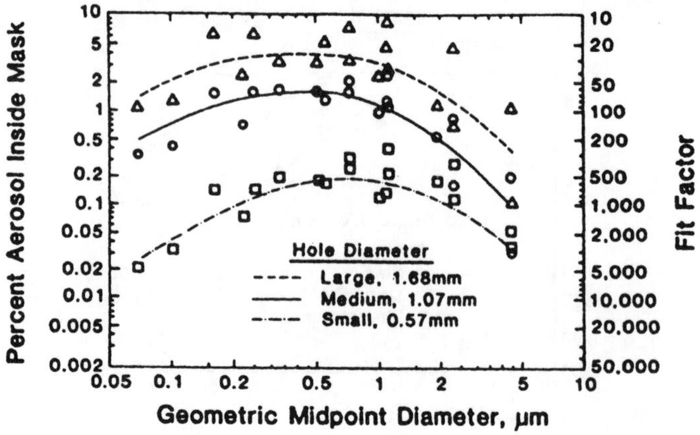

Figure 4 Smooth curves from data showing percent aerosol inside mask for the three hole sizes. (From Reference 1.)

revealed that the air velocities through the medium and large holes were larger than through the small holes. This indicates that higher inertial entry losses may remove more particles larger than approximately 0.9 μm, for the large and medium size holes. These curves show that using a monodisperse fit test aerosol of 0.3 μm or 3.0 μm particles to measure the leakage through a medium size hole would yield a 5:1 ratio in the measured percent leakages. For the large and small hole sizes, the percent leakage ratios would have 3.7:1 and 1.6:1, respectively.

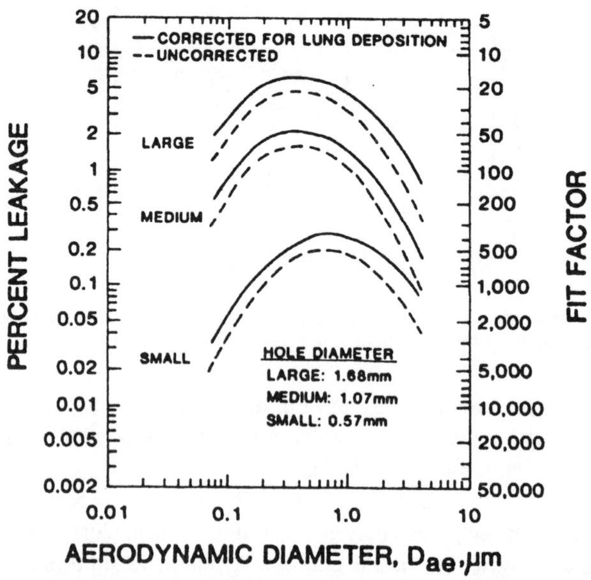

Figure 5 Percent aerosol inside a negative-pressure half-mask respirator for three hole sizes in the facepiece (dashed curves). The same data, corrected for lung deposition, are the best estimate of the actual leakage through the three hole sizes and are designated by the solid curves. (From Reference 2.)

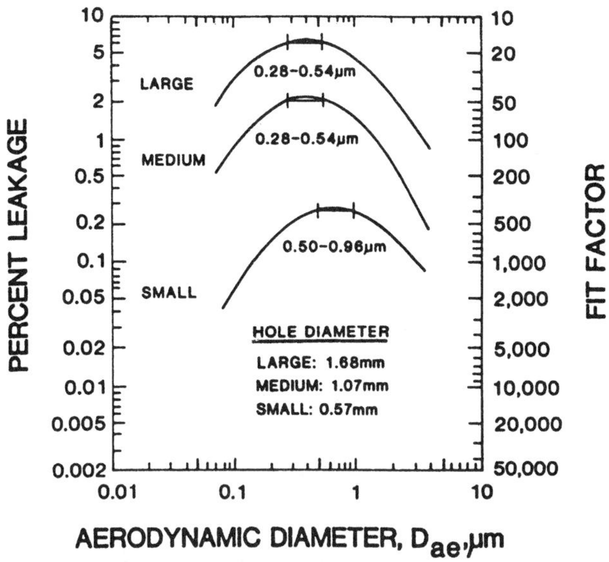

Figure 6 Size ranges of peak leakage for the three hole sizes. (From Reference 2.)

In order to demonstrate the effects of the size distribution and the measurement method on the measured leakage, three aerosol-size distributions with count median diameters of 0.28, 0.6, and 1.1 µm were chosen. Each of the three aerosol size distributions has a geometric standard deviation of 2.0. The middle 97% of the particle count distribution would lie between the boundaries of the particle sizes that were tested, 0.07 to 4.4 µm. In addition, within that middle 97% of each count distribution the number concentration was calculated to give a mass concentration equal to 10 mg/m^3.

The application of the percent leakage results in Figure 6 to the distributions in Figure 7 results in the leakage curves of Figure 8. The curves are constructed from the three commonly used particle measurement methods for detecting facepiece leakage: count, scatter, and mass. In order to obtain the maximum leakage that occurs inside the mask with a polydisperse aerosol regardless of hole size, the fit test aerosol should have a count mean diameter between 0.15 to 0.30 µm and use a light scattering or mass method should be used for detection.

Table 1 lists the fit factors that would be obtained from the different quantitative respirator fit tests by using the data taken from curves in Figures 6 for monodisperse aerosols and from curves in Figure 8 for polydisperse aerosols (Figures 6 and 8). The difference in measured fit factor for different test aerosols and measurement methods can be considerable. The QNFT using the 2.5 µm monodisperse corn oil aerosol and the light-scattering photometer detector gives the highest fit factor and is the least sensitive fit test method of the tests shown, but would be a representative test when the exposure aerosol is composed of large dust particles at sizes approaching 2.5 µm. If the respirator wearer, however, is exposed to arc welding fumes that have a reported mean size of 0.15 µm with a GSD of 1.7, the NaCl test that uses an aerosol with smaller particle would reflect more accurately the higher leakage or smaller fit factor.

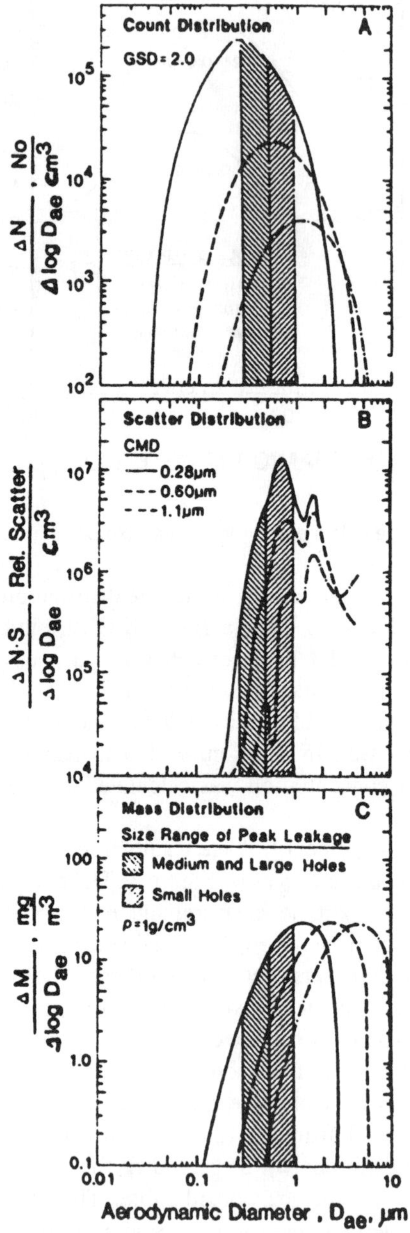

Figure 7 Count, scatter, and mass distributions of three aerosol size distributions with the size ranges of peak leakage accentuated for the three hole sizes. Each count distribution has been calculated to give a mass concentration of 10 mg/m³ within the center 97% of the count distribution. The scattering intensity is designated by the letter "S." (From Reference 2.)

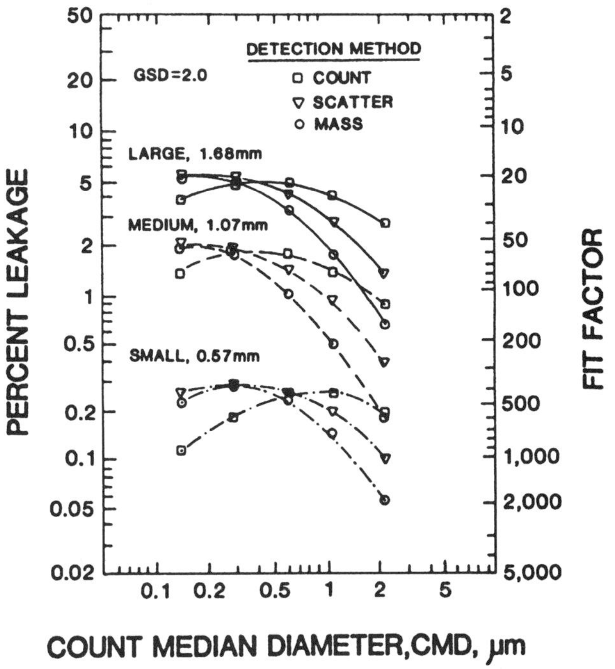

Figure 8 Calculated percent leakage extrapolated to larger and smaller particle sizes for polydisperse aerosols with a geometric standard deviation equal to 2.0. Each count distribution has a mass concentration of 10 mg/m³ within the center 97% of the count distribution. (From Reference 2.)

From Figure 8, the mass leakage into the mask for all three hole sizes will be greatest for an aerosol with a CMD between 0.15 and 0.30 μm. For the small hole size, the percent leakage based on mass with an exposure aerosol with a CMD equal to 0.28 μm would be 0.27%, while for an exposure aerosol with a CMD equal to 1.1 μm the leakage would be 0.14%. The aerosol with a smaller CMD has a greater mass leakage because as the CMD increases, the proportion of larger particles in the size distribution increases, and all three hole sizes in the mask are more effective at reducing the number of particles larger than approximately 1.0 μm entering the mask.

The authors conclude that the maximum leakage through holes in a respirator facepiece occurs approximately between 0.2 and 1.0 μm. Larger and smaller particle sizes do not enter through the leak sites as easily. The count median diameter of a

Table 1 Fit Factors Calculated for Each Hole Size Based on Different Detection Methods and Test Aerosols

Test aerosol/ detection method	CMD (μm)	GSD	Fit factors for hole sizes: Small	Medium	Large
Corn oil/Scatter	0.15	2.0	400	50	19
Corn oil/Count	0.15	2.0	830	70	26
Corn oil/Scatter	2.5	1.0	1250	250	83
Sodium chloride/Mass	0.12	2.0	526	59	20
Silica dust/Mass	0.55	2.0	390	83	29

From Reference 2.

polydisperse aerosol does have an effect on the measured leakage during a quantitative fit test, as does the measurement methods. A 5:1 ratio in leakage can be measured by a light-scattering detection method between two test aerosols with count median diameters (CMD) of 0.28 and 2.2 μm. Likewise, a 4:1 ratio between a count and mass method of detection can be measured when measuring the same polydisperse aerosol with a count mean diameter equal to 2.2 μm. These large ratios indicate that the size distribution of the test aerosol and of the aerosol to which the respirator wearer is exposed, in addition to the measurement method, must be considered to ensure that the measured leakage will reflect the actual leakage into the respirator. The ratio of the fit factors, however, between most of the current quantitative fit test methods, shown in Table 1, are smaller than 2:1. This would indicate that most of the current quantitative fit test methods have approximately the same sensitivity.

The performance of dust filters (replaceable filters and single use) as a function of aerosol particle size and flow rate was evaluated by Hinds et al.[3,4] Respirators equipped with elastomeric facepieces selected for this study were: MSA Comfo II half-mask with dust/mist (type F), fume (type S), high-efficiency particulate air (HEPA) (type H), and paint spray filters; North 7700 half-mask with fume (N7500-7), and HEPA (7500-8) filters; and AO R-5000 half-mask with fume (R56A), and HEPA (R57A) filters. Single use (disposable) respirators selected for this study were 3M 8710, AO R1070, and Gerson 1710. The test chamber had a volume of 107 l. The respirator being tested was mounted on a head form inside the test chamber. The test aerosols were generated from oleic acid with three sizes. The monodisperse aerosols had a size of 7.25 and 11.34 μm, and the size range for the polydisperse aerosol was between 0.1 and 4.0 μm. The aerosol concentration inside the test chamber was less than 300 particles/cm^3. The in-mask sample was collected along the center line of the mouth and the ambient sample was collected a few centimeters left of the mask.

Three types of face seal leaks were studied: natural leak, single or multiple wire, and tube leaks. The filter was blocked when the leak was measured. Up to five tubes with diameters between 0.5 and 3.2 mm were inserted in the facepiece seal. The facepiece was sealed with hot melt glue. Wire diameters of 0.5 and 0.9 mm were selected. The facepiece was sealed with high vacuum grease except within 3 or 4 cm of the wire leak. For the natural leak, the unsealed facepiece was mounted on the head form with a strap tension of 240 g. Both filter penetration and faceseal leakage were measured for each type of leak. For the filter performance run, penetration was measured at seven flow rates: 2, 5, 10, 20, 50, 100, and 150 l/m. The midpoint mass median aerodynamic diameter (MMAD) of test aerosols varied between 0.14, 0.17, 0.23, 0.30, 0.37, 0.51, 0.64, 0.83, 1.04, 1.68, 2.67, 3.65, 7.25, and 11.34 μm. Each leak test was performed the same as a filter test except that measurements were made at seven to nine pressure drops: 0.18, 0.45, 0.94, 1.98, 5.2, 9.4, 12.7, 25.4, and 33.6 mm of water pressure, instead of seven flow rates.

The results indicate that aerosol penetration was strongly dependent on particle size and flow rate for filters. However, the aerosol penetration was less strongly dependent on pressure drop for leaks. The proposition of leak flow relative to filter flow will be different for different flow rates. The percent leak flow rate and filter flow rate of the MSA fume filter is shown in Figure 9. The filter performance of the MSA fume filter at different particle sizes is shown in Figure 10. The filter performance of the 3M 8710 disposable respirator at different particle sizes is shown in

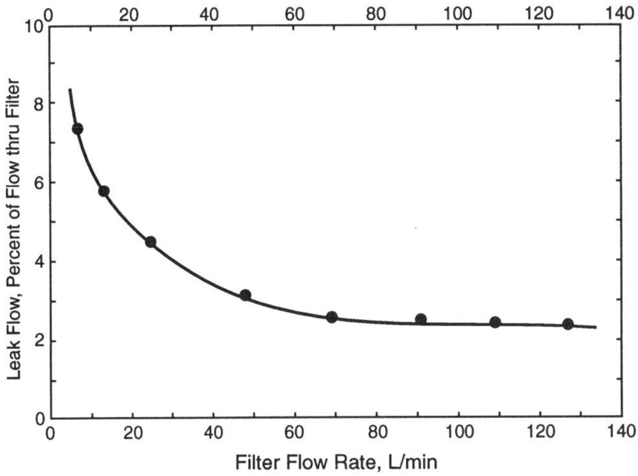

Figure 9 Percent leakage flow rate vs. filter flow rate. MSA Type S filters with two 1.55-mm diameter facial seal leaks. (From Reference 3.)

Figure 11. Filter performance was found to vary significantly between types and between brands within a type. Based on three different filter brands the relative standard deviation for penetration measurement is 65% for the fume filters and 77% for the single use respirators. All filters tested showed an increase in penetration with increasing flow rate suggesting that filtration is dominated by diffusion, interception, and sedimentation mechanisms that are less effective at higher flow rates. Impaction appears less dominating because of the low face velocities selected. Except for the HEPA filter, all filters tested showed a pattern similar to Figures 10 and 11, but there is a level of difference in penetration between different filter types and brands. The

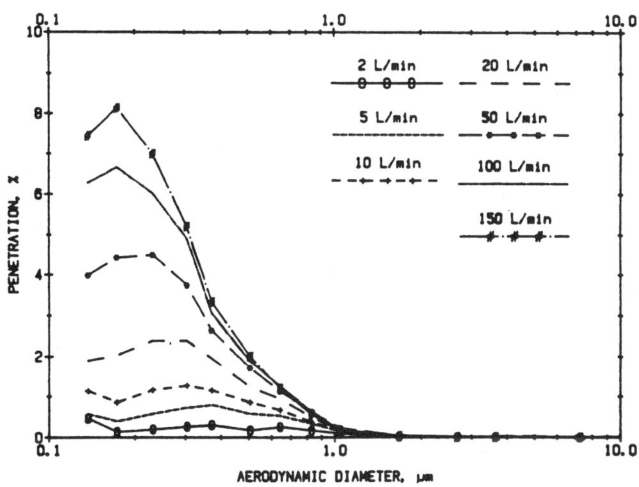

Figure 10 Filter performance, MSA Type S dust, fume, mist cartridge filters, average of lot numbers 1184, 2583, and 3383. (From Reference 3.)

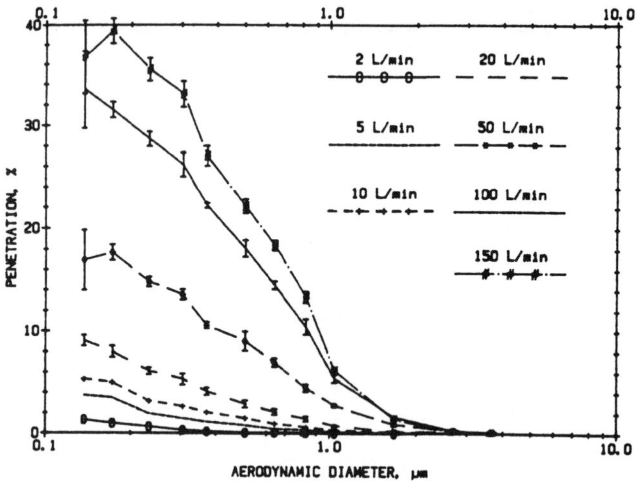

Figure 11 Filter performance, 3M 8710 single-use respirator. Bars on 20 to 150 l/m lines indicate ± standard errors of the mean for each data point. (From Reference 3.)

leak performance of MSA fume filter against tube leak is shown in Figure 12, and the leak performance of MSA fume filter against wire leak is shown in Figure 13. Leaks produced by tubes and wires of different sizes showed similar patterns.

Although air flow rate through leaks depends strongly on leak size (proportional to size raised to the 2.7 power), the shape of the penetration versus particle-size curve varied little with leak size or location. Penetration of aerosols through leaks was approximately 100% for particles having aerodynamic diameters of 0.1 to 1 µm, regardless of leak size or pressure drop. In the size range 1 to 12 µm, as pressure drop increases, penetration first increases due to decreased sedimentation losses in leaks,

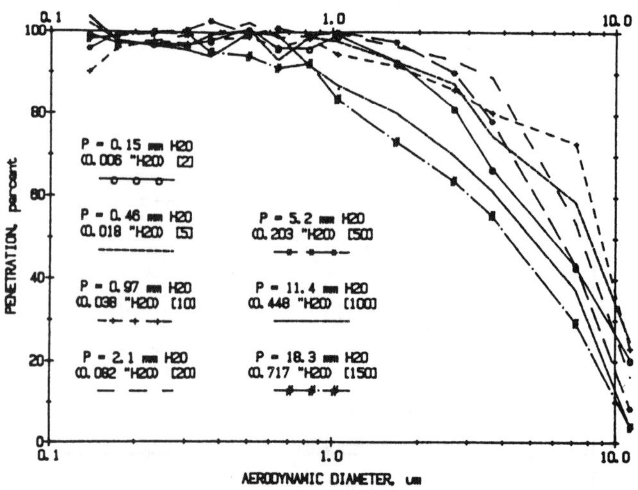

Figure 12 Leak performance, tube leak 1.0 mm ID × 10 mm long. Numbers in brackets represent the equivalent flow rate (l/m) through a pair of MSA type S filters. (From Reference 3.)

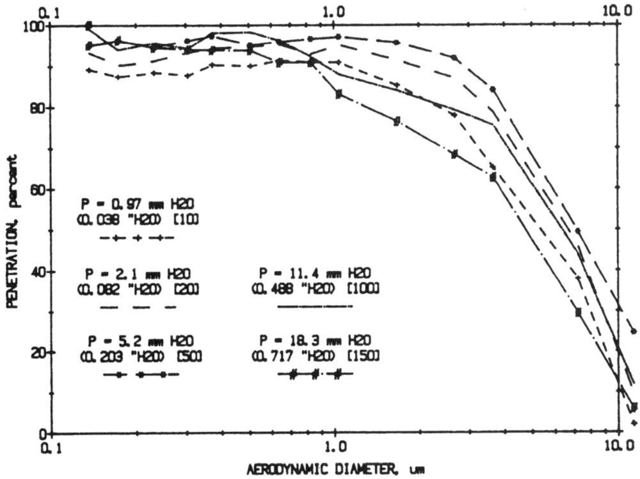

Figure 13 Leak performance, wire leak 0.5 mm-diameter. Numbers in brackets represent the equivalent flow rate (l/m) through a pair of MSA Type S filters. (From Reference 3.)

and then decreases with increasing pressure drop due to increased inlet losses and impaction losses against the face at the leak outlet. Similar penetration curves would be obtained from half-mask or disposable respirators when operated at the same pressure. Due to its design, disposable respirators have a significantly lower pressure drop for the same flow rate.

The combined performance curves for the MSA fume filter or the 3M 8710 disposable respirator (Figures 14 to 16) can be integrated over the work rate to obtain overall penetration as a function of particle size for selected work rates (Figures 17

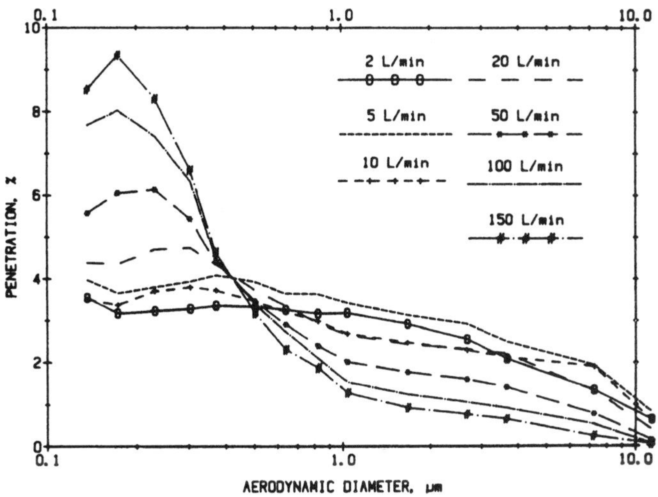

Figure 14 Combined respirator performance, MSA Type S dust, fume, mist cartridge filters (Figure 10) with a 2% leak (Figure 12) at 34.3 l/m. Parameter is total flow. (From Reference 4.)

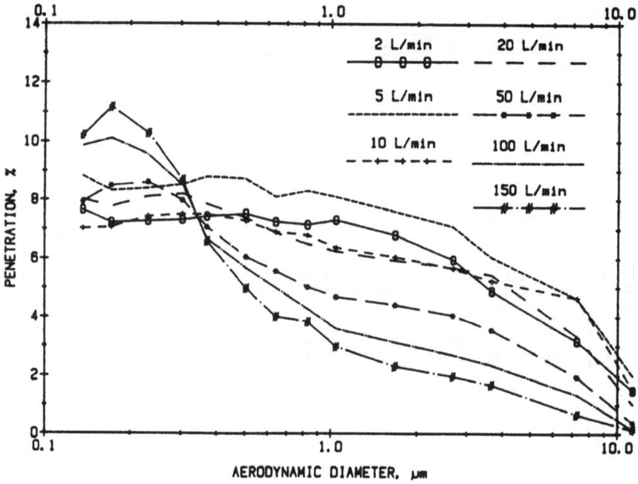

Figure 15 Combined respirator performance, MSA Type S dust, fume, and mist cartridge filters (Figure 10) with a 5% leak (Figure 12) at 34.3 l/m. Parameter is total flow. (From Reference 4.)

and 18). A more general approach is to express the results and data on which they are based in the form of a regression equation:

$$\text{LPen} = 97 - 7.4\ (d_a) \text{ for } 0.1 \leq d_a \leq 12\ \mu m \tag{1}$$

where LPen is penetration in percent and d_a is aerodynamic size of the particle in µm. This equation may be used for facepiece leak prediction with the fraction of each size particle that would enter a mask with a given leakage flow rate, and in the absence of information leak size and pressure drop.

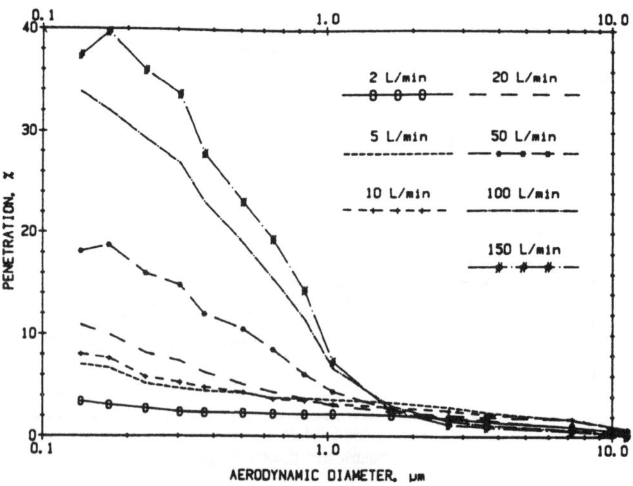

Figure 16 Combined respirator performance, 3M 8710 single-use respirator with a 2% leak (Figure 12) at 34.3 l/m. Parameter is total flow. (From Reference 4.)

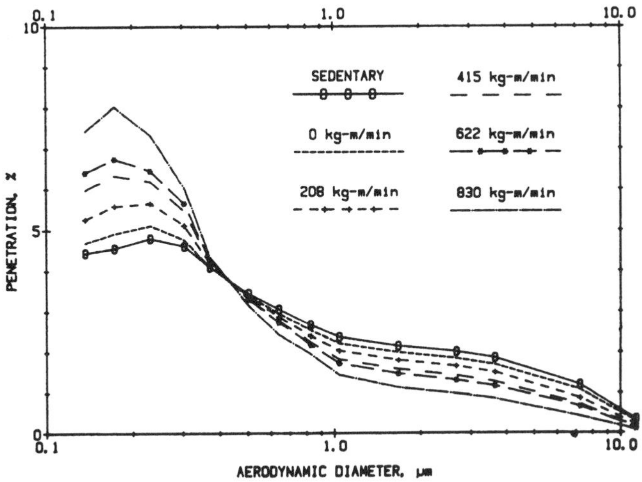

Figure 17 Overall respirator performance at six work rates based on combined respirator performance shown in Figure 14 (MSA Type S filters with a 2% leak at 34.3 l/m). (From Reference 4.)

When pressure drop is included the equation is modified to:

$$\text{LPen}\ (\%) = 90 - 9.8d_a - 7.8\ \ln P - 1.5(\ln P)^2 - 1.4d_a(\ln P) - 0.15d_a(\ln P)^2 \quad (2)$$

where P is the instantaneous pressure drop across the leak in inches of water.

The average predicted filter penetration FPen is given by:

$$\text{FPEN} = (3.49 + 0.014\ \text{WR})\exp[-d_a(3.10 + 0.00127\ \text{WR})] \quad (3)$$

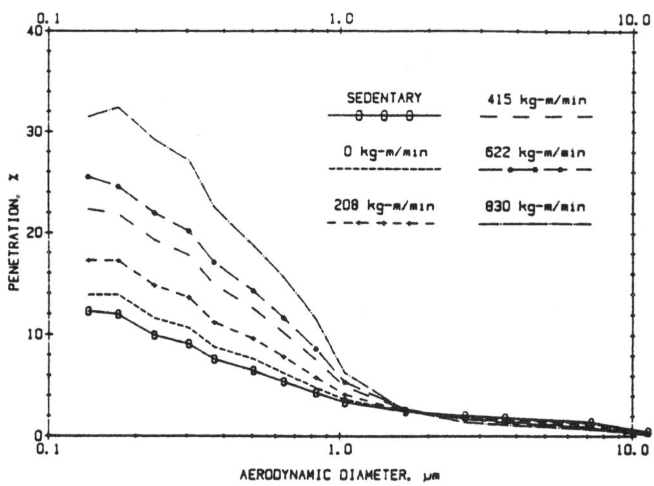

Figure 18 Overall respirator performance at six work rates based on combined respirator performance shown in Figure 16 (3M 8710 respirator with a 2% leak at 34.3 l/m). (From Reference 4.)

Table 2 Correction Factor for Use in the Simplified Method for Estimating Respirator Performance[a]

Work rate, kg-m/min	CF
Sed	1.05
0	1.00
208	0.92
415	0.84
622	0.81
830	0.73

[a] Assumes fit factor was determined at a work rate of 0 kg-m/min. Correction factor can be approximated by CF = 1.0 – 0.00032 (WR).

From Reference 4.

for $0.1 < d_a \leq 1.0$ µm and

$$FPEN = (0.2158 + 0.00025\ WR)\exp[-d_a(0.2192 + 0.000645\ WR)]$$

for $d_a > 1.0$ µm. WR is work rate in kg-m min.

A simplified method is developed to provide a more convenient, but less accurate, estimate of overall respirator performance is shown in the following:

$$TPen = (CF)(F)\ LPen + (1\text{-}F)\ FPen \qquad (4)$$

where TPen is total penetration (a function of particle size and work rate), F is the average fraction that leakage flow rate is out of the total flow rate at 0 kgm-m/min work rate as determined by QNFT test, and CF is a correction factor that depends on the work rate at which the respirator will be used (see Table 2). If accurate filter data are used, the accuracy of this method is about ± 25% of the penetration value for any particle size.

Table 3 presents a comparison between the detailed and the simplified method of calculating mass penetration and protection factor for a half-mask respirator equipped with fume filters that has a fit factor of 50 based on measured QNFT at a work rate of 0 kg-m/min. The respirator may provide in actual use a protection factor ranging from 20 to 81 depending on particle size distribution and work rate.

Table 3 Comparison of Detailed and Simplified Methods of Calculating Mass Penetration and Protection Factor for a Dust, Fume, Mist Respirator with Overall Performance Characterized by Figure 6 (MSA Type S Filters with a 2% Facial Seal Leak at 34.3 l/m)

Aerosol	MMD µm	GSD	WR kg-m/min	Detailed calc. mass pen % (PF)	Simplified calc[a] mass pen % (PF)	Difference %
Lead fume[5]	0.32	2.25	415	4.98 (20)	6.46 (15)	+29.7%
Silica dust, respirable[6]	1.44	1.54	622	1.69 (59)	1.76 (57)	+4.1%
Limestone dust, resuspended[7]	2.1	2.1	208	1.99 (50)	1.97 (51)	–1.0%
Coal mine dust[8]	5.26	2.43	622	1.23 (81)	1.43 (70)	+16.3%

[a] Calculated by Equation 4; LPEN given by Equation 1 and FPEN given by Equation 3.

[5-8] Refer to references at the end of this chapter.

From Reference 4.

The authors state that use of this approach represents a possible alternative to field measurement of worker's actual exposure while wearing a respirator. However, this model has not been validated by experimental measurements in the field because of practical difficulties involved in such measurements.

Both the Hinds and the Willeke studies indicate that the upper limit for the most penetrating aerosol size is 1.0 μm. There are two major sources of leakage through the respirator: facepiece and filter. The penetration through the facepiece is also affected by many factors, such as work rate and environmental conditions. Another factor is donning, which is difficult to control. In order to improve the performance of the respirator, the respirator approval test should limit the penetration of particulate filters against particles larger than 1 μm. Any aerosol used for performing fit testing having a MMAD of larger than 1 μm may not provide the respirator wearer with adequate protection.

II. IDENTIFICATION OF LEAK SITES

While there are studies to measure faceseal leakage, however, there are very few studies to identify the location and shape of respirator faceseal leak sites. Oestenstad et al.[9,10] developed a method using a florescent tracer to identify leak sites. A fluorescent whitening agent, 4-methyl-7-diethyl-amonocoumarin, (MDC) was selected for its optical property as well as for its toxicity. The test aerosol was generated through a nebulizer from an alcohol solution of MDC. It has a mass median aerodynamic diameter (MMAD) of 0.55 μm and a geometric standard deviation of 1.6. The average test chamber aerosol concentration was 38 mg/m^3.

Seventy-three test subjects participated in this study. One brand half-masks consisted of three sizes made by the U.S. Safety were selected for testing. Each test subject performed the standard quantitative fit testing (QNFT) prescribed in the American National Standard Institute Standard Practices for Respiratory Protection, ANSI Z88.2-1980.[11] The exercises performed consist of normal breathing, deep breathing, moving the head side-to-side, moving the head up-and-down, talking, and normal breathing. In order to assess the leaks, photographs were taken before and after the QNFT for comparison purpose.

Leaks were identified according to the location and shape. Leaks can be classified as single, multiple, point (less than 1 cm) or diffused (greater than 1 cm). Leak sites can also be classified as around the nose, on the cheek, under the chin, or a combination of more than one of these sites. Observed leak sites and shapes are shown in Tables 4 and 5. Statistical analysis at the level of $p < 0.0001$ indicated that the proportions of leaks in the nose and nose/chin categories were significantly higher than the null value. The leak site distribution for males and females were similar and were also insignificantly different from the null hypothesis. About 79% of test subjects had faceseal leaks at the nose or multiple leaks which included the nose. About 51% had leaks at the chin or multiple leaks which included the chin, while only about 19% had leaks at the cheek or multiple leaks which included the leak. A total of 110 leaks were observed on 73 test subjects. Statistical analysis at a level of $p < 0.0001$ indicated that proportion of diffuse leaks for all test subjects was significantly greater than 0.5. The proportion of diffuse leaks for males was 0.822

Table 4 Observed Respirator Leak Sites for All Subjects and Gender Subsets

Leak site	All subjects (%)	Gender	
		Male (%)	Female (%)
Nose	24 (32.9)	13 (33.3)	11 (32.4)
Cheek	6 (8.2)	4 (10.2)	2 (5.9)
Chin	6 (8.2)	3 (7.7)	3 (8.8)
Nose and cheek	4 (5.5)	3 (7.7)	1 (2.9)
Nose and chin	26 (35.6)	11 (28.2)	15 (44.1)
Cheek and chin	1 (1.4)	1 (2.6)	0 (0.0)
Nose, cheek, and chin	4 (5.5)	2 (5.1)	2 (5.9)
None detected	2 (2.7)	2 (5.1)	0 (0.0)
Total	73	39	34

From Reference 10.

and for females was 0.630, and only the proportion of diffuse leaks for males was significantly different from the null hypothesis.

The test results indicated that faceseal leaks at the nose and chin are of the greatest importance in affecting leakage on this type of half-mask respirator. Leaks at these sites or multiple leaks included 89% of all the observed leaks. The distribution of leak sites for males and females were similar, but females were found to have significantly fewer diffuse leaks. About 71% of the significant differences in facial dimensions for leaks sites subset were attributed to gender. It indicates that respirators designed for the faces of males may not fit females the same way, even when selected on the basis of facial dimensions.

Of the two dimensions (face length and lip width) used to define the anthropometric test panel, only face length was found to be significantly different in two leak site subsets. There was no correlation between these dimensions and fit factors, which implies that lip width alone may not be good criteria for selection respirators. Based on observed deposition patterns, diffuse leaks were considered to approximately slits and point leaks to approximately round holes.

The limitations of this study cited by the authors were that only one brand of respirator was tested, no determination was made on intraleak variability, and the increase of leak at higher work rates.

III. PREDICATION OF FIT

Based on the model described above,[3,4] Hinds and Bellin[12] developed applications to predict the protection the worker may receive from the quantitative fit testing

Table 5 Observed Leak Site Shapes

Subset	Leak shape		Total
	Point (%)	Diffuse (%)	
All subjects	30 (27.3)	80 (72.7)	110
Male	10 (17.8)	46 (82.2)	56
Female	20 (37.0)	34 (63.0)	54
Black	2 (18.2)	9 (81.8)	11
Asian	3 (23.1)	10 (76.9)	13
White	24 (30.0)	56 (70.0)	80
Other	1 (16.7)	5 (83.3)	6

From Reference 10.

data. Two types of errors were considered: (1) the measurement error due to the attenuation of large test aerosol particles as they traversed faceseal leaks, and (2) the difference in overall performance that resulted from using HEPA filters for testing when fume filters are used in the workplace. The pressure drop of the fume filter may decrease leakage; however, small particles can penetrate the fume filter in greater quantities than the HEPA filter.

The method of fit testing can also affect the results. The sodium chloride (NaCl) fit testing method measures the mass concentration of NaCl inside and outside the respirator facepiece. The NaCl test aerosol has a MMAD of 0.48 to 0.72 μm, and a GSD of 1.8 to 2.0. The corn oil aerosol test method measures aerosol penetration with a forward scattering photometer which is strongly dependent on the particle size. The reported particle size distribution of the corn oil is a count mean diameter (CMD) of 0.25 and a GSD of 1.53. Calculations using the particle size range for performing QNFT recommended by the ANSI Z88.2-1980 standard,[11] resulted in a MMAD of 0.5 to 0.7 μm and a GSD of 2.0 to 2.4. The NaCl instrument would indicate 93 to 96% and the corn oil instrument 77 to 95% of the potential faceseal leakage.

When using the current available fit testing instrument, the system measurement errors of faceseal leakage would usually be less than 5%. However, using the QNFT tests to estimate workplace protection factors may be significant in error for three reasons: (1) difference in breathing rate between test and actual use, (2) the use of a less efficient filter with less resistance may decrease leakage, and (3) the use of a less efficient filter would allow greater filter penetration. The proposed model had considered all of these factors with the additional assumptions that fit factor measured by the QNFT method was representative of the fit achieved in use, and there were no sampling errors during QNFT.

A comparison between measured fit factors and the predicted performance in use for the half-mask respirator, in selected industrial aerosol exposure, is shown in Table 6. The size distribution given in this table represents a compilation of published studies giving aerosol size distributions in terms of either MMAD, GSD, or data for which these values could be easily estimated. All size distributions are lognormal. QNFT was performed at a work rate of 0 kg-m/min with perfect HEPA filters. The predicted overall protection factors were based on a moderate work rate of 415 kg-m/min with fume filters. Arbitrary selected QNFT fit factors of 20 and 50 as measured by corn oil QNFT were used. It was further assumed that during use the worker had the same fit as during the QNFT.

Figure 19 illustrates the QNFT fit factor required to achieve overall protection factors of 10 and 50 at a work rate of 415 kg-m/min. The contour lines represent the required fit factor to achieve overall fit factors of 10 and 50 for a given MMAD and GSD. For large aerosol particle sizes, fit is less important than it is for submicrometer sizes. For most submicrometer particle size distributions, a fit factor greater than 100 is required to achieve an overall fit factor of 50. For size distribution in the shaded region of Figure 19b, an overall fit factor of 50 is not possible because particle penetration through the ordinary fume filter represents more than 2% of exposure aerosol concentration. To achieve an overall fit factor of 50 with fume filters for exposure aerosols having MMADs in excess of 1 μm, the QNFT fit factor must be greater than 85. For a disposable respirator with dust/mist approval, a faceseal leakage of less than 2% is required to achieve an overall protection factor of 5, if the exposure particle size distribution has a MMAD in excess of 1 μm. It should be noted

Table 6 Aerosol Size Distributions for Various Industrial Operations and Predicted Overall Protection Factors for QNFT Fit Factor of 20 and 50

Operation	MMAD, µm	GSD	Predicted PF[a] QNFT FF = 20	QNFT FF = 50
Mining				
Open pit, general environment (ns)[b,c]	2.5	4.7	36	67
Open pit, in cab (2)	1.1	2.4	26	48
Coal mine, continuous miner (27)	4.6	2.5	56	132
Coal mine, continuous miner (8)	15.0	2.9	158	386
Coal mine, continuous miner (50)[c]	17.0	3.1	172	419
Coal mine, other operations (80)[c]	11.5	2.8	122	298
Oil Shale mine (26)	2.8	3.5	39	79
Smelting and foundry				
Lead smelter, sintering (7)[c]	11.0	2.4	130	322
Lead smelter, furnace (8)[c]	3.3	15.7	47	79
Brass foundry, pouring (4)[c]	2.1	10.3	38	64
Brass foundry, grinding (3)[c]	7.2	12.9	52	90
Iron foundry, general environment (1)	2.8	5.1	38	70
Iron foundry, general environment (4)	16.8	4.4	127	290
Be-Cu foundry, furnace (16)	5.0	2.4	59	142
Nuclear fuel fabrication (66)	2.1	1.6	35	85
Nonmineral dust				
Bakery (6)	12.1	4.2	99	222
Cotton gin (5)[c,d]	47.1	2.7	1150	2860
Cotton mill (10)	7.6	4.0	72	158
Swine confinement building (21)[c]	9.6	4.0	83	186
Woodworking, machining, sanding				
Fine mode (6)	1.3	2.7	27	51
Coarse mode (6)	33.1	2.6	687	1710
Wood model shop (9)	7.2	1.4	92	229
Metal fume				
SMA (stick) Welding (ns)[d]	0.38	1.8	17	25
MIG Welding (ns)[d]	0.48	2.3	19	29
Lead fume, (O_2-Nat. gas) (5)	0.37	2.1	17	26
Mist and spray				
Pressroom, ink mist (10)	27.4	4.30	226	529
Spray painting, lacquer (ns)	6.4	3.4	68	152
Spray painting, enamel (ns)	5.7	2.0	67	165
Aerosol spray products (6)[c]	6.4	1.8	76	189
Other				
Forging (ns)[c]	5.5	2.0	65	161
Refinery, fluid catalytic cracker (4)[d]	6.2	2.4	70	171
Cigarette smoke (diluted) (5)	0.4	1.4	17	25
Pistol range (2)	2.6	3.8	37	73
Diesel exhaust (age = 5–600 s) (5)	0.12	1.4	16	21

[a] Predicted protection factors are based on the following assumptions: measured QNFT fit factors are 20 or 50, QNFT test conducted at a work rate of 0 kg-m/min with perfect DFMR filters having average resistance, respirator is properly used at a work rate of 415 kg-m/min with average DFM filters.

[b] Number in parenthesis following the operation name is the number of size distributions on which the data in the table are based; ns = number of size distributions not specified.

[c] Average values for MMAD and GSD used (median values used for all others).

[d] Mass distribution parameters calculated from count distribution data.

From Reference 12.

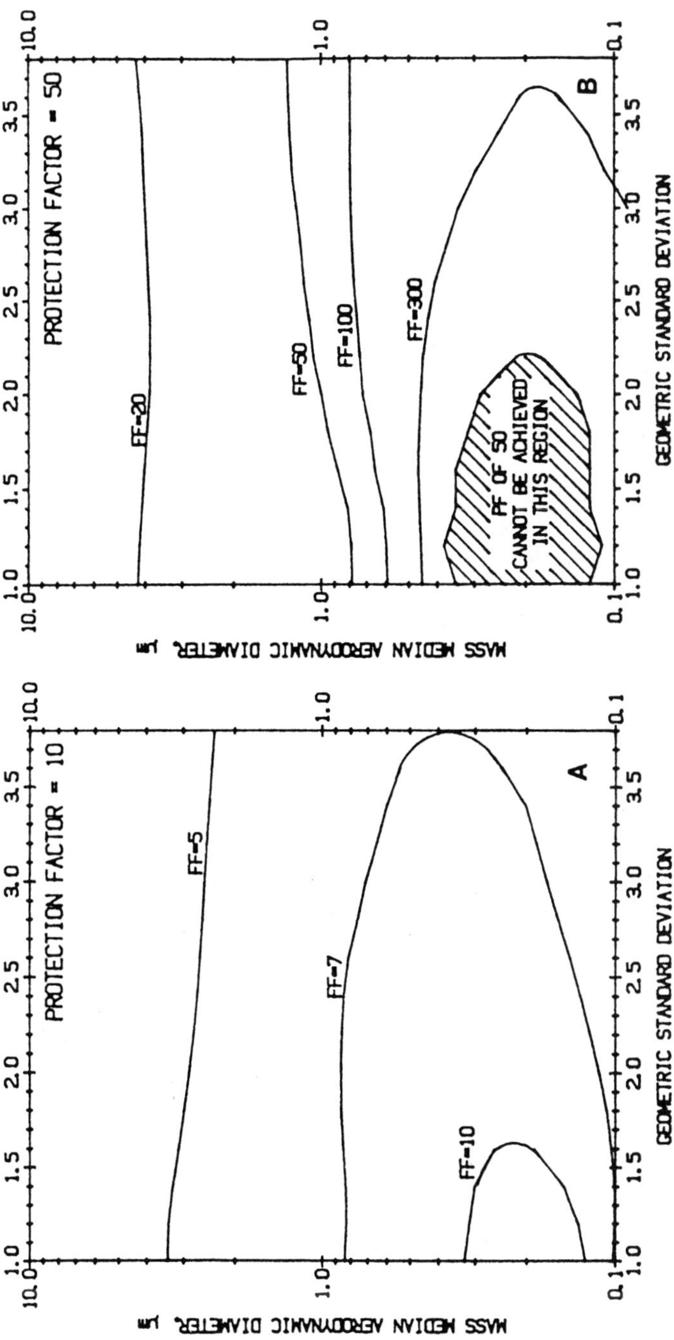

Figure 19 (A) QNFT fit factor required to achieve a protection factor of 10 at a work rate of 415 kg-m/min; (B) QNFT fit factor required to achieve a protection factor of 50 at a work rate of 415 kg-m/min. (From Reference 12.)

that other factors such as environmental factors (e.g., temperature, humidity or air movement), work hours, and body movements could reduce the overall protection factors achieved in the workplace. However, these factors are difficult to include in modeling.

IV. EXHALATION VALVE LEAKAGE

There are three major sources of respirator leakage: filter, faceseal, and exhalation valve. The exhalation valve acts like a check valve: permitting air to flow out of the respirator facepiece, and preventing reverse flow through the valve on inhalation. The testing requirements for the exhalation valve are listed in various subparts of 30 CFR 11.[13] They have the following requirements:

1. Dry exhalation valves and valve seats will be subjected to a suction of 25 mm water-column height while in a normal operating position.
2. Leakage between the valve and valve seat shall not exceed 30 ml/min.

Because it is a static test, it does not provide an indication of the valve leakage to be expected from actual use. As a part of the respirator research studies conducted for NIOSH, the Los Alamos Scientific Laboratory (LASL) developed a system for testing respirator exhalation valves.[14] The test system would be able to perform tests for air as well as aerosol leakage, statically and dynamically, and for instantaneous exhalation resistance during each breathing cycle. The test can be performed with ambient or humid exhalation air and with simulated coughing.

The system consisted of a breathing machine which was designed to simulate a normal breathing cycle. A heater and humidifier were included to control the humidity of exhaled air. The air leakage was measured by a low differential pressure transducer. A cough simulator directed a flow of compressed air through the exhalation valve. A series of solenoid valves controlled the inhalation cycle, exhalation cycle, as well as the simulated coughing. A schematic flow diagram of the system is shown in Figure 20. The test system can also be connected to an aerosol test chamber for aerosol leak testing.

Test results on mushroom type exhalation valves made by American Optical Company indicated that when the valves were tested dry, under 250 Pa (1 in. water column) vacuum, they showed an air leakage to be normally greatest at the beginning of inhalation. The valve seal apparently improved with time causing decreased leakage. This held true for both static and dynamic testing. When two hairs were introduced, the leak increased to 2.83×10^{-7} m^3/s (17 ml/min). The total leakage under dynamic testing was generally small, less than 5×10^{-7} m^3/s (30 ml/min), which met the certification requirement.

The second phase of the valve testing project[15] involved aerosol testing under dynamic test conditions. Seven different types of respirator valves from six manufacturers were tested. The valve and the facepiece combination was sealed onto a mannequin head inside of the aerosol test chamber. The test aerosol was a polydisperse DOP and the exhalation air came from a breathing machine with a work rate of 622 kg-m/min. The exhalation air was heated and humidified at 35°C and 94% relative humidity. The schematic diagram of the test system is shown in Figure 21.

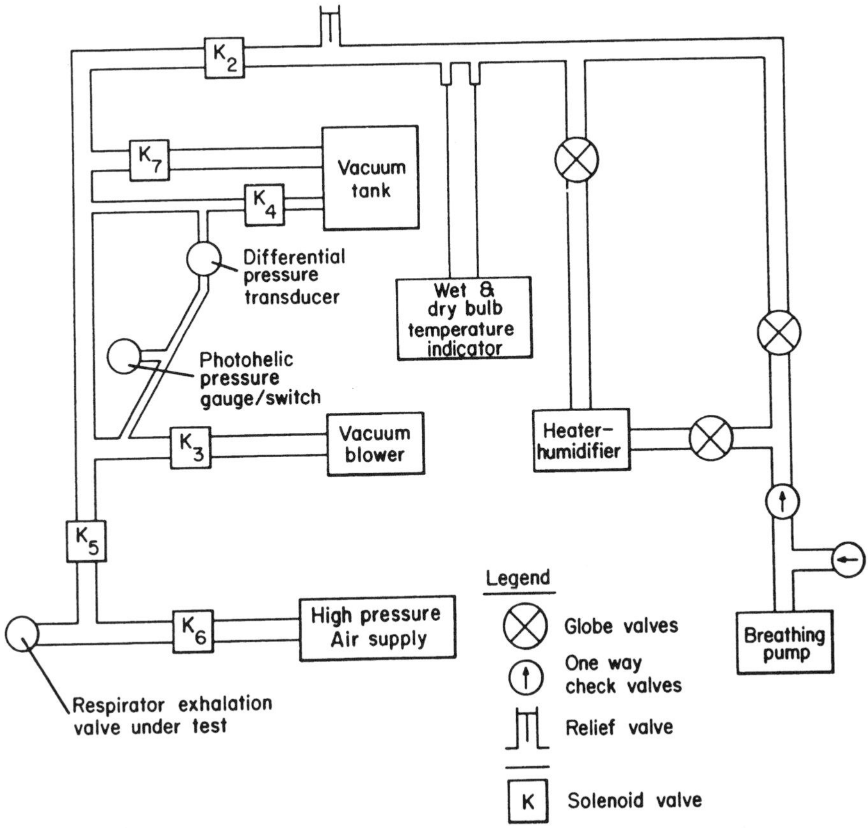

Figure 20 Respirator exhalation valve test system flow schematic. (From Reference 14.)

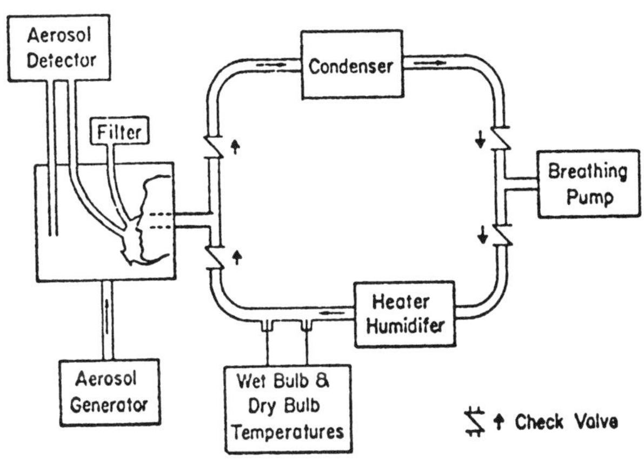

Figure 21 Line diagram of LASL dynamic exhalation valve test system. (From Reference 15.)

Table 7 Summary of Respirator Exhalation Valve Tests

Valve identification	Average air leakage thru valve under static conditions differential pressure of 2.5×10^2 Pa (ml/min)	Average penetration of aerosol into interior of respirator contributed by valve leakage for breathing at work rate of 622 kg-m/min			
		Without valve cover (%)	With valve cover type 1 (%)	With valve cover type 2 (%)	With valve cover type 3 (%)
A	9.2	0.04	<0.01	<0.01	
B	13	0.07	0.05	<0.01	<0.01
C	2.9	<0.01	<0.01		
D	16.6	0.05	<0.01		
E	11.4	<0.01	<0.01		
F	39.8	0.01	<0.01		
G	>50	0.02	<0.01		

From Reference 15.

The test was performed after the device showed less than 0.01% aerosol leak. First, the leak test was performed when the aerosol leakage was measured for 5 min without the exhalation valve cover in place. Then the valve cover was put in place and the whole assembly was tested for 1 min. If the exhalation valve had more than one valve cover, the available valve covers were also tested. In addition, a static test was also performed.

The test results indicated that there was no correlation between static air leakage and dynamic penetration. Valves with very high static air leakage showed very low aerosol leakage. A significant increase in aerosol leakage was measured when the valve cover was removed. The test results are summarized in Table 7. The authors concluded that exhalation valves from approved respirators contributed very little to the total leakage into a respirator facepiece.

Bellin and Hinds[16] developed a system to evaluate the effect of the compromised exhalation valve function on particle size specific aerosol penetration through exhalation valves. Aerosol penetration was measured on the exhalation valve of the MSA Comfo-II half-mask respirator which was sealed to a mannequin. A breathing machine with work rates of 0, 208, 415, and 622 kg-m/min was used to simulate breathing. The stroke rate (frequency) for these work rates was 19.6, 20.6, 22.7, and 29 cycles/min. Aerosol particles having mass median aerodynamic diameter (MMAD) of 0.5, 2, 4, and 8 μm were selected for testing. The 0.51 μm polydisperse aerosol had a geometric standard deviation of 2.1 and a chamber concentration of 12 mg/m³. The other monodisperse aerosols had a concentration of 1 to 20 particles/ml. The test chamber had a size of 109 l. Sampling time varied from 5 min to 1 h. An oleic acid aerosol tagged with uranine (sodium fluorescein) was used for all leakage measurements. The aerosols collected on the filter were extracted and measured for intensity of fluorescence. A schematic diagram of the testing system is shown in Figure 22.

In order to simulate valve defects in a controlled fashion, copper wires having thicknesses of 0.03, 0.08, 0.13, and 0.25 mm were placed in direct contact with the exhalation valve seat. These wires had the same order of magnitude as human hair. Tests were conducted using valves compromised by paint on their exterior surface. Mask performance was measured by simultaneous measurement of aerosol concentration inside and outside the facepiece. Each test condition was repeated five times.

The test results on normal valves indicated that the penetration of submicrometer aerosol was proportional to the work rate. There was a 10-fold increase between the

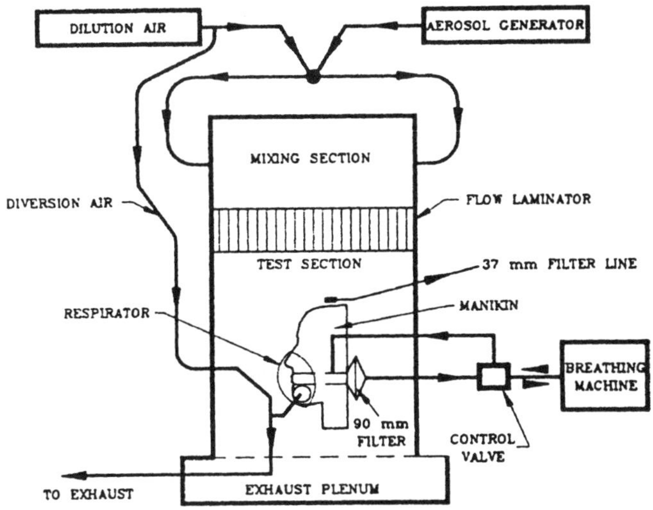

Figure 22 Cross-sectional diagram of manikin and sampling system. (From Reference 16.)

lowest (0 kg-m/min) and the highest (622 kg-m/min) work rate. The results agreed with the LASL study that exhalation valves in good condition allowed very little inward leakage. Clean exhalation valves were challenged with submicrometer aerosols in three flow modes at 0 and 415 kg-m/min work rates. At each work rate the aerosol penetration through the exhalation valve was evaluated under (1) steady flow at the average inhalation flow rate; (2) inhalation only through the mask (i.e., exhalation flow diverted); and (3) normal breathing cycle through the mask. The results indicated that the penetration for the full cycle flow was lower than observed for inhalation-only flow. This was caused in part by the dead space of the mask. The steady flow at the average inhalation flow rate was unreliable as an indicator of valve leakage under cyclic flow. This indicated that the penetration through an exhalation valve at a steady flow was not representative of that at cyclic flow. Cyclic flow was selected for all subsequent tests.

The performance of valves by the presence of fine copper wires on the valve seat was evaluated by submicrometer aerosol under two work rates (0 and 415 kg-m/min) (see Figure 23). Aerosol penetration increased by 100 to 1000 times with the introduction of the 0.25 mm wire. For wire having diameters larger than 0.08 mm, work rate had a minor effect on valve performance compared to the effect of work rate when the valves were clean. The penetration of 0.5 μm MMAD aerosol increased gradually to greater than 1% as wire thickness increased from 0.03 to 0.25 mm. When the wire was not in complete contact with the valve seat, aerosol penetration was highly variable. However, none of the tests indicated a penetration larger than 5%. The penetration of aerosol particles through the leaky valves was highly dependent on particle size. Large particles (1.5 to 2.5 μm MMAD) had half the penetration of the smaller 0.51 μm MMAD aerosol.

The exhalation valve cover provided protection to the valve and prevented the accumulation of foreign materials on the valve and valve seat. The tests indicated that the in place valve cover showed a 1.5 to 2-fold improvement in performance

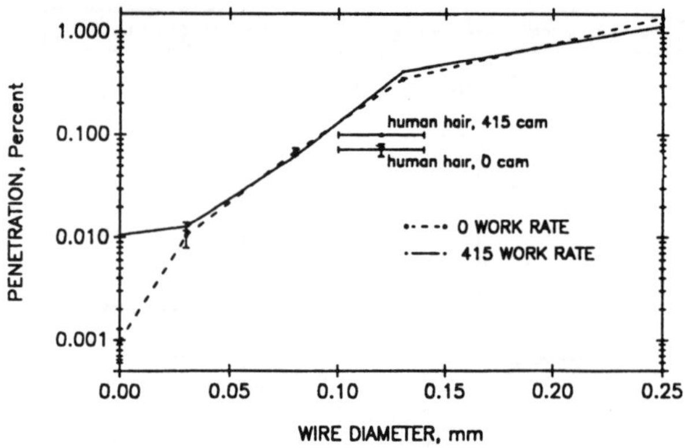

Figure 23 Penetration of submicrometer aerosol through normal and compromised exhalation valves, as a function of wire diameter. (From Reference 16.)

compared to no cover. When an exhalation valve was coated with paint having a thickness of 10 to 20 μm, valve leakage was 1.5% at 0 kg-m/min work rate and 0.75% at 415 kg-m/min for the submicrometer aerosol. The most severe exhalation valve defect was a missing valve or one stuck in the open position. For the missing valve, the penetration fell as particle size increased and reached a plateau at about 30% penetration.

Brueck et al.[17] developed a leakage testing system for exhalation valves which can be used in the laboratory as well as in the field. The system consists of two components: the Respirator Integrity Tester and the Exhalation Valve Tester. The integrity test involves placing the respirator on a soft, pliable medium which simulates the face of a respirator wearer. The test determines the integrity of the respirator by measuring leakage through all sources, including the exhalation valve. If leakage is found to be significant, the Exhalation Valve Tester would distinguish exhalation valve leakage from the potential leak sources. A schematic diagram of the test system is shown in Figure 24.

In the exhalation valve test, the backflow of air through exhalation valves into the respirator is measured at three different negative pressures: 0.5, 1.5, and 2.5 cm water gauge. These pressures correspond to work rates of 208, 415, and 622 kg-m/min. The negative pressure is created by a small pump and the leak flow is measured by a flow sensor. Steady flow was used in this study to produce negative pressures. Exhalation valves were then evaluated for the leakage that occurred after the valve closed in response to the negative pressure. In this study, exhalation valve leakage was first evaluated in new, unused mushroom and flap-type exhalation valves, and then in exhalation valves from regularly used respirators which were exposed to dusts and chemical vapors. A total of 54 new respirator valves and 67 used respirators were tested. Measurements were repeated five to ten times on randomly selected valves for the determination of leakage variation.

The test results indicated that the respirator integrity tester and the exhalation valve tester would measure the same amount of leakage if the only source of leakage into the respirator was the exhalation valve leakage. Figure 25 shows the leak flow rate of new valves tested under three different negative pressures. Most mushroom

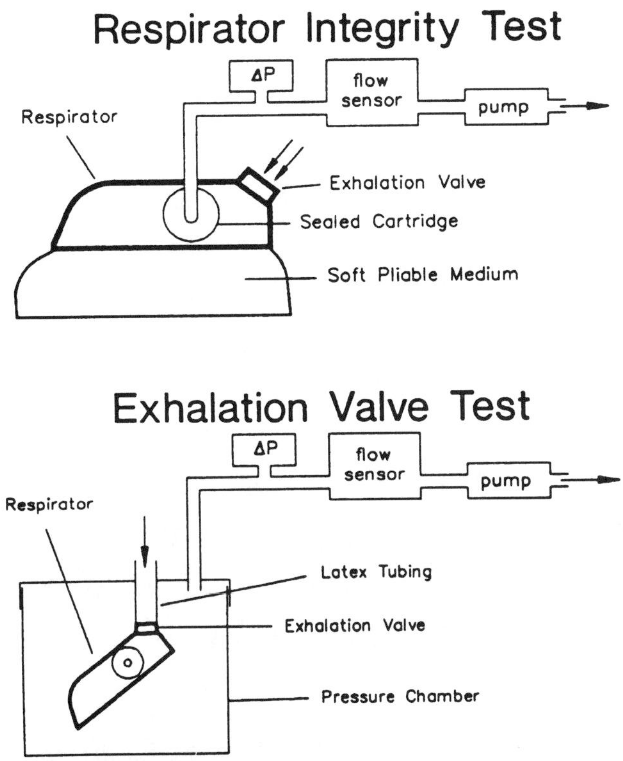

Figure 24 Schematic representation of respirator integrity test and exhalation valve test. (From Reference 17.)

type exhalation valves of Respirator A had leak flow rates less than 10 ml/min. However, the exhalation valve leak rates were much higher for Respirator B, with its flap-type exhalation valves with accordion-like folds. Three of the valves showed leak rates of 100 ml/min or higher. All other valves had leak rates between 10 and 80 ml/min. A leak flow rate in excess of 100 ml/min is considered significant and unsatisfactory. There is a tendency for the leak flow rate through the exhalation valve to increase with an increase in negative pressure.

The test results also indicated that after rinsing and drying, the leakage of exhalation valves was reduced by 40 to 70%. Leak flow through exhalation valves in respirators used in the dusty workplace and the chemical industry showed that greater exhalation valve leakage was found for respirators used in the dusty environment. Upon examination, a light coating of dust was found on the exhalation valve of a majority of the respirators received for testing. Two of the 26 Brand C respirators tested showed valve leakage in excess of 100 ml/min. The remaining 24 respirators showed leak flow between 0.1 and 50 ml/min, with most near 1 ml/min. One of the ten Brand D respirators showed a leak rate in excess of 100 ml/min. The rest had a leak rate between 1 and 20 ml/min. A possible explanation is that dust particles deposited between the valve and the valve seat may prevent the sealing of the valve during inhalation.

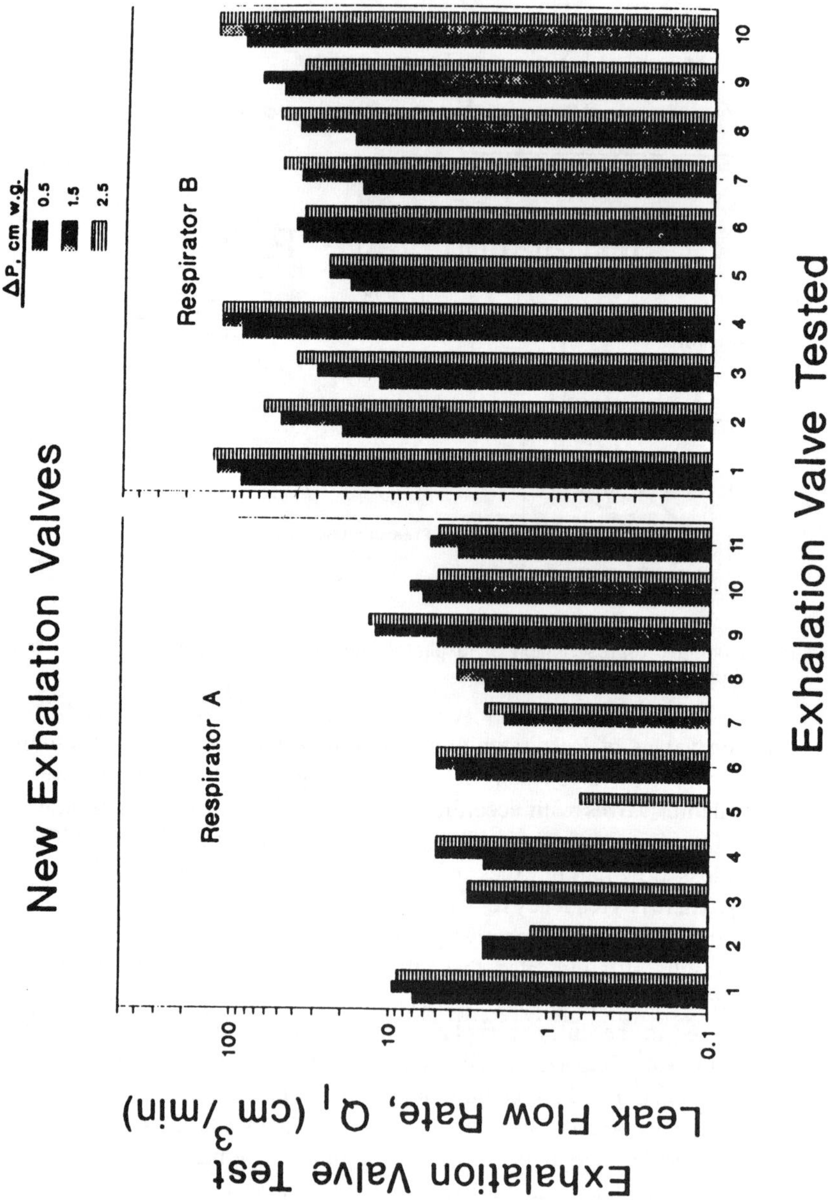

Figure 25 Exhalation valve leak flow rates for new valves. (From Reference 17.)

Table 8 Distribution of Violations of 29 CFR 1910.134

Section	Description	Percent
(a)(1)	Failure to use feasible engineering controls	1.6
(a)(2)	Failure to provide suitable respiratory protection	16.7
(a)(3)	Employee shall use provided respiratory protection	0.7
(b)(1)	Written procedures required	12.2
(b)(2)	Proper respirator selection	3.1
(b)(3)	Employee instruction	6.5
(b)(4)	Individual respirators to be provided	0.2
(b)(5)	Cleaning and disinfection of respirators	7.4
(b)(6)	Proper storage of respirators	7.9
(b)(7)	Inspection during cleaning	3.9
(b)(8)	Maintenance of environmental surveillance	2.1
(b)(9)	Inspection of program	2.7
(b)(10)	Medical screening	3.7
(b)(11)	Approved respirators	5.0
(c)	Selection according to ANSI specifications	0.8
(d)	Quality of supplied air	3.9
(e)(1)	Standard procedures for use of respirators	0.5
(e)(2)	Correct respirator for the job	0.5
(e)(3)	Procedures for use in dangerous atmospheres	3.6
(e)(4)	Random inspections of program	0.7
(e)(5)	Instruction and training in respirator use	5.6
(f)(1)	Maintenance of respirators	2.0
(f)(2)	Inspection of respirators	6.4
(f)(3)	Cleaning and disinfection of respirators	1.0
(f)(4)	Replacement of parts and repair of respirators	0.2
(f)(5)	Storage of respirators	1.1
Total		100.0

From Reference 18.

Exhalation valves from four Brand E respirators used in the chemical industry showed leak rate between 0.6 and 35 ml/min and the exhalation valves of Brand F Respirators had a leak rate between 15 and 30 ml/min. The valve leakage for the same respirator was reduced by 45 to 91% after its valves were rinsed with cold water and dried. These exhalation valve leakage studies indicate that proper maintenance and inspection of respirators is an important factor in minimizing exhalation valve leakage.

III. QUALITY OF RESPIRATOR PROGRAMS

The effectiveness of respiratory protection programs has been evaluated by Rosenthal and Paull using the OSHA compliance data.[18] The data used in analysis were collected during the period from October 1, 1976 to September 30, 1982. Industries having the Standard Industrial Classification (SIC) codes between 2000 to 3999 were selected in this evaluation because respirators are widely used. Citations issued are based on the violation of applicable sections of the OSHA respiratory protection standard, 29 CFR 1910.134. A distribution of violations are shown in Table 8. Respirator violation may also be issued under specific OSHA standards on air contaminants (e.g. asbestos, benzene, lead); however, these types of citations are small compared to the violations issued under 1910.134. The distribution of violations into major categories is shown in Figure 26. The category with the most violations is "inadequate maintenance, cleaning or storage of respirators" which

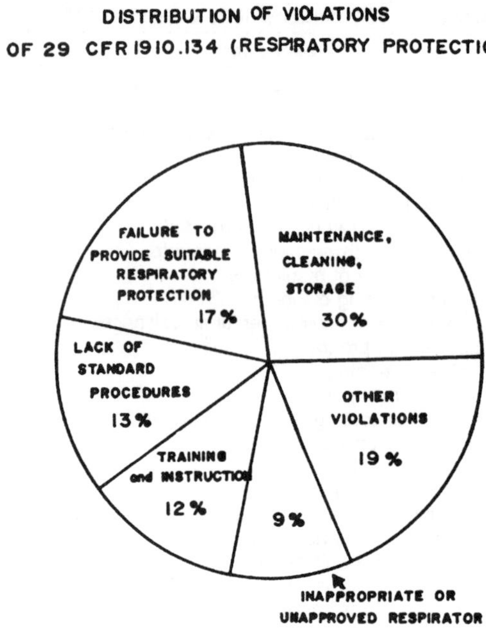

Figure 26 Major categories of respirator violations. Violations are grouped as follows: maintenance, cleaning and storage of respirators [(b)(5-7), (f)(1-5)]; lack of standard procedures [(b)(1), (e)(1)]; training and instruction [(b)(3), (e)(5)]; inappropriate or uncertified respirator [(b)(2), (b)(11), (c), (e)(2)]; failure to provide suitable respirator protection [(a)(2)]. (From Reference 18.)

comprises 30% of all violations. Other major violation categories were "lack of standard operating procedures" (13%), "inadequate training" (13%), and "non-approved or inappropriate respirators" (9%).

The OSHA database only indicates situations in which a citation was actually issued. It may underestimate the occurrence of program deficiencies due to underreporting, mainly when a deficiency was observed but a decision was made on site, or at the area or regional level, not to issue a citation. Another point which leads to underreporting is that respirator citations will not be issued unless the air contaminant concentration exceeds the applicable OSHA permissible exposure limit. Al-

Table 9 Statistics on Violations Cited in Manufacturing Health Inspections

FY year	Number of inspections	Number (%) of inspections with violations of: 1910.134	1910.134 A Only and 1910.1000	1910.134 (Program deficiencies)[a]
1977	6983	798 (11.4)	201 (2.9)	597 (8.5)
1978	7932	991 (12.5)	225 (2.8)	766 (9.7)
1979	7930	1027 (13.0)	191 (2.4)	836 (10.5)
1980	8667	1153 (13.3)	213 (2.5)	940 (10.8)
1981	7840	930 (11.9)	173 (2.2)	757 (9.7)
1982	7356	912 (12.4)	123 (1.7)	789 (10.7)
Totals	46708	5811 (12.4)	1126 (2.4)	4685 (10.0)

[a] Estimated, see text for explanation.

From Reference 18.

Table 10 Statistics on Health Inspections with Violations of 1910.1000

FY Year	Inspections with violations of 1910.1000	Number (%) of inspections with violations of 1910.1000 and: 1910.134	1910.134 (b)-(g)	1910.134 (a)(1) Only	1910.134 (a)(2) Only	1910.134 and (a)(1) (a)(2) Only
1977	605	393 (65.0)	196 (32.4)	47 (7.8)	148 (24.4)	2 (0.3)
1978	697	501 (71.9)	281 (40.3)	42 (6.0)	176 (25.3)	2 (0.3)
1979	607	467 (76.9)	280 (46.1)	31 (5.1)	155 (25.5)	1 (0.2)
1980	721	508 (70.5)	302 (41.9)	15 (2.1)	188 (26.1)	3 (0.4)
1981	567	396 (69.8)	224 (39.5)	12 (2.1)	156 (27.5)	4 (0.7)
1982	375	307 (81.9)	186 (49.7)	17 (4.5)	103 (27.5)	1 (0.3)
All yrs.	3567	2572 (72.1)	1469 (41.1)	164 (4.6)	926 (26.0)	13 (0.4)

From Reference 18.

though positive bias may result from an incorrect citation, the fraction of citations which are vacated by subsequent hearing is small.

During the 6-year period studied, OSHA conducted approximately 8000 health inspections in the manufacturing industries each fiscal year. In these inspections a total of approximately 20,000 violations of 29 CFR 1910.134 were cited. In inspections in which respirator violations were cited, the mean number of such violations cited was 3.4. Table 9 lists the number of inspections with at least one violation of 1910.134, during each fiscal year and for all years combined. For the 6-year period studied, 12.4% of the inspections resulted in citations for one or more sections of 1910.134. Since most over exposures to air contaminants are cited under OSHA air contaminants standard, 1910.1000, the maximum number of such cases can be approximated by the number of inspections in which both citations for both 1910.1000 and either 1910.134 (a)(1) or 1910.134 (a)(2) were issued (see column 4 of Table 9).

Respiratory protection program deficiencies present the greatest potential risk to health when there are high environmental levels of air contaminants. The subgroup of inspections for which an overexposure to a toxic substance listed in 1910.1000 was cited is shown in Table 10. With respect to this subset, 72.1% of the inspections resulted in citation for at least one violation of 1910.134, with 41.2% resulting in citations for specific program deficiencies.

Deficiencies in respiratory protection program were also cited by several investigators. Toney et al.[19] conducted a study in the paint spray operations. They found that 82% of the respirators in use for organic solvents protection were not approved, 35% of the wearers had "unacceptable" fits, and 28% of the respirators were used in "unacceptable " conditions. Another study conducted by Mahon et al.[20] on foundries indicated that 32% of the respirators used for protection against metal fumes and 7.5% of the respirators used for protection against silica were not MSHA/NIOSH approved.

REFERENCES

1. Holton, P. M., Tackett, D. Y., and Willeke, K., Particle size-dependent leakage and losses of aerosol in respirators, *Am. Ind. Hyg. Assoc. J.*, 48, 848, 1987.
2. Holton, P. M., Tackett, D. Y., and Willeke, K., The effect of aerosol size distribution and measurement method on respirator fit, *Am. Ind. Hyg. Assoc. J.*, 48, 838, 1987.

3. Hinds, W. C. and Kraske, G., Performance of dust respirators with facial seal leaks. I. Experimental, *Am. Ind. Hyg. Assoc. J.*, 48, 836, 1987.
4. Hinds, W. C. and Bellin, P., Performance of dust respirators with facial seal leaks. II. Predictive model, *Am. Ind. Hyg. Assoc. J.*, 48, 842, 1987.
5. Japuntich, D. A. and Johnson, B. A., Characteristics of lead fume aerosols generated under different conditions by an oxygen-natural gas welding torch, in *Aerosols in the Mining and Industrial Work Environments*, Vol. 2, Marple, V. A. and Liu, B. Y. H., Eds., Ann Arbor Science, Ann Arbor, MI, 1983, 513.
6. Welker, R. W., Eisenberg, W., and Semmler, R. A., Mine particulate characterization, in *Aerosols in the Mining and Industrial Work Environments*, Vol. 2, Marple, V. A. and Liu, B. Y. H., Eds., Ann Arbor Science, Ann Arbor, MI, 1983, 455.
7. Lilenfeld, P., Current mine dust monitoring developments, in *Aerosols in the Mining and Industrial Work Environments*, Vol. 2, Marple, V. A. and Liu, B. Y. H., Eds., Ann Arbor Science, Ann Arbor, MI, 1983, 733.
8. Tomb, T. F., Treaftis, H. N., and Gero, A. J., Characteristics of underground coal mine dust aerosols, in *Aerosols in the Mining and Industrial Work Environments*, Vol. 2, Marple, V. A. and Liu, B. Y. H., Eds., Ann Arbor Science, Ann Arbor, MI, 1983, 395.
9. Oestenstad, R. K., Perkins, J. L., and Rose, V. E., Identification of faceseal leak sites on a half-mask respirator, *Am. Ind. Hyg. Assoc. J.*, 51, 280, 1990.
10. Oestenstad, R. K., Dillon, H. K., and Perkins, L. L., Distribution of faceseal leak sites on a half-mask respirator and their association with facial dimensions, *Am. Ind. Hyg. Assoc. J.*, 51, 285, 1990.
11. ANSI, Standard practices for respiratory protection, ANSI Z88.2-1980, American National Standards Institute (ANSI), New York, 1980.
12. Hinds, W. C. and Bellin, P., Effect of facial-seal leaks on protection provided by half-mask respirators, *Appl. Ind. Hyg.*, 3, 158, 1988.
13. Code of Federal Regulations, Mine Safety and Health Administration, Labor: 30 CFR 11, 11.102-2 and 11.140-10, Government Printing Office, Washington, DC, 1993.
14. Held, B. J., Revoir, W. H., Davis, T. O., Pritchard, J. A., Lowry, P. L., Hack, A. L., Pritchard, C. P., Geoffrion, L. A., Wheat, L. D., and Hyatt, E. C., Los Alamos National Laboratory: Respirator Studies for the National Institute for Occupational Safety and Health, July 1, 1973 through June 30, 1974. LA 5805-PR, pp. 10–14, December 1974.
15. Douglas, D. D., Revoir, W. H., Lowry, P. L., Pritchard, J. A., Richards, C. P., Hack, A. L., Wheat, L. D., Geoffrion, L. A., Bustos, J. M., Davis, T. O., and Hesch, P. R., Los Alamos National Laboratory: Respirator Studies for the National Institute for Occupational Safety and Health, July 1, 1974 through June 30, 1975. LA 6386-PR, Darrel Douglas, Project Manager, pp. 18–25, August 1976.
16. Bellin, P. and Hinds, W. C., Aerosol penetration through respirator exhalation valves, *Am. Ind. Hyg. Assoc. J.*, 51, 555, 1990.
17. Brueck, S., Lehtimake, M., Krishnan, U., and Willeke, K., Method development for measuring respirator exhalation valve leakage, *Appl. Occup. Environ. Hyg.*, 7, 174, 1992.
18. Rosenthal, F. S. and Paull, J. M., Quality of respirator programs: An analysis from OSHA compliance data, *Am. Ind. Hyg. Assoc. J.*, 46, 709, 1985.
19. Toney, C. R. and Barnhart, W. L., Performance evaluation of respiratory protective equipment used in paint spraying operations, NIOSH (DHEW/NIOSH-76-177), Cincinnati, OH, 1976.
20. Mahon, R. D., Morrison, J. H., and Weller, L. A., Survey of Personal Protective Equipment Used in Foundries, NIOSH (DHEW-NIOSH-80-100), Cincinnati, OH, 1980.

INDEX

A

B

D

E

G

H

I

K

L

M

N

S

T